With thanks
for all the
collaboration

George S.

Peter Lees
Department of Veterinary Basic Sciences
The Royal Veterinary College
Hawkshead House
Hawkshead Lane
North Mymms
Hatfield, Herts AL9 7TA
United Kingdom

Experimental Design Techniques in Statistical Practice:

a practical software-based approach

"Talking of education, people have now a-days" (said he) "got a strange opinion that every thing should be taught by lectures. Now, I cannot see that lectures can do so much good as reading the books from which the lectures are taken. I know nothing that can be best taught by lectures, except where experiments are to be shewn. You may teach chymestry by lectures — You might teach making of shoes by lectures!"

James Boswell: *Life of Samuel Johnson, 1766* (1709-1784)

The direction in which education starts a man will determine his future life.

Plato: *The Republic, Book V*

(427-347 BC)

THE AUTHORS

Bill Gardiner

Dr Bill Gardiner graduated with first class honours in mathematics at the University of Strathclyde and later with a Ph.D. in statistics and mathematical modelling. He has devoted his professional life to high quality teaching of statistics to under-graduates across a range of disciplines and also to postgraduate MSc students in industrial mathematics. He has provided consultancy support for over 16 years to biologists, chemists and health providers. Much of his work has involved the use of experimental designs within practical problems which, with his teaching and consultancy experience, culminated in the writing of this book together with George Gettinby. He currently lectures at Glasgow Caledonian University, where he emphasises the importance and practical benefits of statistical data analysis as a fundamental and integral part of data interpretation for students of all disciplines. He has also recently published work on unbalanced experimental designs and two introductory statistics books for bioscientists and chemists. This present textbook covers both elementary and advanced methods and reflects Dr Gardiner's successful teaching style in practice, where the emphasis is on using software as an aid to statistical analysis and decision-making.

George Gettinby

George Gettinby is Professor and Chairman in the Department of Statistics and Modelling Science at the University of Strathclyde, Glasgow. He graduated in applied mathematics at Queens University Belfast and obtained a DPhil for his thesis on mathematical and statistical modelling at the University of Ulster. He subsequently qualified as a chartered statistician. Over the last twenty years he has been an adviser to international agencies, industry and governmental bodies on the design and analysis of studies using experimental design methods. His research and teaching interests have focused on the use of statistical and mathematical models for the study of the environment and the control of diseases. His industrial interests have centred around the research, development and manufacture of human and animal medicines. In recent years he has taught many short courses for industry that promote the use of statistical methods for assessing the quality of products. He is a member of the Royal Statistical Society, Professional Statisticians in Industry and the UK Medicines Commission. In 1997 he was elected a Fellow of the Royal Society of Edinburgh.

Experimental Design Techniques in Statistical Practice:

a practical software-based approach

William P. Gardiner, Bsc, PhD
Lecturer in Statistics
Glasgow Caledonian University

and

George Gettinby, BSc, Dphil, CStat
Professor of Statistics and Modelling Science
University of Strathclyde
Glasgow

Horwood Publishing
Chichester

First published in 1998 by
HORWOOD PUBLISHING LIMITED
International Publishers
Coll House, Westergate, Chichester, West Sussex, PO20 6QL
England

British Library Cataloguing in Publication Data
A catalogue record of this book is available from the British Library

ISBN 1-898563-35-7

Printed and bound in Great Britain by MPG Books Ltd, Bodmin, Cornwall

Table of Contents

Authors' Preface

Statistical data analysis techniques are increasingly being relied upon to translate experimental data into useful knowledge. Familiarity with statistical techniques is necessary if scientists, industrial personnel, and researchers are to design experiments to obtain the most relevant data for the specified objectives and if they are to use the data collected to best advantage in the advancement of the knowledge of new phenomena, processes, and products.

The development of experimental design principles is generally attributed to Sir Ronald Fisher who was concerned with agricultural research in the 1920s for the improvement of yield. Further development of these initial principles was provided by innovators such as Frank Yates and George Box, most notably through their contributions to agriculture and industrial experimentation. The areas in which experimental designs can be applied have continued to expand. Recently, the increased awareness of Taguchi methods for quality improvement has led to an increase in the use of statistics and experimental design in industry, though not necessarily all industry. New areas of application continue to be developed through increased awareness of the benefits of experimental design and through improved software facilities.

This text will provide an introduction to the diverse subject area of experimental design. Despite its length, it does not attempt to be an all embracing guide to all aspects of experimental designs. There are numerous other excellent texts available for that purpose. The important principles that we instill are that stating the aims and objectives, planning, and data analysis go hand in hand and must be seen as part of the overall investigative process. Inefficient designs represent wasted effort and so ensuring that these three principles are addressed is vital to efficient design and analysis.

A particular purpose of this book is to develop in the reader an appreciation and understanding of experimental designs and to equip the reader with the ability to use experimental designs in a practical way. The book illustrates experimental designs using practical problems from a wide range of application areas. We have de-emphasised the mathematical aspects underpinning experimental designs to concentrate more on the practicality of design usage and the subsequent data handling. We believe this approach will provide the reader with a greater feel and stronger understanding of how to use experimental designs in practice. The data handling is explained from both exploratory data analysis and inferential data analysis aspects through the provision of detailed solutions. This enables the reader to develop a sound understanding of how to analyse data and of the role statistical methods can play within both the design and interpretational aspects of experimentation. We concur with the trend of including more exploratory data analysis in Statistics teaching to enable data to be explored visually and numerically for inherent patterns. This aspect of data analysis has been incorporated in all the illustrations. Each chapter also contains simple, practical, and applicable problems for the reader to attempt to provide additional illustrations of the concepts and data analysis principles described. Summary solutions to selected problems are presented at the end of the text.

The use of statistical software packages has increased markedly in recent years, with inefficient and inappropriate usage still a frequent occurrence. With this in mind, we have decided to base all aspects of the design illustrations presented on the use of software output to reflect the wide availability of statistical software for data handling. It must be appreciated, however, that software is only a tool to aid data presentation for analysis purposes. The investigator must use the output created to interpret the data with respect to the specified experimental objectives. The emphasis we have placed on software usage has enabled the calculation aspects of the data analysis procedures to be kept to a minimum. This has allowed us to focus more on describing the underlying principles and methods with which to expedite the data analysis. Software usage also enables the data to be viewed and interpreted from graphical, summary, and inferential perspectives, so providing a more comprehensive base for applicable data analysis than calculation methods alone can provide. It is this philosophy which is the basis of the practical illustrations within this book.

Numerous statistical software packages are available and it was difficult to decide which to include in the book. We chose to use both Minitab and SAS as they are simple to use, ubiquitous, and compatible in most of their operations with the operational features of Windows software. Minitab is widely available throughout educational institutions while SAS is used extensively in industry. We acknowledge that this decision may not suit everyone but the data presentation and analysis principles have been presented in a way that can be readily transferred to other software packages. Detailed information and explanations of software operation are provided in appendices at the end of each chapter for both Minitab and SAS to provide the reader with full information on how to use the software to obtain the illustrated statistical output.

Acknowledgements

We wish to express our sincere thanks to Dr Bill Byrom, Professor Stuart Reid, and Dr Xuerong Mao for reviewing the manuscript. Their many constructive suggestions and helpful criticisms improved the structure and explanations provided in the text. We also wish to thank the many journals and publishers who have graciously granted us permission to reproduce materials from their publications. Thanks are also due to our publisher, Ellis Horwood MBE, for the enthusiasm he showed for the project and the assistance he has provided in the preparation of the manuscript. Finally and most importantly, we must express our appreciation and heartfelt thanks for the support of our families, Moira, Debbie and Greg, and Ruth, Michael, and Peter, through the long running saga of text development, preparation, review, and final typing. This book is dedicated to them and we hope they have as many happy hours reading it as we did preparing it!

W. P. Gardiner
Glasgow Caledonian University
w.gardiner@gcal.ac.uk

G. Gettinby
University of Strathclyde
g.gettinby@stams.strath.ac.uk

January 1998

Glossary

Accuracy The level of agreement between replicate determinations of a measurable property and a reference or target value.

Aliasing The sharing of contrast expressions and sum of squares of separate factorial effects.

Alternative hypothesis A statement reflecting a difference or change in the level of a response as a result of experimental intervention, denoted by H_1 or H_A or AH.

Analysis of Variance (ANOVA) The technique of separating, mathematically, the total variation within experimental measurements into sources corresponding to controlled and uncontrolled components.

BIBD Balanced Incomplete Block Design.

Blocking The grouping of experimental units into homogeneous blocks to remove an extraneous source of response variation.

Boxplot A data plot comprising tails and a box from lower to upper quartile separated in the middle by the median for detecting data spread and patterning together with the presence of outliers.

CCD Central Composite Design.

CI Confidence interval, an interval or range of values which contains an unknown parameter with a specified probability.

COD CrossOver Design.

Contrasts Corresponds to linear combinations of treatments for specific treatment comparisons, also the underpinning components of two-level designs.

Confounding The design technique for blocking a factorial experiment where information on certain treatment effects is sacrificed as they are indistinguishable from the block effects.

CRD Completely Randomised Design.

CV Coefficient of variation, a dimensionless quantity which is a measure of the relative precision of replicated experimental data.

Decision rule Mechanism for using test statistic or p value to decide whether to accept or reject the null hypothesis in inferential data analysis.

Defining contrast A treatment effect confounded with the blocks or fractions in Fractional Factorial Designs.

Defining relation The complete specification of sacrificed treatment effects in a Fractional Factorial Design expressed as the set of factors equal to the identity column I.

Descriptive statistics The graphical presentation and calculation of summary statistics for experimental data.

df Degrees of freedom, number of independent measurements that are available for estimation, generally corresponds to number of measurements minus number of parameters to estimate.

Diagnostic checking An analysis tool for assessment of the assumptions associated with an inferential data analysis procedure.

Dispersion The level of variation within collected data corresponding to the way data cluster around their "centre" value.

Dotplot A data plot of recorded data where each observation is presented as a dot to display its position relative to other measurements within the data set.

EDA Exploratory data analysis, visual and numerical mechanisms for presenting and analysing data to help gain an initial insight into the structure and patterning prevalent within the data.

Error Deviation of a response measurement from its true value.

Estimation Methods of estimating the magnitude of an experimental effect within an investigation.

Experiment A planned inquiry to obtain new information on a measurable, or observable, outcome or to confirm results from previous studies.

Experimental design The experimental structure used to generate practical data for interpretative purposes.

Experimental plan Step-by-step guide to experimentation and subsequent data analysis.

Experimental unit An experimental unit is the physical experimental material to which one application of a treatment is applied, e.g. manufactured product, water sample, food specimen, subject.

FD Factorial Design.

FFD Fractional Factorial Design.

Fixed effects The treatments to be tested correspond to those specifically chosen for investigation or to the only ones associated with an investigation.

Heteroscedastic Data exhibiting non-constant variability as the mean changes.

Homoscedastic Data exhibiting constant variability as the mean changes.

Inferential data analysis Inference mechanisms for testing the statistical significance of collected data through weighing up the evidence within the data for or against a particular outcome.

Interaction The joint influence of treatment combinations on a response which cannot be explained by the sum of the individual factor effects.

Location The centre of a data set which the recorded responses tend to cluster around, e.g. mean, median.

LS Latin Square design.

Main effects Independent factor effects reflecting the change in a response as a result of changing the factor levels.

MD Mixture Design.

Mean The arithmetic average of a set of experimental measurements.

Median The middle observation of a set of experimental measurements when expressed in ascending order of magnitude.

Mixed model experiment An experiment in which factors of both fixed and random effect type appear.

Model The statistical mechanism where an experimental response is explained in terms of the factors controlled in the experiment.

Model I experiment An experiment in which the treatments, or factors, tested all correspond to fixed effects.

Model II experiment An experiment in which the treatments, or factors, tested all correspond to random effects.

Multiple comparison procedures Statistical procedures for pairwise treatment comparison which provide an understanding of how detected treatment differences are occurring.

ND Nested Design.

Non-parametric procedures Methods of inferential data analysis, often based on ranking, which do not require the assumption of normality for the measured response.

Normal (Gaussian) The most commonly applied population distribution in Statistics, the assumed distribution for a measured response in parametric inference.

Null hypothesis A statement reflecting no difference between observations and target or between sets of observations as a result of experimental intervention, denoted H_0 or *NH*.

OAD Orthogonal Array Design.

Observation A measured or observed data value from a study or an experiment.

OFAT One-factor-at-a-time experimentation.

OLS Ordinary least squares, a parameter estimation technique used within regression modelling to determine the best fitting relationship for a response *Y* in terms of one or more experimental variables.

Orthogonality The property of a design matrix whereby the inner product of any pair of columns is zero.

Orthogonal polynomials Specific treatment contrasts which can assess for evidence of trend effects in quantitative treatments.

Outlier A recorded response measurement which differs markedly from the majority of the data collected.

***p* Value** The probability that a calculated test statistic value could have occurred by chance alone, provides a measure of the probability that the level of treatment difference detected has occurred purely by chance, compared to significance level.

Paired sampling A design principle where experimental material to be tested is split into two equal parts with each part tested on one of two possible treatments.

Parameters The terms included within a response model which require to be estimated and assessed for their statistical significance.

Parametric procedures Methods of inferential data analysis based on the assumption that the measured response data conform to a normal distribution.

PBD Plackett-Burman Design.

Power The probability of correctly rejecting a false null hypothesis, power = 100[1 − P(Type II error)], often set at 80%.

Power analysis An important part of design planning to assess suitability of design structure for its intended purpose.

Precision The level of agreement between replicate measurements of a measurable property.

Protocol An outline of the study approach specifying objectives, sampling strategy, power analysis, and details of planned data analysis.

Quality assurance (QA) Procedures concerned with monitoring of laboratory practice, manufacturing practice, and measurement reporting to ensure quality of reported measurements.

Quality control (QC) Mechanisms for checking that reported measurements are free of error and conform to acceptable accuracy and precision.

Quantitative data Physical measurements of a study outcome conforming to a validated scale system.

Random effects The treatments to be tested represent a random sample from larger population.

Random error Causes response measurements to fall either side of a target affecting data precision.

Randomisation Reduces the risk of bias in experimental results, concerned with selection of experimental units for use within an experiment and run order of experiments.

Range A simple measure of data spread.

Ranking Ordinal number corresponding to the position of a measurement value when measurements are placed in ascending order of magnitude.

Repeatability A measure of the precision of a method expressed as the agreement attainable between independent determinations performed by a single individual using the same instrument and techniques in a short period of time.

Replication The concept of repeating experimentation to produce multiple measurements of the same response to enable data accuracy and precision to be estimated.

Reproducibility A measure of the precision of a method expressed as the agreement attainable between determinations performed in different locations.

Residuals Estimates of model error, determined as the difference between the recorded observations and the model's fitted values.

Resolution The ability of a two-level or three-level Fractional Factorial Design to provide independent factor effect estimates of the main components of interest.

Response The characteristic measured or observed in a study.

RMD Repeated Measures Design.

Robust statistics Data summaries which are unaffected by outliers and spurious measurements.

RSM Response Surface Methods.

Sample A set of representative measurements of a measurable or observable outcome.

Sample size estimation Integral part of design planning, ensure sufficient measurements are collected to enable study objectives to be properly assessed.

Screening experiments Simple to implement and analyse multi-factor experiments for the early stages of projects to identify important factors.

Significance level The probability of rejecting a true null hypothesis, P(Type I error), typically set at 5% or 0.05.

Skewness Shape measure of data for assessing lack of symmetry.

SNK Student-Newman-Keuls multiple comparison procedure.

SPC Statistical Process Control.

SPD Split-Plot Design.

SQC Statistical Quality Control.

Standard deviation A magnitude dependent measure of the absolute precision of replicate experimental data.

Systematic error Causes response measurements to be in error affecting data accuracy.

Test statistic A mathematical formulae which provides a measure of the evidence that the study data provide in respect of acceptance or rejection of the null hypothesis, numerically estimable using study data.

Taguchi methods Experimental design and analysis techniques initially pioneered by Genichi Taguchi for quality improvement of products and processes.

Transformation A technique of re-coding data so that the non-normality and non-constant variance of reported data can be corrected.

Treatment The controlled effect being assessed in an experiment for its influence on a measurable or observable outcome.

Type I error (False positive) Rejection of a true null hypothesis, P(Type I error) = significance level.

Type II error (False negative) Acceptance of a false null hypothesis, P(Type II error) = 1 – power/100.

Variability The level of variation present within collected data, also consistency and spread.

1

Introduction

1.1 INTRODUCTION

The origins of contemporary experimental design are generally attributed to Sir Ronald Fisher who published seminal work on statistical principles in the 1920s. The impact of his work was to become apparent in the late 1930s when terms such as **statistics**, **experimental design**, **treatment effect**, **randomisation**, **Analysis of Variance**, and **significance** and were to become recognised as synonymous with the efficient planning and analysis of data over a wide range of subject areas. In a famous address to the Indian Statistical Congress, Fisher was reported to have said:

> "To consult a statistician after an experiment is finished is often merely to ask him to conduct a *post mortem* examination. He can perhaps say what the experiment died of."

Fortunately, since then, there has been a prolific increase in the use of statistical methods and an appreciation of the merits of planned data collection and analysis. The methods have become the servant of the research and development community for the design and manufacture of new useful products. The methods have underpinned the discovery of fundamental knowledge and contributed to our understanding of the life sciences and social behaviour. Indeed, very few disciplines have made progress in the twentieth century and not been influenced by the principles of experimental design. In particular, experimental design has become the cornerstone of good statistical practice and internationally adopted by regulatory authorities and statutory bodies concerned with the safe and effective development of new processes and products. It should be realised that the term **experiment** is open to a very broad interpretation and covers any type of study, trial, or investigation where data are to be collected and assessed. It is not confined to the narrow interpretation of a laboratory experiment!

Despite these achievements the subject is still largely unexplored and under used. Many practitioners seldom get to implement the full range of rich techniques developed over the last 80 years by distinguished contributors such as Jerzy Neyman, Egon Pearson, Karl Pearson, Maurice Kendall, George Snedecor, John Tukey, Frank Yates, George Box, and Genichi Taguchi. This is either because of resource constraints or the benefits of the methods are not sufficiently promulgated. This comes at a time when data are increasingly generated and stored in large quantities using high speed computer systems and when there is a demand for the rapid assessment of complex multi-factor studies. Often the investigator will not be an expert in mathematics or statistics but someone from a completely different discipline with an interest in the application of experimental design to his or her problem domain.

Statistical practice covers a large number of statistical methods, and the key challenge to the modern day practitioner is to design the collection of data so that it can be effectively converted into useful knowledge. Experimental design does exactly that. Using the methods is like using a diagnostic test to detect disease or an electron microscope to reveal unknown cell structures. Central to the methods described within this text is the identification of an underlying mathematical model which explains variations affecting a single measured response. In particular, the model asserts that

data = controlled variation + random variation

where random variation is commonly referred to as **experimental error**. From the model, statistical analysis procedures can be constructed which enable hypotheses to be tested that are relevant to the study objectives.

Most of these statistical tests have become readily accessible with the advent of computers and statistical software for the processing of data. In this chapter, it therefore seems appropriate to commence with an introduction to those software packages which are adopted throughout the text to illustrate the application of the methods of analysis from which the inferences are finally drawn. Thereafter, graphical presentations and numerical summaries will be discussed as they are important to all of the analysis methods which follow. Graphical procedures enable the data to be explored visually while use of numerical summaries provides the means of quantitatively condensing the data. These materials and methods will set the scene for comparing the means and variability between two groups of data as outlined in Chapter 2.

The formal extension of the methods to more efficient designs for the collection and analysis of data follows in the remaining chapters where the methods to be discussed make use of inferential procedures and, in particular, the **Analysis of Variance** (ANOVA) technique. Inference procedures provide the techniques for objective interpretation based on using the collected data to weigh up the evidence for or against the study findings. As the more advanced and recent techniques are covered in the final chapters, and two-level design techniques within Fractional Factorial Designs (FFD), Orthogonal Array Designs (OAD), and Taguchi Methods (TM) are described, the emphasis moves full circle back to the use of graphical plots and numerical summaries. The advantages of combining exploratory and inferential analysis approaches lies in their ability to provide a comprehensive view of the data from which both subjective and objective interpretations can be derived.

1.2 INFORMATION ON THE STATISTICAL SOFTWARE

Statistical software packages, such as SAS[1], Minitab[2], SPSS[3], S-Plus[4], GLIM[5], BMDP[6], and Statgraphics[7], are widely used for data handling. They contain extensive provision of commonly used graphical and statistical analysis routines which are the backbone of statistical data analysis. After a little familiarisation, they are simple to use and their wide coverage of routines enables a variety of data presentation and analysis methods to be available to an investigator. In particular, SAS is considered an industrial standard while Minitab is widely available throughout educational institutions. Such software is available across many computing platforms though most are now implemented within the personal computer environment under Windows. Spreadsheets, such as Microsoft Excel[8], come into this category though their provision of default statistical procedures is currently limited.

We have chosen to make use of the statistical software packages SAS and Minitab. Other software, such as S-Plus and GLIM, could equally be used but we believe SAS and Minitab are best as they are simple to use, ubiquitous, and compatible in most of their operations with the operational features of Windows software. The data presentation and analysis principles will be presented in a way that can be readily transferred to other software packages.

A major feature of the statistical data analysis illustrations is that information on how the output was obtained within the software will be provided in appendices to the chapters to guide the reader through software usage. Output editing has also been used on occasions to enable the outputs to be better presented than would initially have been the case. Nevertheless, it is intended that once an investigator has identified an appropriate design and used such to collect response data, it will be possible to expedite the analysis using the software illustrations.

1.2.1 SAS

The information presented in this text refers to SAS Release 6.11 where entry will result in a VDU screen similar to that of Fig. 1.1. The menubar at the top of the screen displays the Menu procedures available in SAS. The toolbar buttons below cover many of the same procedures and by resting the mouse pointer on a button (without clicking), a short description of the procedure is displayed with the status bar at the bottom of the screen showing a fuller definition. SAS can be operated using various windows, these being Program Editor, Log, Output, and Graph. The Program Editor window is active on entry and is where SAS programs are entered before execution. The Log window is used to display program execution information. The Output and Graph windows are hidden on entry and are only activated when program execution requires that numerical information be generated (Output) and high-resolution data plots be produced (Graph).

The **File** option, in Fig. 1.1, contains access to file opening, saving, and window printing while the **Edit** option provides access to SAS's copy and paste facilities for copying programs, output, and plots as well as facilities for modification of screen styling. **View** provides movement between plots in the Graph window. **Locals** is used to submit and retrieve programs when the Program Editor window is active and provides graph editing when the Graph window is active. **Globals** and **Options** enable aspects of SAS's default operation to be customised while the **Window** option allows for movement between windows. As usual with windows type software, extensive on-line help and tutorial support can be accessed through the **Help** menu.

The default statistical analysis features of SAS can be accessed by either entering and executing sequences of SAS statements in the Program Editor window or using the SAS/ASSIST menu interface. All aspects of SAS operation within this text will be illustrated using the former approach. In the appendices to each chapter, sample SAS programs will be provided to demonstrate how output illustrated was, or could have been, generated. Data entry in SAS will be described by means of direct input through the Data step within a sequence of SAS statements. Data entry from an ASCII file is also possible as is spreadsheet data entry through SAS/INSIGHT. Data variables must be of one type, either numerical or alpha-numeric. Missing value code is given by a . (dot). Appendix 1A provides a brief introductory outline of the operational features of SAS.

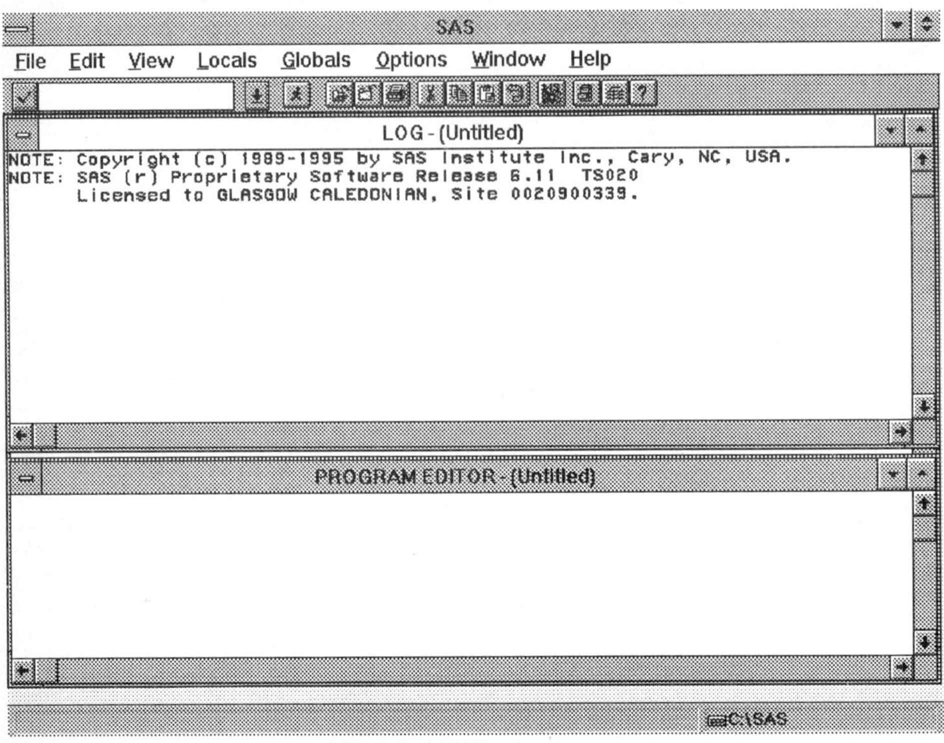

Fig. 1.1 — SAS entry screen

Saving of programs of SAS code and the output which has been created is based on the menu commands **File ➤ Save As** with subsequent saving using only **File ➤ Save**. Programs are generally saved using the .sas extension. Use of **File ➤ Open** enables previously saved programs or output to be retrieved for modification and re-execution but only in the Program Editor window. Data transfer from SAS is available through copying a data print in the Output window, transferring it to the Program Editor window, editing the window to remove unwanted material, and copying the window to the clipboard for export to other Windows-based software packages. Data transfer by means of ASCII files is also possible as is use of Dynamic Data Exchange (DDE). DDE enables data to be entered in one software system, e.g., SAS, and received automatically in another, e.g., Excel.

1.2.2 Minitab

The information presented in this text refers to Minitab Release 11.2 where entry will result in the VDU screen shown in Fig. 1.2. Minitab is operated using a sequence of Windows for storage of data and printing of the presentational elements of data analysis. The **Data** window, which is the active window on entry, is the spreadsheet window for data entry. The **Session** window, also shown on entry, is for entering session commands and displaying, primarily, numerical information. The **Info** window, which is hidden on entry, contains a compact overview of the data and number of observations in the worksheet displayed in the Data window. The **History** window, also hidden on entry, displays all session commands produced during a Minitab session. The **Graph** window, which will only be activated when a data plot is requested, displays the professional graphs produced by Minitab. When using Minitab, the Data, Session, and Graph windows are the most commonly accessed.

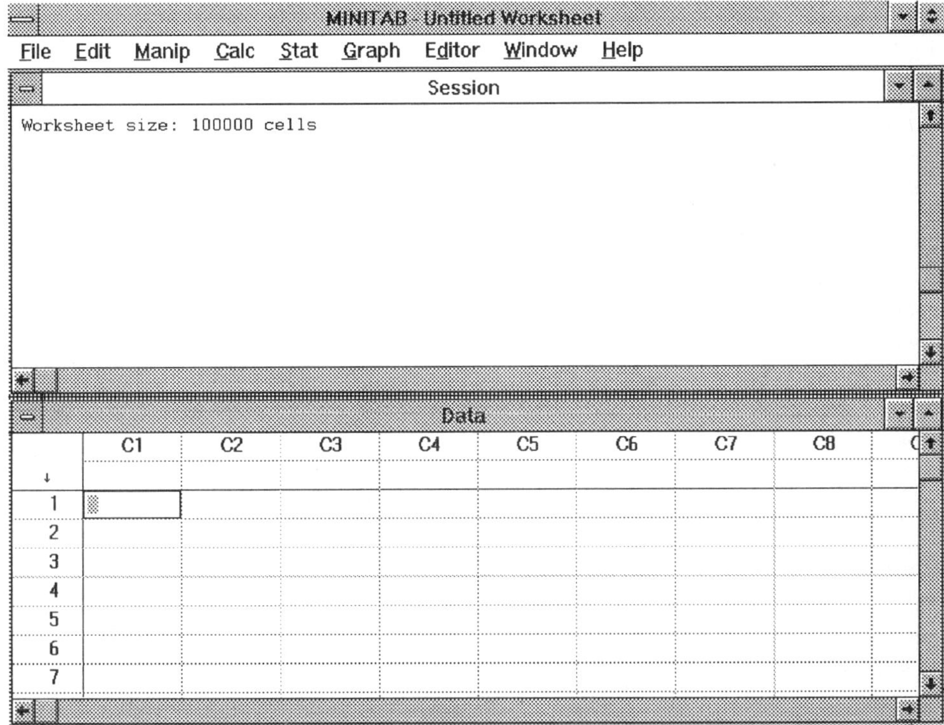

Fig. 1.2 — Minitab entry screen

The menubar at the top of the screen specifies the menu procedures available within Minitab. Irrespective of active window, these menu procedures mostly remain unaltered. The **File** menu contains access to worksheet opening and saving, graph storage and retrieval, output file creation, and file printing. The **Edit** menu provides access to Minitab's copy and paste facilities for copying output and graphs to the clipboard for onward transfer, and moving cells in the data window. The **Manip** and **Calc** menus provide access to data manipulation and calculation features with **Manip** also accessing data printing in the Session window. Data summary and inferential procedures are accessed through the **Stat** menu and choice of an appropriate sub-menu associated with the required data analysis. The routines available are comprehensive and cover most statistical data analysis procedures from basic statistics (Basic Statistics), incorporating descriptive statistics and two sample inference procedures, through to ANOVA procedures (ANOVA) for formal one factor and multi-factor experimental designs. Experimental design construction and analysis components for two-level multi-factor experiments are also available (DOE). The **Graph** menu, as the name suggests, provides access to Minitab's extensive plotting facilities incorporating simple *X-Y* plots (Plot), boxplots (Boxplot), dotplots (Character Graphs ➤ Dotplot), interval plots (Interval Plot), and normal plots (Normal Plot). The **Editor** menu changes with the active window. For Session, it enables session command language and fonts to be modified while for Data, it provides spreadsheet editing facilities. When the Graph window is active, graph editing facilities are available under **Editor**. Movement between windows can be achieved through use of the **Window** menu while comprehensive on-line help is provided through the **Help** menu.

Minitab can be operated using either session commands or menu commands. We will only use the latter for each illustration of Minitab usage. A summary of the conventions adopted for explanation of the menu procedures in Minitab is shown in Table 1.4 in Appendix 1A. Only those features of Minitab appropriate to the statistical techniques explored will be outlined though this represents only some of the operational potential of Minitab in terms of data handling and presentation.

In Minitab, data entry is best carried out using the spreadsheet displayed in the Data window though session commands could equally be used. Data are entered down the columns of the spreadsheet (see Fig. 1.2) where C1 refers to column 1, C2 column 2, C3 column 3, and so on. Unlike the Microsoft Excel spreadsheet, it is not possible to mix numerical and alpha-numeric data types in a column in the Minitab spreadsheet. Only one type of data can therefore be entered in a column with labelling (up to eight characters) achieved using the empty cell immediately below the column heading. Missing value code is given by a * symbol.

Once data have been entered, it is always important to check them for accuracy and then save them in a worksheet file (.mtw extension) on disc. When first saving data, we use the menu commands **File ➤ Save Worksheet As** and fill in the presented dialog boxes accordingly. For subsequent savings, if data are edited or added to, we would use the menu commands **File ➤ Save Worksheet** which will automatically up-date the data file. The **File ➤ Open Worksheet** menu command enables previously saved worksheets to be retrieved. The **Edit ➤ Copy** facility provides a means of exporting data or output from Minitab to other windows-based software via the clipboard when operating the software packages simultaneously. Data interchange with SAS is possible by use of ASCII data files or through DDE links. Minitab can also save data in a Microsoft Excel format (.xls extension) enabling data interchange with Excel. Saving of graphics is available through the **File ➤ Save Window As** facility when the **Graph** window is active.

1.3 SUMMARISING EXPERIMENTAL DATA

To interpret study data, it is necessary to first present them in summary form to provide the basis for analysis and interpretation. All of the analysis methods to be described in this text will be directed to handling quantitative data and will be approached from two angles: use of **descriptive statistics** and use of **inferential statistics** including **statistical tests** and **estimation**. Descriptive statistics cover graphical presentations (simple data plots) and the evaluation of relevant summary statistics (mean, standard deviation, *CV*). Such components are often used as the principal aspects of **exploratory data analysis** (**EDA**) when analysing and interpreting study data.

1.3.1 Graphical Presentations

Data gathered from experiments can be difficult to understand and interpret in their raw form. **Data plots** represent simple pictorial representations which provide for concise summarising of data in simple and meaningful formats. The use of graphics to assess data has been common practice for over a century. In fact, Florence Nightingale, in the 1850s, advocated the use of graphical methods to interpret patient data. The widespread availability of powerful statistical and graphical software has made graphical analysis of data more accessible.

Which graphical presentations to use, however, depend on the study objective(s), the study design, the nature and level of measurement of the response(s), and the amount of data collected. Several forms of graphical presentation exist with histogram, boxplot, dotplot, scatter diagram (*X*-*Y* plot), standard error plot, normal plot, control charts, time series plots, interaction plots, and quantile-quantile (*Q*-*Q*) plots representing some of those in general use. Several of these pictorial data presentations will be used within this text.

One of the most commonly used and simple to interpret data plots is the **dotplot**. In such a plot, each response measurement is presented separately enabling each measurement's position to be displayed as illustrated in Fig. 1.3. The plot consists of a horizontal axis covering the range of measurements with each measurement specified by a dot, or suitable symbol, at the requisite value on the axis. A dotplot is particularly useful when comparing measurements from two or more data sets as later illustrations will show.

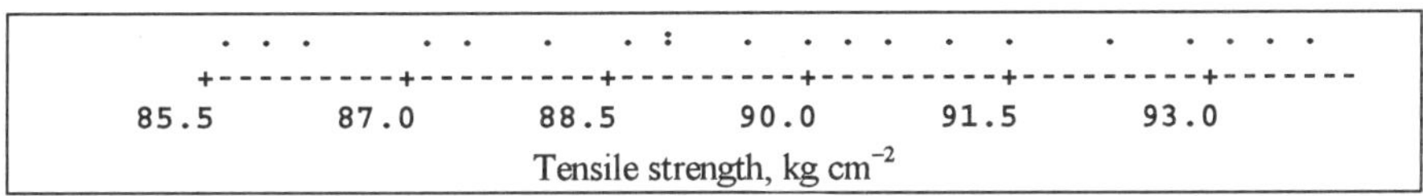

Fig. 1.3 — Illustration of data dotplot

Another useful plot within exploratory data analysis is a **boxplot** shown in Fig. 1.4. Boxplots illustrate the spread and patterning of data and are useful for identifying outliers (see Section 1.5). The plot corresponds to a box, based on the lower quartile (Q_1, 25% of data below) and upper quartile (Q_3, 25% of data above), where the vertical crossbar inside the box marks the position of the median, the value which splits the data in half (Q_2, 50% of data below and 50% of data above) (see Box 1.1). Lines, or **whiskers**, are used to connect the box edges to the minimum and maximum values in the data set.

In respect of using boxplots to assess spread and patterning, the illustrations in Fig. 1.4 show three likely cases. In Set 1, the median lies near the middle of the box and the tails are of roughly equal length indicative of **symmetric** data where mean $\approx$ median. **Right-skew** data are represented by Set 2 where the median lies close to the lower quartile with the plot exhibiting a short left tail and long right tail with mean greater than median. Set 3 corresponds to **left-skew** data with median close to upper quartile and the related boxplot showing a long left tail and short right tail and mean less than median.

Fig. 1.5 provides further illustration of skewness. It shows that right-skew data tend to concentrate at the lower end of their range of measurement though a few larger measurements may also appear. Hence, the long tail to the right and the specification that the data are right-skew. Left-skew data exhibit the opposite trend with most data concentrated at the upper end of the range of measurement with a long left tail appearing.

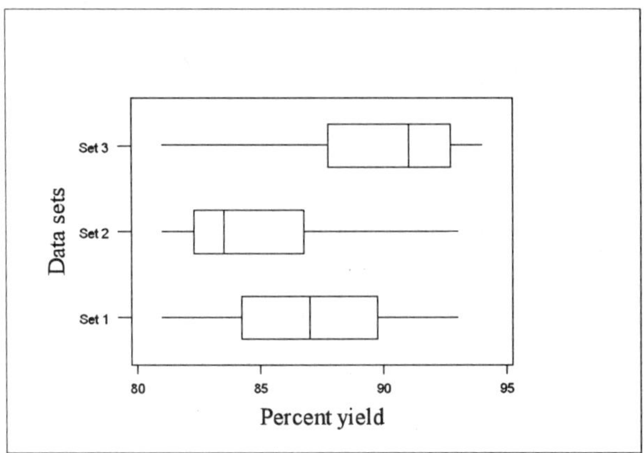

Fig. 1.4 — Illustration of data boxplots

A numerical summary of data skewness is provided by the statistic

$$\frac{n\sum(x-\bar{x})^2}{(n-1)(n-2)s^3} \qquad (1.1)$$

where n is the number of data, x represents each data value, $\bar{x}$ and s are numerical summaries referring to the sample mean and sample standard deviation, respectively (see Section 1.3.2). Negative values of expression (1.1) suggest left-skew data, positive values right-skew, and values near zero indicate symmetrical data.

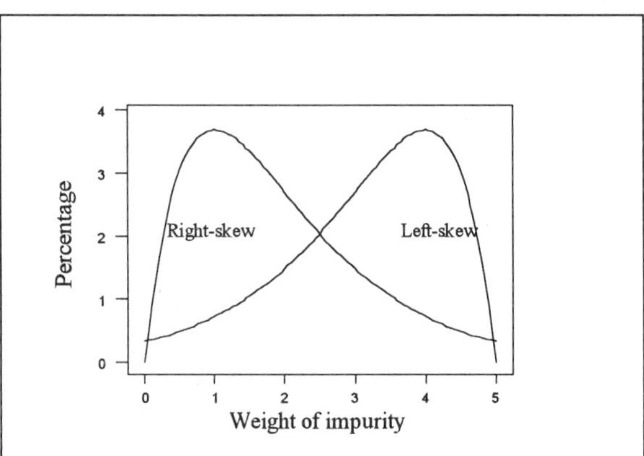

Fig. 1.5 — Illustrations of skew-type data

1.3.2 Numerical Summaries

Plotting data is the first step in analysis. The second step requires determination of numerical summary measures which are thought to be "typical" of the measurements collected. Two basic forms of measure are generally used: a **measure of location** and a **measure of variability**.

The former is defined as a single measurement value specifying the position of the "centre" of the data set, i.e. a datum the data tend to cluster around. The

arithmetic mean, or **average**, and the **median** are two commonly used measures of location. Derivation of both is explained in Box 1.1. Comparison of a sample mean to a reference measurement or target specification can be used to assess data **accuracy** with similarity suggesting good accuracy and dissimilarity inaccuracy.

Box 1.1 — Summary Measures of Location

Mean The mean corresponds to the sum of the measurements divided by the number n of measurements collected and is denoted by $\bar{x}$, i.e.

$$\bar{x} = \sum x / n \quad . \tag{1.2}$$

Median The median refers to the value which splits the data 50:50 with 50% of the measurements below and 50% above. It is denoted by Q_2 (the second quartile) and can be obtained by arranging the measurements in order of magnitude from smallest to largest and finding the value which divides the ordered data into two equal halves. For a sample of n measurements, this will correspond to the $[(n + 1)/2]$th ordered observation when n is odd, and is the average of the $(n/2)$th and $(n/2 + 1)$th ordered observations when n is even.

A measure of data variability is used to provide a numerical summary of the level of variation present in the measurements in respect of how the data cluster around their "centre" value. Variability is often also referred to as **spread**, **consistency**, or **precision**. Variation measurements in common usage for summarising data include **range**, **standard deviation**, and **coefficient of variation**, the computation of which are described in Box 1.2. Low variability measures signify closely clustered data exhibiting low spread, good consistency, or high precision. In contrast, high values are indicative of wide dispersion in the data and are reflective of high variability or imprecise data.

The reason behind dividing by $(n - 1)$ in the standard deviation calculation (1.3) lies with the concept called **degrees of freedom**. It is known that the sum of the deviations of each measurement from their mean will always be equal to zero, i.e. $\sum(x - \bar{x}) = 0$. If we know $(n - 1)$ of these deviations, this constraint means the nth deviation must be fixed. Therefore, in a sample of n observations, there are $(n - 1)$ independent pieces of information, or degrees of freedom, available for standard deviation estimation. Hence, the use of a divisor of $(n - 1)$ in the calculation. Squaring the standard deviation (1.3) provides the **variance** of the data. The standard deviation is a measure of the **absolute precision** of data, the value of which depends on the magnitude of the data. The *CV* provides a measure of the **relative precision** of data and is a measure of data variability which is independent of units. In this sense, the *CV* is more useful when comparing different data sets which may differ in magnitude or scale of measurement. The *CV* is also referred to as the **relative standard deviation** (RSD).

Example 1.1 Table 1.1 provides data on the weights, in g, of 20 pieces of scrap metal from a metal fabrication process.

Table 1.1 — Scrap metal weights for Example 1.1

2.83	2.68	2.57	2.73	2.56	2.62	2.38	2.45	2.35	2.43
2.37	2.62	2.41	2.69	2.78	2.47	2.35	2.52	2.63	2.76

Box 1.2 — Summary Measures of Variability

Range This statistic is specified as the difference between maximum and minimum measurements, i.e. maximum – minimum. It is simple to compute but ignores most of the available data and can be influenced by atypical values.

Standard deviation This provides a summary of variability which reflects how far the measurements deviate from the mean measurement. The sample standard deviation, for n measurements, is expressed as

$$s = \sqrt{\frac{\Sigma(x-\bar{x})^2}{n-1}} = \sqrt{\frac{\Sigma x^2 - \frac{(\Sigma x)^2}{n}}{n-1}} \qquad (1.3)$$

where the numerator expression is referred to as the corrected sum of squares and cannot be negative.

Coefficient of variation (CV) This is simply the standard deviation expressed as a percentage of the mean, i.e.

$$CV = 100(s/\bar{x})\% \quad . \qquad (1.4)$$

Standard error of the mean This is a measure of the precision of the sample mean as an estimate of the population mean, small values implying good precision. It is expressed as

$$se(\bar{x}) = s/\sqrt{n} \quad . \qquad (1.5)$$

Exercise 1.1 Consider the weight data in Example 1.1. We will illustrate the calculation of relevant summary statistics both manually and using SAS.

Manual derivation: We have $n = 20$ weight measurements. Using these measurements, we first calculate Σx as 2.83 + 2.68 + ... + 2.63 + 2.76 = 51.2 and Σx^2 as $2.83^2 + 2.68^2 + ... + 2.63^2 + 2.76^2 = 131.5236$.

Using $\Sigma x = 51.2$ and $n = 20$, the mean (1.2) becomes $\bar{x} = 51.2/20 = 2.56$ g.

Ordering the measurements with respect to magnitude provides the sequence

2.35	2.35	2.37	2.38	2.41	2.43	2.45	2.47	2.52	2.56
2.57	2.62	2.62	2.63	2.68	2.69	2.73	2.76	2.78	2.83

As $n = 20$ is even, the median will be the average of the 20/2 = 10th and (20/2 + 1) = 11th ordered observations. As these are 2.56 and 2.57 respectively, then $Q_2 = 2.565$ g which is almost identical to the mean indicating symmetric data.

As the maximum is 2.83 and the minimum is 2.35, the range will be 2.83 – 2.35 = 0.48 g.

Using the summations found initially, the standard deviation (1.3) becomes

$$s = \sqrt{\frac{131.5236 - 51.2^2/20}{20-1}} = \sqrt{0.023768} = 0.1542 \text{ g} \quad .$$

The CV expression (1.4) provides 100(0.1542/2.56) = 6.023% which is low indicating small variability and good consistency in the weight measurements.

The standard error expression (1.5) becomes $0.1542/\sqrt{20} = 0.0345$ g. This figure is relatively low further reinforcing the conclusion of small variability and good consistency in the weight measurements.

SAS generation: The summaries generated by SAS's PROC MEANS procedure for the weight measurements are presented in Display 1.1. Routine procedures in most software can generate more statistical summaries than we would normally require. Only a subset of these are generally used to summarise a data set. The figures presented in Display 1.1 for the mean, range, standard deviation, CV, and standard error agree with those determined manually and so provide the same data interpretation. □

Display 1.1

```
Analysis Variable : Weight
 N      Mean    Minimum    Maximum      Range    Std Dev          CV    Std Error
-------------------------------------------------------------------------------
20    2.5600     2.3500     2.8300    0.4800     0.1542    6.0223         0.0345
-------------------------------------------------------------------------------
```

Exercise 1.1 has been included primarily to illustrate manual derivation of summary statistics. All future illustrations of data handling will be oriented towards using software (SAS or Minitab) to provide the summary measures as good statistical practice generally requires calculations to be undertaken using dedicated software. Obviously, investigations can often involve collection of more than one sample of data. Exploratory comparison of summaries is straightforward in these cases with means compared on the basis of differences and variabilities compared on basis of multiplicative difference with a ratio of 2:1 or more generally indicative of important differences in data variability.

1.4 THE NORMAL DISTRIBUTION WITHIN DATA ANALYSIS

The **normal**, or **Gaussian, distribution** is the most important of all the probability distributions which underpin statistical principles. It is at the heart of parametric inference procedures where it is assumed that the measured response is normally distributed. A knowledge of the shape of the distribution is therefore useful for the understanding of inferential data analysis. The normal distribution function is specified by the expression

$$f(x) = \frac{1}{\sqrt{2\pi}\sigma} e^{-(x-\mu)^2/2\sigma^2} \tag{1.6}$$

where μ (*mu*) is the population mean, σ (*sigma*) is the population standard deviation, and e represents the exponential constant. Graphically, this function describes a bell-shaped curve symmetrical about the mean μ with the area under the curve equalling one unit. This distribution is found useful in describing, either exactly or approximately, many different responses such as germination time of seedlings, lead pollution levels in fauna, pH of soil, concentration of albumin in blood sera, and noise pollution.

Fig. 1.6 provides a diagrammatic representation of expression (1.6) for a mean μ of 100 and three values of standard deviation σ. For σ equal to 5, the curve is sharply peaked with short tails signifying that data with such a pattern would be representative of a highly precise experiment (low variability, good precision). Increasing variability flattens the peak and lengthens the tails corresponding to data with greater variability and less precision. An important property of all normal distributions is that 68% of the values lie between $\mu - \sigma$ and $\mu + \sigma$, 95% of values lie between $\mu - 1.96\sigma$ and $\mu + 1.96\sigma$, and 99% of values lie between $\mu - 2.5758\sigma$ and $\mu + 2.5758\sigma$. Data sampled from any of the populations exhibited in Fig. 1.6 are described as **symmetrical** and generally have similar mean and median.

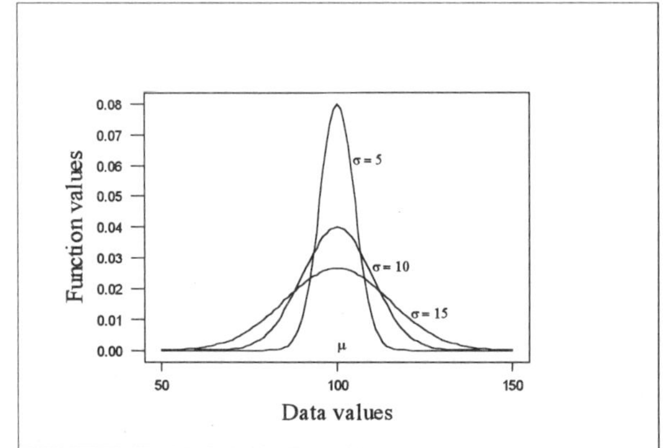

Fig. 1.6 — Normal distribution with the same mean $\mu = 100$ and different standard deviations $\sigma = 5$, 10, and 15

Our interest in the normal distribution lies in its use as the assumed distribution for the measured response and the role this plays in the underlying theory of parametric inferential data analysis procedures.

1.5 OUTLIERS IN DATA

Summary statistics like the mean and standard deviation can be sensitive to **outliers**, atypical measurement values which differ from the majority of measurements. Inclusion of outliers can influence data interpretation while exclusion may be justified on the basis of knowledge of why such data values differ from the majority. However, inclusion of outliers is often necessary in order to assess the collected data fully and also because assessors monitoring experimental analyses may not accept unexplained omission of data measurements, e.g. the Food and Drug Administration practice in the US for clinical trial data. In the interests of good statistical practice, outliers can only be omitted under exceptional circumstances when there is justification. The point is that we should be aware of how to detect outliers and of their effect on the interpretation of the data.

Graphical detection of outliers can be based on the use of **boxplots** (see Section 1.3.1). When software is used to generate the boxplot, an outlier is generally represented by a "*" beyond the whiskers. Detection of an outlier uses hidden **inner** and **outer fences** located, respectively, at a distance $1.5(Q_3 - Q_1)$ and $3(Q_3 - Q_1)$ below the lower quartile and above the upper quartile. Observations between the inner and outer fences of the boxplot are considered **possible**, or **suspect**, outliers while values beyond the outer fences are classified as **probable**, or **highly suspect**, outliers.

Provided data can be assumed normally distributed, numerical detection of outliers can be based on evaluating the z score

$$z = (\text{value} - \bar{x})/s \tag{1.7}$$

for each measurement and comparing it to ±2.698 which is the equivalent z score to the inner fence values in a boxplot. These scores have a normal distribution with mean 0 and standard deviation 1. Consequently, only 1% lie beyond ±2.5758 and only 5% lie beyond ±1.96. If a z score lies beyond ±2.698, then the corresponding measurement can be considered a **possible** outlier. The probability of observing such a z value is as small as 0.007.

Example 1.2 Table 1.2 contains measurements of the pH of gastric juice samples from 15 different patients within a gastroenterological investigation.

Table 1.2 — pH measurements of gastric juice specimens for Example 1.2

4.49	4.34	2.40	3.95	3.74	3.50	4.12	4.20
4.09	3.56	3.78	3.94	4.80	4.63	4.05	

Exercise 1.2 Checking the data of Example 1.2, we see that the measurement 2.40 appears very different from the rest which lie between 3.5 and 4.8. We need, therefore, to assess whether this measurement is an outlier and to investigate the effect it has on the data summaries.

Detection of outlier: A boxplot of the pH data presented in Fig. 1.7 clearly highlights this datum as a possible outlier (presence of * symbol). The remainder of the measurements are similar as shown by the compactness and near symmetry of the boxplot. The measurement of 2.40 has a z score (1.7) of

$$z = (2.40 - 3.973)/0.571 = -2.75$$

within the range of it being classified as a possible outlier. Both graphical and numerical detection methods clearly specify that the 2.40 pH measurement appears a possible outlier.

Summary statistics: Summaries of the pH data with and without the possible outlier are presented in Table 1.3. Inclusion of the value 2.40 decreases the mean marginally, due to the outlier being an observation with a low value. The standard deviation is higher with the outlier because it is effectively increasing the range of reported pH measurements. For this case, only the standard deviation is markedly affected by the outlier suggesting more variability in the data than is actually occurring in the majority. □

Table 1.3 — pH summaries for Exercise 1.2

	Mean	Median	Standard deviation	Minimum	Maximum
With outlier	3.973	4.05	0.571	2.4	4.8
Without outlier	4.085	4.07	0.385	3.5	4.8

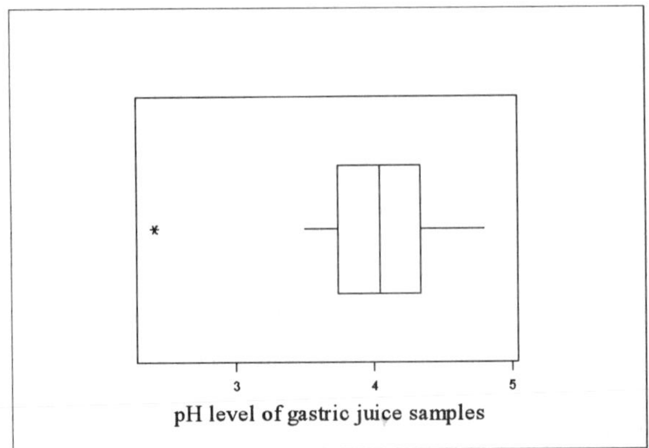

Fig. 1.7 — Boxplot of gastric juice pH measurements for Exercise 2.2 with * denoting possible outlier

From Exercise 1.2, we can see that both outlier detection methods arrive at the same conclusion on the unusual pH observation of 2.40. The differences shown in the summaries highlight the effect outliers can have and the importance of detecting such values prior to analysis. In most cases, and in particular if there is more than one outlier in a data set, the boxplot and fences method is better for detection purposes because it is unaffected by the presence of atypical observations. Presence of an outlier can inflate the standard deviation which, in turn, affects the z score method. Dixon's Q test and Grubb's test provide formal statistical tests for outliers (Miller & Miller 1993, Snedecor & Cochran 1980) while use of robust statistics such as the **trimmed mean** (mean of mid 90% of data) can enable data containing outliers to be summarised accounting for the presence of atypical data values.

Notes

1. SAS (Statistical Analysis System) is a registered trademark of the SAS Institute, Inc., SAS Campus Drive, Cary, NC 27513, USA.
2. Minitab is a registered trademark of Minitab, Inc., 3081 Enterprise Drive, State College, PA 16801, USA.
3. SPSS (Statistical Package for the Social Sciences) is a registered trademark of SPSS, Inc., 444 N. Michigan Avenue, Chicago, Illinois 60611, USA.
4. S-Plus is a registered trademark of StatSci Europe, Osney House, Mill Street, Oxford OX2 0JX, UK.
5. GLIM is a registered trademark of NAG Ltd, Wilkinson House, Jordan Hill Road, Oxford ON2 8DR, UK.
6. BMDP is a registered trademark of SPSS, Inc., 444 N. Michigan Avenue, Chicago, Illinois 60611, USA.
7. Statgraphics is a trademark of Statistical Graphics Corporation, Inc., 2115 East Jefferson Street, Rockville, Maryland 20852, USA.
8. Microsoft Excel is a registered trademark of the Microsoft Corporation, One Microsoft Way, Redmond, WA 98052-6399, USA.

1A Appendix: Introductory Software Information

Data entry

Data entry in software generally requires columns of data to be entered where the columns correspond to the response and character or numerical codes define the factors associated with the sample observations. Both SAS and Minitab can use character sample coding though certain of Minitab's procedures require numeric coding. An explanation of Minitab operation will be by use of menu procedures and to this end, for which Table 1.4 provides a brief summary of the conventions adopted throughout the chapter Appendices. In this introductory Appendix, all the information provided will relate to the scrap weight study of Example 1.1.

Table 1.4 — Minitab conventions

Item	Convention
Menu command	The menu to be chosen is specified in bold with the first letter in capital form, e.g. **Calc** for access to the calculation facilities and **Stat** for access to the statistical analysis facilities.
Emboldened text	Bold text corresponds to either the text to be typed by the user, e.g. **Species**, the menu option to be selected, or the button to be checked within a dialog box window.
Menu instructions	Menu instructions are set in bold with entries separated by a pointer. For example, Select **Stat ➤ ANOVA ➤ Balanced ANOVA** means select the Stat menu, open the ANOVA sub-menu by clicking the ANOVA heading, and choose the Balanced ANOVA procedure by clicking the Balanced ANOVA heading. This will result in the Balanced ANOVA dialog window being displayed on the screen. When text such as "for *Variables*, select **Weight** and click **Select**" is presented, this means click the specified Weight label appearing in the variables list box of the sub-menu window and click the Select button. This procedure specifies that the Lead data are to be used in the routine selected. When text such as "select the **From last value** box and enter **1**" is presented, this means click the box specified From last value and enter the emboldened information in the box. This enables the entered information to be used within the Minitab procedure being implemented. When text such as "for *Statistic*, select **Standard deviation**" is presented, this means select the Standard deviation option or check the labelled box appropriately to specify that the information generated by this choice is to be included in the output created.

SAS: Data entry in SAS is achieved through use of DATA step statements, use of the SAS/ASSIST menu, or the spreadsheet within SAS/INSIGHT. The former involves entering sequences of SAS statements in the PROGRAM EDITOR window and executing them. For the data associated with Example 1.1, the data entry program reads as follows:

```
DATA CH1.EX11;
    INPUT WEIGHT @@;
    LINES;
    2.83  2.68  2.57  2.73  2.56  2.62  2.38  2.45  2.35  2.43
    2.37  2.62  2.41  2.69  2.78  2.47  2.35  2.52  2.63  2.76
;
PROC PRINT DATA = CH1.EX11 NOOBS;
    VAR WEIGHT;
RUN;
```

The DATA step begins with a DATA statement and can include any number of data manipulation statements. The element CH1.EX11 specifies that the DATA step is to create a SAS data set EX11 in the directory referenced by the library reference CH1 which corresponds to the location to be used for data storage. Library reference set-up must be implemented before program execution and is based on the following option selection which will set up the disk in a: drive as the storage location:

> Select **Globals** ➤ **Access** ➤ **Display Libraries** ➤ **New Library** ➤ in *Library Field*, enter **CH1** ➤ select the **Directory to Assign** box and enter **a:** ➤ **Assign** ➤ **Close**.

The INPUT statement specifies the variable name for the response with @@ indicating that the data are being entered in a continuous stream. The LINES statement specifies that data are to be entered in the data file as in-stream data, i.e. within the sequence of presented statements.

Checking of data entry can be achieved using PROC PRINT with the NOOBS element being included to prevent observation numbers from being printed. The VAR statement indicates the response variable to be printed. The RUN statement specifies that the procedure statements prior to it are to be submitted for execution.

Data entry from an ASCII file can be achieved by a simple modification of the illustrated data entry step. For the weight data in an ASCII file EX11.DAT on a disk in drive a:, the following would create from the data file a SAS data set EX11 in the directory referenced by the library reference CH1.

```
DATA CH1.EX11;
      INFILE 'A:EX11.DAT';
      INPUT Y $ WEIGHT;
RUN;
```

Selection of **Locals** ➤ **Submit** will submit a sequence of SAS statements for execution. The Output window will open and a data print should be provided. If execution is not successful, return to the Program Editor window (**Window** ➤ **PROGRAM EDITOR**) and select **Locals** ➤ **Recall text** to retrieve the executed program. The Log window will contain an explanation of why the program did not execute and using this, the program can be edited and re-submitted.

Minitab: Entry of response data in Minitab can be through either the Data window or use of Session commands in the Session window. The former involves entering the data into the columns C1, C2, etc. in the Data window where each cell will contain one observation and every column corresponds to a variable (see Fig. 1.2). For the scrap weight study of Example 1.1, the weight responses would be entered into column C1 as one observation per row of the column. Once data are entered, they require to be checked for accuracy. In Minitab, the **Manip** menu, as shown below, provides access to the data printing facilities with printing being provided in the Session window. The menu procedures for checking the weight data entry would be as follows:

> Select **Manip** ➤ **Display Data** ➤ for *Columns, constants, and matrices to display*, select **Weight** and click **Select** ➤ click **OK**.

Data plots

SAS: The simple dotplot of one sample data, comparable to that shown in Fig 1.3, could be generated using the SAS statements

```
PROC GPLOT DATA = CH1.EX11;
      PLOT Y*WEIGHT / HAXIS = AXIS1 VAXIS = AXIS2;
```

```
        AXIS1 LABEL = (F = CENTB J = C 'Weight') VALUE = (H = 0.8)
              LENGTH = 40;
        AXIS2 LENGTH = 5 VALUE = NONE LABEL = NONE STYLE = 0
              MAJOR = NONE;
RUN;
```

The above statements are typical of the structure of SAS procedure statements. The procedure to be invoked is specified within the PROC statement with GPLOT indicating that a high-resolution plot is required using the data contained in the SAS data file EX11. The PLOT statement specifies Y plotted against X (Y*X), Y referring to a character label which is the same for all observations, with options appearing after the slash symbol for customising the plot, H referring to the horizontal axis and V to the vertical axis. The AXIS elements provides means of customising the plot axes in respect of label font (F), position of the label (J), label size (VALUE), axis length (LENGTH), type of axis points label (LABEL), and nature of axis tick marks (MAJOR). For the AXIS2 statement, which refers to the Y axis, STYLE = 0 prevents an axis line from appearing while MAJOR = NONE prevents tick marks from being produced. Selection of **Locals** ➢ **Submit** will submit such SAS statements for execution with the Graph window opening to provide the requested data plot.

Boxplot production is not available by direct means in SAS though the PROC UNIVARIATE procedure can provide such, as well as extensive summary statistics, a stem-and-leaf plot, and a normal plot. Interactive (menu driven) exploratory data analysis facilities can also be accessed through SAS/INSIGHT.

Minitab: In Minitab, dotplot production, such as that illustrated in Fig. 1.3, uses the following menu selection:

Select **Graph** ➢ **Character Graphs** ➢ **Dotplot** ➢ for *Variables*, select **Weight** and click **Select** ➢ click **OK**.

while boxplot production, such as illustrated in Fig. 1.7, uses the following menu selections:

Select **Graph** ➢ **Boxplot** ➢ for *Graph 1 Y*, select **Weight** and click **Select**.
Select **Options** ➢ **Transpose X and Y** ➢ click **OK** ➢ click **OK**.

Data summaries

SAS: Numerical summaries of response data can be generated using the PROC MEANS procedure (see Display 1.1) as follows:

```
PROC MEANS DATA = CH1.EX11 MAXDEC = 4 N MEAN MIN MAX RANGE
          STD CV STDERR;
     VAR WEIGHT;
RUN;
```

MAXDEC defines the number of decimal places for the summaries requested while N, MEAN, MIN, MAX, and RANGE are self-explanatory. STD defines standard deviation, CV specifies coefficient of variation, and STDERR defines standard error. The VAR statement specifies the response to be summarised. Selection of **Locals** ➢ **Submit** will submit such SAS statements for execution with the Output window opening to provide a table of the requested data summaries.

Minitab: Menu commands for default data summary production, similar to that shown in Display 1.1, are given below. The **Graphs** option within this procedure can also be used to generate a high resolution dotplot and boxplot of the data.

> Select **Stat** ➤ **Basic Statistics** ➤ **Descriptive Statistics** ➤ for *Variables*, select **Weight** and click **Select** ➤ click **OK**.

2

Inferential Data Analysis for Simple Experiments

2.1 INTRODUCTION

Industrial and scientific experiments can often be simple in nature involving comparison of, for example, two production processes, two advertising strategies, tensile strengths of two nickel-titanium alloys, toxic chemical removal by two waste management strategies, and animal growth using two different animal feeds. To carry out such assessment appropriately requires the use of basic statistical methods, such as **graphical presentations** and **numerical summaries** as discussed in Chapter 1, and **inferential procedures** in order to gain as much information as possible from the measurements against the background of the inherent variability present in them.

Inference procedures are numerous, providing mechanisms for dealing with a wide variety of experimental situations. They can be conveniently split into two families: **parametric procedures** and **non-parametric procedures**, the difference between them being (1) in the assumptions underpinning them, (2) the type of data to which they are most suited, (3) the way in which the data are used, and (4) their associated power. Parametric tests typically use quantitative continuous data and are generally based on the assumption of normality for the measured response. By contrast, non-parametric procedures are less restrictive in terms of assumptions and in the type of data to which they can be applied, with data being ranked or put into classes. This results in non-parametric procedures generally having lower power than comparable parametric ones.

The inferential statistics and estimation aspects cover those statistical procedures (t tests, F tests, confidence intervals) which are used to help draw objective inferences and derive parameter estimates from data. They enable assessment of the statistical significance of an experimental objective to be carried out by weighing up the evidence within the data, and using it to assert whether the data agree or disagree with certain hypotheses. Emphasis throughout this chapter and others will be on illustrating how descriptive and inferential methods can be used in tandem to interpret data.

2.2 BASIC CONCEPTS OF INFERENTIAL DATA ANALYSIS

Inferential data analysis by means of **hypothesis** or **significance testing** is based on the principle of modelling a response as an additive model of the explanatory factors and an error term. The testing aspect assesses how the variation explained by the controlled components differs from that associated with the error. If controlled variation exceeds error variation markedly, the evidence within the data appears to be pointing to a particular effect being the reason for the difference, and that the effect is potentially of practical significance. If controlled variation does not exceed error, then the differences are likely to be random and not due to an attributable cause. Significance testing of study data therefore assesses whether result differences are likely to be significant due to identifiable differences, or are merely likely to be due to random variations in the measurements. Parametric inference procedures, which cover many of the procedures in this text, are based on the assumption that the error component is normally distributed. It is this assumption which provides the foundation for the underlying theory of parametric inference and estimation procedures.

Significance testing of experimental data rests on the construction of **statistical hypotheses** which describe the likely responses to the study objective being assessed. Two basic hypotheses exist, the **null** and the **alternative** (or **research**). The null hypothesis, denoted H_0 or *NH*, is almost always used to reflect no difference either between the observations and their target or between sets of observations. The alternative, denoted H_1, H_A, or *AH*, describes the difference or change being tested for. These hypotheses are often expressed in terms of the population parameter being tested, μ for mean and σ^2 for variability, though it is feasible to express hypotheses in simpler ways as Example 2.1 will show.

A further aspect to consider, particularly in small sample studies, is that of the **nature of the test**. This reflects whether the question to be assessed concerns a general directional difference (bi-directional) or a specific directional difference (uni-directional). The former occurs when the not equal to symbol ($\neq$) is present in the alternative hypothesis H_1 (two-sided alternative, two-tail test). Presence of less than ($<$) or greater than ($>$) symbols in the alternative hypothesis corresponds to specific directional difference (one-sided alternative, one-tail test) as it reflects that the study objective specifies assessing whether a decrease or increase in the response appears present within the measured data.

Example 2.1 A new, simpler analytical procedure for determining copper (Cu) levels in blood serum has been proposed. The new procedure is to be compared with the standard procedure to examine if the method is comparable in respect of both accuracy and precision of reported measurements. As the investigation is to assess both mean (accuracy) and variability (precision), hypotheses relevant to each case need to be constructed.

Mean: The null hypothesis reflects no difference and is expressed as H_0: no difference in the mean Cu measurement with procedure, i.e. $\mu_{new} = \mu_{standard}$. Assessment of difference in Cu measurement between the two procedures reflects a general directional difference corresponding to a two-sided alternative of the form H_1: difference in mean Cu value, i.e. $\mu_{new} \neq \mu_{standard}$.

Variability: As with mean assessment, the associated null hypothesis corresponds to no difference in the variabilities and so is presented as H_0: no difference in variability of Cu measurements with procedure, i.e. $\sigma^2_{new} = \sigma^2_{standard}$. The question posed is concerned with assessing if there is a difference in the Cu values. This reflects a general directional difference requiring specification of a two-sided alternative of the form H_1: difference in the variability of Cu measurements, i.e. $\sigma^2_{new} \neq \sigma^2_{standard}$. □

Presence of the equality sign when specifying the null hypothesis does not infer numerical equality of the observations. It simply means that the population parameters appear similar enough to reflect no detectable difference of importance. Hypotheses for specific difference testing (one-tail tests) may be expressed differently as H_0: $\mu_A \leq \mu_B$ and H_1: $\mu_A > \mu_B$ though practical definition of the hypotheses is unchanged. Another form of null hypothesis is H_0: $\mu_A - \mu_B = d$ if assessment of a specified level of difference of d between the two population means were appropriate.

The second stage of statistical inference concerns the determination of a **test statistic**, a simple formula based on the sample data. This provides a measure of the evidence that the study data provide in respect of acceptance or rejection of the null hypothesis. Test statistic formulae depend on study structure, study objective(s), and corresponding statistical theory of the inference procedure. Large values generally indicate that the observed difference is unlikely to have occurred by chance and so the null hypothesis is less likely is to be true. Small values provide the opposite interpretation. Significance testing is a decision making tool, with a decision necessary on whether to accept or reject the null hypothesis using the evidence from the study data.

Such a decision could be in error because, in inferential data analysis, we can never state categorically that a decision reached is true but only that it has not been demonstrated to be false. Errors in significance testing are referred to as **Type I error**, or **false positive**, and **Type II error**, or **false negative**, with their influence on the decision reached summarised in Table 2.1. The probability of a Type I error is generally specified as α, the **significance level of the test**. Given that significance level is a measure of error probability, all statistical testing is based on the use of low significance levels such as 10%, 5%, and 1%, with **5%** the generally accepted default. If we reject the null hypothesis at the 5% level, we specify that the difference tested is significant at the 5% level and that there is only a 5% probability that the evidence for this decision was due to chance alone. In other words, the evidence points to the detected difference being statistically important which will, hopefully, also reflect practical importance. In prospective studies, it is good practice to choose α **before** undertaking the data collection. For this reason, the significance level of the test will be specified in the study protocol (see Section 2.7).

Table 2.1 — Errors in inferential data analysis

		Decision reached	
		accept H_0	reject H_0
True	H_0 true	no error	Type I
result	H_0 false	Type II	no error

Implementation of all statistical inference procedures for the objective assessment of data is based on minimising both Type I and Type II errors. In practice, the probabilities of Type I and Type II errors, denoted α and β respectively, are inversely related such that, for a fixed sample size, decreasing α causes β to increase. Generally, we must specify the significance level of the test α and select a test statistic which maximises the **power** of the test $(1 - \beta)$.

In scientific journals, inferential results are often reported using the ***p* value**, the probability that the observed difference could have occurred by chance alone. It provides a measure of the weight of evidence in favour of acceptance of the null hypothesis, large values indicative of strong evidence and small values indicative of little or no evidence. Often, it is expressed in a form such as $p < 0.05$ (null hypothesis rejected at 5% significance level) meaning that there is, on average, a 1 in 20 chance that a difference as large as the one detected was due to chance alone. In other words, there is little or no evidence in favour of acceptance of the null hypothesis and the evidence points to the existence of a significant and meaningful difference. Another definition of p is that it is the lowest significance level at which H_0 can be rejected.

Since both the test statistic and p value provide measures of acceptance of the null hypothesis, we can use either as a means of constructing general rules for deciding whether to accept or reject the null hypothesis in inferential data analysis. Such rules are explained in Box 2.1 and are applicable to most practical applications of inferential data analysis.

Box 2.1 — General Decision Rule Mechanisms for Inferential Data Analysis

Test statistic and critical value approach This approach involves comparing the computed test statistic with a **critical value** which depends on the inference procedure, the significance level of the test, the type of alternative hypothesis (one- or two-sided), and, in many cases, the number of measurements. The latter is generally reflected in the **degrees of freedom** of the test statistic corresponding to the amount of independent sample information available for parameter estimation. Critical values for inference procedures are extensively tabulated as evidenced in the Tables in Appendix A.

For a two-sided alternative (two-tail test), the decision rule is generally

$$\textit{critical value 1} < \textit{test statistic} < \textit{critical value 2} \Rightarrow \textit{accept } H_0.$$

For a one-sided alternative (one-tail test) with H_1 containing the inequality $<$, e.g., $\mu_1 < \mu_2$, we use

$$\textit{test statistic} > \textit{critical value} \Rightarrow \textit{accept } H_0$$

while if H_1 contains the inequality $>$, e.g., $\mu_1 > \mu_2$, we use

$$\textit{test statistic} < \textit{critical value} \Rightarrow \textit{accept } H_0 .$$

p value and significance level approach An estimate of the p value generally appears as part of statistical software output. The associated decision rule, which is dependent only on the significance level of the test, is

$$\textit{p value} > \textit{significance level of the test} \Rightarrow \textit{accept } H_0.$$

Acceptance of the null hypothesis H_0 indicates that the evidence within the data suggests that they appear near target or that the groups compared appear similar to one another. It does not mean that full equality of observations has been demonstrated. Initially, both test statistic and p value approaches will be outlined though, if software is used to provide inferential results, only one of these decision approaches need be used.

Software generation of p values generally provides figures rounded to a given accuracy, e.g., to three decimal places. A value of $p = 0.000$, for instance, does not mean that the p value is exactly zero. Such a result should be interpreted as meaning $p < 0.0005$ and would specify that the null hypothesis H_0 can be rejected at the 0.05% significance level.

One drawback of classical hypothesis testing, as highlighted by Kay (1995), is that inference procedures provide no information about the magnitude of an experimental effect. They simply provide means of detecting evidence in support of or against the null hypothesis. By contrast, estimation through calculation of a **confidence interval** (CI) provides a means of estimating the magnitude of an experimental effect enabling a more specific practical interpretation of the data to be forthcoming.

A confidence interval is defined as a range of response values within which we can be reasonably certain that the experimental effect being estimated will in fact lie. They are constructed in additive form

$$\text{estimate} \pm \text{critical value} \times \text{measure of variability} \qquad (2.1)$$

for location effects such as mean or difference of means, and in multiplicative form

$$\text{function of critical value} \times \text{variability measure} \qquad (2.2)$$

for variability measures. The additive nature of CIs for location measures stems from the requisite critical values being based on symmetric distributions (normal and t) resulting in the interval being symmetrical about a centre value. Confidence intervals for variability measures, by contrast, use critical values from right-skew distributions (χ^2 and F) and so this leads to intervals which are not symmetrical around a centre value. The level of a CI, expressed as $100(1 - \alpha)$ percent, provides a measure of the confidence (degree of certainty) that the experimental effect being estimated lies in the constructed interval. In practice, such levels are customarily set at one of three values: 90% ($\alpha = 0.1$), 95% ($\alpha = 0.05$), or 99% ($\alpha = 0.01$), comparable to the three significance levels associated with inferential data analysis.

Having discussed the principles underpinning statistical inference, it is important to consider how statistical significance relates to practice. Statistical analysis may result in a conclusion stating that a "statistically significant" difference has been detected. This difference may not be large enough, however, to claim that an important interpretable effect has been found. Significance of an effect on a statistical basis, therefore, does not necessarily relate directly to practical importance. Such a decision ultimately rests with the investigator based on their expertise, understanding of the problem, and the size of difference which they consider to be appropriate for practical significance.

2.3 INFERENCE METHODS FOR TWO SAMPLE STUDIES

Two sample experiments are particularly useful when wishing to compare, for example, a new production process with the present process or two methods of feeding animals in respect of rate of growth. Such a structure, as shown in Table 2.2, is often referred to as a "treatment" versus "control" design as its basis lies in subjecting one group to a treatment and using the other to act as the control. As with most simple inference, there are inference and estimation procedures for means and variability. Only small sample tests (sample size $n < 30$) will be described.

Table 2.2 — Two sample design structure

Sample 1	x	x	x	x	.	.	.	.
Sample 2	x	x	x	x	.	.	.	.

x denotes an observation

Example 2.2 As part of a study into lead pollution and its absorbance by seaweed, two species of fucoid seaweed, *Fucus serratus* and *F. spiralis*, were investigated. The former grows at the lower shore and is covered by the sea for most of its life while the latter is generally found at the top of the shore where it is uncovered by the tide for most of its life. The purpose of the investigation was to assess whether lead levels differ between the two species both in terms of the average recorded and the associated variability in lead levels. Samples from 10 plants at each location were taken and assessed for levels of lead, measured as μg Pb g^{-1} dry weight. The results are presented in Table 2.3.

Table 2.3 — Lead measurements for Example 2.3

Fucus serratus	10.0	15.0	21.1	7.1	11.0	10.0	21.2	20.0	12.3	17.6
Fucus spiralis	13.1	18.1	20.0	7.5	14.5	20.0	11.2	16.4	20.0	10.5

2.3.1 Hypothesis Test for Difference in Mean Responses

Two sample studies provide two sets of comparable sample information. Difference of means testing is based on assessing whether the population means are similar (H_0) or deviate markedly from each other (H_1). Box 2.2 provides a summary of the components of this inference procedure.

The check of ratio of variances enables a decision to be made on whether or not to assume equal population variances. The ratio value of 3, described in Box 2.2, stems from the fact that the variances can differ by as much as a factor of three, and the equality of variability assumption is still valid for equal sample sizes. This simple check on data variability ensures its effect is properly accounted for in test statistic derivation. The numerators in expressions (2.3) and (2.5) correspond to the standard error of the difference between the mean measurements providing a measure of the precision of this difference estimate.

The Student's t distribution first appeared in an article published by William S. Gossett in 1908. Gossett worked at the Guinness Brewery in Dublin but the company did not permit him to publish the work under his own name so he chose to use the pseudonym "Student". The form of the t distribution exhibits a comparable symmetric pattern to that displayed in Fig. 1.6 for the normal curve. Critical values in Table A.1 are presented as $t_{\alpha,df}$ where α is related to the significance level and nature of the test, and df is the degrees of freedom of the test statistic.

The test statistic decision rule approach for alternative hypothesis format 1 in Box 2.2 is often expressed as

$$|t| < t_{\alpha/2,df} \Rightarrow \text{accept } H_0$$

where $|t|$ refers to the absolute value of test statistic (2.3) or (2.5) irrespective of sign and divisor 2 in $\alpha/2$ is used to signify the two possibilities defined for the mean in the associated alternative. The basic premise of this test procedure is that small values of the test statistic will imply acceptance of the null hypothesis and large values rejection. The former is indicative of comparable population means (no significant evidence of difference) and the latter is indicative of different population means (significant evidence of difference). With the p value, acceptance of the null hypothesis is associated with large p values and rejection with small p values corresponding to the same basic interpretation as the test statistic approach.

Box 2.2 — Two Sample t Test

Assumptions Two independent random samples and response variable approximately normally distributed.

Hypotheses The null hypothesis always reflects no difference, i.e. H_0: no difference between the two populations ($\mu_1 = \mu_2$). The alternative hypothesis can be expressed as either two-sided (1) or one-sided (2 and 3):

1. H_1: difference between the two populations ($\mu_1 \neq \mu_2$), includes both $\mu_1 < \mu_2$ and $\mu_1 > \mu_2$,
2. H_1: population 1 lower ($\mu_1 < \mu_2$),
3. H_1: population 1 higher ($\mu_1 > \mu_2$).

Exploratory data analysis (EDA) Assess data plots and summaries to obtain an initial feel for the information within the response data.

Test statistic The form of two sample t test to use depends on whether the **ratio of larger sample variance to smaller sample variance** is below or above 3.
ratio less than 3: For this case, the test statistic, referred to as the **pooled t test**, is based on the Student t distribution and is expressed as

$$t = \frac{\bar{x}_1 - \bar{x}_2}{s_p\sqrt{\frac{1}{n_1} + \frac{1}{n_2}}} \tag{2.3}$$

with $df = n_1 + n_2 - 2$ degrees of freedom where $\bar{x}_1$ and $\bar{x}_2$ are the sample means, and n_1 and n_2 the number of observations in the samples. The term s_p defines the pooled estimate of standard deviation

$$s_p = \sqrt{\frac{(n_1 - 1)s_1^2 + (n_2 - 1)s_2^2}{n_1 + n_2 - 2}} \tag{2.4}$$

which is a weighted average of the sample standard deviations s_1 and s_2 and provides an estimate of the variability in the response based on information from both samples.
ratio exceeds 3: For this case, the form of the test statistic, referred to as the **separate-variance t test** or **Welch's test**, is given by

$$t = \frac{\bar{x}_1 - \bar{x}_2}{\sqrt{\frac{s_1^2}{n_1} + \frac{s_2^2}{n_2}}} \tag{2.5}$$

with degrees of freedom approximated by

$$df = \frac{\left(s_1^2 / n_1 + s_2^2 / n_2\right)^2}{\frac{\left(s_1^2 / n_1\right)^2}{n_1 - 1} + \frac{\left(s_2^2 / n_2\right)^2}{n_2 - 1}} \tag{2.6}$$

rounding down to the nearest integer.

Decision rule Testing at the 100α percent significance level can be operated in either of two ways if software produces both test statistic and p value.
test statistic approach: This approach depends on which of the three forms of alternative hypothesis is appropriate and uses the t distribution critical values displayed in Table A.1.

1. two-sided alternative H_1: $\mu_1 \neq \mu_2$ (two-tail test), $-t_{\alpha/2,df} < t < t_{\alpha/2,df} \Rightarrow$ accept H_0,
2. one-sided alternative H_1: $\mu_1 < \mu_2$ (one-tail test), $t > -t_{\alpha,df} \Rightarrow$ accept H_0,
3. one-sided alternative H_1: $\mu_1 > \mu_2$ (one-tail test), $t < t_{\alpha,df} \Rightarrow$ accept H_0.

p value approach: p value > significance level $\Rightarrow$ accept H_0 where significance level must be expressed in decimal and not percent format.

Exercise 2.1 We now want to use the collected lead data of Example 2.2 to carry out a full statistical analysis of the data with respect to the study objectives.

Assumptions: The lead levels measured are approximately normally distributed.

Hypotheses: Construction of the requisite hypotheses is relatively straightforward based on the planned study objectives. The null hypothesis for the comparison of mean lead levels is simply H_0: no difference in mean lead levels of the two species, i.e. $\mu_{serr} = \mu_{spir}$, while the two-sided alternative hypothesis is H_1: difference in the mean lead level of the two species, i.e. $\mu_{serr} \neq \mu_{spir}$. For variability, the two hypotheses are comparable to those specified for the mean namely, H_0: no difference in variability of lead levels between the two species, i.e. $\sigma^2_{serr} = \sigma^2_{spir}$, and H_1: difference in the variability of the lead levels of the two species, i.e. $\sigma^2_{serr} \neq \sigma^2_{spir}$.

EDA: Display 2.1 contains an initial plot of the lead data collected. From the plot, we can see that each data set is relatively similar with both covering a similar range of values. The *F. serratus* (Serr) data have a grouping near the top end of its range of measurements whereas the *F. spiralis* (Spir) data tend to be more evenly spread out.

Display 2.1

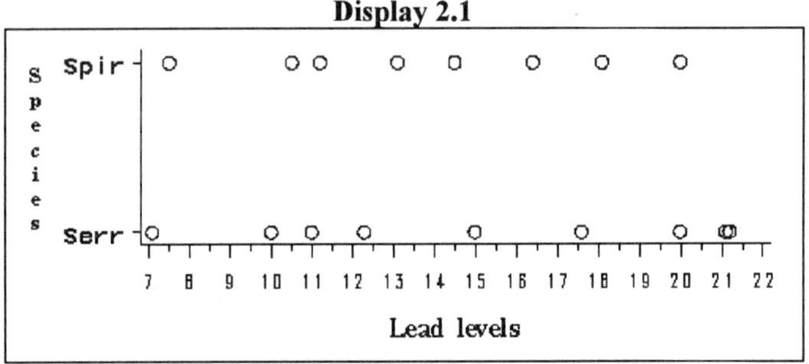

Corresponding summary statistics are provided in Display 2.2. The means differ by 0.6 μg Pb g^{-1} dry weight in favour of *F. spiralis*. The *CV*s are 35.6% and 29.7% for *F. serratus* and *F. spiralis* showing minor difference in variability, but also indicating large spread in both sets of measurements, as may be expected given the nature of the investigation. The summaries back-up the data plot information, suggesting little or no evidence of a substantive difference between the two species. Initial impressions therefore suggest that there appears little difference in the lead levels of the two species.

Test of difference of means: Given the ratio of variances of $5.1737^2/4.4870^2 = 1.33$, we must use the pooled t test (2.3). Display 2.2 contains the associated test result based on SAS's PROC TTEST procedure. The pooled t test statistic is given as −0.2771 (column 'T', row 'Equal') with associated p value of 0.7849 (column 'Prob>|T|', row 'Equal'). It should be noted, however, that the p value generated by SAS corresponds to that for a two-tail test so if one-tail testing is necessary, the SAS generated p value requires to be halved.

For the *test statistic approach*, the test statistic (2.3) is $t = -0.2771$. Based on the test being two-tail with significance level of 5% ($\alpha = 0.05$) and degrees of freedom $df = 10 + 10 - 2 = 18$, the critical value from Table B.2 is $t_{0.025,18} = 2.101$. As the test statistic lies between −2.101 and 2.101, it is obvious we should accept the null hypothesis and conclude that there appears to be insufficient evidence of a statistical difference in the lead levels between the two species.

For the *p value approach*, the p value is given as 0.7849. Based on testing at the 5% ($\alpha = 0.05$) significance level, we can see that the p value exceeds the significance level, specified as 0.05, leading to acceptance of H_0 as per the test statistic approach. □

Display 2.2

```
Variable: Lead
Species    N       Mean    Std Dev     Std Error    Minimum    Maximum
-----------------------------------------------------------------------
Serr      10    14.5300     5.1737        1.6361     7.1000    21.2000
Spir      10    15.1300     4.4870        1.4189     7.5000    20.0000

Variances        T        DF    Prob>|T|
----------------------------------------
Unequal    -0.2771      17.6      0.7850
Equal      -0.2771      18.0      0.7849

For H0: Variances are equal, F' = 1.33   DF = (9,9)   Prob>F' = 0.6783
```

It is also possible to modify the two sample t test to look at a specified difference D_0 between the sample means. This requires that the numerator of the test statistics (2.3) and (2.5) be modified to read ($\bar{x}_1 - \bar{x}_2 - D_0$). Use of software to provide the inferential elements generally entails modifying one of the data sets to suit the difference for which we are testing.

2.3.2 Confidence Interval for Difference in Mean Responses

Hypothesis testing provides a method for weighing up the evidence for or against the null hypothesis. Often, estimation of the effect being assessed would be beneficial to understand more about the data collected. Construction of a CI for the difference between two means will provide such an assessment. The determination of this is explained in Box 2.3. The interval can also be used as an alternative to two-tail testing since if the interval contains zero, we can accept the null hypothesis H_0.

Box 2.3 — Confidence Interval for Difference in Means

Assumptions As two sample t test (see Box 2.2).

Confidence interval A 100(1 – α) percent confidence interval for the difference between two sample means is expressed as

$$\bar{x}_1 - \bar{x}_2 \pm t_{\alpha/2,df}\,se(\bar{x}_1 - \bar{x}_2) \qquad (2.7)$$

where df refers to degrees of freedom and $se(\bar{x}_1 - \bar{x}_2)$ is the standard error of the difference between the means which measures the precision of the difference estimate, i.e. denominator of test statistic (2.3) or (2.5) as appropriate.

Exercise 2.2 For the lead data of Example 2.2, we will now illustrate construction and interpretation of a 95% confidence interval for the difference in mean lead levels.

Assumptions: As stated in Exercise 2.1.

Confidence interval: A pooled t test (2.3) was used in Exercise 2.1 to generate the required test statistic, so the necessary 95% confidence interval (2.7) will be based on df = 18 and standard error $se(\bar{x}_1 - \bar{x}_2) = s_p\sqrt{(1/n_1 + 1/n_2)}$ where s_p can be found using expression (2.4). From the summary information in Display 2.2, we have n_1 = 10, s_1 = 5.174, n_2 = 10, and s_2 = 4.487. Expression (2.4) is therefore

$$s_p = \sqrt{\frac{(10-1)\text{x}(5.174^2)+(10-1)\text{x}(4.487^2)}{10+10-2}} = 4.8427 \quad .$$

Together with the critical value $t_{0.025,18} = 2.101$, we generate the confidence interval

$$14.53 - 15.13 \pm (2.101)\times(4.8427)\times\sqrt{(1/10 + 1/10)} = (-5.15, 3.95)\ \mu\text{g Pb g}^{-1}\ \text{dry weight.}$$

Based on the limits derived, the 95% confidence interval for the difference between the means, *F. serratus* – *F. spiralis*, is (−5.15, 3.94) μg Pb g^{-1} dry weight. This implies that we are 95% confident that the true difference between the means lies between −5.15 and 3.95 μg Pb g^{-1} dry weight. Because the interval contains zero, there appears no difference in lead levels between the two species and both have comparable lead levels as also shown in Exercise 2.1. □

2.3.3 Hypothesis Test for Variability

In addition to testing and estimation of location effects in two sample studies, assessment of the precision (variability) of the measurements can also be of interest. This can be achieved through a test of the equality of the sample variances called a **variance ratio test** as summarised in Box 2.4. The test can be used to formalise the decision on which form of t test, expression (2.3) or expression (2.5), to use for comparison of the means.

The F distribution is a skewed statistical distribution similar to the right-skew pattern illustrated in Fig. 1.5. The test statistic in equation (2.8) can alternatively be expressed in the form F = smaller/larger but this leads to a more complex decision rule process. By evaluating the test statistic using expression (2.8), the numerical value generated will never be less than 1 and it is this result that enables the decision rule for alternative hypotheses 2 and 3 to be expressed similarly.

Exercise 2.3 Referring to the lead pollution study of Example 2.2, a part of the experimental objective was to assess whether the variability in reported results was different between the species.

Assumptions: As stated in Exercise 2.1.

Hypotheses: The null hypothesis is, as usual, reflective of no difference. Thus, we have H_0: no difference in the variability of the reported lead measurements, i.e. $\sigma^2_{\text{serr}} = \sigma^2_{\text{spir}}$. Since the test is to assess for difference in variability, the alternative will be two-sided and specified as H_1: variability of reported lead measurements differs with species, i.e. $\sigma^2_{\text{serr}} \neq \sigma^2_{\text{spir}}$.

Variances test: Based on the sample standard deviations presented in Display 2.2, the variance ratio test statistic (2.8) becomes $F = 5.1737^2/4.4870^2 = 1.33$ as shown in the last line of Display 2.2. Given $n_{\text{lsv}} = 10 = n_{\text{ssv}}$, the degrees of freedom are $df_1 = n_{\text{lsv}} - 1 = 9$ and $df_2 = n_{\text{ssv}} - 1 = 9$. Using a 5% significance level and the information that we are conducting a two-tail test, Table A.3 provides the critical value $F_{0.025,9,9} = 4.03$. Since 1.33 is less than 4.03, we accept H_0 and conclude that there appears to be no statistical difference in the variability in lead levels between the two species. This also confirms the viability of using the pooled t test in Exercise 2.1.

Display 2.2 also provides a p value for the test of 0.6783 (Prob>F') which would generate an identical conclusion. Again, SAS only provides the p value for a two-tail test of variability with the presented p value requiring to be halved for use in a one-tail test. □

Box 2.4 — Variance Ratio F Test

Assumptions As two sample t test (see Box 2.2).

Hypotheses The null hypothesis again reflects no difference but this time with respect to data variability, i.e. H_0: no difference in variability between the two populations ($\sigma_1^2 = \sigma_2^2$). The alternative hypothesis can again be specified in one of three forms:

1. H_1: difference in variability ($\sigma_1^2 \neq \sigma_2^2$),
2. H_1: variability lower in population 1 ($\sigma_1^2 < \sigma_2^2$),
3. H_1: variability higher in population 1 ($\sigma_1^2 > \sigma_2^2$).

Test statistic The associated test statistic can be simply constructed as

$$F = \frac{\text{larger sample variance (lsv)}}{\text{smaller sample variance (ssv)}} \tag{2.8}$$

based on an F distribution with (df_1, df_2) degrees of freedom where $df_1 = n_{lsv} - 1$, $df_2 = n_{ssv} - 1$, and n_{lsv} and n_{ssv} are the respective sample sizes for the larger and smaller sample variance estimates.

Decision rule Decision rule approaches, for testing at the 100α percent significance level, conform to the same principles as previously outlined.

test statistic approach: This approach again depends on which form of alternative hypothesis is appropriate and uses the critical values displayed in Table A.3.

1. $F < F_{\alpha/2, df_1, df_2} \Rightarrow$ accept H_0,
2. $F < F_{\alpha, df_1, df_2} \Rightarrow$ accept H_0,
3. As 2.

p value approach: As for two sample t test (see Box 2.2).

In conclusion, the analyses pertaining to the lead study of Example 2.2 point to there being no evidence to suggest a statistical difference in the lead levels of the two species with respect to the mean ($p > 0.05$), or to the variability ($p > 0.05$). It appears that the two species absorb lead in equal quantities though we would need to know what level is detrimental to growth to properly investigate the phenomena being studied.

2.3.4 Confidence Interval for the Ratio of Two Variances

In addition to statistically testing variability differences, it can be useful to consider construction of a confidence interval for the ratio of two variances to provide an estimate of the variability differences between the samples. As the confidence interval concerns a variability characteristic, it is multiplicative in nature and not evenly spread around the estimated ratio. The $100(1-\alpha)$ percent confidence interval for the ratio of two variances is expressed as

$$\frac{s_1^2}{s_2^2} F_{1-\alpha/2, df_1, df_2} < \frac{\sigma_1^2}{\sigma_2^2} < \frac{s_1^2}{s_2^2} F_{\alpha/2, df_1, df_2} \tag{2.9}$$

where, in this instance, df_1 and df_2 refer to the degrees of freedom for sample 1 and sample 2 respectively.

2.3.5 Non-parametric Methods

In certain cases within two sample studies, study data may not conform to the normality assumption which underpins parametric inference and estimation procedures. The measurement scale may not have an underlying normal distribution. This sometimes happens when the data are percentages, or when the data are numerical rating scores or categorical data. To deal with these occurrences, we need to have alternative means of carrying out inferential data analysis. **Non-parametric**, or **distribution-free**, techniques provide such alternatives. The **Mann-Whitney test**, also called **Wilcoxon Rank Sum test**, and the **Ansari-Bradley test** are non-parametric alternatives to the two sample t test and variance ratio F test, respectively. Confidence interval estimation for the difference in medians can also be considered. Fuller details on non-parametric inference and estimation procedures can be found in Daniel (1990), Siegel & Castellan (1988), and Zar (1996).

2.4 INFERENCE METHODS FOR PAIRED SAMPLE STUDIES

In studies based on two independent samples, results may be influenced by variation in the experimental material. For example, comparing two procedures for measuring the lead content of plant specimens may detect differences in the results. This difference may be due to procedure differences, but could equally be as a result of variations between the sets of specimens assessed by each procedure. In other words, a detected difference in lead content measurement may be due to procedure or specimens tested or a mixture of both. In such a case, we should try to account for specimen differences by splitting them in half and allocating a separate half of each to each procedure. By so doing, specimen variation can be accounted for, enabling any differences detected to be most likely attributable to procedure or treatment differences, the assessment of which is the primary objective of the study.

This form of study design, illustrated in Table 2.4, is generally referred to as **paired comparison testing**. The design eliminates a source of extraneous variation by making the pairings chosen as similar as possible with respect to a confounding variable. Such studies are special cases of two sample studies and involve observations on two treatments being collected in pairs under as near identical conditions as possible. Tests and estimation procedures in respect of mean difference and difference in variability will be outlined.

Table 2.4 — Paired comparison design structure

Data pairing	1	2	3	4	.	.	.	.
Treatment 1	x	x	x	x	.	.	.	.
Treatment 2	x	x	x	x	.	.	.	.

x denotes an observation

Example 2.3 A study was conducted to compare two methods of determining the percentage titanium content in nickel-titanium alloys used in electrical components. Ten samples of the alloy were selected from the manufacturing process and were analysed by both methods. The results collected are presented in Table 2.5. It has been suggested that method A tends to produce lower but equally variable titanium content measurements.

Table 2.5 — Titanium content data for Example 2.4

Sample	1	2	3	4	5	6	7	8	9	10
Method A	16.35	17.48	16.26	17.18	15.13	16.75	15.46	16.89	16.57	17.05
Method B	16.61	17.59	16.14	17.02	15.26	16.97	15.46	17.05	16.71	17.28

2.4.1 Hypothesis Test for Mean Difference in Responses

This parametric procedure is based on collecting the data in pairs from like material and using the difference between the recorded measurements corresponding to each matched pair as the basis of the procedure. No detectable difference between the pairings would be considered reflective of little obvious difference between the treatments. To test this difference statistically, we use the **paired comparison *t* test**. This is essentially a one sample t test of the observation differences as summarised in Box 2.5.

Box 2.5 — Paired Sample t Test

Difference As paired comparison studies are based on measurement of a response from like samples, we require to first specify the requisite difference D. This is most often expressed as

$$D = \text{treatment } 1 - \text{treatment } 2$$

though order of determination is unimportant.

Assumptions Differences approximately normally distributed.

Hypotheses Again, the null hypothesis corresponds to no difference between the treatments, i.e. H_0: no difference between the two treatments (mean difference $\mu_D = 0$). As with previous testing, three forms of alternative hypothesis can be considered:

1. H_1: difference between treatments (mean difference $\mu_D \neq 0$),
2. H_1: treatment 1 lower (mean difference $\mu_D < 0$),
3. H_1: treatment 1 higher (mean difference $\mu_D > 0$).

Exploratory data analysis (EDA) Use data plots and difference summaries to obtain an initial picture of the data information.

Test statistic The associated test statistic, based on the Student's t distribution, is expressed as

$$t = \frac{\overline{D}}{s_D/\sqrt{n}} \quad (2.10)$$

where $\overline{D}$ is the mean difference, s_D is the standard deviation of the differences, n is the number of data pairings, and $n-1$ is the associated degrees of freedom.

Decision rule As for two sample t test (see Box 2.2).

As with two independent samples, this test procedure can also be used to assess a specific level of difference D_0 rather than simply a mean difference of zero. To account for this, we would require to alter the numerator of the test statistic (2.10) to read ($\overline{D} - D_0$) to enable relevant assessment to occur.

Exercise 2.4 We want to illustrate assessment of the two specified objectives within Example 2.3. The variability aspect will be discussed more fully in Exercise 2.6.

Difference: For this example, the difference D for titanium measurements will be determined as

$$D = \text{Method A} - \text{Method B}$$

based simply on the order of the data presentation.

Assumption: Differences in titanium measurements are approximately normally distributed.

Hypotheses: The mean aspect requires assessment of a specific difference. This can be based on testing a null hypothesis of no difference, i.e. H_0: no difference in the mean titanium content ($\mu_D = 0$), against a one-sided alternative expressed as H_1: mean titanium content lower with method A ($\mu_D < 0$).

EDA: The plot in Display 2.3, indicates a comparable trend in the reported measurements. However, for most samples tested, it appears that method A provides lower measurements. The only exceptions to this are samples 3 and 4 which exhibit an opposite effect and sample 7 which shows no difference. Variability difference between the two sets of measurements appears negligible as they provide a comparable trend for each paired sample.

Display 2.3

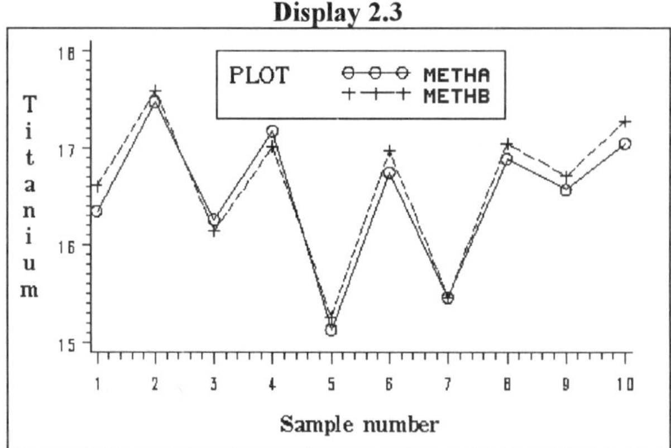

A dotplot of the method differences D is shown in Display 2.4. The plot amply illustrates that there are more negative differences than positive. Such a trend provides further evidence that method A appears to be returning lower measurements compared to method B.

Summaries of the method differences generated by the PROC MEANS procedure in SAS are shown in Display 2.5. The differences range from −0.26 to 0.16 with a negative mean difference. The standard error is less than half this value suggesting that the trend in the differences is mostly negative and that this trend could be of statistical significance. All initial analyses point to the same conclusion of lower method A results but comparable variability.

Display 2.4

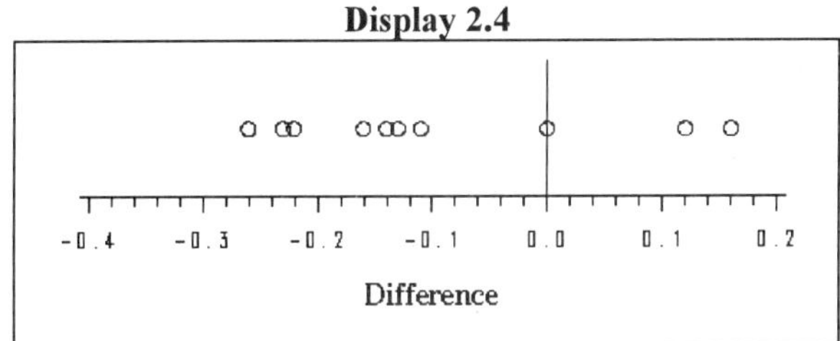

Display 2.5

Minimum	Maximum	Mean	Std Error
-0.260	0.160	-0.097	0.046

t test for mean difference: Display 2.6 contains the paired comparison t test result from PROC MEANS in such a form as to enable either decision rule to again be considered. Since the alternative hypothesis is one-sided with inequality <, this means use of either $t > -t_{\alpha,df}$ or $p >$ significance level.

Display 2.6

T	Prob>|T|
-2.117	0.0634

The paired sample test statistic (2.10) is quoted as $t = -2.177$ ('T'). The p value quoted ('Prob>|T|') is that for a two-tail test so for this one-tail test, we require to halve the presented value to give $p = 0.0317$. Based on the 5% critical value of $t_{0.05,9} = 1.833$, we have t less than -1.833 indicating rejection of H_0. As the p value is less than the significance level of 0.05, this also specifies reject H_0. Both procedures again provide the same conclusion, that there appears sufficient evidence within the collected data to suggest method B records higher titanium content measurements compared to method A ($p < 0.05$). □

2.4.2 Confidence Interval for Mean Difference

For paired studies, a confidence interval (CI) for the mean difference can be used to provide an estimate of the magnitude of the difference being tested for. The CI is comparable to that associated with the mean in one sample inference and is summarised in Box 2.6.

Box 2.6 — Confidence Interval for Mean Difference

Assumptions As for paired comparison t test (see Box 2.5).

Confidence interval The $100(1 - \alpha)$ percent confidence interval for a mean difference is expressed as

$$\overline{D} \pm t_{\alpha/2,n-1}(s_D/\sqrt{n}) \tag{2.11}$$

where $\overline{D}$, n, and s_D are as defined in Box 2.5, and $s_D/\sqrt{n}$ refers to the standard error of the mean difference, a measure of the precision of the mean difference estimate.

Exercise 2.5 Referring to Example 2.3, we will now illustrate construction and interpretation of the confidence interval for the mean titanium difference for the two methods.

Difference: As stated in Exercise 2.4.

Assumption: As stated in Exercise 2.4.

Confidence interval: Derivation of the confidence interval (2.11) is feasible in SAS using PROC MEANS by including the CLM option as Display 2.7 illustrates. The CLM option refers to confidence limits for the mean. The interval of (−0.201, 0.007) g contains zero but only just and is mostly covering negative values. This provides further evidence that the two methods are differing in their reported measurements with method A, in general, lower. □

Display 2.7

```
Lower 95.0% CLM  Upper 95.0% CLM
--------------------------------
         -0.201            0.007
--------------------------------
```

2.4.3 Hypothesis Test for Variability

In paired studies, interest may also lie in comparing the variability of the treatment measurements. For example, in a study comparing a new analytical procedure with a reference procedure using paired specimens, we may wish to assess whether the precision (variability) of the reported measurements from each procedure is comparable so we can conclude equality of precision irrespective of any differences in accuracy of the reported measurements.

The variance ratio F test (2.8) for two independent samples is inappropriate in this case as the sample variance estimates s_1^2 and s_2^2 may be correlated because the measurements are based on a natural relationship between the samples tested. In such a case, a special form of t test, summarised in Box 2.7, can be used for testing variability in paired sample studies. The test is based on the t test for correlation and is referenced, in a modified form, in Snedecor and Cochran (1980).

Box 2.7 — Variance Test for Paired Sample Studies

Hypotheses As for variance ratio F test (see Box 2.4).

Test statistic The form of the test statistic for testing paired sample variability is

$$t = \frac{\left(s_1^2 - s_2^2\right)\sqrt{n-2}}{2s_1 s_2\sqrt{1-r^2}} \tag{2.12}$$

based on the t distribution with $n - 2$ degrees of freedom. s_1 and s_2 are the standard deviations of each treatment group, and r defines the correlation (degree of linear association) between the two data sets.

Decision rule As for two sample t test (see Box 2.2).

The correlation measure r within test statistic (2.12) is included to provide a measure of the similarity in the trend of the measurements recorded for each pairing. Correlation measures the level of association between experimental variables and is part of regression procedures. A fuller description of correlation can be found in Ott

(1993) and Zar (1996). If each pair of observations are of comparable magnitude or follow a similar trend, it will be close to +1 while a value near 0 would be indicative of little similarity in the measurements. An alternative way of computing (2.12) is provided by

$$t = \frac{(F-1)\sqrt{n-2}}{2\sqrt{F(1-r^2)}} \tag{2.13}$$

where F is the ratio s_1^2/s_2^2 of sample 1 variance over sample 2 variance, irrespective of the magnitude of the variances unlike the variance ratio test statistic (2.8).

Exercise 2.6 Referring to the titanium data of Example 2.3, part of the experimental objective was to assess whether the variability (precision) of the measurements was similar with method. This will require the variances of each method to be tested using the test statistic (2.12).

Hypotheses: The variability of the paired titanium measurements is based on assessing a null hypothesis of H_0: no difference in variability of titanium measurements ($\sigma_A^2 = \sigma_B^2$) against a two-sided alternative of H_1: variability in titanium measurement differs with method ($\sigma_A^2 \neq \sigma_B^2$).

t test for variances: The necessary summaries underpinning this test are presented in Display 2.8. From this, we have $s_A = 0.744$, $s_B = 0.766$, $n = 10$, and $r = 0.98198$. Using expression (2.18), the test statistic becomes

$$t = \frac{\left(0.744^2 - 0.766^2\right)\sqrt{10-2}}{2(0.744)(0.766)\sqrt{1-0.98198^2}} = \frac{-0.093960}{0.215407} = -0.44$$

Display 2.8

```
            Variable          Std Dev
            ----------------------
            MethA              0.744
            MethB              0.766
            ----------------------

            Correlation Analysis
        2 'VAR' Variables:   MethA    MethB
  Pearson Correlation Coefficients / N = 10
                                   MethB
               MethA             0.98198
```

The critical value, from Table A.1, is $t_{0.025,8} = 2.306$ based on a two-tail test at the 5% ($\alpha = 0.05$) significance level with 10 − 2 = 8 degrees of freedom. Since $|t|$ is less than 2.306, we must accept the null hypothesis and conclude that, on the basis of the data collected, it appears that the variability in titanium measurement does not differ with method used ($p > 0.05$). □

Summarising the titanium measurement analysis of Exercises 2.4 to 2.6, we can conclude that there appears evidence that method A provides lower ($p < 0.05$) but equally variable ($p > 0.05$) measurements suggesting a difference only in method accuracy. The 95% confidence interval for the mean difference of (−0.201, 0.007) g shows that the majority of the trend within the data is commensurate with method A providing lower values, in general. The results reached suggest that titanium measurement may be affected by method of measurement which, ideally, should not be the case. Knowledge of the target titanium content of the manufactured alloy could have been useful.

2.4.4 Non-parametric Methods

As with two sample studies, the data collected within paired sample studies may not conform to normality necessitating use of alternative procedures with which to statistically analyse the data. The **Wilcoxon signed-ranks procedure** or **sign test** can be used in place of the paired comparison *t* test (Daniel 1990, Siegel & Castellan 1988, Zar 1996) where comparison is based on the median difference and not the difference in medians, as occurred in two independent sample testing. Confidence interval estimation on the basis of the median difference is also possible.

2.5 SAMPLE SIZE ESTIMATION IN DESIGN PLANNING

In study design, knowledge of how many measurements to take is important in order to ensure that sufficient data are collected to satisfy appropriate assessment of the experimental objectives. Sample size estimation is part of **power analysis** and should be considered when planning the study, though it is often overlooked. Power analysis enables a planned design structure and related inference procedure to be assessed in respect of their ability to detect specified treatment differences when present within the data. This ability is specified as the probability of the inference procedure rejecting the null hypothesis when the null hypothesis is not true, i.e. 1 − P(Type II error).

Many studies are typically based on a power of **at least 80%** with studies exhibiting low power making little sense as it is unlikely they will be capable of properly answering the study objectives. Thus such a study would fail on the basis of **statistical conclusion validity**. This is one of four aspects of statistical practice which strongly influence the validity of study findings (see Section 2.6). If power analysis suggests inadequate data collection, there is an opportunity to re-design the study to increase the power and consequently its usefulness.

Estimation of sample size in two sample studies requires the investigator to consider likely study outcomes and to set conditions under which they wish to operate the inference procedure. Specification of study outcomes corresponds to setting a difference d in response means (practical difference of interest) which quantifies the smallest level of treatment difference that is considered relevant from a practical perspective, i.e. a difference that will specify attainment of the experimental objective(s). The conditions to be set for operation of the statistical test of inference include setting the 100α percent significance level of the test and its power. Simple approximation methods exist to help estimate an acceptable sample size in two sample studies.

2.5.1 Sample Size Estimation for Two Sample Studies

In two sample studies, we can consider sample size estimation for either equal-sized samples or unequal-sized samples, although the latter is generally not preferred. Both estimation procedures are based on similar concepts and methods which we will now explain.

Example 2.4 A study is planned into the determination of nitrogen (N_2) in a particular type of soil by Kjeldahl digestion at two temperature settings to assess if temperature affects measurement accuracy and precision. Equal numbers of soil samples are required for each regimen but the investigator is unsure of how many to use. The investigator has suggested that a mean difference of 0.4% N_2 would be of importance to provide evidence of a difference in the nitrogen measurements recorded at the two temperatures. Past such studies suggest a standard deviation estimate of 0.33% N_2 would be appropriate. A statistical test for comparison of mean N_2 content is planned to be carried out at the 5% significance level with power of at least 80%.

2.5.1.1 Equal Sample Sizes

For a planned two-tail test, the investigator must first specify the likely difference d between the two treatment means which would reflect, in their opinion, evidence of a practical difference. Using this difference, sample size n for each group can be estimated using the expression

$$n = 2(z_\beta + z_{\alpha/2})^2/ES^2 \tag{2.14}$$

where $\beta = 1 -$ Power/100, z_β is such that $P(z > z_\beta) = \beta = P(\text{Type II error})$, $z_{\alpha/2}$ refers to the 100α percent two-tail critical z value for the proposed inference test (see Table A.4), and z corresponds to the standard normal distribution, i.e. equation (1.6) with μ set at 0 and σ set at 1. The ES term is specified as d/σ and refers to **effect size** where σ represents the likely level of variability for the data to be recorded based on past knowledge.

For expression (2.14) to operate, the investigator must pre-specify the terms based on their requirements for the operation of the planned study and the related statistical test. Using these specifications, we estimate n from equation (2.14) and decide if such an estimate is practical. The approximate sample sizes generated by this procedure can often be larger than would be feasible within a practical study. Modification of the pre-specified constraints could reduce this to more appropriate levels so consideration of the practicality of the estimate provided is important before using it as the basis of any study. For one-tail testing, z_α replaces $z_{\alpha/2}$ in expression (2.14).

Exercise 2.7 Consider the planned study outlined in Example 2.4. It is based on equal sample sizes. How many soil samples need to be collected?

Experiment information: The planned study is designed to compare two temperature settings using a two sample structure and to assess whether reported measurements differ with temperature (two-tailed test). Equal sample sizes are required for each method and so we will use expression (2.14) to estimate sample size.

Parameter determination: From the information provided, we know $d = 0.4$, $\sigma = 0.33$, and $ES = 0.4/0.33 = 1.2121$. Power specification of at least 80% implies $\beta = 1 - (80/100) = 0.20$. From $\beta = P(z > z_\beta)$, Table A.4 provides $z_\beta = z_{0.20} = 0.8416$. The associated $z_{\alpha/2}$ term (two-tail test, 5% significance level) is $z_{0.025}$ which, from Table A.4, is specified as 1.96.

Evaluation of number of specimens: With all the parameters known, expression (2.14) becomes

$$n = 2\times(0.8416 + 1.96)^2/1.2121^2 = 10.685 \approx 11,$$

truncated (rounded upwards) to nearest integer. Estimation suggests 11 soil samples per temperature regimen would be advisable. This should provide sufficient nitrogen data for the investigator to be 80% certain of detecting a difference of 0.4% N_2, if it exists, between the two temperature settings using a 5% significance level test. □

2.5.1.2 Unequal Sample Sizes

For unequal sample sizes, estimation can be based on assuming that one sample is to be a proportion m of the other, i.e. assume $n_2 = mn_1$. In this situation, expression (2.20) is modified to

$$n_1 = (1 + 1/m)(z_\beta + z_{\alpha/2})^2/ES^2 \tag{2.15}$$

to provide an estimate of the size of sample 1. Using $n_2 = mn_1$ provides the estimate of sample size for the second sample. For $m > 1$, i.e. $n_2 > n_1$, power is only minimally affected by the imbalance in sample sizes but for $m < 1$ and n_2 small, power can diminish markedly as m decreases. Unequal sample sizes may be necessary when certain treatments are expensive in industry or large numbers of untreated patients are unethical in a medical study.

2.5.2 Sample Size Estimation for Paired Sample Studies

Similar estimation and power analysis concepts can be applied if a paired sample study is planned. Their basis once again depends on the investigator's expectation of the likely treatment difference d for practical effect. In paired sample studies, sample size estimation can be achieved through use of the expression

$$n = (z_\beta + z_{\alpha/2})^2/ES^2 \tag{2.16}$$

where $ES = d/\sigma_D$ and σ_D is an estimate of the variability of the paired differences, $\beta = 1 - \text{Power}/100$, and z_β and $z_{\alpha/2}$ are as defined earlier with z_α replacing $z_{\alpha/2}$ if the planned inference is to be one-tailed. Again, the estimate provided is only a guideline figure with consideration of its feasibility required before the planned study is carried out.

Example 2.5 A water engineer plans to compare two similar methods for determining the number of particles of effluent in water specimens collected at a particular location in a water treatment plant. The water specimens are to be divided equally with one half randomly assigned to each method, i.e. a paired experiment is to be conducted. The engineer wishes to examine whether there is a bias in the measurements quoted. It is believed that a mean difference in particle numbers of 0.85 would provide evidence of bias in the method measurements. From past studies, the standard deviation of the differences in the particulate count is expected to be 1.25, approximately. The associated paired sample inference is planned to be carried out at the 5% significance level with a power of approximately 90%.

Exercise 2.8 How many water specimens are necessary for the planned study described in Example 2.5?

Experiment information: The planned experiment is to compare particulate count methods for counting the number of effluent particles in water specimens using a paired experimental set-up. We will use expression (2.16) to estimate the number of water specimens n necessary to satisfy the study objective of a general directional difference (two-tailed test).

Parameter determination: From the information provided, we know $d = 0.85$, $\sigma_D = 1.25$, and $ES = 0.85/1.25 = 0.68$. The power constraint of 90% provides $\beta = 1 - (90/100) = 0.1$ and $z_\beta = z_{0.1} = 1.2816$ (see Table A.4). The proposed two-tail test at 5% significance level means $z_{\alpha/2} = z_{0.025} = 1.96$ (see Table A.4).

Estimation of number of specimens: With all the parameters known, expression (2.16) provides the sample size estimate of

$$n = (1.2816 + 1.96)^2/0.68^2 = 22.7 \approx 23,$$

truncating to nearest integer, suggesting 23 water specimens would be necessary. However, this may appear excessive because of the amount of experimentation it would entail. Increasing the level of difference deemed important or reducing the difference variability estimate can help to reduce the sample size estimate. Using $d = 0.95$, for example, produces $n = 19$; while $\sigma_D = 1.15$ with $d = 0.95$ results in $n = 16$, the latter being possibly most realistic. This highlights not only the importance of correctly specifying the level of difference reflecting the study objective but also the need to modify expectations to enable acceptable sample size estimates to be provided. □

Guidelines for power estimation in tests concerning means and variances are also produced by the British Standards Institution (BSI) and the International Standards Organisation (ISO). Zar (1996) describes an alternative trial and error approach to sample size estimation while Lipsey (1990) discusses graphical approaches to power and sample size estimation.

2.6 VALIDITY AND GOOD STATISTICAL PRACTICE

Section 2.5 has addressed one of the most important issues for designing experiments if statistical conclusions are to be valid. In addition to **statistical conclusion validity**, good statistical practice requires

(i) **internal validity** whereby the internal details of the study have been correctly observed such as proper randomisation of the experimental units to groups and the maintenance of similar environmental conditions for different groups during the study,

(ii) **external validity** whereby reasonable precautions have been taken to ensure that the study findings will extend to the real world, and

(iii) **construct validity** whereby the investigator checks that the study objectives are answering the correct questions.

The findings from an industrial study based on the analysis of data from only one manufacturing plant may not apply to all of the company's manufacturing plants due to changes in climate, management, labour force, etc. and so the study could fail to be externally valid. This often happens in studies and it is one reason why unsatisfactory results are obtained when the findings of an experimental study are applied on a wide scale. In a medical study, the objective may be to provide treatment that improves the general demeanour of patients with a particular condition. The investigator may decide to collect biochemical measurements in control and treated subjects to demonstrate a difference. However, it may not be obvious that such clinical measurements contribute to an improvement of the general demeanour of the subjects and so the study may fail on the basis of construct validity. In addition, if the studies have insufficient data to provide statistical conclusion validity and are not properly monitored, then the expense of collecting data and carrying out analyses will have been wasted. At the design stage, a good investigator should check that the study meets the four validity requirements. Lipsey (1990) gives an excellent overview of the role of design sensitivity for statistical practice.

2.7 PROTOCOLS

For all studies, it is necessary to prepare a protocol or study plan prior to commencing data collection. This should include the following:

- *A statement of the study objective. Some studies may have more than one objective and so it will be important to identify the primary objective if sample size calculation is to be undertaken. Secondary objectives can be pursued but they may not always be achievable.*

- *A statement of the target population and how the experimental units are to be sampled from the target population so that they are representative. In particular, in medical studies certain subjects may not be appropriate because they are receiving concurrent medication and so the protocol needs to specify the inclusion-exclusion criteria.*

- *The experimental design should be stated and a detailed description of how the experimental units are to be randomised to groups should be provided.*

- *The start and end dates of the study should be provided and details of any planned interim analyses along with the significance levels to be used.*

- *There should be statistical consideration of the sample size requirements based on power calculations and a description of the statistical methods.*

- *A description should be provided on how the data are to recorded, analysed and quality controlled. In particular, details of the software packages to be used for data handling should be provided.*

Once the study is completed, a final report should be prepared in which a reviewer can follow the flow of raw data to summary data to statistical analysis and conclusions. In some industries, the final report will be scrutinised by auditors and so it is important that there is a proper audit trail from raw data to final report. The report should include details of the statistical design used and appropriate checks on the adequacy of the statistical model. It should be clear how the results lead to the final claims, and all statistical methods and materials should be clearly documented.

2.8 COMMONLY OCCURRING DESIGN MISTAKES

Despite the best of intentions, studies often go wrong due to design mistakes. This makes the cost of product development and basic research unnecessarily expensive and studies ineffective. Some of the commonly occurring mistakes are:

1. *The sample size is too small.* This occurs all too frequently. The temptation is to save on costs of experimental material and the effort involved in data collection which leads to experimental designs which have insufficient power. It is false economy as in many cases, it may be better not to undertake the study at all.
2. *The sample size is too large.* This is increasingly a problem with modern computer recording systems. The more data collected for analysis the more powerful are the statistical methods of analysis with the results that small orders of differences may be statistically significant when they are insignificant in practical terms. Fortunately, in recent years, most product development has been market-driven and market forces often set what order of difference must be achieved if a new product is to be profitable.
3. *There is a 'placebo' effect.* Often one of the groups in the analysis will be designated the placebo group to only receive a placebo treatment. Unfortunately, if the experimental units are people or are assessed by people, the recording of the data may lead to an apparent treatment effect. This makes it more difficult to detect a real effect in the treatment group. Allowance can be made for a placebo effect by revising the power calculation for sample size so that the sample size is increased or by using a run-in period prior to the start of the study.
4. *There are too many statistical tests.* A basic principle of good statistical practice is to carry out as few statistical tests as possible. This avoids the risk of finding differences that do not exist in the population being studied.
5. *Experimental units are not randomly assigned to groups.* This often happens because investigators feel it is not really important. Non-randomisation of experimental units can result in patterns in the response data which can invalidate the study inferences. Random assignment of experimental units is therefore a key feature of efficient study design.
6. *Introduction of bias and confounding.* Bias and confounding often arise unwittingly. Bias can arise as a result of data being omitted after the study has been completed, e.g., omission of atypical data values with no justification. Confounding may arise when experimental units in treatment groups are managed under different conditions, e.g., discovering that subjects assigned to one treatment group were managed by a different clinician to the other treatment group. Proper attention to how the trial is to be conducted, and the correct choice of experimental design can avoid bias and confounding.
7. *The investigator insists on using a complex statistical analysis.* This only makes a simple problem difficult. The methods of analysis should be kept as simple as possible and complicated statistical techniques avoided.
8. *Assumptions not valid.* Invalid assumptions affect the sensitivity of the inferential tests and can increase the true significance level of the statistical conclusion reached. For example, the response data might not be normal but bimodal. However, the tests described are often considered to be 'robust' to minor shortfalls in the assumptions and so the inferences are usually safe.

It can be seen, therefore, that study design is vital if the investigation is to provide meaningful conclusions. It can often be necessary to compromise between what is ideal and what is achievable within the constraints surrounding the investigation. The important point is that study design and choice of data analysis procedures go hand in hand and that inadequate planning will generate inappropriate data which cannot be validly interpreted even by the use of sophisticated statistical analysis. The "square peg into a round hole" syndrome of collect data and consider how to analyse after does not constitute good statistical practice and will never provide good interpretable study data.

PROBLEMS

2.1 In a production facility, part of the production process involves depositing a coating on the surface of the product. Two ways of applying the coating are possible. Concern has been expressed about the level of occurrence of abnormal amounts of coating. A simple study is planned to assess whether there is a difference in both the level and variability of coating deposition of the two methods. Construct relevant hypotheses.

2.2 A study is to be carried out on the lengths of a particular chromosome in healthy females aged from 30 to 40 and in similarly aged females suspected of having a genetic abnormality. Equal-sized random samples are to be selected from both groups and the "ratio of long arm to short arm" of the chromosome determined for each individual selected. It has been asserted that this "ratio" is lower and more variable in females with the genetic abnormality. Construct relevant hypotheses for the study objectives.

2.3 As part of a study of diabetes mellitus, 20 mice were randomised equally between a control and treatment group, the latter receiving a drug treatment in their diet for a one week period. At the end of the treatment period, the mice were given a glucose injection and 45 minutes later had their blood glucose, in μg l^{-1}, measured as shown. Is there evidence, at the 5% significance level, that the mice receiving the drug treatment have higher and less variable blood glucose levels?

Control	48.6	53.4	49.5	48.8	51.2	46.7	52.5	51.2	50.4	48.7
Treatment	55.3	52.3	54.5	57.8	51.8	51.2	53.8	55.9	56.7	50 3

2.4 Batches of a chemical ingredient from two different suppliers were delivered to a manufacturing plant. As part of the quality control of incoming raw material, ten samples of the material supplied by each supplier were selected and the level of manganese, in μg l^{-1}, measured. The results were:

Supplier 1	558	643	623	612	589	576	582	619	645	574
Supplier 2	584	613	623	606	596	615	603	593	584	597

Do the manganese levels present in the supplied raw material differ with supplier in terms of both average and variability?

2.5 Elongation measurements on manufactured steel were made on steel treated in two different ways. Treatment A involved using aluminium plus calcium while treatment B only used aluminium. Measurements, in mm, were made on 24 specimens, 12 given treatment A and 12 given treatment B. The recorded data were as follows:

Treatment A	4.45	4.54	3.87	5.76	4.26	4.65	4.23	5.03	4.85	5.45	4.54	3.95
Treatment B	4.02	3.45	4.21	4.65	4.32	3.76	3.54	4.08	4.31	3.87	4.07	3.59

It is conjectured that the addition of calcium can improve elongations and decrease the variability in elongation. Test these assertions.

2.6 To asses the effect of washing on the breaking extension of cotton thread, a company carried out a simple quality control check. Eight lengths of thread were each cut into two halves. Each half was then assigned randomly to either a washing or control group, the latter to receive no washing. Each tested piece of thread was then measured for percent breaking extension as follows:

Piece	1	2	3	4	5	6	7	8
Washed	13	19	14	11	10	14	12	14
Control	12	17	12	8	9	16	10	16

Is there any statistical evidence that washing affects extensibility and variability in extensibility of the cotton thread?

2.7 An experiment was conducted to investigate the phenolase activity in a particular variety of potato. Twelve potato specimens were split in half with one half of each specimen assigned randomly to each treatment group. One group was to be physically wounded and the other not, the latter acting as controls. The purpose of the investigation was to assess whether phenolase activity was higher and more variable in wounded potatoes. Cross section samples were taken and the level of phenolase activity, in absorbance min^{-1} per mg protein, recorded. Carry out the relevant assessment.

Sample Number	1	2	3	4	5	6	7	8	9	10	11	12
Control	0.133	0.019	0.012	0.056	0.025	0.075	0.013	0.021	0.102	0.124	0.046	0.063
Wounded	0.192	0.024	0.018	0.057	0.024	0.078	0.023	0.023	0.107	0.123	0.048	0.067

2.8 Twelve elderly patients suffering from rheumatoid arthritis took part in a study into the benefits of the administration of a therapeutic procedure on the level of wrist angular movement. Prior to treatment, measurements were made on the angular movement, in degrees, of the right hand. Each patient then received the active ingredient for a specific period after which measurements were again made. The data collected are presented below. Does administration of the therapeutic procedure improve angular movement where improvement corresponds to higher and less variable measurements?

Patient	1	2	3	4	5	6	7	8	9	10	11	12
Before	21.61	24.19	14.56	18.61	28.17	15.24	12.39	22.47	20.03	16.86	13.87	19.56
After	21.37	25.59	15.78	19.43	28.17	16.78	12.12	23.03	21.56	16.23	15.14	19.87

2.9 Prior to use in experiments, a bacteriology laboratory carries out a check on the pH of distilled water used to reconstitute dehydrated media. Two methods of storing the distilled water are to be tested with each method requiring an independent random sample of specimens in equal numbers. However, the investigator is unsure of how many samples to use for each method but wishes the sample sizes to be the same. A difference in mean pH of 0.1 would, in the investigator's opinion, be indicative of a important difference between the storage regimes. Previous data suggest a variability in pH of 0.05 could be assumed for both storage methods. Given that the associated statistical test is to be carried out at the 5% significance level with power of 95%, estimate the required sample sizes to meet the experimental objectives. Is the number suggested adequate?

2.10 A trading standards officer was concerned with the number of complaints being received on the length of life of a light bulb manufactured by a particular company. To test customer complaints, a small test study of samples of light bulbs from this company against a branded competitor is to be carried out using equal numbers of bulbs from each. The trading standards officer is unsure of how many bulbs to use in the test. A difference in mean lifetime of 50 hours was thought to be indicative of significantly lower quality of bulb. Previous data suggest a variability in lifetime of 60.3 hours could be assumed for both companies' bulbs. The associated statistical test is planned to be carried out at the 5% significance level with power of at least 80%. Based on these constraints, show that the required sample size estimate to meet the experimental objectives is 18.

2.11 A study is to be conducted to compare two diets in respect of their effect on the weight gain of pigs. A randomly selected groups of pigs is to be tested using a paired sample design structure though the investigator is unsure of how many pigs to select. It is suspected that a weight gain difference of 1.36 kg would provide sufficient evidence to suggest that diets differ in their influence on weight gain. A standard deviation of weight gain differences of 1.76 kg has been assumed from past studies. The paired comparison t test is planned to be carried out at the 5% significance level with power approximately 85%. How many pigs are necessary for this experiment?

2.12 A water quality laboratory is concerned with assessing two methods of determining chromium levels in river water: atomic absorption spectrophotometry (AAS) and complexometric titration (CT). Water specimens are to be collected and divided into two equal parts with one half randomly assigned to each method. The laboratory manager wishes to assess whether AAS produces higher measurements on average. A mean difference of 0.6 ppm between the paired observations would, in the manager's opinion, provide evidence to back-up this assertion. The number of samples to select is unknown though it is proposed to carry out the associated paired comparison t test at the 5% significance level with approximate power of 90%. Assuming an estimate of standard deviation of the differences in chromium levels of 0.85 ppm, estimate the number of water samples that would need to be taken to satisfy the stated experimental objective.

2A.1 Appendix: Software Information For Two Sample Studies

Data entry

Data entry for studies using two independent samples generally requires two columns of data to be entered: one for the response and one for the character or numerical codes corresponding to each sample. Both SAS and Minitab can use character sample coding though certain of Minitab's procedures require numeric coding for the samples.

SAS: Data entry in SAS is achieved through use of the DATA step. For Example 2.2, this is as follows:

```
DATA CH2.EX22;
    INPUT SPECIES $ LEAD @@;
    LINES;
    SERR 10   SERR 15   SERR 21.1   SERR 7.1   SERR 11   SERR 10
    SERR 21.2   SERR 20   SERR 12.3   SERR 17.6   SPIR 13.1   SPIR 18.1
    SPIR 20   SPIR 7.5   SPIR 14.5   SPIR 20   SPIR 11.2   SPIR 16.4
    SPIR 20   SPIR 10.5
;
RUN;
```

The DATA step begins with a DATA statement and can include any number of program statements. The element CH2.EX22 specifies that the DATA step is to create a SAS data set EX22 in the directory referenced by the library reference CH2. This library reference is usually the disk in a:, as described in Appendix 1A. The INPUT statement specifies the variable names for the samples and response, $ signifying a character variable for the SPECIES sampled and @@ indicating that the data will be entered in a continuous stream. The LINES statement specifies that data are to be entered in the data file as in-stream data, i.e. within the sequence of presented statements. The RUN statement specifies that the procedure statements prior to it are to be submitted for execution with **Locals ➤ Submit** used for this purpose.

Checking of data entry can be achieved using PROC PRINT. In the statements below, the NOOBS element is included to prevent observation numbers from being printed while use of BY enables the data to be printed by sample grouping, in this case the seaweed species.

```
PROC PRINT DATA = CH2.EX22 NOOBS;
    VAR LEAD;
    BY SPECIES;
RUN;
```

Minitab: Entry of response data into column C1 requires to be by sample grouping with codes entered into column C2 to distinguish the sample groupings. For Example 2.2, this means lead response will be entered in two blocks of ten observations starting with *F. serratus*. Codes entry into column C2 uses the **Calc ➤ Make Patterned Data** menu followed by the **Manip ➤ Code ➤ Numeric to Text** menu to convert the codes into character codes reflecting the species sampled. Numeric codes entry is also necessary for certain default graphical procedures. The menu procedures for sample codes entry for Example 2.2 would be as follows:

Select **Calc ➤ Make Patterned Data ➤ Simple Set of Numbers ➤** for *Store patterned data in*, enter **Specs** in the box ➤ select the **From first value** box and enter **1** ➤ select the **To last value** box and enter **2** ➤ ensure the **In steps of** box has entry **1** ➤ select the **List each value** box and enter **10** ➤ ensure the **List the whole sequence** box has entry **1** ➤ click **OK**.

Select **Manip ➤ Code ➤ Numeric to Text ➤** for *Code data from columns*, select **Specs** and click **Select** ➤ select the **Into columns** box and enter **Species** ➤ select the first box of **Original values** boxes, enter **1**, press **Tab**, enter **Serr**, press **Tab**, enter **2**, press **Tab**, and enter **Spir** ➤ click **OK**.

To check data entry, the following menu selection can be used:

Select **Manip** ➤ **Display Data** ➤ for *Columns, constants, and matrices to display,* select **Lead** and click **Select** and select **Species** and click **Select** ➤ click **OK**.

For Example 2.2, either of these forms of data entry would result in a data structure as follows:

Lead	Species	Lead	Species	Lead	Species	Lead	Species
10	Serr	10	Serr	13.1	Spir	20	Spir
15	Serr	21.2	Serr	18.1	Spir	11.2	Spir
21.1	Serr	20	Serr	20	Spir	16.4	Spir
7.1	Serr	12.3	Serr	7.5	Spir	20	Spir
11	Serr	17.6	Serr	14.5	Spir	10.5	Spir

Data plot

SAS: The simple data dotplot of lead response for each species in Display 2.1 was generated using the SAS statements

```
PROC GPLOT DATA = CH2.EX22;
    PLOT SPECIES*LEAD / HAXIS = AXIS1 VAXIS = AXIS2;
    AXIS1 LABEL = (F = CENTB J = C 'Lead levels') VALUE = (H = 0.8)
        LENGTH = 40;
    AXIS2 LABEL = (F = CENTB A = –90 R = 90 J = C H = 0.8 'Species')
        LENGTH = 8;
RUN;
```

The above are typical of the structure of SAS procedure statements. The procedure to be invoked is specified within the PROC statement with GPLOT indicating that a high resolution plot is required using the data contained in the SAS data file EX22. The PLOT statement specifies Y plotted against X (Y*X) with options appearing after the slash symbol for customising the plot, H referring to the horizontal axis and V to the vertical axis. The AXIS elements provide a means of customising the plot axes in respect of label font (F), position and angle of the label (J, A, R), label size (VALUE), and axis length (LENGTH).

Minitab: In Minitab, dotplot production uses the following menu selections based on the numeric sample coding only:

Select **Graph** ➤ **Character Graphs** ➤ **Dotplot** ➤ for *Variables*, select **Lead** and click **Select** ➤ select **By variable**, click the empty box, select **Specs**, and click **Select** ➤ click **OK**.

Data summaries and test statistic

SAS: Numerical summaries of response data by sample and the associated t test statistic can be generated using the PROC TTEST procedure (see Display 2.2) as follows:

```
PROC TTEST DATA = CH2.EX22;
    CLASS SPECIES;
    VAR LEAD;
RUN;
```

The CLASS statement identifies the required sample code variable. The VAR statement specifies the response to be summarised for each sample and to be tested through the two sample *t* test.

Minitab: Menu commands for default data summary production by sample are as follows:

Select **Stat** ➤ **Basic Statistics** ➤ **Descriptive Statistics** ➤ for *Variables*, select **Lead** and click **Select** ➤ select **By variable**, click the empty box, select **Species**, and click **Select** ➤ click **OK**.

The **Graphs** option within this procedure can also be used to generate a data dotplot and/or boxplot. Test statistic generation uses a different menu selection procedure as follows, though only the *t* test statistic can be evaluated:

Select **Stat** ➤ **Basic Statistics** ➤ **2-Sample t** ➤ ensure **Samples in one column** selected ➤ select the **Samples** box, select **Lead** and click **Select** ➤ for *Subscripts*, select **Species** and click **Select** ➤ for *Alternative*, select **not equal** ➤ **Assume equal variances** ➤ click **OK**.

2A.2 Appendix: Software Information For Paired Sample Studies

Data entry

Data entry for studies with related samples generally requires three columns of data to be entered: one for the sample number, one for the first set of paired data, and one for the second.

SAS: Data entry in SAS for Example 2.3 was achieved through use of the DATA step as follows:

```
DATA CH2.EX23;
     INPUT SAMPNO METHA METHB DLABEL $ @@;
     DIFF = METHA–METHB;
     LINES;
     1 16.35 16.61 D   2 17.48 17.59 D   3 16.26 16.14 D   4 17.18 17.02 D
     5 15.13 15.26 D   6 16.75 16.97 D   7 15.46 15.46 D   8 16.89 17.05 D
     9 16.57 16.71 D   10 17.05 17.28 D
;
RUN;
```

The INPUT statement specifies the variable names for the each of the necessary variables, $ signifying a character variable for the DLABEL variable, and @@ indicating that the data will be entered in a continuous stream. The DIFF statement specifies calculation of the difference *D* on which paired sample analysis is based.

Checking data entry using PROC print is again necessary to ensure that the correct data have been entered. This can be achieved as follows:

```
PROC PRINT DATA = CH2.EX23 NOOBS;
     VAR SAMPNO METHA METHB DIFF;
RUN;
```

Minitab: Entry of response data into the first three columns requires to be by sample number. Once response data for each method have been entered, the difference D for paired sample analysis can be calculated as follows:

> Select **Calc** ➤ **Calculator** ➤ for *Store result in variable*, enter **Diff** ➤ select the **Expression** box and enter **MethA – MethB** ➤ click **OK**.

Data printing for checking purposes can be achieved with the menu selection:

> Select **Manip** ➤ **Display Data** ➤ for *Columns, constants, and matrices to display*, select **SampNo** and click **Select**, select **MethA** and click **Select**, select **MethB** and click **Select**, and select **Diff** and click **Select** ➤ click **OK**.

For Example 2.3, either of these forms of data entry would result in a data structure as follows:

SampNo	MethA	MethB	Diff	SampNo	MethA	MethB	Diff
1	16.35	16.61	−0.26	6	16.75	16.97	−0.22
2	17.48	17.59	−0.11	7	15.46	15.46	0.00
3	16.26	16.14	0.12	8	16.89	17.05	−0.16
4	17.18	17.02	0.16	9	16.57	16.71	−0.14
5	15.13	15.26	−0.13	10	17.05	17.28	−0.23

Data plot
SAS: The simple data plot of the titanium response for each method tested in Display 2.3 was generated using the SAS statements

```
PROC GPLOT DATA = CH2.EX23;
    PLOT METHA*SAMPNO METHB*SAMPNO / HAXIS = AXIS1
        VAXIS = AXIS2 OVERLAY LEGEND = LEGEND1;
    SYMBOL1 C = BLACK V = CIRCLE I = JOIN L = 1;
    SYMBOL2 C = BLACK V = PLUS I = JOIN L = 3;
    AXIS1 LABEL = (F = CENTB J = C 'Sample number') VALUE = (H = 0.8)
        LENGTH = 40;
    AXIS2 LABEL = (F = CENTB A = –90 R = 90 J = C 'Titanium')
        VALUE = (H = 0.8) LENGTH = 15;
    LEGEND1 FRAME LABEL = (F = CENTB)
        POSITION = (TOP CENTER INSIDE) DOWN = 2;
    RUN;
```

Again, PROC GPLOT was invoked to produce the high resolution plot illustrated using the data stored in the SAS data file EX23. The PLOT statement specifies Y plotted against X (Y*X) for both method results with the options appearing after the slash symbol for customising the plot. Use of H refers to the horizontal axis, V to the vertical axis, OVERLAY for superimposing two X-Y plots on top of one another, and LEGEND specifying how the plot legend is to presented. The SYMBOL statements specify how the points and lines within the plot are to be presented, V for type of symbol, I for how the points are to be joined, and L for the line type. The AXIS elements provide the means of customising the plot axes in respect of label font (F), positioning (J, A, R), label size (VALUE), and axis length (LENGTH). The LEGEND1 statement specifies how the plot legend is to be displayed with DOWN indicating position of points definition in column format.

Minitab: In Minitab, multiple plot production uses the following menu selections:

Select **Graph** ➤ **Plot** ➤ for *Graph 1 Y*, select **MethA** and click **Select** ➤ for *Graph 1 X*, select **SampNo** and click **Select** ➤ for *Graph 2 Y*, select **MethB** and click **Select** ➤ for *Graph 2 X*, select **SampNo** and click **Select**.

Select **Frame** ➤ **Multiple Graphs** ➤ for *Generation of Multiple Graphs*, select **Overlay graphs on the same page** ➤ click **OK**.

Select **Frame** ➤ **Axis** ➤ select the **Label 1** box and enter **Sample number** ➤ select the **Label 2** box and enter **Titanium** ➤ click **OK**.

Select **Annotation** ➤ **Line** ➤ for *Points 1*, enter **SampNo MethA** ➤ select the **Points 2** box and enter **SampNo MethB** ➤ click **OK** ➤ click **OK**.

Difference plot and summaries

SAS: A simple dotplot (Display 2.4) and numerical summaries (Display 2.5) of the method differences can be generated using the PROC GPLOT and PROC MEANS procedures as follows:

```
PROC GPLOT DATA = CH2.EX23;
    PLOT DLABEL*DIFF / HAXIS = AXIS1 VAXIS = AXIS2 HREF = 0;
    AXIS1 LABEL = (F = CENTB J = C 'Difference') VALUE = (H = 0.8)
        LENGTH = 40 ORDER = (-0.4 TO 0.2 BY 0.1);
    AXIS2 LENGTH = 5 VALUE = NONE LABEL = NONE STYLE = 0
        MAJOR = NONE;
PROC MEANS DATA = CH2.EX23 MAXDEC = 3 MIN MAX MEAN STDERR;
    VAR DIFF;
RUN;
```

The HREF element in the PLOT statement requests that a line be inserted in the plot at difference value 0 for reference purposes. For the AXIS1 statement, ORDER specifies where the X axis markers are to be placed between -0.4 and 0.2 in steps of 0.1. For the AXIS2 statement, which refers to the Y axis, STYLE = 0 prevents an axis line from appearing while MAJOR = NONE prevents tick marks from being produced. In the PROC MEANS statement, MAXDEC defines the number of decimal places for the summaries requested and STDERR defines the standard error. The VAR statement specifies the response to be summarised.

Minitab: Menu commands for the differences dotplot and summaries creation are as follows:

Select **Graph** ➤ **Character Graphs** ➤ **Dotplot** ➤ for *Variables*, select **Diff** and click **Select** ➤ click **OK**.

Select **Stat** ➤ **Basic Statistics** ➤ **Descriptive Statistics** ➤ for *Variables*, select **Diff** and click **Select** ➤ click **OK**.

Test and confidence interval information

SAS: The PROC MEANS procedure, listed below, performs the t test calculation (see Display 2.6) and confidence interval (see Display 2.7) calculation for a paired samples study.

```
PROC MEANS DATA = CH2.EX23 MAXDEC = 3 T PRT;
    VAR DIFF;
PROC MEANS DATA = CH2.EX23 ALPHA = 0.05 MAXDEC = 3 CLM;
```

```
    VAR DIFF;
RUN;
```

The T and PRT elements in PROC MEANS determine the *t* test statistic and associated *p* value. In the second use of PROC MEANS, CLM requests printing of the lower and upper confidence limits with ALPHA = 0.05 specifying that the 95% limits be provided.

Minitab: Use of **Stat** procedures provides the same statistical information as PROC MEANS. The necessary menu commands are as follows:

Select **Stat** ➢ **Basic Statistics** ➢ **1-Sample t** ➢ for *Variables*, select **Diff** and click **Select** ➢ select **Test mean** and ensure **0** is the entry in the box ➢ for *Alternative*, select **not equal** ➢ click **OK**.

Select **Stat** ➢ **Basic Statistics** ➢ **1-Sample t** ➢ for *Variables*, select **Diff** and click **Select** ➢ select **Confidence interval** and ensure **Level** set at **95.0** ➢ click **OK**.

Variability test information

SAS: Use of PROC MEANS, as listed below, produces the data summaries necessary for calculation of the paired samples variability test (see Display 2.8). The STD element in PROC MEANS requests determination of the standard deviation. Use of PROC CORR provides the means of determining the correlation coefficient *r* for the paired data, with NOPROB suppressing the printing of *p* values for the inferential test of correlation and NOSIMPLE suppressing the printing of summary statistics for the two variables. By default, the parametric Pearson correlation coefficient is calculated.

```
PROC MEANS DATA = CH2.EX23 MAXDEC = 3 STD;
    VAR METHA METHB;
PROC CORR DATA = CH2.EX23 NOPROB NOSIMPLE;
    VAR METHA METHB;
RUN;
```

Minitab: The **Calc** and **Stat** procedures provide for the same information. The associated menu commands would be as follows:

Select **Calc** ➢ **Column Statistics** ➢ for *Statistic*, select **Standard deviation** ➢ select the **Input variable** box, select **MethA**, and click **Select** ➢ click **OK**.

Select **Calc** ➢ **Column Statistics** ➢ for *Statistic*, select **Standard deviation** ➢ select the **Input variable** box, select **MethB** and click **Select** ➢ click **OK**.

Select **Stat** ➢ **Basic Statistics** ➢ **Correlation** ➢ for *Variables*, select **MethA** and click **Select**, and select **MethB** and click **Select** ➢ click **OK**.

3

One Factor Designs

3.1 INTRODUCTION

In Chapter 2, methods for exploratory and inferential data analysis for simple two sample experimentation were introduced. This simple form of experimental design can be easily extended to many samples using **One Factor Designs** where the effect of different levels of a controllable factor, or treatment, on a measured response can be investigated. One factor structures are simple to implement in practice and represent a more efficient form of investigation which enables better use of information on how treatments affect a response to be forthcoming without necessarily increasing the experimental effort.

Illustrations of one factor investigations include comparison of the noise levels in three similar digital computer circuits, comparison of four protective treatments for wear on optic lenses, comparison of different apple mildew fungicides taking account of systematic variations in the environment, and comparison of the yield of a product using five different catalysts accounting for five different experimental conditions. The factors planned to be tested in such experiments may be classified as either **qualitative**, e.g. protective treatments and computer circuits, or **quantitative**, e.g. temperature, pressure, and quantity of raw material. As the illustrations show, the design plan involves testing different levels of treatment to ascertain if they influence the level of a measured outcome and if they do, how this influence is occurring.

The experimental designs to be discussed in this text are those commonly used as the basis of studies across a wide range of application areas including industrial processes, product development, marketing, biological sciences, chemical sciences, veterinary science, nursing research, and psychology to name but a few. All such experiments are constructed to investigate a specific experimental objective. Such an objective could include ascertaining the best fungicide for treating apple mildew, the best medium for advertising a product, and the best combination of chemicals for most active fertiliser. Data from such experiments are analysed using the many and varied techniques of **Analysis of Variance** (**ANOVA**, **AoV**).

Pioneering work on the principles of ANOVA was carried out by Sir Ronald Fisher at the Rothamsted Experimental Station in the 1920s and 1930s within agricultural experimentation for the assessment of different fertilisers on crop yield. Hence, the use of terminology such as treatment, plot, and block in experimental design explanations. This pioneering work brought to the fore the three basic principles of experimentation: **randomisation**, **replication**, and **blocking**, as well as providing the base for the derivation of many fundamental statistical analysis techniques. Experimental designs have since been extended and are now universally recognised as a rigorous approach appropriate for industrial and scientific investigations. In particular, in the physical, engineering, social, and environmental sciences, they are used to make comparisons and in the medical sciences, they are often the basis of claims of efficacy and safety.

Randomisation is fundamental to ensuring data collection is free of accountable trends and patterns which reflect unwanted causal variation. It represents the procedure undertaken to ensure that each experimental unit has an equal chance of being assigned to different treatments, or treatment combinations. Randomisation can reduce the risk of bias and can remove the correlation between the errors in order to validate the assumptions required to undertake inferential data analysis. It can also be appropriate when considering the order in which independent runs of the experiment are to be performed.

Replication refers to the number of times experimentation is to be repeated to provide adequate data for analysis. It is tied to the expectation of size of difference we wish to detect and the principle of power analysis (see Sections 2.5 and 3.6). Essentially, it corresponds to the number of times each treatment or treatment combination is tested. Replication is particularly useful when assessing data accuracy and precision, an important aspect of data handling in analytical laboratories. Provision of replication enables experimental error to be estimated, improves the precision of estimated parameters, and can reduce the level of noise or unexplained variation in the response data. Single replication, referring to one observation per treatment combination, may be acceptable if experimental constraints such as time and cost dictate that few experiments can be carried out. Experiments based on single replication need careful attention and will be a feature of the methods described in Chapters 7 to 11.

Blocking represents a second measure which, in conjunction with randomisation, can remove the effect of unwanted causal variation. It corresponds to the grouping of experimental units into homogeneous blocks with each treatment tested within each block. Batches of raw material, or samples of test material, are typical of blocking factors as batch or sample tested could differ in their effect on the experimental response. Blocking is therefore an extension of the pairing principle (see Section 2.4) to account for an identifiable source of response variation which is thought to be important but which we do not need to fully investigate. Inclusion of a blocking factor can increase the sensitivity of detecting significant treatment effect if such an effect exists. To illustrate blocking, consider the comparison of the level of employee experience on the number of paperwork errors across a number of sections in a large organisation. The main factor of interest, the treatment, is the employee experience factor with the section factor, the blocks, simply included to account for an effect which could influence the response as the level of such work may vary with section. Through such a design, we can examine if experience helps to minimise paperwork errors across the sections and, if it does, in what way.

ANOVA methods are in effect a generalisation of two sample procedures. They essentially attempt to separate, mathematically, the total variation within the experimental measurements into sources corresponding to the elements controlled within the experiment, the factors studied, and the elements not controlled, the unexplained variation or experimental error. From this, the variation due to controlled factors can be compared to that of the unexplained variation, large ratios signifying a significant explainable effect. These variance ratios, known as F ratios in honour of Sir Ronald Fisher, provide the primary statistical test element of experimental designs.

Classical experimental designs are based on the specification of a **model for the measured response** expressed as the sum of the variation corresponding to controlled components (treatment variation) and the variation corresponding to uncontrolled components (unexplained or error variation). It is hoped that the effect of the controllable components is far greater than that of the uncontrolled components, thus enabling conclusions on how the factors affect the response to stand out. In experimental design application in industry, the effect of the controlled components is often referred to as the signal and the effect of error as noise, with signal that is large relative to noise indicative of potentially important treatment effect.

The designs to be outlined in this chapter are all comparative and enable treatments or the effects of different experimental conditions to be compared. They are structured in a deliberate way to obtain appropriate response measurements and to conform to the associated statistical analysis procedures. Discussion will centre on how to analyse design data using exploratory analysis, ANOVA techniques, and diagnostic checking to gauge the level of influence of the factors studied. Only balanced designs, where all observations are present, will be discussed though the concepts introduced can be extended to cater for unbalanced designs where observations may be missing. However, balanced designs are to be much preferred. The emphasis throughout will be on the use of software output from both SAS and Minitab as the basis of the data presentation, analysis, and interpretation using similar principles to those of Chapter 2. Often, an experiment may need to be conducted to investigate the influence, either independently or in combination, of many factors on a measurable response. This concept of **multi-factor experimentation**, which comes within the domain of Factorial Designs, will be discussed more fully in Chapter 5 and developed further in subsequent chapters.

3.2 COMPLETELY RANDOMISED DESIGN

The **Completely Randomised Design** (**CRD**) is one of the simplest design structures on which practical studies can be based. It is so called because experimental units are randomly assigned to treatment groups, often referring to different levels of a factor, and because interest lies in examining if any difference exists in the responses associated with each treatment group. No other influencing elements beyond the treatment effect are accounted for making this design a simple extension of the two sample study structure introduced in Section 2.3.

Illustrations where such a design would be appropriate are numerous. In a study into three different marketing strategies for a new product, we may be interested in assessing if there is a difference between the strategies and if so, which appears best for the advertising campaign. A simple project into cadmium levels in soil may be concerned with comparing the levels in four areas of a particular site to see if the areas sampled differ with respect to cadmium level and, if they do, which areas contain more or less cadmium. In the removal of toxic wastes in water by a polymer, interest may lie in assessing how varying temperature affects waste removal to ascertain which temperature setting appears best suited to the task.

3.2.1 Design Structure

The layout for a CRD comprising k treatments is shown in Table 3.1. It can be seen from this structure that each treatment is tested the same number of times providing n replicate measurements for each treatment and a **balanced** design structure. Unequal replication would be represented by columns of differing length. This form of study structure is such that no other factors other than the treatments are accounted for.

Table 3.1 — Completely Randomised Design (CRD) structure

	Treatments 1	2	.	.	k
replicate 1	x	x	.	.	x
2	x	x	.	.	x
.	.	.	.	.	.
.	.	.	.	.	.
n	x	x	.	.	x

x denotes an observation

Example 3.1 A chemical engineer conducted a study into a new polymer planned to be used in removing toxic substances from water. Experiments were conducted at five different temperatures to assess if temperature affected the percentage of impurities removed by the treatment using water samples contaminated with the same level of impurity. The collected responses are given in Table 3.2. This study has therefore been based on a CRD structure using five temperature settings (k = 5) with six observations per setting (n = 6) and a response of percent impurity removed.

3.2.2 Model for the Measured Response

Data collected from a CRD structure are assumed to be influenced by only two components: the treatment tested and the unexplained variation (error or noise); the former corresponding to the variation effects capable of being controlled by the investigator and the latter to the variation not properly controlled. Error variation can encompass misrecording of data, instrument error, calibration error, environmental differences between samples, and contaminated solutions.

Table 3.2 — Data for percent impurity removed in toxic removal study of Example 3.1

Temperature	I	II	III	IV	V
	49	47	40	36	55
	51	49	35	42	60
	53	51	42	38	62
	53	52	38	39	63
	52	50	40	37	59
	50	51	41	40	61

Reprinted from Milton, J. S. & Arnold, J. C. (1990) *Introduction to probability and statistics.* 2nd ed. McGraw-Hill, New York with the permission of The McGraw-Hill Companies.

Response model specification for a CRD is based on this split of response variation leading to a recorded observation X_{ij} in the CRD structure being expressed as

$$X_{ij} = \mu_j + \varepsilon_{ij} = \mu + \tau_j + \varepsilon_{ij} \tag{3.1}$$

assuming additivity of effects where $i = 1, 2,..., n$, and $j = 1, 2,..., k$. The expression $\mu_j + \varepsilon_{ij}$ defines the **means model** while $\mu + \tau_j + \varepsilon_{ij}$ corresponds to the **effects model**.

The term μ (*mu*) refers to the grand mean, assumed constant for all k treatments, while μ_j defines the population mean for the jth treatment. The effect of treatment j, denoted τ_j (*tau*), is assumed constant for all units receiving treatment j but may differ for units receiving a different treatment. The experimental error, denoted ε_{ij} (*epsilon*), explains the response variation due to uncontrolled random sources of variation.

In general, τ_j is defined as a fixed component meaning that we assume we are experimenting on all possible treatments for such an investigation. Under such an assumption, we say that the experiment is a **fixed effects**, or **Model I**, experiment. Treatments may also be classified as **random effects** if the treatments tested represent a selection or sample of all potential treatments for such an experiment. We will only consider fixed effects treatments in the illustrations of design structures within this chapter and Chapter 4. Model specification, therefore, indicates how we think the response is constructed, in this case through additivity of effects, with the expectation that the treatment effect explains considerably more of the response variation than does error.

To illustrate the model specification, consider an experiment into four teaching methods and their effect on student performance in an assessment. Students are to be randomly assigned in groups of ten to each teaching method, and after a period of instruction given a short test. In this study, only the teaching method is being controlled and as we wish to compare four methods of instruction, a CRD would be a suitable design basis. Since student performance is to be the measured response, then the model we would specify for this study would be

$$\text{student performance} = \mu + \tau_j + \varepsilon_{ij}$$

where μ is the overall mean and $i = 1, 2,..., 10$ (ten students per teaching method), $j = 1, 2, 3, 4$ (four teaching methods), and the τ_j term defines the contribution of the jth teaching method to student performance. Analysis of data from such an experiment would be based on measuring the magnitude of the teaching method effect relative to error effect, a large value being indicative of teaching method differences.

3.2.3 Assumptions

The experimental error component within response model (3.1) is generally **assumed** to be normally distributed, with the variance of this error requiring to remain constant within and across all treatments, i.e. comparable levels of response variability for each treatment. This can be succinctly summarised as $\varepsilon_{ij} \sim \text{NIID}(0, \sigma^2)$ where NIID means normal independent identically distributed and σ^2 refers to the error variance. Small departures from these assumptions may not seriously affect the analysis results from this design structure resulting in the analysis procedures being classified as **robust**.

Exercise 3.1 The first step in the analysis of the toxic removal study of Example 3.1 is specification of the response model.

Response model: As the experimental structure is that of a CRD, the response model will be

$$\text{percent impurity removed} = \mu + \tau_j + \varepsilon_{ij}$$

where i = 1, 2,..., 6 (six measurements per temperature) and j = I, II, III, IV, V (five temperatures). The term τ_j represents the contribution of temperature to percent impurity removed while ε_{ij} refers to the random error between water samples within a treatment.

Assumptions: For this CRD based study, we assume $\varepsilon_{ij} \sim \text{NIID}(0, \sigma^2)$ implying percent impurity removed is normally distributed and variability in this response similar for all temperatures.

3.2.4 Exploratory Data Analysis

As with previous illustrations of data handling, it is always advisable to examine data plots and summaries to gain insight into what they may be specifying about the experimental objectives.

Exercise 3.2 Exploratory analysis of the percent impurity removed data is the second stage of the analysis of the data collected in the toxic removal study of Example 3.1.

EDA: Display 3.1 contains SAS generated dotplots for each temperature. The output highlights that the amount of impurity removed appears to vary markedly with temperature with temperatures I, II, and V providing high responses. Three groupings of temperature appear to be emerging, these being (III, IV), (I, II), and V. Consistency, approximated by length of the sequence of points, shows minor difference with temperature I least and III and V looking higher though similar to each other.

Display 3.1

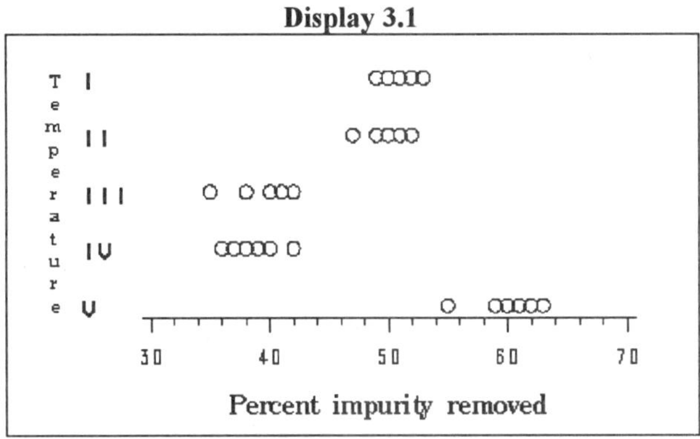

Display 3.2 provides the numerical summaries of the impurity data by temperature. The means (Mean) show temperature differences with temperature V highest. The means further confirm the temperature groupings highlighted in the data plot and suggest that choice of temperature can influence the polymer's ability to remove toxic waste. The coefficient of variation (*CV*) values suggest different levels of variability in percent removed according to temperature suggesting unequal variability of response with temperatures tested. □

Display 3.2

Analysis Variable : Imprem

Temp	N Obs	Minimum	Maximum	Mean	Std Dev	CV
I	6	49.00	53.00	51.33	1.63	3.18
II	6	47.00	52.00	50.00	1.79	3.58
III	6	35.00	42.00	39.33	2.50	6.36
IV	6	36.00	42.00	38.67	2.16	5.59
V	6	55.00	63.00	60.00	2.83	4.71

3.3 ANOVA PRINCIPLE FOR THE COMPLETELY RANDOMISED DESIGN

The next phase of data analysis after EDA involves determining whether treatment differences, possibly detected in the exploratory analysis, can be substantiated to be statistically significant. This is where the **ANOVA principle** comes into play by providing the mechanism for test statistic construction and derivation based on the response model specification and statistical theory. The general procedure involves specification of hypotheses, derivation of an ANOVA table and treatment test statistic, and decision on which hypothesis best reflects the evidence which the data are providing. A fuller description of the underlying theory of experimental designs can be found in Montgomery (1997) and Winer *et al.* (1991).

3.3.1 Hypotheses

Hypotheses associated with the CRD experimental structure reflect a test of no treatment difference versus treatment difference. This results in a null hypothesis specification of

H_0: no difference amongst means of treatments tested
($\mu_1 = \mu_2 = \ldots = \mu_k$, i.e. $\tau_1 = \tau_2 = \ldots = \tau_k = 0$)

and an alternative hypothesis specification of

H_1: difference amongst means of treatments tested
(at least one μ_j different, i.e. at least one $\tau_j \neq 0$).

These hypotheses are general hypotheses in respect of treatment differences and do not provide any information as to how a treatment difference, if detected, is reflected within the response data. Follow-up analysis would therefore be necessary (see Section 3.4).

3.3.2 ANOVA Table

The **ANOVA principle** within a CRD structure involves splitting response variation into two parts, treatment and error, sometimes referred to as the within treatment variation, corresponding to the sources of variation within the specified response model (3.1). This variation splitting is achieved through the determination of sum of square (*SS*) terms which reflect the level of response variation associated with a particular source. The main *SS* term is *SS*Total corresponding to total response variation in the collected data which, based on the additive response model (3.1), can be split into two parts thus

$$SSTotal\ (SST) = SSTreatment\ (SSTr) + SSError\ (SSE) \quad . \tag{3.2}$$

For historical reasons, the calculation aspects associated with this splitting of response variation is generally summarised in tabular form using an **ANOVA table**. The ANOVA table for a balanced CRD experiment is given in Table 3.3 where the mean square (*MS*) terms represent variance components which are part of the statistical theory associated with experimental designs. For an unbalanced CRD with n_j replicates for treatment j, the form of ANOVA table is unaltered though the calculation formulae must be modified to account for the unequal number of measurements across the treatments. This involves replacing n by n_j and kn by Σn_j with the error degrees of freedom becoming $(\Sigma n_j - k)$.

Table 3.3 — ANOVA table for a balanced Completely Randomised Design

Source	*df*	*SS*	*MS*
Treatments	$k-1$	$SSTr = \sum_{j=1}^{k}\frac{T_j^2}{n} - \left(\sum_{i=1}^{n}\sum_{j=1}^{k} X_{ij}\right)^2 \Big/ kn$	$MSTr = SSTr/df$
Error	$k(n-1)$	$SSE\ (= SST - SSTr)$	$MSE = SSE/df$
Total	$kn-1$	$SST = \sum_{i=1}^{n}\sum_{j=1}^{k} X_{ij}^2 - \left(\sum_{i=1}^{n}\sum_{j=1}^{k} X_{ij}\right)^2 \Big/ kn$	

k is the number of treatments, T_j is the sum of the responses for treatment j, n is the number of observations recorded for each treatment, X_{ij} is the experimental response for unit i within treatment j, *df* refers to degrees of freedom, *SS* refers to sum of squares, *MS* refers to mean square

3.3.3 Treatment Test Statistic

Based on the statistical theory underpinning the CRD design structure, the **treatment effect test statistic** is specified as the variance ratio

$$F = MSTr/MSE \tag{3.3}$$

with degrees of freedom $df_1 = k - 1$ and $df_2 = k(n-1)$, i.e. treatment degrees of freedom and error degrees of freedom.

To decide whether the evidence within the data leads to acceptance or rejection of the null hypothesis, we can apply either of two **decision rules** provided software producing p values has been used. The **test statistic approach** is based on the same form of decision rule as that of the F test for variability in two independent samples (see Section 2.3.3) except that the degrees of freedom for critical value determination become $[k-1, k(n-1)]$, i.e. [treatment degrees of freedom, error degrees of freedom]. If the test statistic is less than the critical value (see Table A.3) at the 100α percent significance level, then we accept the null hypothesis of no difference between the treatments, i.e. test statistic < critical value $\Rightarrow$ accept H_0, and we conclude no significant treatment differences detected. The ***p* value approach** follows from software use and operates as in previous illustrations, i.e. p value > significance level $\Rightarrow$ accept H_0.

Exercise 3.3 In Exercise 3.2, we carried out some exploratory analysis of the impurity data of Example 3.1 and found evidence of temperature differences. As the study basis was a CRD, we now need to carry out the associated main statistical test in order to assess whether the detected differences are statistically significant.

Hypotheses: The basic hypotheses we are going to test are H_0: no difference in mean percent impurity removed with temperature tested and so temperature settings produce similar results against H_1: difference in mean amount removed according to temperature tested and so at least one temperature setting returns different values.

ANOVA table: Display 3.3 presents the ANOVA information for the impurity data generated by SAS's PROC ANOVA procedure. The presented ANOVA information does not conform to the general ANOVA table shown in Table 3.3 though all the necessary *df*, *SS*, and *MS* terms are provided. The output shows that *SS*Total = *SS*Temperature + *SS*Error with the *MS* terms derived as the ratio *SS*/*df* for both temperature and error effects as per the theory.

Test statistic: Equation (3.3) specifies that the treatment *F* test statistic is the ratio of the treatment and error mean square terms. The figure of $F = 97.15$ in the 'F Value' column corresponding to the row 'Temp' in Display 3.3 is the numerical estimate of this test statistic for the impurity data, i.e. 481.867/4.96. As we have both a test statistic and a *p* value, we can use either approach to assess the specified temperature effect hypotheses.

Display 3.3

```
Dependent Variable: Imprem
Source            DF    Sum of Squares    Mean Square    F Value    Pr > F
Model              4     1927.46666667   481.86666667      97.15    0.0001
Error             25      124.00000000     4.96000000
Corrected Total   29     2051.46666667

     R-Square               C.V.           Root MSE        Imprem Mean
     0.939555           4.652728         2.22710575        47.86666667

Source            DF       Anova SS       Mean Square    F Value    Pr > F
Temp               4   1927.46666667     481.86666667      97.15    0.0001
```

For the *test statistic approach*, we know that $F = 97.15$. As $k = 5$ and $n = 6$, the degrees of freedom of the test statistic are $df_1 = k - 1 = 4$ and $df_2 = k\times(n - 1) = 25$. The critical value from Table A.3, for the 5% significance level, is $F_{0.05,4,25} \approx 2.76$ using linear interpolation to approximate the missing critical value. Since test statistic exceeds the critical value, we must reject the null hypothesis and conclude that there appears sufficient evidence to suggest that temperature settings differ statistically in their effect on percent impurity removed ($p < 0.05$).

For the *p value approach*, the column 'Pr > F' in Display 3.3 provides the *p* value for the temperature *F* test as 0.0001. This low value also indicates rejection of the null hypothesis and therefore implies the same conclusion of temperature difference as reached above ($p < 0.05$). □

3.4 FOLLOW-UP ANALYSIS PROCEDURES

The treatment effect *F* test in the CRD, expression (3.3), only allows for statistical checking as to whether there is evidence of no difference or difference between the treatments. It provides no information as to how treatment differences, if detected, are occurring. Further analysis, therefore, is necessary to enhance the conclusion and to pinpoint how treatments may be differing. Techniques for doing this include graphical approaches such as **main effects plot** and **standard error plot** and inferential approaches such as **multiple comparisons**, **linear contrasts**, and **orthogonal polynomials**. Generally, several should be considered.

3.4.1 Main Effects Plot

This represents a simple line plot of the response means for each treatment. Fig. 3.1 illustrates a Minitab generated main effects plot for the impurity removal data of Example 3.1. The plot shows a distinct quadratic trend and provides further visual evidence of the differing effects of temperature on percent impurity removed. The temperature groupings are further emphasised. Optimal temperature appears to be temperature V with III and IV obviously worst. A drawback of using such a plot to pinpoint treatment differences lies in its inability to assess variability around the treatment mean response which could lead to a misleading interpretation of treatment differences.

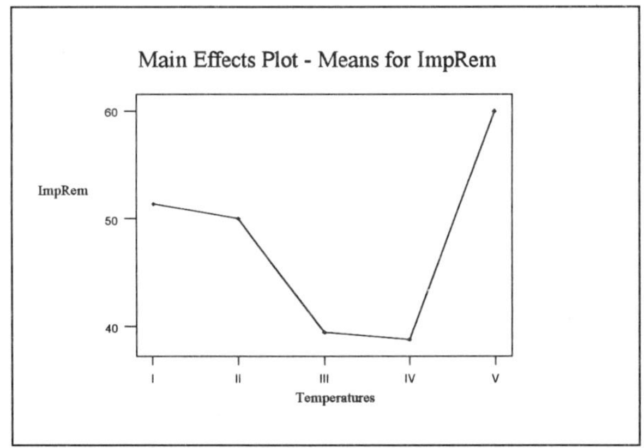

Fig.3.1 — Main effects plot for mean percent impurity removed for temperatures tested

3.4.2 Standard Error Plot

This represents a graphical display of treatment means while also accounting for variability in response around this mean value. The plot is best produced using the expression

$$\text{mean response} \pm 2\sqrt{(MSE/n)} \tag{3.4}$$

to produce a two standard error plot where MSE is the error mean square in the ANOVA table, n is the number of observations per treatment, and $\sqrt{(MSE/n)}$ defines the standard error of the mean (see Box 1.1). Use of two standard errors means the plot is essentially a graphical presentation of approximate 95% confidence intervals for each of the treatment group means.

Its use as a method of pinpointing treatment differences is only really appropriate provided most of the response variation is due to solely treatment differences, i.e. treatment differences are the primary cause of response differences. Overlap of treatment intervals suggests no evidence of statistical difference between the compared treatments, whereas non-overlap of treatment intervals would be suggestive of evidence of statistically significant treatment differences. Fig. 3.2 shows a Minitab generated two standard error plot for the impurity removal data of Example 3.1.

The plot highlights the significant differences between temperatures (I, II), (III, IV), and V where there is no interval overlap. The statistical similarity between temperatures I and II, and III and IV are clearly seen as the intervals show a strong degree of overlap. Based on this, we can state that there appears evidence that level of impurity removed is strongly affected by temperature at which treatment takes place with the temperature effects grouping neatly into three distinct groups.

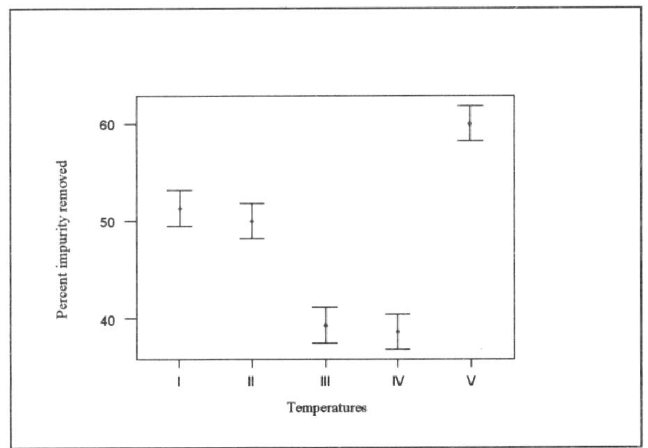

Fig. 3.2 — Two standard error plot for mean percent impurity removed for temperatures tested

3.4.3 Multiple Comparisons

Often after a significant treatment effect is suggested by the treatment effect F test, we should compare each treatment against every other treatment to help identify treatment differences occurring within the response data. This involves carrying out ***post hoc*** or ***a posteriori*** comparisons of all treatment pairings in the form of independent tests.

In a study involving k treatments, we have $c = k(k - 1)/2$ possible pairwise comparisons. Using a 100α percent significance level for each of the c comparisons means that the overall probability of at least one pairwise comparison incorrectly indicating a significant difference, i.e. committing at least one Type I error, can grow to approximately $1 - (1 - \alpha)^c$. This type of error is generally referred to as the **experimentwise Type I error rate** or **overall error rate**. For example, a study with $k = 4$ treatments provides six ($c = 6$) possible pairwise comparisons. If $\alpha = 0.05$ (5% significance level) is used for each of the six comparisons, this results in an experimentwise Type I error rate of as high as 26.5%. Thus, there is a high probability of committing at least one Type I error over the six comparisons carried out, if all tests are to be carried out at the 5% significance level.

For this reason, when multiple comparisons are required, it is necessary to adjust the significance level of each comparison, the **comparisonwise Type I error rate**, to ensure that the experimentwise error rate will be satisfied. This corresponds to **Bonferroni's inequality** which states that the experimentwise error rate should be less than or equal to the sum of the individual comparisonwise error rates. In the illustration discussed, to ensure an experimentwise error rate of 5%, we would need to set the comparisonwise error rate of 100α percent so that the expression $100[1 - (1 - \alpha)^6] = 5$ is satisfied, i.e. $\alpha = 1 - (0.95)^{1/6}$. This results in $\alpha = 0.0085$ so choice of comparisonwise error rate of 0.85% should ensure an experimentwise error rate of 5% for all six treatment comparisons.

To account for the presence of Type I error when comparing many treatments, a **multiple comparison** procedure should be used. Such pairwise contrasts include **Fisher's Least Significant Difference** (LSD), **Tukey's Honestly Significant Difference** (HSD), **Duncan's Multiple Range Test**, and the **Student-Newman-Keuls** (SNK) procedure. These procedures are based on carrying out all possible pairwise treatment comparisons in a fashion which restrains the overall significance level, or experimentwise error rate, of such comparisons to 100α percent thus accounting for Bonferroni's inequality. They all work on the same principle though some are more conservative whereas others are more liberal. Conservative procedures can make attaining significant differences more difficult and often result in certain treatment differences being quoted as insignificant when they really are significant (false negative). Liberal procedures provide the opposite effect and can result in significant differences being quoted for treatment differences which are not significant (false positive).

We will consider only the **Student-Newman-Keuls procedure** which is one of several available in SAS through the MEANS option in PROC ANOVA and PROC GLM (see Appendix 3A). This procedure is based on the use of the Studentised range statistic and focuses upon a series of ranges rather than a collection of differences using the position of each treatment mean, in an ordered list, as its base. The implementation steps of the SNK procedure are outlined in Box 3.1.

Box 3.1 — Student-Newman-Keuls Multiple Comparison Procedure

Hypotheses The null hypothesis to be tested in this form of multiple comparison is H_0: no difference between compared treatments (means similar) with alternative hypothesis specifying treatment difference.

Test statistics The numerical difference between each pair of treatment means becomes the test statistic for each pairwise comparison.

Critical values Obtain, for the 100α percent significance level, the least significant range

$$W_r = q_{\alpha,r,errordf}\sqrt{\frac{MSE}{n}} \tag{3.5}$$

where $r = j + 1 - i$ is the number of steps the treatments in positions i and j are apart in the ordered list of means, *errordf* is the degrees of freedom for the error component, $q_{\alpha,r,errordf}$ is the 100α percent critical value from the tables of the Studentised range statistic (see Tables A.5 and A.6), *MSE* is the error mean square, and n is the number of observations per treatment.

Decision rule If the numerical difference in treatment means for treatments in positions i and j, specified as $|\overline{T}_{(i)} - \overline{T}_{(j)}|$, is less than the associated critical value W_r, we accept H_0 and declare that the evidence suggests that the treatments being compared do not appear to differ statistically.

Tables of the Studentised range statistic do not always specify the q values for all possible error degrees of freedom, especially those above 20. Entries not specified in a table can be estimated through either linear or harmonic interpolation. Alternatively, the next lower degrees of freedom may be used leading to a conservative use of the procedure.

The results from this comparison are often summarised in a simple graphical form where pairs of treatment means that do not differ significantly share the same underline. For example, based on a treatment order of A, C, D, and B, the summary A C D B would indicate that treatments A and C and treatments D and B are not significantly different from each other (underlined) but that the two groups of like treatments appear to differ statistically from each other (no underlining). If compared treatments in positions i and j are based on unequal numbers of observations (unbalanced CRD), critical value calculation must be modified to account for the imbalance. This can be achieved by replacing MSE/n in expression (3.5) by $(MSE/2)(1/n_i + 1/n_j)$.

Exercise 3.4 For the toxic removal study of Example 3.1, we know that temperature effects differ statistically. Figs 3.1 and 3.2 provided graphical evidence of the temperature differences. We will now carry out the SNK multiple comparison procedure at the 5% (α = 0.05) significance level following the guidelines provided in Box 3.1 to provide the statistical evidence of treatment differences accounting for design structure and number of comparisons carried out. The design structure provides us with n = 6 and k = 5, and the ANOVA information in Display 3.3 provides $errordf$ = 25 and MSE = 4.96.

Hypotheses: The null hypothesis we will test is H_0: no difference in removal rate between temperature settings being compared.

Test statistics: The impurity means and table of differences in means, the test statistics, are presented in Table 3.4. Temperature order is shown to be IV, III, II, I, and V.

Table 3.4 — Test statistics (differences in means) for multiple comparison of Exercise 3.4

	IV	III	II	I	V
Mean	38.67	39.33	50.00	51.33	60.00
IV	38.61 -	0.66	11.33 *	12.66 *	21.33 *
III	39.33	-	10.67 *	12.00 *	26.67 *
II	50.00		-	1.33	10.00 *
I	51.33			-	8.67 *
V	60.00				-

* denotes statistically significant difference at the 5% significance level

Critical values: The corresponding 5% critical values of W_r, from evaluation of expression (3.5), are shown in Table 3.5. The Studentised range q values associated with error degrees of freedom 25 are not specified in Table A.5. The values presented in Table 3.5 for q have been estimated using simple linear interpolation. To illustrate multiple comparison interpretation, we will look at temperature comparisons for temperature setting IV only. All other comparisons are based on similar reasoning.

Table 3.5 —5% critical values for SNK illustration in Exercise 3.4

r	2	3	4	5
$q_{0.05,r,25}$	2.915	3.523	3.890	4.158
W_r	2.65	3.20	3.54	3.78

IV versus III: The test statistic for this comparison, from Table 3.4, is 0.66. These temperatures are in positions $i = 1$ (1st) and $j = 2$ (2nd) in the ordered list so the critical value for this comparison corresponds to $r = 2 + 1 - 1 = 2$ (two steps apart), i.e. use $W_2 = 2.65$ from Table 3.5. As the test statistic is less than the critical value, we accept H_0 and indicate that there appears no evidence of a statistical difference in removal rate between temperatures IV and III ($p > 0.05$).
IV versus II: The comparison test statistic from Table 3.4 is 11.33. These temperatures are in positions $i = 1$ (1st) and $j = 3$ (3rd) in the ordered list. The associated critical value corresponds to $r = 3 + 1 - 1 = 3$ (three steps apart), i.e. use $W_3 = 3.20$ from Table 3.5. As the test statistic exceeds the critical value, we reject H_0 and conclude that there appears sufficient evidence to imply a statistical difference in removal rate associated with these two temperatures ($p < 0.05$). A symbol "*" (significant at the 5% level) is inserted in Table 3.4 at the corresponding test statistic to indicate this result. By similar reasoning, the difference between temperatures IV and I (positions $i = 1$ and $j = 4$) can also be shown to be statistically significant ($p < 0.05$).
IV versus V: The necessary test statistic from Table 3.4 is 21.33. These temperatures are in positions $i = 1$ and $j = 5$ in the ordered list so the critical value to use for this comparison corresponds to $r = 5 + 1 - 1 = 5$ (five steps apart), i.e. use $W_5 = 3.78$ from Table 3.5. As the test statistic exceeds the critical value, we reject H_0 and conclude that there appears sufficient evidence to suggest a statistical difference in removal rate between temperatures IV and V ($p < 0.05$). Again, a symbol "*" is inserted in Table 3.4 to indicate a statistical difference has been detected.

From all the temperature comparisons, as summarised in Table 3.4, we can deduce that the grouping feature highlighted earlier is statistically important and that each temperature grouping differs in its effect on removal rate on a statistical basis. In summary form, we would set out these findings as $\underline{\text{IV III}}$ $\underline{\text{II I}}$ V.

The corresponding SAS generated multiple comparison information is presented in Display 3.4 for information. The results shown agree with the interpretation of three statistically different temperature groups. The W_r values quoted are comparable to those shown in Table 3.5 though they have been based on using Lagrangian or harmonic interpolation, a form of polynomial interpolation, for determination of the associated q values. □

3.4.4 Linear Contrasts

A linear contrast represents a linear combination of treatment means (or totals) which is of interest to the investigator to explore. It is useful for specific comparisons between treatments or groups of treatments where such **planned**, or ***a priori***, comparisons are decided upon prior to experimentation, e.g. control procedure versus modifications of the control or chemically based drugs versus homeopathic remedies for hay fever. The general form of a linear contrast L is

$$L = c_1\mu_1 + c_2\mu_2 + \dots + c_k\mu_k = \sum c_j\mu_j \tag{3.6}$$

where μ_j is the mean of treatment j and the c_j terms represent constants where usually $\Sigma c_j = 0$. Multiple comparisons, described in Section 3.4.3, are simply pairwise linear contrasts.

Display 3.4

```
        Student-Newman-Keuls test for variable: Imprem

NOTE: This test controls the type I experimentwise error rate under
the complete null hypothesis but not under partial null hypotheses.

              Alpha= 0.05   df= 25   MSE= 4.96

 Number of Means            2         3         4         5
 Critical Range     2.6482017 3.2027595 3.5368342 3.7762909

Means with the same letter are not significantly different.

        SNK Grouping               Mean       N  Temp
                   A             60.000       6  V
                   B             51.333       6  I
                   B
                   B             50.000       6  II
                   C             39.333       6  III
                   C
                   C             38.667       6  IV
```

In a three treatment CRD, for example, the general form of the linear contrast (3.6) will be

$$L = c_1\mu_1 + c_2\mu_2 + c_3\mu_3$$

with specific contrasts based on this general form. For instance, choosing $c_1 = 0$, $c_2 = 1$, and $c_3 = -1$ provides the contrast $L = \mu_2 - \mu_3$ for comparison of the average response of treatments 2 and 3 while choosing $c_1 = 1$, $c_2 = -½$, and $c_3 = -½$ provides the contrast $L = \mu_1 - (\mu_2 + \mu_3)/2$ which compares the average response of treatment 1 against the average for treatments 2 and 3 combined. This contrast could also be written as $L = 2\mu_1 - (\mu_2 + \mu_3)$ which still satisfies the coefficient property though the numerical estimate of L will double in size. This scaling effect does not affect the statistical interpretation of contrast results. The steps associated with analysis of linear contrasts, the test aspect of which can be derived in SAS using the CONTRAST option in PROC GLM, are briefly summarised in Box 3.2.

If the confidence interval contains 0 or the F test statistic is less than the requisite critical value at the 100α percent significance level, the contrast in question is reported to be not statistically significant, resulting in the conclusion that the evidence within the response data suggests no detectable difference between the contrast elements. If the confidence interval does not contain 0 and the F test statistic exceeds the associated critical value, we conclude that the evidence points to a detectable difference between the contrast elements which requires appropriate contextual interpretation. The $100(1 - \alpha)$ percent confidence interval and F test therefore provide the same conclusion, but the confidence interval approach is often preferable since it generates a numerical estimate of the effect being assessed thus providing a measure of the contrast effect.

Box 3.2 — Linear Contrasts

Hypotheses The null hypothesis for a contrast is expressed as H_0: $L = 0$, i.e. the effect being assessed by the specified contrast is not significant.

Estimate of L To estimate L, the mean responses for the specified treatments are substituted into the appropriate contrast expression and a numerical estimate for the contrast calculated.

Confidence interval for contrast The $100(1 - \alpha)$ percent confidence interval for a contrast L is given by

$$\text{estimate of } L \pm t_{\alpha/2,errordf}\sqrt{\left(\frac{\Sigma c_j^2}{n}\right)MSE} \tag{3.7}$$

where $t_{\alpha/2,errordf}$ is the $100\alpha/2$ percent value for error degrees of freedom (*errordf*) in Table A.1, c_j are the contrast coefficients, n is the number of observations for each treatment, and MSE is the error mean square. Generally, a 95% confidence interval is used for linear contrasts.

F test approach Based on the estimate of L, we determine the sum of squares of the specified contrast as

$$SS\text{Contrast} = n(\text{estimate of } L)^2/(\Sigma c_j^2) \tag{3.8}$$

The associated test statistic is given by

$$F = SS\text{Contrast}/MSE \tag{3.9}$$

based on degrees of freedom $df_1 = 1$ and $df_2 = errordf$.

If more than one linear contrast is planned, some adjustment to the significance level associated with the confidence intervals and tests may be necessary to ensure that the overall error rate is maintained at 100α percent. As with multiple comparisons and assuming 95% confidence intervals are required, this means ensuring α satisfies the expression $1 - (1 - \alpha)^c = 0.05$ where c is the number of linear contrasts planned.

If response measurement is based on unequal numbers of observations (unbalanced CRD), evaluation must be modified to account for the imbalance. This can be achieved by replacing $\Sigma c_j^2/n$ in expression (3.7) by $\Sigma(c_j^2/n_j)$ and $n/\Sigma c_j^2$ in expression (3.8) by $\Sigma(n_j/c_j^2)$.

Exercise 3.5 For the toxic removal study of Example 3.1, suppose it was of interest to compare temperatures I and II against temperature V. This comparison can be expressed in contrast form as $L = (\text{I} + \text{II})/2 - \text{V}$.

Hypotheses: The null hypothesis associated with this contrast is specified as H_0: $L = 0$, i.e. mean removal rate of temperature V not statistically different from the combined average effect of temperatures I and II.
Estimate of L: The contrast of $L = (\text{I} + \text{II})/2 - \text{V}$ can be estimated as $(51.33 + 50.00)/2 - 60.00 = -9.335$ using the temperature means from Display 3.2. The negative value of this estimate implies that temperature V appears to have a higher mean removal rate compared to the other two temperatures, as also highlighted in the exploratory analysis of Exercise 3.2.
Confidence interval: We know that *errordf* = 25, $n = 6$, and *MSE* = 4.96 (from Display 3.3). The 95% confidence interval (3.7), using a critical value of $t_{0.025,25} = 2.06$ from Table A.1, is given by

$$-9.335 \pm (2.06)\sqrt{\frac{((1/2)^2 + (1/2)^2 + (-1)^2)}{6}(4.96)} = -9.335 \pm 2.294$$

which results in an interval (−11.629, −7.041). As this interval does not contain 0, we conclude that the mean removal figure associated with temperature V appears to differ significantly from that of the other two temperatures ($p < 0.05$). The large negative estimate of L and the range covered by the interval provide further evidence of the much higher removal rate associated with temperature V.
F test: The SAS output in Display 3.5 provides the F test statistic for the planned temperature contrast. The test statistic (3.9) is presented as 70.25 with p value estimated as 0.0001. Both provide support for the conclusion reached through the confidence interval approach. □

Display 3.5

Contrast	DF	Contrast SS	Mean Square	F Value	Pr > F
(I, II) versus V	1	348.44444444	348.44444444	70.25	0.0001

3.4.5 Orthogonal Polynomials

A further procedure for assessing treatment differences is that of **orthogonal polynomials**. They represent special forms of treatment contrasts which are particularly appropriate when a quantitative experimental factor has been tested at evenly spaced intervals over a specified scale, e.g. three temperature levels set 5 degrees apart. They enable the response data to be assessed for evidence of the presence of linear or polynomial (quadratic, cubic, etc.) trend effects across the quantitative treatments by splitting the treatment effect into its trend components and assessing which, if any, is present within the measured response data. Exploratory analysis of factor plots or main effect plots may help indicate trend possibilities.

For k such treatments, it is possible to extract from the treatment sum of squares (*SSTr*) polynomial effects up to order $(k - 1)$. For example, for $k = 4$ temperature levels (four treatments), $k - 1 = 3$ so a significant treatment effect could contain linear (order 1) and/or quadratic (order 2) and/or cubic (order 3) trends. Orthogonal polynomials can be constructed, as with linear contrasts, using treatment means or totals. The mechanism, based on treatment means, is explained in Box 3.3.

3.4.6 Treatment Effect Estimation

For the model I experiment (treatments classified as fixed effects), least squares estimation generates unbiased estimators of the terms μ, μ_j, and τ_j of the response model (3.1). These are shown in Table 3.6. Using the τ_j estimators, we can obtain estimates of the treatment effects for each treatment relative to the overall mean response which can help to indicate optimum factor levels, if any exist. Positive estimates would be appropriate for maximisation of response while negative estimates would be appropriate for minimisation of response.

Box 3.3 — Orthogonal Polynomials

Estimation of trend effect This involves calculating an average trend effect estimate as $\Sigma c_j \overline{X}_j$ where c_j represents the tabulated coefficients of the associated orthogonal polynomial (see Table A.7) and $\overline{X}_j$ is the mean of treatment j.

Test statistic Using this estimate of trend effect, we derive the sum of squares of the trend effect as

$$SS(\text{Trend effect}) = n(\Sigma c_j \overline{X}_j)^2 / \Sigma c_j^2 \qquad (3.10)$$

where n is the number of measurements per treatment and the values for c_j and Σc_j^2 are tabulated for combinations of trend type and number of factor levels (see Table A.7). Using expression (3.10), we then generate the trend test statistic as

$$F = SS(\text{Trend effect})/MSE \qquad (3.11)$$

based on degrees of freedom $df_1 = 1$ and $df_2 = errordf$. The usual decision rule procedures for F tests then applies to assess the statistical significance of the associated trend effect.

Table 3.6 — Least squares estimators for model components in the Completely Randomised Design

Parameter	Point estimator	Variance of estimator
μ	$\overline{G}$	$\sigma^2/(kn)$
τ_j	$\overline{T}_j - \overline{G}$	$\sigma^2(k-1)/(kn)$
μ_j	$\overline{T}_j$	σ^2/n

$\overline{G}$ is the overall mean, $\overline{T}_j$ is the mean of treatment j, σ^2 is the estimate of error variance

Using the *MSE* term from the ANOVA table as an estimate of σ^2 (the error variance) and the associated error degrees of freedom (*errordf*), it becomes relatively straightforward to calculate confidence intervals for these parameter estimates by using the t distribution and the associated variance estimate. The form of confidence interval for each estimated parameter is given by

$$\text{estimate} \pm t_{\alpha/2,errordf}\sqrt{(\text{variance of estimator})} \qquad (3.12)$$

If the confidence interval for any treatment effect estimate contains 0, then it suggests that the effect being considered appears to have no statistically significant influence on the recorded response. Additionally, if the confidence interval for treatment effects is large, then it suggests a weakness in the design structure reflecting either too few observations or incomplete explanation of response influences.

Exercise 3.6 To illustrate effect estimation, consider the toxic removal study of Example 3.1.

Treatment effect estimation: From the design structure and data collected, we have $\overline{G}$ = 47.87, *errordf* = 25, $t_{0.025,25}$ = 2.06, σ^2 = 4.96 (= *MSE*), k = 5, and n = 6. Treatment means are shown in Display 3.2. 95% temperature effect confidence intervals for the five tested temperatures are

I: (1.78, 5.14)% II: (0.45, 3.81)% III: (−10.22, −6.86)%
IV: (−10.88, −7.52)% V: (10.45, 13.81)%

All intervals differ from 0, providing further indication that each temperature tested has some influence on level of impurity removed while the confidence limits indicate which have greatest effect. The intervals for I and II, and III and IV overlap, showing again the similarity of each pair of temperatures. The greatest effect on impurity removed occurs with temperature V. However, all intervals cover a large range of values suggesting that, though the study provides interesting results in respect of temperature, it is perhaps not providing full information on all possible factors affecting the treatment process. □

3.4.7 Model Fit

This can be checked by looking at the ratio of error sum of squares to total sum of squares, i.e. (*SSE*/*SST*)%. If this is large, say greater than 30%, it implies that the suggested model for the experimental response may not be explaining all the variability in the response data indicating there may be missing factors not included in the experiment which may be affecting the response. If this ratio is small, say less than 10%, it may suggest acceptable model fit.

For the toxic removal study of Example 3.1, this ratio is estimated as 6%. This percent value is low indicating that the response model of percent impurity removed as a function solely of temperature appears to be an acceptable description of the problem. This, however, does not imply that temperature is the only factor governing the level of impurity removed. It only shows that temperature setting chosen is an important component of the treatment process but by no means the only one.

3.5 DIAGNOSTIC CHECKING OF MODEL ASSUMPTIONS

Model assumption checks are an important diagnostic tool within the data analysis procedures associated with the CRD, as well as with most other design structures. They can be carried out using **residual (error) analysis**, where the **residuals** are estimates of the model errors corresponding to the difference between the observed data and the corresponding predicted values from the response model. The model assumptions in question are responses equally variable for each treatment tested and response data normally distributed.

Such analysis can, in addition, assess model applicability and may show that insignificant factors in terms of mean response may be important in their effect on response variability. Invalid assumptions can affect the sensitivity of the statistical tests, and may increase the true significance level of the testing carried out. This results in potentially incorrect parameter estimates and invalid conclusions. Most experimental data generally satisfy these assumptions, and so conclusions drawn from ANOVA analysis are usually acceptable. If either variability or normality is inappropriate, we can explore transformations (re-scaling) of the experimental data to enable variability and normality assumptions to be more acceptable (see Section 3.8) or we can use alternative non-parametric procedures to analyse the measured response data (see Section 3.7).

3.5.1 Graphical Checks

Graphical means of assessing model assumptions provide a visual check method. How this can be carried out is summarised in Box 3.4.

Box 3.4 — Graphical Methods of Diagnostic Checking

Equal variability Plot residuals against each factor. The plot should generate a horizontal band of points equally distributed around 0. Patterns in the plot, e.g. different lengths of column, reflect non-constant variance (heteroscedasticity) indicative that a variance stabilising transformation of the data may be required while bunching will indicate inappropriate response model. Outliers, corresponding to atypical residuals, appear as values lying beyond $\pm 3\sqrt{MSE}$. Their presence can affect the data interpretations and so it is necessary to check the source of such values carefully to determine why the associated measurements have resulted in an unusual residual value.

Model validity Plotting residuals against model predictions (fitted values) for the response model enables the validity of the response model to be assessed. Random patterning would be indicative of "acceptable" model fit across the full range of measured responses. Patterns or trends in the plot would be indicative of inappropriate model specification.

Normality Normality of the measured response can be checked by examining a normal probability plot of the residuals (residuals against normal scores) where the normal scores are determined using the position value of each residual and the standard normal distribution (expected value of order statistics assuming normality for the collected data). Departure from linearity may indicate response non-normality and extreme points suggest outliers.

Examples of likely treatment residual plots are shown in Fig. 3.3. Fig. 3.3A illustrates a plot with columns of equal length and no discernible pattern indicative that equal variability across all temperature treatments appears acceptable. Fig. 3.3B, on the other hand, shows differing column lengths implying that, for the related response data, equal variability for all time treatments is in doubt.

Fig. 3.4 provides two illustrations of a normal plot of residuals presented as residuals against normal scores. The latter refer to data we would expect to obtain if a sample of the same size were selected from a standard normal distribution (normal, mean 0, standard deviation 1). Ideally, linearity of trend in such a plot provides evidence of normality of response as illustrated in Fig. 3.4A. Non-linearity of trend, such as the S-shape pattern in Fig. 3.4B, would be suggestive of non-normal data.

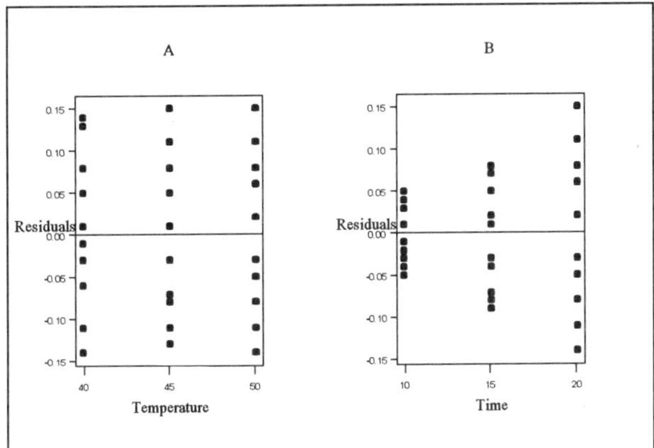

Fig. 3.3 — Illustration of treatment residual plots

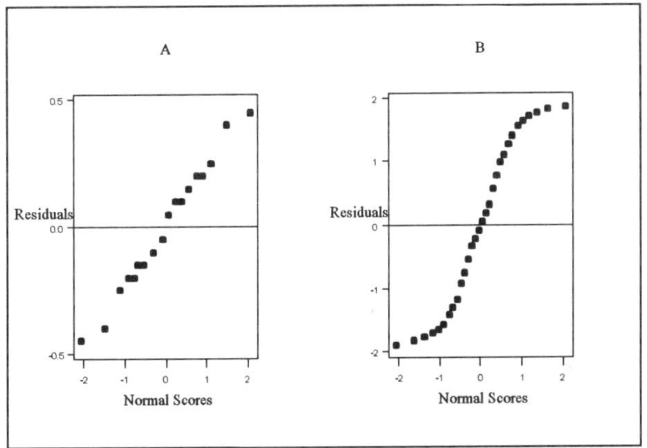

Fig. 3.4 — Illustration of normal plots of residuals

Exercise 3.7 The final part of the analysis of the toxic removal study of Example 3.1 is the diagnostic checking aspect based on the residuals (measurements – predicted values) and fits generated by the fitted response model (see Exercise 3.1 for model specification).

Diagnostic checking: Plotting the residuals against temperature settings is the first step in the diagnostic checking of the reported data. The plot in Display 3.6 shows a columnar pattern because of the code values used to specify the temperature factor. All columns are reasonably similar with only minor differences in column lengths. This minor difference in variability was hinted at within the exploratory analysis presented in Exercise 3.2.

The fits plot for this study is not included as, for a CRD, it simply represents the same as the treatment plot except that column re-ordering occurs so it can provide no additional diagnostic information for such a design structure. The normal plot, presented in Display 3.7, looks reasonably linear indicative that normality for the percent impurity response appears acceptable.

The outcome of the diagnostic checking appears to be that the reported toxic removal data may not fully conform to equal variability but do appear to be normally distributed. A variance stabilising transformation of such data may be worthy of consideration in future experiments when such a response is measured. □

Display 3.6

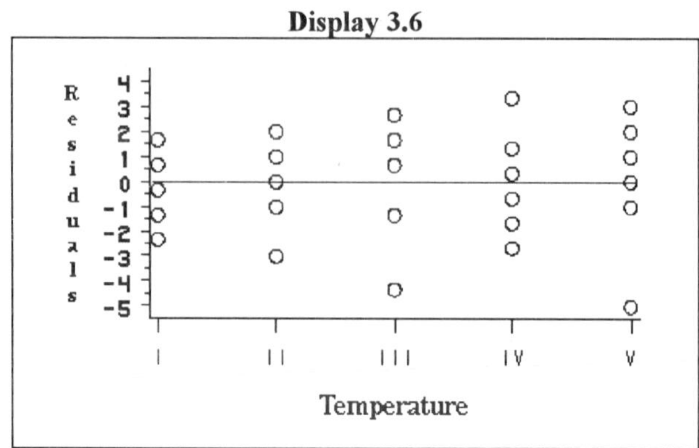

Display 3.7

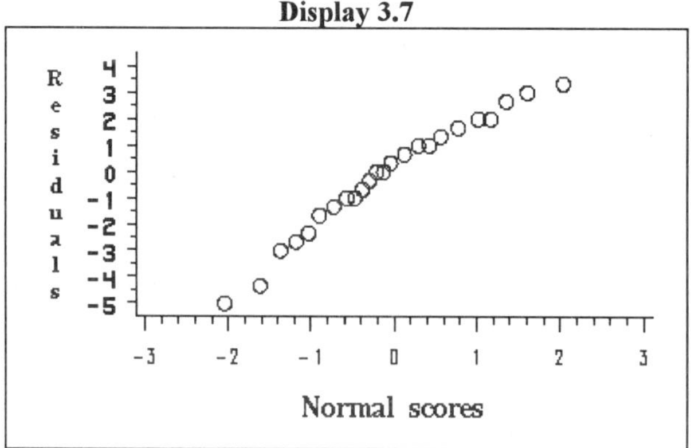

From Exercises 3.1 to 3.7, we can see that the steps in the analysis associated with a CRD experimental structure involve model specification, exploratory analysis, inferential data analysis through treatment effect test and follow-up to identify specific treatment differences, and diagnostic checking. Through these steps, a full and comprehensive analysis of study data can be forthcoming. On the basis of such an approach, we can say that the toxic removal experiment of Example 3.1 shows difference between the temperatures ($p < 0.05$) with temperature V providing significantly different, and higher, measurements from the other four temperatures ($p < 0.05$). Temperature groupings of I and II, and III and IV have emerged providing two temperature settings which perform similarly within each group ($p > 0.05$) but which show performance differences between the groups ($p < 0.05$). There is also a suggestion that transformation of the response may be worth considering in future studies.

A further diagnostic check is to plot the collected response data against model predictions. Such a plot should exhibit a pattern of points lying on a 45° line passing through the origin. Deviations from this line signify data variability problems.

3.5.2 Statistical Checks

Graphical checks provide a visual means of assessing whether or not the data conform to the model assumptions. Formal statistical mechanisms for such assessment can also be adopted though graphical is generally sufficient.

3.5.2.1 Testing Equality of Variability

A formal statistical test for assessing equality of variability across many treatments is provided by **Bartlett's test of homogeneity** which can statistically validate the equal variance assumption central to ANOVA procedures. The operational elements of this test are summarised in Box 3.5 with application most relevant for One Factor Designs. Minitab can be used to calculate this test statistic (see Appendix 3A).

Box 3.5 — Bartlett's Test of Homogeneity

Hypotheses The null hypothesis assessed by this inference procedure is H_0: equal variability across all treatments, i.e. $\sigma_1^2 = \sigma_2^2 = \ldots = \sigma_k^2$.

Test statistic The test statistic is expressed as

$$B_c = B/C \tag{3.13}$$

where

$$B = k(n-1)\log_e[(\Sigma s_j^2)/k] - 2(n-1)\Sigma\log_e(s_j)$$

$$C = 1 + \frac{1}{3(k-1)}\left(\frac{k}{n-1} - \frac{1}{k(n-1)}\right)$$

using k to represent the number of treatments, n the number of measurements per treatment, and s_j the standard deviation of the jth treatment. Test statistic (3.13) follows an approximate χ^2 distribution with $k - 1$ degrees of freedom, requisite critical values for which are presented in Table A.2.

If B_c is small, it suggests equality of variability with the opposite interpretation for large values. One drawback of the test is that it can be sensitive to data departure from normality. For the toxic removal study of Example 3.1, test statistic (3.13) was 1.89 with p value of 0.756, too low to suggest significant difference in the variability of removal rate with temperature. Thus, graphical check and test result provide comparable conclusions.

3.5.2.2 Testing Normality

The "straightness" of the normal plot of residuals can be measured by the correlation coefficient of the points in the plot. This coefficient, and thus the normality of response data, can be tested by using a special form of correlation test where high correlation near +1 will be indicative of normality. This special test is referred to as the **Ryan-Joiner test** and is similar to the Shapiro-Wilks test of normality. It is based on the principle underlying the normal plot, that linearity is indicative of normality and that an objective measure of linearity can be found in the correlation coefficient. Box 3.6 contains an outline of the operation of this test procedure which is generally conducted at the 10% significance level. Appendix 3A contains details of how this can be obtained in both SAS and Minitab.

Essentially, the test is based on the premise that if experimental data conform to normality, the data within the normal plot should exhibit a strong positive linear trend. This strength can be measured using the correlation between the two sets of points within the plot. Positive linearity will be appropriate if the correlation coefficient is near to +1 highlighting why the test statistic and critical value decision rule is altered to be such that large values of the test statistic become indicative of acceptance of the null hypothesis of data normality and not rejection as would usually be the case. For the normal plot in Display 3.7, the correlation coefficient is 0.98291. The corresponding 10% critical value, for $N = 30$, is $R_{0.1,30} = 0.9707$ leading to acceptance of normality for the response data of percent impurity removed agreeing with the graphical check result in Exercise 3.7.

Box 3.6 — Ryan-Joiner Test

Hypotheses The hypotheses tested by this procedure are H_0: data are normally distributed against H_1: data not normally distributed.

Test statistic The Ryan-Joiner test statistic is specified as

$$R = \text{correlation of the residual and normal scores within the normal plot of residuals} \quad (3.14)$$

Decision rule

test statistic approach: If R exceeds the critical value $R_{\alpha,N}$ at the 100α percent significance level, N being the total number of observations, then we accept H_0 with Table A.8 providing $R_{\alpha,N}$.

p value approach: As in all previous illustrations.

3.6 POWER ANALYSIS IN DESIGN PLANNING

In Section 2.5, the **power analysis** aspect of experimentation was introduced and its importance to statistical conclusion validity explained. When using formal design structures, the same concepts and principles apply in order to ensure the best design commensurate with the stated experimental objectives is being implemented. In experimental designs, when testing a null hypothesis of no difference between treatments, the ability to correctly support the alternative hypothesis that at least one treatment differs is of fundamental importance. Without this ability, the treatment effect statistical test will lack power (probability of detecting a true alternative hypothesis) and be unable to adequately answer the question being investigated. We must therefore be sure that the statistical test planned can detect significant treatment differences if such exist.

In one factor experiments, power analysis can be approached in a number of ways. Most are based on using the non-central F distribution characterised by the numerator and denominator degrees of freedom of the relevant F test statistic (treatment f_1, error f_2), and a third parameter λ referred to as the **non-centrality parameter**. For a CRD, we have $f_1 = k - 1$ and $f_2 = k(n - 1)$ as shown in the general ANOVA table displayed in Table 3.3. The non-centrality parameter λ provides a measure of the departure from the null hypothesis exhibited by the data. Table A.9

provides power estimates for the non-central $F(f_1, f_2, \lambda)$ distribution for planned tests at the 5% significance level.

To use these tables in power analysis, we require to specify a **minimum detectable difference** δ between the treatments which, if detected, would provide the investigator with practical proof of treatment differences. This difference essentially reflects the smallest numerical difference we expect to detect between two population means. Using this pre-defined estimate, we compute a parameter ϕ (*phi*)

$$\phi = \sqrt{\frac{nES^2}{2k}} \tag{3.15}$$

which is related to the non-centrality parameter λ by the expression $\lambda = \phi^2(f_1 + 1)$. In expression (3.15), n is the number of observations per treatment, $ES = \delta/\sigma$ refers to **effect size** based on σ^2 defining the error mean square estimate (MSE), and k is the number of treatments.

Examining Table A.9, we can see that the larger the value of ϕ then the larger the power, and from expression (3.15) we detect that ϕ can increase by increasing n, increasing δ, reducing k, or decreasing the error mean square estimate σ^2. Specification of these parameters is required before the power analysis procedure can be implemented to assess the viability of a proposed CRD experiment.

3.6.1 Power Estimation

For power estimation, the parameters n, δ, k, σ^2, and the significance level (100α percent) of the planned statistical test of treatments require to be specified by the investigator. Using these pre-set values, we evaluate ϕ from (3.15) and use Table A.9 for the characteristics combination $f_1 = k - 1$, $f_2 = k(n - 1)$, and ϕ to estimate the power of the treatment effect test at the 5% significance level. As in Section 2.5, a power estimate in **excess of 80%** is normally desirable.

Exercise 3.8 Consider a CRD investigation involving comparison of four treatments using a total of five replicate measurements for each treatment. We wish to estimate the power of the treatment effect F test, at the 5% significance level, for a minimum detectable difference of 6.75 to assess if it exceeds the usual target of 80%. Past data suggest an error mean square estimate of 7.05 can be assumed.

Information: This is a CRD experiment with $k = 4$ treatments and $n = 5$ measurements per treatment. From this information, we have $f_1 = 4 - 1 = 3$ and $f_2 = 4(5 - 1) = 16$. We also know that the minimum detectable difference of interest is $\delta = 6.75$, the estimate of error mean square $\sigma^2 = 7.05$, and that the treatment effect test is to be carried out at the 5% ($\alpha = 0.05$) significance level. From the information provided, $ES = 6.75/\sqrt{7.05} = 2.5422$.

Estimate of ϕ: Using expression (3.15), we estimate ϕ as

$$\sqrt{\frac{5 \times (2.5422)^2}{2 \times 4}} = \sqrt{4.039} = 2.01 \quad .$$

Power estimation: An extract from Table A.9 based on $f_1 = 3$ and $f_2 = 16$ is presented in Table 3.7. As $\phi = 2.01$ is near the tabulated entry of 2.0, we could assume power is approximately 86%. As this is in excess of the 80% target, the proposed design structure appears acceptable. □

Table 3.7 — Extract from Table A.9 for Exercise 3.8 on power estimation

ϕ	0.5	1.0	1.2	1.4	1.6	1.8	2.0	2.2	2.6	3.0
Power%	10.1	28.8	40.3	53.0	65.6	76.7	85.6	91.8	98.0	99.7

3.6.2 Sample Size Estimation

Power analysis calculations are most often reversed to estimate the size of the sample (n) required in an experiment to ensure a treatment effect F test power of at least 80%. In other words, we want to design the study so that the sample size ensures that the experiment has a sufficient chance of detecting significant treatment differences if such are present. This ensures statistical conclusion validity. The estimation procedure is based on trial and error using an initial guess for n and refining this until a power close to the desired figure is reached.

Example 3.2 Before being used in a medicinal preparation, chemical raw material is checked for purity. For chemical powders, for instance, powder samples are heated and the residual ash content, which represents the impurities, weighed. A processing company wishes to compare, through a CRD structure, the purities of a chemical powder that it receives from four regular suppliers. There is a belief that a treatment difference of at least 4.24% would suggest that purity differed with supplier sufficiently to suggest quality difficulties with the supplied raw material. A batch of the powder received from each supplier is to be randomly selected. However, the quality assurance (QA) manager is unsure as to how many measurements to make on each selected batch but believes that six will be sufficient. The manager wishes the test for supplier differences to be carried out at the 5% significance level and to have a power of at least 80%. Past quality information suggests an error mean square estimate of 5.54 would be appropriate.

Exercise 3.9 For the planned supplier quality study in Example 3.2, how many measurements should be made on each sampled batch?

Information: We have $k = 4$ suppliers ($f_1 = 3$), minimum detectable difference of $\delta =$ 4.24%, significance level of test of 5% ($\alpha = 0.05$), error mean square estimate of $\sigma^2 =$ 5.54, and power of at least 80%. From this information, we have $ES = 4.24/\sqrt{5.54} =$ 1.8014.

Estimation of n: For $n = 6$, the figure initially suggested by the QA manager, we have $f_2 = 4\times(6-1) = 20$ and expression (3.15) becomes

$$\phi = \sqrt{\frac{6\times(1.8014)^2}{2\times 4}} = 1.56 \quad .$$

Consulting Table A.9 for $f_1 = 3$ and $f_2 = 20$, power is estimated to be approximately 65.4% which is lower than the required power. We must increase n and check the power again.

Trying $n = 7$, we have $f_2 = 24$ and $\phi = 1.69$ which provides a power estimate of approximately 74.4% (Table A.9, $f_1 = 3$, $f_2 = 24$) which is still too low for adequate power. Using $n = 8$, we have $f_2 = 28$, $\phi = 1.80$, and power of approximately 81.4% (Table A.9, $f_1 = 3$, $f_2 = 28$). This satisfies the power stipulation and so it can be recommended that each batch of material supplied be tested using at least eight measurements. □

Most experimenters neglect power analysis within the planning phase of an experiment partly because they do not give enough thought to formulating a minimum treatment difference which, if detected, would provide scientific evidence in support of the conclusion. Its omission is often also because experimenters do not appreciate the importance of power and sample size estimation as part of design planning to ensure the statistical tests to be applied are capable of detecting significant effects if they exist. Other mechanisms of carrying out power analysis include the use of operating characteristic curves (Lipsey 1990) and the specification of treatment effects for all treatments in conjunction with use of Table A.9. Retrospective power analysis after data collection and analysis can also be informative (Zar 1996).

3.7 NON-PARAMETRIC ALTERNATIVE TO THE ANOVA BASED TREATMENT TESTS

As with one and two sample studies, data collected from experiments based on the CRD structure may not conform to the underlying assumptions of parametric ANOVA methods, i.e. equal variability and normality. Use of a non-parametric alternative to the ANOVA F test, expression (3.3), and the SNK multiple comparison, expression (3.5), may therefore be necessary in order to be able to carry out relevant inferential data analysis.

3.7.1 The Kruskal-Wallis Test of Treatment Differences

The non-parametric alternative to the treatment F test for a CRD structure is called the **Kruskal-Wallis test** and represents a simple extension of the non-parametric two sample Mann-Whitney test to the case of k samples. It is also referred to as **analysis of variance by ranks**. Application of this test generally occurs if normality for the measured response cannot be safely assumed or if the response data consist of ranks. Box 3.7 contains an outline of the analysis methods for such a case.

In the test statistic calculation, significant differences in the total ranks provides evidence of treatment differences as the position of the measurements associated with each treatment is differing. Information on the median responses for each treatment should also be presented to translate this positional result into a practical interpretation. The test statistic (3.16) can be modified easily to accommodate an experiment based on unequal numbers of observations per treatment (unbalanced CRD) by replacing N by $\sum n_j$ and n by n_j, $j = 1, 2,..., k$. If there is a substantial number of tied observations, test statistic (3.16) can be adjusted by including a correction factor based on the extent of tied observations present (Daniel 1990). Inclusion of this factor marginally inflates the test statistic. Expression (3.16) could also be computed by ranking the response data and carrying out ANOVA on the ranks to provide

$$H = SS\text{Treatment}/[kn(kn + 1)/12] \qquad (3.17)$$

Box 3.7 — Kruskal-Wallis Test for Treatment Differences

Assumptions Measured data constitute k random samples and the underlying populations are identical except for location differences in at least one.

Hypotheses The hypotheses related to the Kruskal-Wallis test are essentially the same as those presented for parametric inference (see Section 3.3.1) namely, H_0: no difference between treatments (median treatment effects same) and H_1: treatments differ (at least one median treatment effect different).

Exploratory data analysis (EDA) Exploratory analysis of the actual data is again advisable to gain initial insight into what they indicate in respect of the experimental objective.

Test statistic To determine the test statistic, the first step involves pooling the observations into one large group and ranking the pooled data in ascending order, smallest to largest, with tied observations given an average rank. The ranks associated with each of the k treatments are then summed to generate the rank totals C_1 to C_k. From these, we use

$$H = \frac{12}{N(N+1)} \sum_{j=1}^{k} \frac{C_j^2}{n} - 3(N+1) \tag{3.16}$$

to calculate the Kruskal-Wallis test statistic where N is the total number of measurements, k is the number of treatments, and n is the number of observations associated with each treatment.

Decision rule The test statistic (3.16) follows an approximate χ^2 distribution with $k - 1$ degrees of freedom enabling Table A.2 to be used for critical value determination. As with most inference procedures, low values of the test statistic ($H <$ critical value) or high p values ($p >$ significance level) will be indicative of acceptance of H_0 and the conclusion that the treatments tested do not appear to differ.

3.7.2 Multiple Comparison Associated with the Kruskal-Wallis Test

The Kruskal-Wallis test, as with the ANOVA F test, only provides a statistical test of general treatment differences. To pinpoint how the treatments differ if a difference is detected, we need to employ a non-parametric follow-up. Pairwise Mann-Whitney tests could be considered, but testing all possible treatment pairings this way could increase the probability of committing at least one Type I error, i.e. raise the probability of rejecting a true null hypothesis in at least one of the treatment comparisons. It is therefore best to use a non-parametric multiple comparison for the same reasons as were discussed when describing the background to ANOVA based multiple comparisons in Section 3.4.3. Such a procedure is available and is called **Dunn's procedure** which, like the Student-Newman-Keuls procedure, provides a mechanism for treatment comparison which ensures that the experimentwise error rate (overall significance level) for all pairwise tests is constrained to 100α percent. An outline of the calculation and decision elements of this non-parametric follow-up is provided in Box 3.8.

Box 3.8 — Dunn's Non-parametric Multiple Comparison

Hypotheses The null hypothesis to be tested in this multiple comparison is H_0: no difference between compared treatments (medians similar). The alternative specifies difference between the compared treatments.

Test statistics The numerical differences between the average ranks $\overline{R}$ for each pair of treatments become the test statistics for the pairwise comparisons.

Critical value The critical value necessary for comparison with the difference test statistics is expressed as

$$W_{ij} = z_{\alpha/[k(k-1)]}\sqrt{\frac{kn(kn+1)}{12}\left(\frac{2}{n}\right)} \tag{3.18}$$

where 100α percent is the experimentwise error rate (significance level), k is the number of treatments, z refers to the $100\alpha/[k(k-1)]$ percent critical z value from Table A.4, and n is the number of observations per treatment.

Decision rule If the numerical difference in the average ranks for treatments i and j ($|\overline{R}_i - \overline{R}_j|$) is less than the specified critical value (3.18), we accept H_0 and declare that the evidence suggests that the compared treatments appear not to differ significantly.

In most CRD based experiments, the number of measurements associated with the compared treatments should, ideally, be the same. If the number of observations in the compared treatments differs, then the term $2/n$ in expression (3.18) can be modified to $(1/n_i + 1/n_j)$ to accommodate the differing numbers of observations associated with treatments i and j. Adjustment of the critical value (3.18), if there are large numbers of tied observations, can be considered but the effect on the value produced is only minor.

In a study involving k treatments, there are $k(k-1)/2$ possible treatment comparisons. As each comparison is essentially a two-tail test and assuming an overall significance level of 100α percent, then $100\alpha/2$ percent has to be assigned equally to each comparison if Bonferroni's inequality is to be adopted (see Section 3.4.3). With $k(k-1)/2$ comparisons possible, this results in the z value being determined at the $100\alpha/[k(k-1]$ level. Hence, the use of the $z_{\alpha/[k(k-1)]}$ term in expression (3.18). Choice of experimentwise error rate, 100α percent, is also based partly on k, the number of treatments tested, with the general rule being to select 100α percent larger than that customarily used in inference, i.e. use 10% ($\alpha = 0.1$), 15% ($\alpha = 0.15$), or even 20% ($\alpha = 0.2$).

3.7.3 Linear Contrasts

Frequently, we may also require to test a planned comparison (contrast) of the treatments as illustrated in section 3.4.4 for ANOVA based procedures. An equivalent non-parametric contrast exists and can be expressed in terms of the k average ranks $\overline{R}$ as

$$L = c_1\overline{R}_1 + c_2\overline{R}_2 + \ldots + c_k\overline{R}_k = \Sigma c_j\overline{R}_k \tag{3.19}$$

based on the constraint that the constants c_j should generally be such that $\Sigma c_j = 0$. The mechanisms of non-parametric contrasts are summarised in Box 3.9 with similar reasoning applied to the interpretation of the calculated confidence interval as for ANOVA based linear contrasts.

Box 3.9 — Non-parametric Linear Contrasts

Hypotheses As Box 3.2.

Estimate of L To estimate L, the mean ranks for the specified treatments are substituted into the appropriate contrast expression and a numerical estimate calculated.

Confidence interval for contrast An approximate $100(1 - \alpha)$ percent confidence interval estimate for a non-parametric contrast, similar to the contrast confidence interval (3.7), is expressed as

$$\text{estimate of } L \pm \chi^2_{\alpha,k-1}\sqrt{\left(\frac{\Sigma c_j^2}{n}\right)\frac{kn(kn+1)}{12}} \tag{3.20}$$

based on the previous definitions of k, n, and c_j.

3.8 DATA TRANSFORMATIONS

The underlying equal variance (homoscedastic) and normality assumptions required by parametric inference procedures may not always be satisfied in a study. Often, study data may be inappropriate for normality to be assumed (skewed, count data), or they may be such that the variances associated with the treatments are not constant and seem to vary with the magnitude of the treatment mean (heteroscedastic). Data re-scaling, through application of a simple transformation, is particularly useful in these cases to enable the non-normality, non-constant variance, and non-additivity to be corrected before implementing inferential data analysis which may depend on these specifications being valid for the analysed data. Re-scaling essentially changes the scale of the response data with the transformed metric becoming the basis of the inferential analysis. Several transformations exist with some of the most commonly used illustrated in Box 3.10.

Often, knowledge of the type of data collected dictates the transformation to apply. For skew data, the logarithm transformation, for which $\log_{10}$ is also possible, can produce symmetric data with this type of transformation often used in laboratory experimentation to re-scale factor settings for the purpose of results presentation, e.g., concentration of a solution and density of organisms in a growth experiment. The square root transformation is useful when it is necessary to make the group variances independent of the means. It often occurs when data are in the form of numbers of independent events in a fixed period of time or spatial region corresponding to a Poisson random variable, e.g., radiation counts and number of abnormal cells in viral cultures. In both logarithm and square root transformations, it is also possible to select an optimal constant c so that the applied transformation, either $\log(X + c)$ or $\sqrt{(X + c)}$, is optimised.

Box 3.10 — Commonly Applied Data Transformations

Logarithm If the standard deviation is proportional to the mean for each treatment group ($\sigma \propto \mu$), i.e. standard deviation over mean is approximately constant (data exhibit heteroscedasticity), then we can use the logarithmic transformation $\log_e(X)$ or $\log_e(X + 1)$ if the data contain zeros to avoid occurrence of $\log_e(0)$.

Square root If the experimental data are such that the variance is proportional to the mean for each treatment grouping ($\sigma^2 \propto \mu$, Poisson data), i.e. variance over mean approximately constant, then the square root transformation $\sqrt{X}$, or $\sqrt{(X + 0.5)}$ if the data are low and contain zeros, should be applied.

Arcsine (angular) When the data are in the form of probabilities, proportions, or percentages (binomial data), we can apply the arcsine transformation $\sin^{-1}(\sqrt{X})$ where X lies between 0 and 1.

Reciprocal For each data grouping, if the standard deviation is proportional to the square of the mean ($\sigma \propto \mu^2$), then we can use the reciprocal transformation $1/X$ to transform the data.

Box-Cox generalised This corresponds to a generalised form of data transformation provided by the Box-Cox family of power transformations $(X^\lambda - \lambda)/\lambda$, $\lambda \neq 0$. Here, the data are used to determine the maximum likelihood estimate of λ which provides best transformation expression.

Data appropriate for the arcsine transformation include, for example, the proportion of cells inhibited by toxic substances and the percentage of respiring bacteria in sewage samples. If small numbers of observations have been collected, the range can be used in preference to the standard deviation in assessing if a transformation of experimental data is necessary. The Box-Cox transformation was suggested by Box & Cox (1964) and is most appropriate for single data sets. Unlike the others presented in Box 3.10, it represents a re-scaling method whereby the response data as a whole generate the necessary function for transformation.

Data analysis, when data transformation is necessary, is carried out on the transformed data with the inferences reached extrapolated to the original experimental investigation for problem context inference. For estimation procedures, in particular, re-transformation of the results to the original data scale should be carried out. Example 3.3 which follows will be used to illustrate the application of data transformations.

Example 3.3 Consider the data presented in Table 3.8 which correspond to tube conductivity measurements for three different coatings of colour picture tubes. The measurements are presented as joules per second per metre per Kelvin. The summaries for the collected data are also presented in Table 3.8 and show that the standard deviations are not constant for all coatings implying that the response data appear heteroscedastic. We need, therefore, to assess which transformation to apply to the collected data to enable the property of similar variability (homoscedasticity) across all coatings to be satisfied.

Table 3.8 — Data for conductivity in the picture tube study of Example 3.3

Coating	A		B		C	
Original data	139	137	269	250	63	60
	134	142	256	266	59	63
	144	136	260	264	61	62
Mean	138.6667		260.8333		61.3333	
Standard deviation	3.7771		6.9976		1.6330	
Scaled data	2.143	2.137	2.430	2.398	1.799	1.778
	2.127	2.152	2.408	2.425	1.771	1.799
	2.158	2.134	2.415	2.422	1.785	1.792
Mean	2.1418		2.4163		1.7873	
Standard deviation	0.0116		0.0118		0.0114	

Ratio check: We first check the ratio of standard deviation to mean for each coating. This provides values of 0.0272, 0.0268, and 0.0266 for coatings A, B, and C respectively. These values are roughly constant so the logarithmic transformation would appear appropriate. We should therefore transform the data using $\log_{10}(X)$.
Application of transformation: Applying this logarithm transformation results in the re-coded data presented in Table 3.8. The summary figures for these coded data show that variability, as measured through the standard deviation, is now very similar across all tested coatings. □

We can see from the scaled data of Example 3.3 how transforming data can improve certain properties of the data, so enabling the assumptions of a planned inference procedure to be more readily satisfied. Re-coding of data before analysis is not always necessary within experimental designs though use of diagnostic checking, as illustrated, can point to the need for data transformation in subsequent comparable experiments.

PROBLEMS

3.1 As part of an investigation into pollution levels in seaweed, five specimens from each of four species of seaweed were sampled. The specimens were examined in the laboratory where the levels of cadmium present, in $\mu g\ g^{-1}$ dry weight, was measured as presented below. Interest lay in ascertaining whether cadmium levels differed between the species and, if so, whether particular species absorb more or less cadmium compared to others. Also of interest was a comparison of the mean cadmium level of *Fucus vesiculosus* with the mean of the other three species.

Pelvetia	*Fucus vesiculosus*	*Ascophyllum*	*Laminaria*
2.04	2.17	1.93	1.94
1.97	2.25	2.08	2.02
2.06	2.11	2.03	1.93
2.01	2.07	2.05	1.85
1.95	2.14	2.13	1.97

3.2 Tensile strength measurements from ten randomly selected bars produced within three high-temperature alloy castings were made. The tensile strength measurements collected, in lb in^{-2}, are presented below. Carry out a full and comprehensive statistical analysis of these data.

Casting	A		B		C	
	88.0	89.0	85.9	86.0	94.2	93.8
	88.0	86.0	88.6	91.0	91.5	92.5
	94.8	92.9	90.0	89.6	92.0	93.2
	90.0	89.0	87.1	93.0	96.5	96.2
	93.0	93.0	85.6	87.5	95.6	92.5

3.3 γ-BHC is an insecticide widely used to treat structural timber. It has been suggested that increased use of γ-BHC has resulted in the decline of bat populations. The following data show the weights, in g, of 24 bats of a particular species six weeks after they had been assigned to one of three roosts containing wood shavings with low, medium, and high concentration of γ-BHC. Analyse these data and draw conclusions.

Concentration	Low		Medium		High	
	61	49	46	48	52	43
	72	62	49	37	35	39
	58	57	36	34	37	40
	50	49	35	45	42	45

3.4 In an investigation to compare the effects of three medications, three groups of six randomly selected dogs were administered the three medications and a fourth group of dogs was given a placebo. Treatments A and B both contain one unit of active ingredient while treatment C contains 1½ units of active ingredient and the placebo zero units. The time, in minutes, till an observed response appeared was recorded as follows:

Placebo		A		B		C	
10.8	10.3	9.1	9.7	9.6	9.5	8.9	9.3
9.5	10.2	9.7	9.0	9.8	8.8	9.9	9.0
11.0	10.3	10.1	10.3	10.5	9.4	8.8	9.5

Analyse these data to investigate for evidence of differences in response time amongst groups. Of particular interest was the comparison of the medication with the placebo and the comparison of the different levels of the medication administered. Carry out these comparisons and draw relevant conclusions.

3.5 A food analyst wished to set up a CRD based experiment to assess the butyric acid content of four commercially produced butter-based biscuits. The analyst is unsure of how many samples of each biscuit type to test but is prepared to assume that if a minimum difference of 0.7 mg/ml in the mean butyric acid content could be detected, it would provide sufficient evidence to conclude that butyric acid content differed with biscuit type. The general significance test is planned to be carried out at the 5% significance level with power of at least 80%. An error mean square estimate of 0.14 is available from past comparable studies. The analyst initially believes four biscuits of each type is a reasonable number of samples for assessment. Is four sufficient or are more necessary?

3.6 As part of a study of diabetes mellitus, mice are to be randomly assigned, in equal numbers, to one of four experimental groups, a control group and three treatment groups representing different drugs added to their diet for a one week period. At the end of the treatment period, the mice are to be given a glucose injection and, after 45 minutes, have their blood glucose, in $\mu g\ l^{-1}$, measured. The study director is unsure of how many mice to assign to each experimental group but is of the opinion that a minimum detectable difference of 2.55 $\mu g\ l^{-1}$ would be indicative of significant treatment effect. The associated statistical test of treatment differences is planned to be carried out at the 5% significance level with an approximate power of at least 85%. An error mean square estimate of 2.3 can be assumed. How many mice per treatment group would satisfy these constraints?

3.7 Three analytical laboratories are to be asked to perform a standard form of chemical analysis on sample material aimed at determining if they report, on average, comparable measurements. Each laboratory is required to carry out six replicate analyses on supplied material with the resultant data to be analysed according to a CRD structure. Suppose the investigator wished to test, at the 5% significance level, for a minimum detectable difference of 2 units which would be reflective of significant laboratory differences. From previous comparable quality assurance studies, an estimate of mean square error of 1.82 can be assumed. Show that the power of the test is approximately 87%.

3.8 An excessive amount of ozone in the air is an indicator of high levels of air pollution. A monitoring study is planned whereby ten air volumes are to be collected from each of four locations in an industrial region in order to monitor ozone levels. The environmental health officer setting up the study believes that a minimum detectable difference of 0.048 ppm in ozone would be sufficient to indicate location differences. To be confident of his conclusion, he has decided to carry out the appropriate location test at the 5% significance level based on a power of at least 80%. Assuming an experimental error estimate of 0.0006, confirm that ten air volumes is adequate. Can this number be reduced while still retaining sufficient power for the planned location test?

3A Appendix: Software Information For Completely Randomised Design

Data entry
Data entry for CRDs generally requires two columns of data to be entered: one for the response and one for the character or numerical codes corresponding to each treatment.

SAS: Data entry in SAS is achieved through use of the DATA step with the PROC PRINT procedure providing a data check. The statements for Example 3.1 were as follows:

```
DATA CH3.EX31;
    INPUT TEMP $ IMPREM @@;
    LINES;
    I 49   I 51   I 53   I 53   I 52   I 50
    II 47  II 49  II 51  II 52  II 50  II 51
    III 40  III 35  III 42  III 38  III 40  III 41
    IV 36  IV 42  IV 38  IV 39  IV 37  IV 40
    V 55   V 60   V 62   V 63   V 59   V 61
;
PROC PRINT DATA = CH3.EX31 NOOBS;
```

```
        VAR IMPREM;
        BY TEMP;
    RUN;
```

Statement explanation is provided in Appendices 1A and 2A.1.

Minitab: Entry of response data into column C1 requires to be by treatment. For Example 3.1, this means impurity response will be entered in five blocks of six observations starting with temperature I. Codes entry into column C2 uses the **Calc ➤ Make Patterned Data** option followed by the **Manip ➤ Code ➤ Numeric to Text** option to convert the codes into character codes reflecting the temperature level labelling. The menu procedures for temperature codes entry for Example 3.1 would be as follows:

> Select **Calc ➤ Make Patterned Data ➤ Simple Set of Numbers ➤** for *Store patterned data in*, enter **Temp** in the box ➤ select the **From first value box** and enter **1** ➤ select the **To last value** box and enter **5** ➤ ensure the **In steps of** box has entry **1** ➤ select the **List each value** box and enter **6** ➤ ensure the **List the whole sequence** box has entry **1** ➤ click **OK**.
>
> Select **Manip ➤ Code ➤ Numeric to Text ➤** for *Code data from columns*, select **Temp** and click **Select** ➤ select the **Into columns** box and enter **Temps** ➤ select the first box of **Original values** boxes, enter **1**, press **Tab**, enter **I**, press **Tab**, enter **2**, press **Tab**, enter **II**, and continue in this way until all five conversion codes are entered ➤ click **OK**.

For Example 3.1, either of these forms of data entry would result in a data structure as follows:

ImpRem	Temp	ImpRem	Temp	ImpRem	Temp	ImpRem	Temp
49	I	51	II	40	III	40	IV
51	I	52	II	41	III	55	V
53	I	50	II	36	IV	60	V
53	I	51	II	42	IV	62	V
52	I	40	III	38	IV	63	V
50	I	35	III	39	IV	59	V
47	II	42	III	37	IV	61	V
49	II	38	III				

Data plot

SAS: The simple data dotplot of impurity response for each temperature setting in Display 3.1 was generated using the SAS statements below with Appendix 2A providing explanation of the statements.

```
PROC GPLOT DATA = CH3.EX31;
    PLOT TEMP*IMPREM / HAXIS = AXIS1 VAXIS = AXIS2 VREVERSE;
    AXIS1 LABEL = (F = CENTB J = C 'Percent impurity removed')
        VALUE = (H = 0.8) LENGTH = 30;
    AXIS2 LABEL = (F = CENTB A = –90 R = 90 J = C H = 0.7 'Temperature')
        LENGTH = 10 STYLE = 0 MAJOR = NONE;
RUN;
```

Minitab: In Minitab, dotplot generation can be obtained using **Graph ➢ Character Graphs ➢ Dotplot** (see Appendix 2A.1) or from within the **Descriptive Statistics** procedure (see ***Data summaries***).

Data summaries

SAS: Numerical summaries of response data by treatment can be generated using the PROC MEANS procedure (see Display 3.2) as follows:

```
PROC MEANS DATA = CH3.EX31 MAXDEC = 2 MIN MAX MEAN STD CV;
    CLASS TEMP;
    VAR IMPREM;
RUN;
```

Statement explanation is provided in Appendices 1A and 2A.

Minitab: Menu commands for default data summary production by treatment are as follows:

Select **Stat ➢ Basic Statistics ➢ Descriptive Statistics ➢** for *Variables*, select **Imprem** and click **Select ➢** select **By variable**, click the empty box, select **Temps**, and click **Select**.

Select **Graphs ➢ Dotplot of data ➢** click **OK ➢** click **OK**.

ANOVA information

SAS: The PROC ANOVA procedure, listed below, performs analysis of variance for all balanced experimental designs (see Display 3.3).

```
PROC ANOVA DATA = CH3.EX31;
    CLASS TEMP;
    MODEL IMPREM = TEMP;
    MEANS TEMP / SNK;
RUN;
```

The CLASS statement identifies *all* the design factors. The MODEL statement specifies the response model with response to the left of the equals sign and factors to the right. The MEANS statement is optional and produces a multiple comparison of the treatment means where SNK specifies that this comparison be based on the Student-Newman-Keuls procedure (see Display 3.4).

Unfortunately, PROC ANOVA does not provide for contrast estimation and residual data recording. This requires use of the PROC GLM procedure, shown below, which follows a similar syntax but has the two necessary option statements for contrast test derivation and diagnostic information storage. The CONTRAST statement must specify the factor being tested with –0.5 –0.5 0 0 1 representing the contrast coefficients for the five temperatures (see Display 3.5). Residual (R) and fits (P) storage is provided within the OUTPUT statement with OUT = specifying the data set for storage which must be a new data set. The P = FITS element of the OUTPUT statement stores the predicted values in the variable FITS while R = RESIDS stores the residuals, estimated as observed value minus predicted value, in the variable RESIDS.

```
PROC GLM DATA = CH3.EX31;
    CLASS TEMP;
    MODEL IMPREM = TEMP;
```

```
        MEANS TEMP / SNK;
        CONTRAST '(I, II) versus V' TEMP –0.5 –0.5 0 0 1;
        OUTPUT OUT = CH3.EX31A P = FITS R = RESIDS;
RUN;
```

Minitab: Minitab's Balanced ANOVA procedure will provide the same ANOVA information as PROC ANOVA and PROC GLM but does not have a facility for multiple comparison and contrast evaluation. The necessary menu commands are as follows:

Select **Stat** ➤ **ANOVA** ➤ **Balanced ANOVA** ➤ for *Responses*, select **Imprem** and click **Select** ➤ select the **Model** box, select **Temps**, and click **Select**.
Select **Storage** ➤ **Fits** ➤ **Residuals** ➤ click **OK** ➤ click **OK**.

Diagnostic checking
SAS: Factor residual plots can be produced using the PROC GPLOT procedure as specified below where VREF in the PLOT statement simply enables a horizontal reference line to be inserted on the plot at 0 on the residuals axis (see Display 3.6).

```
PROC GPLOT DATA = CH3.EX31A;
        PLOT RESIDS*TEMP / HAXIS = AXIS1 VAXIS = AXIS2 VREF = 0;
        AXIS1 LABEL = (F = CENTB J = C 'Temperature') VALUE = (H = 0.8)
                LENGTH = 30;
        AXIS2 LABEL = (F = CENTB A = –90 R = 90 J = C H = 0.8 'Residuals')
                LENGTH = 10;
RUN;
```

Producing a normal plot of residuals (see Display 3.7) is somewhat complicated in SAS. It is first necessary to generate the normal scores using PROC RANK and the BLOM estimation method, BLOM referring to the normal scores estimation method of Blom (1958). This procedure computes the normal scores from the ranks of the residuals by means of the expression $\Phi^{-1}(r_i - 3/8)/(n + 1/4)$ where Φ^{-1} is the inverse cumulative normal function, r_i is the rank of the ith ordered residual, and n is the number of nonmissing observations. The resulting normal plot is generated using PROC GPLOT, while PROC CORR is used to generate the Ryan-Joiner test statistic.

```
PROC RANK DATA = CH3.EX31A NORMAL = BLOM OUT = CH3.EX31A;
        VAR RESIDS;
        RANKS NSCORES;
PROC GPLOT DATA = CH3.EX31A;
        PLOT RESIDS*NSCORES / HAXIS = AXIS1 VAXIS = AXIS2;
        AXIS1 LABEL = (F = CENTB J = C 'Normal scores') VALUE = (H = 0.8)
                LENGTH = 30;
        AXIS2 LABEL = (F = CENTB A = –90 R = 90 J = C H = 0.8 'Residuals')
                LENGTH = 10;
PROC CORR DATA = CH3.EX31A NOPROB NOSIMPLE;
        VAR RESIDS NSCORES;
RUN;
```

PROC UNIVARIATE could also be used. This procedure can generate a normal probability plot of data (PLOT) as well as a test for normality (NORMAL) based on the

Shapiro-Wilks test for sample sizes less than or equal to 2000. The code structure would be as follows:

```
PROC UNIVARIATE DATA = CH3.EX31A PLOT NORMAL;
      VAR RESIDS;
RUN;
```

Minitab: In Minitab, there is a simple macro which can produce diagnostic plots of residuals which include the fits plot and normal plot. The menu commands below would generate these plots based on the storage of residuals (RESI1) and fits (FITS1) when the Balanced ANOVA procedure was implemented. As indicated in the text, the fits plot would be sufficient for a CRD because it and the factor residual plot are identical apart from positioning of the columns with respect to the X axis. Residual plots can also be generated from the **Stat ➤ ANOVA ➤ Balanced ANOVA** procedure by choosing the **Graph** option and selecting the residual plots required.

Select **Stat ➤ ANOVA ➤ Residual Plots ➤** for *Residuals*, select **RESI1** and click **Select ➤** for *Fits*, select **FITS1** and click **Select ➤** click **OK**.

Minitab also has facilities for generation of both statistical checks of model assumptions. Bartlett's test can be generated through the menu selection

Select **Stat ➤ ANOVA ➤ Homogeneity of Variance Test ➤** for *Response*, select **ImpRem** and click **Select ➤** for *Factors*, select **Temps** and click **Select ➤** ensure confidence level is set at **95% ➤** click **OK**.

The Ryan-Joiner test, and a normal plot of the residuals, are available through the menu selection

Select **Stat ➤ Descriptive Statistics ➤ Normality Test ➤** for *Variable*, select **RESI1** and click **Select ➤** for *Tests for Normality*, select **Ryan-Joiner ➤** click **OK**.

4

One Factor Blocking Designs

4.1 INTRODUCTION

The CRD structure for testing different treatments represents the first stage in the development of a study design from the simple two sample case. In this structure, the experimental error associated with the response model arises from the difference between responses of experimental units within treatments, i.e. the within group variation. The influence of this error may be reduced further by blocking the experiment and obtaining measurements on each treatment in every block. Introduction of a blocking factor, or factors, can help to reduce the level of unexplained variation thereby improving the sensitivity of the treatment effect analysis and increasing the likelihood of detecting true treatment differences if they exist within the response data. Blocking concepts are not restricted to One Factor Designs but can also be considered for higher order designs such as the ones discussed in later chapters.

To illustrate the principles underlying blocking, consider a study into the effect on product yield of four modifications to production using five batches of raw material. Comparison of production modifications is the main purpose of the study. However, variation between batches could markedly affect product yield and so each treatment should be considered across each batch of raw material. The first batch of raw material could be divided into four equal parts to form one block with the four modifications randomly assigned to the four parts. The next batch would be divided similarly to form the next block and this would continue until all five batches of raw material had been divided and the production modification schemes assigned accordingly, as summarised in Table 4.1.

Table 4.1 — Possible design layout for product yield experiment

Modification		A	B	C	D
	1	Part 3	Part 2	Part 1	Part 4
	2	Part 4	Part 3	Part 2	Part 1
Batch	3	Part 1	Part 2	Part 4	Part 3
	4	Part 2	Part 4	Part 3	Part 1
	5	Part 3	Part 1	Part 2	Part 4

4.2 RANDOMISED BLOCK DESIGN

The **Randomised Block Design (RBD)** structure is essentially an extension of the paired sample structure described in Section 2.4 and is such that each block provides a response measurement for every treatment. The blocks used in such a structure generally correspond to homogeneous groups, the units of which are similar to one another, e.g. batches of material (see Table 4.1) or age groups of subjects. Assignment of the treatments is randomised within the blocks in a RBD structure, i.e. blocking represents a restriction on randomisation. This restriction means the RBD structure should not be mistaken for a CRD structure or a two factor Factorial Design which will be described in Chapter 5.

4.2.1 Design Structure

The experimental layout for a RBD structure comprising k treatments, or factor levels, and n blocks is shown in Table 4.2 where each treatment is tested on each block of material. The structure is such that there are k experimental units in each block providing a total of kn experimental units as in the CRD. The difference between the CRD and the RBD lies in the inclusion of the blocking factor thought to influence the response but not requiring full assessment. Again, the structure is such that only one treatment factor is being investigated.

Table 4.2 — Randomised Block Design structure

	Treatments	1	2	.	.	k
	1	X_{11}	X_{12}	.	.	X_{1k}
Blocks	2	X_{21}	X_{22}	.	.	X_{2k}
	.	.	.	.	.	.
	.	.	.	.	.	.
	n	X_{n1}	X_{n2}	.	.	X_{nk}

X_{ij} is the measurement made on the experimental unit within block i receiving treatment j

Example 4.1 A study was undertaken into the relative effectiveness, as measured by sales, of four packaging methods for a grocery item. It was thought that the type of shop in which the item was sold could affect sales so the shop factor was used as a blocking element based on using four similar shops of each type. The number of items sold in the study week, to the nearest hundred, are shown in Table 4.3. Also of particular interest was the comparison of sales for method A, the presently used method, with the other three methods tested.

Table 4.3 — Data for sales in packaging method study of Example 4.1

		Packaging method			
		A	B	C	D
	I	22	14	15	14
	II	20	10	20	19
Shop type	III	11	11	14	10
	IV	10	8	8	10
	V	16	17	14	9

4.2.2 Model for the Measured Response

For a recorded observation X_{ij} within a RBD experiment, the suggested **response model** structure is

$$X_{ij} = \mu_{ij} + \varepsilon_{ij} = \mu + \beta_i + \tau_j + \varepsilon_{ij} \tag{4.1}$$

where μ_{ij}, ε_{ij}, and τ_j are as for the CRD (see Section 3.2.2). The expression $\mu_{ij} + \varepsilon_{ij}$ is the **means model** with $\mu + \beta_i + \tau_j + \varepsilon_{ij}$ describing the **effects model**. β_i defines the effect of block i assumed constant for all units within block i but may be different for units within different blocks and is often regarded as a fixed effect but may be a random effect. This additive response model again specifies that response variation is explained by the effect of the controlled factors, blocks and treatments, plus the effect of uncontrolled factors, error. Again, the hope is that controlled variation explains considerably more of the variability in the measured response than does error with the treatment effect contributing most to controlled variation.

Using the illustration on production yield, then the response model we would specify for this study would be

$$\text{yield} = \mu + \beta_i + \tau_j + \varepsilon_{ij}$$

where $i = 1, 2, \ldots, 5$ (five batches) and $j = \text{A, B, C, D}$ (four production modifications). The β_i term reflects the effect of the ith batch on yield while τ_j defines the contribution of the jth modification to production yield. Analysis of data from such an experiment would be based on primarily measuring the magnitude of the modifications effect relative to error effect, a large value being indicative of factor differences.

4.2.3 Assumptions

The inherent **assumptions** of the RBD response model are primarily associated with the error component of the response model. It is assumed that the response variability for all treatments is equal, the blocks are independent of the treatments, and responses are normally distributed, i.e. $\varepsilon_{ij} \sim \text{NIID}(0, \sigma^2)$.

Exercise 4.1 First stage of the analysis of the sales study of Example 4.1 is, as usual, specification of response model for the data collected.

Response model: As the design structure is that of a RBD, then the response model will be

$$\text{sales} = \mu + \beta_i + \tau_j + \varepsilon_{ij}$$

where i = I, II, III, IV, V (five shop types) and j = A, B, C, D (four packaging methods). The term β_i defines the shop type effect, τ_j is the packaging effect, and ε_{ij} defines the random error effect. In addition, we assume $\varepsilon_{ij} \sim \text{NIID}(0, \sigma^2)$ and so sales data are assumed normally distributed and that variability in these data is assumed similar for all four methods tested. □

4.2.4 Exploratory Data Analysis

As with previous illustrations, initial examination of data plots and summaries with respect to the treatments tested should be considered to gain insight into what they convey about the differences between treatment groups.

Exercise 4.2 Exploratory analysis represents the second stage in the analysis of the sales data of Example 4.1.

EDA: Exploratory analysis will be based on the SAS plots and summaries presented in Display 4.1. The plot highlights similar sales for each method with no obvious difference between them. Consistency of sales level differs slightly with packaging method but range of sales covered by all the methods is wide. The means show B and D similar with A possibly highest. The *CV*s are similar and highlight the wide variation in the recorded product sales. Initial impressions suggest packaging method may not influence sales. □

Display 4.1

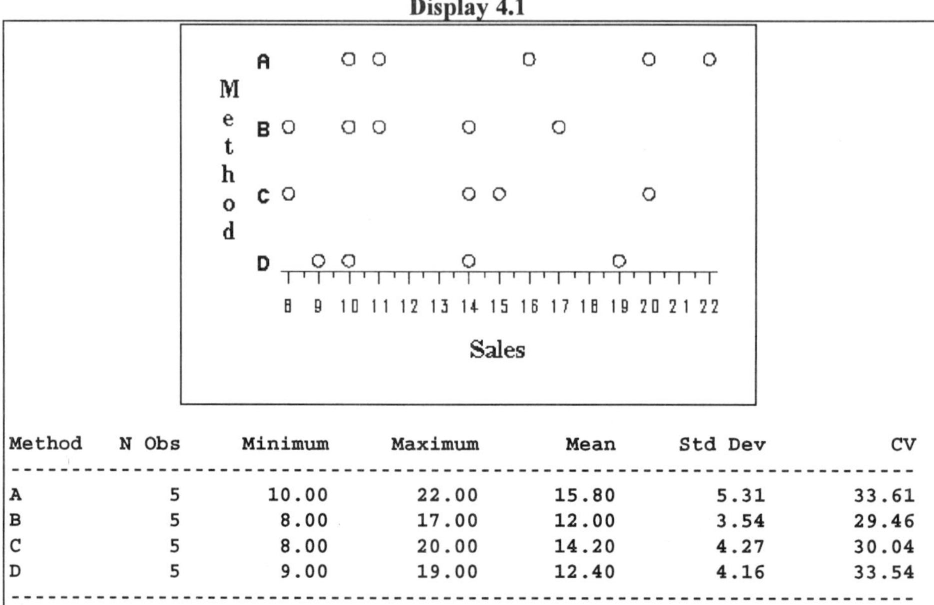

Method	N Obs	Minimum	Maximum	Mean	Std Dev	CV
A	5	10.00	22.00	15.80	5.31	33.61
B	5	8.00	17.00	12.00	3.54	29.46
C	5	8.00	20.00	14.20	4.27	30.04
D	5	9.00	19.00	12.40	4.16	33.54

4.3 ANOVA Principle for the Randomised Block Design

As in the CRD, we need to determine if treatment differences are of statistical significance. Again, the ANOVA principle provides the background to this aspect of the data analysis enabling test statistic construction and derivation based on the response model specification and the underlying statistical theory. The general procedure involves specification of hypotheses, derivation of an ANOVA table and treatment test statistic for testing which hypothesis better reflects the evidence provided by the data.

4.3.1 Hypotheses

Hypotheses for a RBD experiment can be specified in the same way as those for the CRD. This means H_0: no treatment difference versus H_1: treatment difference, i.e. at least one treatment mean differs in level of response.

4.3.2 ANOVA Table

As with the CRD, application of the **ANOVA principle** involves decomposing response variation into the sources of variation specified in the response model (4.1) namely, blocks, treatments, and error. Again, this decomposing of response variation into component parts is based on the determination of sum of squares (*SS*) terms for the controlled and uncontrolled components within the response model. From this, the total sum of squares can be expressed as

$$SS\text{Total } (SST) = SS\text{Blocks } (SSBl) + SS\text{Treatment } (SSTr) + SS\text{Error } (SSE) \qquad (4.2)$$

The calculation aspects of this process are summarised in the **ANOVA table** displayed in Table 4.4 where the mean square (*MS*) terms represent estimates of the expected mean squares associated with the theory underpinning experimental designs. The structure of the ANOVA table mirrors that associated with a CRD structure (see Table 3.3) though an additional controlled component, the block effect, has been incorporated.

Table 4.4 — ANOVA table for a Randomised Block Design

Source	*df*	*SS*	*MS*
Blocks	$n-1$	$SSBl = \dfrac{\sum_{i=1}^{n} B_i^2}{k} - \left(\sum_{i=1}^{n}\sum_{j=1}^{k} X_{ij}\right)^2 \Big/ kn$	$MSBl = SSBl/df$
Treatments	$k-1$	$SSTr = \dfrac{\sum_{j=1}^{k} T_j^2}{n} - \left(\sum_{i=1}^{n}\sum_{j=1}^{k} X_{ij}\right)^2 \Big/ kn$	$MSTr = SSTr/df$
Error	$(k-1)(n-1)$	$SSE\ (= SST - SSBl - SSTr)$	$MSE = SSE/df$
Total	$kn-1$	$SST = \sum_{i=1}^{n}\sum_{j=1}^{k} X_{ij}^2 - \left(\sum_{i=1}^{n}\sum_{j=1}^{k} X_{ij}\right)^2 \Big/ kn$	

n is the number of blocks, B_i is the sum of responses for block i, k is the number of treatments, X_{ij} is the experimental response for block i and treatment j, T_j is the sum of responses for treatment j, *df* defines degrees of freedom, *SS* defines sum of squares, *MS* defines mean square

4.3.3 Test Statistics

Determination of a treatment effect is based, as in the CRD, on the ratio of treatment mean square (between treatment variation) to error mean square (within treatment variation) while the block effect is based on the ratio of block mean square (between block variation) to error mean square. From the statistical theory underlying the RBD design structure, the **treatment effect test statistic** is expressed as the variance ratio

$$F = MSTr/MSE \tag{4.3}$$

with degrees of freedom $df_1 = k-1$ and $df_2 = (n-1)(k-1)$, i.e. treatment degrees of freedom and error degrees of freedom. This is the same test statistic as used in the CRD (see Section 3.3.2). The **block effect test statistic** is specified as

$$F = MSBl/MSE \tag{4.4}$$

with degrees of freedom $df_1 = n-1$ and $df_2 = (n-1)(k-1)$.

To decide whether the evidence within the data supports acceptance or rejection of the null hypothesis, either of the two standard **decision rules** can be considered provided software producing p values has been used for ANOVA table determination. The **test statistic approach** is based on the same form of decision rule as previously, i.e. test statistic < critical value $\Rightarrow$ accept H_0 and no treatment differences detected. The ***p* value approach** depends on use of software and operates as previously, i.e. p value > significance level $\Rightarrow$ accept H_0.

Formal testing of the block effect is not always necessary in a RBD experiment but qualitative assessment of its influence is advised. This can be achieved by carrying out a **block effect check** using the p value for the blocking factor. If the p value is less than 0.1, then use of blocking has been beneficial to explaining the response variation while if greater than 0.1, inclusion of the blocking factor has not been particularly useful.

Exercise 4.3 In Exercise 4.2, we carried out exploratory analysis on the sales data of Example 4.1 and found little evidence of differences due to packaging method. We ought to carry out the associated main statistical test to confirm that the method differences are insignificant on a statistical basis.

Hypotheses: The null and alternative hypotheses for the method effect within this RBD experiment are H_0: no difference in mean sales with packaging method against H_1: difference in mean sales with packaging method.

ANOVA table: The SAS output presented in Display 4.2 contains the ANOVA information for the sales data and conforms to the general pattern shown in Table 4.4 with response variation split into shop type, packaging method, and error components. The mean square terms are again derived as the ratio *SS*/*df* for each of the shop type, packaging method, and error effects.

Display 4.2

```
Source            DF   Sum of Squares   Mean Square    F Value    Pr > F
Model              7     230.30000000   32.90000000       3.28    0.0345
Error             12     120.50000000   10.04166667
Corrected Total   19     350.80000000

R-Square                C.V.              Root MSE            Sales Mean
0.656499            23.30043            3.16885889           13.60000000

Source            DF        Anova SS    Mean Square    F Value    Pr > F
Shoptype           4    184.30000000    46.07500000       4.59    0.0177
Method             3     46.00000000    15.33333333       1.53    0.2580
```

Test statistic: Expression (4.3) specifies the method F test statistic as the ratio of *MS*Method to *MSE*. The value of 1.53 in the 'F Value' column for Method in Display 4.2 has been evaluated as this ratio, i.e. 15.333/10.042. Again, both test statistic and p value approaches are available for testing treatment differences statistically.

For the *test statistic approach*, we know that $F = 1.53$. As $k = 4$ and $n = 5$, the degrees of freedom of the test statistic are $df_1 = k - 1 = 3$ and $df_2 = (k-1)\times(n-1) = 12$. The critical value from Table A.3, for the 5% significance level, is $F_{0.05,3,12} = 3.49$. Since test statistic is less than the critical value, we must accept the null hypothesis and conclude that there appears insufficient evidence to suggest that product packaging method affects sales ($p > 0.05$).

Using the *p value approach*, we have the p value for the method effect F test as 0.2580 from the column 'Pr > F' for Method in Display 4.2. It is greater than the significance level of 5% ($\alpha = 0.05$) implying acceptance of H_0 and so the conclusion is the same as that obtained using the test statistic approach ($p > 0.05$).

Block effect: Though the objectives of this study do not require analysis of the shop type (blocking) effect, it is advisable to assess its effectiveness in order to confirm whether blocking has been beneficial. From Display 4.2, the shop type F test statistic is 4.59, i.e. ratio 46.075/10.042, with p value quoted as 0.0177. This implies that there appears evidence of a substantial difference in sales between the shops chosen and that blocking has been beneficial to the explanation of the variation in sales. □

4.4 FOLLOW-UP ANALYSIS PROCEDURES FOR RANDOMISED BLOCK DESIGNS

As with the CRD, the treatment F test within a RBD study is simply a general test of treatment differences with further analysis through **standard error plot** (see Section 3.4.2), **multiple comparison** (see Section 3.4.3), or **treatment effect estimation** (see Section 3.4.6) necessary to identify specific treatment differences if they exist. As no treatment effect was detected in the sales study of Example 4.1, there is no need to consider a follow-up. If treatment effects are detected, at least one of these follow-up procedures should be implemented to pinpoint the source of treatment differences. Confidence intervals for treatment means can also be considered. Pre-planned comparisons through **linear contrasts** (see Section 3.4.4) should always be carried out irrespective of treatment effect significance. The final check in RBD analysis is **diagnostic checking** of the residuals, as explained in Section 3.5.

Exercise 4.4 For the sales study of Example 4.1, no statistical difference in sales with packaging method was detected. This result would suggest that no further analysis of sales data need be considered. This is untrue. A planned comparison of packaging method A versus the other three methods was also of interest and this comparison must be assessed. In addition, the final aspect of diagnostic checking must also be undertaken.

Linear contrast: The comparison of A versus others can be expressed in contrast form as

$$L = \mathrm{A} - (\mathrm{B} + \mathrm{C} + \mathrm{D})/3.$$

The null hypothesis associated with this contrast is specified as H_0: $L = 0$, i.e. mean sales for method A not statistically different from mean sales of the other three methods.

The contrast L can be estimated as

$$15.80 - (12.00 + 14.20 + 12.40)/3 = 2.933$$

using the sales means from Display 4.1. The positive value of this estimate implies that packaging method A appears to have a higher mean sales compared to the other three tested methods suggesting that customers may prefer the present product packaging.

We know that *errordf* = 12, n = 5, and MSE = 10.042 (from Display 4.2). The 95% contrast confidence interval (3.7), using a critical value of $t_{0.025,12} = 2.179$ from Table A.1, is given by

$$2.933 \pm (2.179)\sqrt{\frac{(1^2 + (-1/3)^2 + (-1/3)^2 + (-1/3)^2)}{5}(10.042)} = 2.933 \pm 3.566$$

which results in an interval (−0.633, 6.499). As this interval contains 0, we conclude that the mean sales associated with method A appears not to differ significantly from mean sales of the other three methods ($p > 0.05$). The interval contains zero but only just and mostly covers positive values. This trend in the contrast difference suggests that packaging method A appears to generate higher sales than the other methods, but not sufficiently high enough to be statistically significant.

The SAS output in Display 4.3 provides the contrast F test statistic (3.9). The test statistic is presented as $F = 3.21$ with p value estimated as 0.0983. Both lead to acceptance of the null hypothesis at the 5% significance level and the conclusion that, statistically, mean sales for method A appear not to differ from those of the other methods ($p > 0.05$). However, such a procedure cannot provide the information on effect size that the confidence interval can.

Display 4.3

Contrast	DF	Contrast SS	Mean Square	F Value	Pr > F
A versus others	1	32.26666667	32.26666667	3.21	0.0983

Model fit: The ratio (*SSE*/*SST*)% for the sales data is 34.4%. This is high suggesting that though inference is possible for the collected data, we do not have a full explanation of what affects product sales. Other factors such as time of study, type of product, competition, location of shops, and customer base of shops could all, in some way, affect sales.

Diagnostic checks: Lastly, we require to analyse response model residuals. The plot of residuals (error estimates) against the methods in Display 4.4 shows that column length appears to differ with method primarily due to two possible outliers associated with method B. This indicates that the equal variability assumption may not be acceptable.

Display 4.4

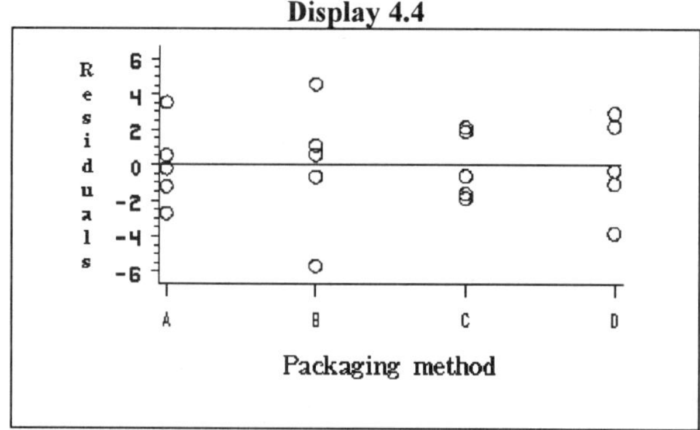

The model fits residual plot can be considered for a RBD as the plot is not, as with the CRD, a simple re-specification of the treatment residual plot. Display 4.5 contains this plot for the sales data. The pattern and trends in the plot look random which suggest the model is reasonable over the range of sales data collected.

Display 4.5

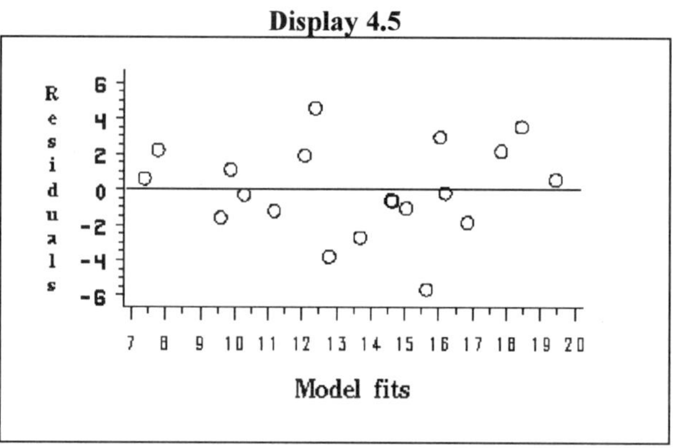

The normal plot of residuals in Display 4.6 exhibits a relatively straight line suggestive that normality of sales data appears acceptable. The Ryan-Joiner test statistic (3.14) is $R = 0.99262$ and since $R_{0.1,20} = 0.9600$, this provides further evidence ($p > 0.1$) that the assumption of normality for the sales response appears acceptable. □

Display 4.6

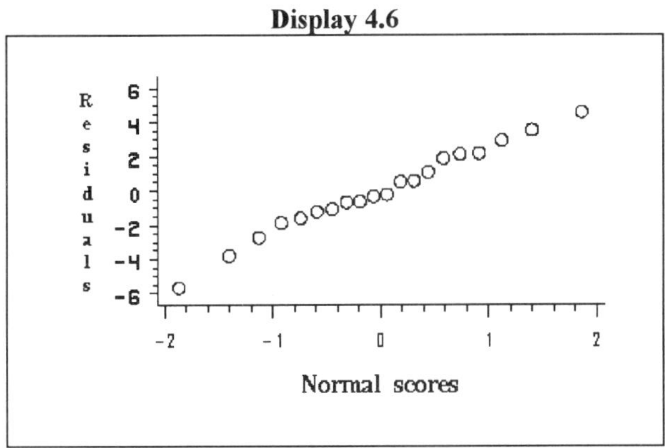

Summarising the analyses of the sales data, we can conclude that the packaging method chosen does not appear to affect sales ($p > 0.05$) suggesting that customers will buy the product irrespective of the way it is presented. It appears also that method A generates higher sales though not statistically so ($p > 0.05$). This effect may be worth investigating in a future study. The outcome of the diagnostic checking appears to be that the equal variability assumption may not be satisfied but that the normality assumption appears to be by the sales data obtained over the study period. However, model fit suggests that the full picture of how sales are affected is not yet fully explained.

In summary, we can see that the steps in the analysis associated with a RBD experimental structure are model specification, exploratory analysis, inferential data analysis through treatment effect test and the identification of specific treatment differences, and diagnostic checking. Using these steps, which are the same as those necessary for a CRD, enables a full and comprehensive statistical and contextual analysis of the collected data to be provided.

4.5 ADDITIONAL ASPECTS OF BLOCKING DESIGNS

4.5.1 Power Analysis

Power analysis can be considered for experiments planned around the RBD structure. The calculations are as those explained in Section 3.6 except that the error degrees of freedom become $(k - 1)(n - 1)$. Sample size estimation essentially becomes estimation of the number of blocking levels, if only one observation is planned for each block × treatment combination. When more than one observation is planned to be collected, sample size estimation will provide a value for number of blocking levels times number of replicate measurements.

4.5.2 Missing Observations

In experiments, an observation may be missing due to loss of experimental material or data recording. When this occurs within a RBD experiment, the block and treatment effects become non-orthogonal. It should be noted that missing data lessen the reliability of the inferences reached. Two ways of dealing with missing observations in the RBD structure are available.

We can estimate the missing value in such a way as to minimise the error mean square and include the estimate with the remaining data for analysis purposes to provide an **approximate analysis**. For one missing observation, we estimate it by calculating

$$X' = \frac{nB' + kT' - G'}{(n-1)(k-1)} \tag{4.5}$$

where B' is the sum of the response for the block containing the missing observation, T' is the sum of the responses for the treatment containing the missing observation, and G' is the overall total of all available observations. In the ANOVA table, the error degrees of freedom and total degrees of freedom are each reduced by one to compensate for the loss of information. Appropriate modification to account for missing observation(s) is adopted for use of multiple comparisons and contrasts. Estimation of more than one missing observation is also possible using an iterative approach.

The framework of the **General Linear Model (GLM)** multiple regression based estimation procedures can be considered as an alternative to estimating the missing value. It provides for what is called an **exact analysis** (Montgomery 1997). The GLM approach adjusts the treatment effect to account for the effect on the blocks of the design imbalance. When more than one missing observation occurs, the GLM approach should be considered.

4.5.3 Efficiency

The efficiency of a RBD structure relative to a CRD structure for the same experimental units can be calculated as

$$E = \left[\frac{(n-1)MSBl + n(k-1)MSE}{(kn-1)MSE}\right]100\% \tag{4.6}$$

E greater than 100% suggests that the RBD is more efficient in comparing treatment means and blocking has provided a more sensitive test of treatment effects while E less than 100% implies that there has been a loss in sensitivity as the RBD structure has reduced the precision of the treatment comparisons. This can be confirmed by looking at the associated p value for the blocking effect in the ANOVA table as indicated in Section 4.3.3 and highlighted in Exercise 4.3. For the sales data study of Example 4.1, E is 175.5% further emphasising the usefulness of blocking on shop type.

4.6 NON-PARAMETRIC ALTERNATIVE TO ANOVA BASED TREATMENT TESTS

It may be necessary in a RBD, as occurred with CRD experimentation, to replace ANOVA as the inferential method if normality for the response cannot be safely assumed. As such, we would need to use appropriate non-parametric approaches to replace the ANOVA F test (4.3) and the SNK multiple comparison (3.5) when the latter is required.

4.6.1 Friedman Test of Treatment Differences

For RBD based experiments, the non-parametric alternative to the treatment F test is provided by the **Friedman test**. The test is based on data ranking, as was the Kruskal-Wallis non-parametric alternative, with ranks replacing the recorded measurements in the test statistic calculation. A summary of the analysis components based on the Friedman test for treatment difference is presented in Box 4.1 where the test is concerned with testing for difference in medians. As with the Kruskal-Wallis test, information on median treatment responses should also be reported to provide practical interpretation of significant treatment effects. As with other non-parametric inference procedures for design structures, test statistic (4.7) can be adjusted to account for tied observations though adjustment does not greatly change the test statistic in most cases (Daniel 1990).

Box 4.1 — Friedman Test for Treatment Differences

Assumptions Data constitute k random samples of size n and there is no interaction between blocks and treatments.

Hypotheses The hypotheses related to this test are as those of the Kruskal-Wallis test, i.e. H_0: no difference between treatments (median treatment responses same) and H_1: treatments differ (at least one median treatment response differs).

Exploratory data analysis (EDA) Exploratory analysis of the data is again advisable to gain initial insight into what they indicate in respect of the experimental objective.

Test statistic To determine the test statistic, the first step requires the observations within each block to be ranked in ascending order of magnitude, smallest to largest, with tied observations given an average rank. The resultant ranks associated with each of the k treatments are then summed to generate the rank totals C_1 to C_k. From these, we use

$$S = \frac{12}{kn(k+1)} \sum_{j=1}^{k} C_j^2 - 3n(k+1) \tag{4.7}$$

to calculate the Friedman test statistic where k is the number of treatments and n is the number of blocks.

Decision rule This test statistic, as with the Kruskal-Wallis test statistic (3.16), approximately follows a χ^2 distribution with $k - 1$ degrees of freedom and so Table A.2 can be used to provide the necessary critical values. Again, low values of the test statistic ($S <$ critical value) or high p values (p value $>$ significance level) will lead to acceptance of the null hypothesis H_0 and the conclusion that the evidence suggests that the treatments do not appear to differ.

4.6.2 Multiple Comparison Associated with Friedman's Test

Common to the previous illustrations of designs, the main test of inference can only provide information on general treatment differences. Use of a follow-up is necessary to help understand how the treatment differences are occurring if differences are detected. The non-parametric multiple comparison for use in association with the Friedman test, described in Box 4.2, differs marginally in its operation from Dunn's procedure (see Section 3.7.2) though the underlying philosophy and application are similar.

Box 4.2 — Multiple Comparison for Use with Friedman Test

Hypotheses The null hypothesis to be tested in this multiple comparison is as previously, i.e. H_0: no difference between compared treatments (medians similar).

Test statistics The numerical differences between the rank totals R for each pair of treatments become the test statistics for the pairwise comparisons.

Critical value The critical value for comparison with each test statistic is specified as

$$W = z_{\alpha/[k(k-1)]}\sqrt{\frac{nk(k+1)}{6}} \tag{4.8}$$

where n is the number of blocks in the design structure, and 100α percent, k, and z are as defined for Dunn's procedure (see Box 3.8).

Decision rule If the numerical difference in rank totals for treatment i and j ($|R_i - R_j|$) is less than the critical value (4.8), we accept H_0 and declare that the evidence suggests the compared treatments do not appear to differ.

As with Dunn's procedure, application of Bonferroni's inequality leads to use of $z_{\alpha/[k(k-1)]}$ for expression (4.8). Choice of experimentwise error rate of 100α percent is again important and largely depends on k, the number of treatments tested. Generally we should select 100α percent larger than generally used in inferential data analysis, i.e. use 10% ($\alpha = 0.1$), 15% ($\alpha = 0.1$), or even 20% ($\alpha = 0.1$), with larger values chosen for larger k.

4.7 INCOMPLETE BLOCK DESIGNS

Experimental design models can be considered particular cases of the **General Linear Model** (GLM). So far, we have discussed complete or balanced designs which contain a full data set and which can be easily analysed using statistical packages. However, when observations are missing by choice or through necessity, the design is no longer complete and there can be difficulties with the implementation within a package and the subsequent statistical analysis. GLM estimation procedures offer an alternative method of obtaining sums of squares, the ANOVA table, and parameter estimation using similar principles to those of multiple regression involving dummy variables.

Sometimes, we may be forced to design an experiment in which we must sacrifice some block-treatment combinations in order to perform the experiment. For example, suppose a comparison is to be made of the potencies of three different batches of the same drug. The analyses are to be carried out on the same day to help eliminate any effect the day of analysis may have. There are three analysts available on each day and suppose that each analyst can manage just two analyses per day due to the complexity of the analytical procedure. It would be possible to complete the comparison of the three batches in a single day if each analyst examines just two of the three possible batches. One possible design structure is displayed in Table 4.5 where carrying out the analyses on a single day could help to eliminate time, storage, and other factors which may affect the experimental outcome. In the design structure proposed, the analysts represent the treatments and the batches correspond to the blocks with the number of treatments per block less than the number of treatments being tested.

Table 4.5 — Possible design layout for drug potency experiment

		Analyst		
		A	B	C
	1	x	x	-
Batch	2	x	-	x
	3	-	x	x

x denotes an analysis measurement

4.7.1 Balanced Incomplete Block Design

The design layout shown in Table 4.5 illustrates a special case of an incomplete block design in which a degree of balance has been retained in the design. This form of design structure is referred to as a **Balanced Incomplete Block Design (BIBD)**. For the general BIBD, we define the following terms:

t the number of treatments b the number of blocks
k the number of treatments per block r the number of repetitions of each treatment
$\lambda = r(k - 1)/(t - 1)$ the number of times each pair of treatments appears together within the design, λ always integer valued, $\lambda < r < b$.

For balancing, the number of blocks b should ideally be the number of ways of choosing k treatments from a total of t treatments. This is equal to $t!(t - k)!/k!$ where $t!$ is the multiple of all the integers from t down to 1, i.e. $t! = t\times(t - 1)\times(t - 2)...3\times2\times1$ and $0! = 1$. Clearly, the total number of observations is bk which equals rt. Note that a BIBD does not exist for all possible combinations of t, k, and r. Cochran & Cox (1957) provide an extensive list of Incomplete Block Designs.

Referring to the example on comparison of drug potencies, we have $t = 3$, $b = 3$, and $k = 2$. From the total number of observations of $rt = 6$, we can deduce that $r = 2$ and so each analyst must conduct two analyses. Hence, $\lambda = 2\times(2 - 1)/(3 - 1) = 1$ confirming that each pair of analysts appear together only once within the proposed design structure.

Example 4.2 A production manager wishes to investigate the effect that experience on the assembly line has on the time, in minutes, to complete a particular assembly task. If it is found that experience is an important factor, an appropriate training programme could be developed. The manager has decided to experiment on two factors, experience and task, with the latter assigned as a blocking factor. Ten tasks have been selected for the study. Five categories of experience ranging from least to most are to be used. However, raw material and assembly line constraints dictate that only three runs of each task can be carried out at any one time.

Design structure: The planned study is to be based on $t = 5$ treatments (experience categories) using $b = 10$ blocks. The constraint of only three runs of a task at any one time means $k = 3$ and the ideal RBD structure must be incomplete. The study will involve collection of $bk = 30$ observations which, given that $bk = rt$, means $r = 6$ and each treatment requires to be tested six times. Thus, $\lambda = 6\times(3 - 1)/(5 - 1) = 3$ indicating that each pair of treatments must appear together in the treatment combinations three times.

Design construction involves permuting each feasible three treatment combination across the 10 tasks in such a way that ensures a different combination is associated with each block and that the necessary design properties are satisfied. A possible structure is displayed in Table 4.6 based on randomisation of the order of the treatment combinations. The structure shown clearly demonstrates all the necessary properties for a BIBD. □

Table 4.6 — Possible Balanced Incomplete Block Design structure for experience study of Example 4.2

		Level of experience				
		A	B	C	D	E
Task	1	x	-	x	-	x
	2	x	x	-	x	-
	3	-	x	-	x	x
	4	x	-	-	x	x
	5	x	x	x	-	-
	6	-	x	x	x	-
	7	x	-	x	x	-
	8	-	-	x	x	x
	9	x	x	-	-	x
	10	-	x	x	-	x

x denotes an assembly time measurement

4.7.2 ANOVA Principle

ANOVA analysis for a BIBD is best performed using the **General Linear Model** (GLM) estimation procedure. The basic structure of the data analysis follows that used for the RBD, i.e. exploratory analysis, inferential analysis, follow-up analysis, and diagnostic checking.

As with the RBD, response variation, measured through *SS*Total, is partitioned into three components. For a BIBD, these are blocks, treatments adjusted for blocks, and error resulting in *SS*Total being expressed in the form

$$SS\text{Blocks } (SSBl) + SS\text{Treatment adjusted for blocks } (SSTra) + SS\text{Error } (SSE) \quad (4.9)$$

The *SSTra* term is an adjusted sum of squares accounting for the missing observations at certain block-treatment combinations as each treatment occurs within a different set of blocks. This modification enables the treatment and block effects to be properly separated and thus independently estimable.

The ANOVA table for a BIBD is shown in Table 4.7. Ratios of mean squares of effect to error again provide the requisite block and treatment test statistics. When using SAS's PROC GLM procedure for generation of ANOVA information, the Type

III sums of squares and corresponding test statistics and p values should be used for statistical checking of the block and treatment effects. Type III *SS*s make adjustment for blocks in the derivation of the treatment sum of squares and ensure each block has equal weighting (SAS/STAT User's Guide).

Table 4.7 — ANOVA table for a Balanced Incomplete Block Design

Source	*df*	*SS*	*MS*	*F*
Blocks	$b-1$	*SSBl*	-	-
Treatments (adjusted for blocks)	$t-1$	*SSTra*	*MSTra* = *SSTra*/*df*	*MSTra*/*MSE*
Error	$bk-t-b+1$	*SSE*	*MSE* = *SSE*/*df*	
Total	$bk-1$	*SST*		

b is the number of blocks, t is the number of treatments, k is the number of treatments tested at each block level, *df* defines degrees of freedom, *SS* defines sum of squares, *MS* defines mean square

4.7.3 Follow-up Analysis Procedures

As with the CRD and RBD, if a significant treatment effect is found from the treatment F test, we can use a modified form of the **Student-Newman-Keuls procedure** to compare each treatment against every other treatment using estimated treatment means to take account of the incomplete nature of the data set. The procedure requires each treatment mean to be re-estimated to account for the missing block × treatment combinations and is summarised in Box 4.3. This re-estimation is performed automatically in SAS using the LSMEANS statement in PROC GLM.

Box 4.3 — Student-Newman-Keuls Procedure for Balanced Incomplete Block Designs

Re-estimation of treatment means The method of re-estimation of the treatment means uses

$$\text{treatment mean} = \overline{G} + \tau_j = \overline{G} + (kT_j - B_{(j)})/(t\lambda) \qquad (4.10)$$

to calculate the new mean value where $\overline{G}$ is the overall mean of the data set, τ_j is the estimated effect of treatment j, T_j is the sum of the observations for treatment j, and $B_{(j)}$ is the sum of the block observation totals for blocks containing treatment j.

Test statistics As previously (see Box 3.1) except that the re-estimated treatment means are used.

Critical value The relevant critical value for comparison with the pairwise treatment differences is given by

$$W_r = q_{\alpha,r,errordf}\sqrt{\frac{k(MSE)}{t\lambda}} \qquad (4.11)$$

Decision rule See Box 3.1.

To show how T_j, $B_{(j)}$, and the re-estimated treatment mean are determined for each treatment, consider the analyst illustration outlined previously. Table 4.8 illustrates the steps in this process. Note that the value for $B_{(j)}$ represents a combination of block totals which will vary for each treatment within the experiment.

Other follow-up mechanisms include confidence intervals for treatment effects and adjusted treatment means, and use of linear contrasts of the treatment means.

Table 4.8 — Block and treatment totals derivation for drug potency based on a Balanced Incomplete Block Design structure

Analyst	A	B	C	Block totals
1	x	x	-	B_1
Batch2	x	-	x	B_2
3	-	x	x	B_3
T_j	T_1	T_2	T_3	
$B_{(j)}$	$B_1 + B_2$	$B_1 + B_3$	$B_2 + B_3$	
τ_j	$(kT_1 - B_{(1)})/(t\lambda)$	$(kT_2 - B_{(2)})/(t\lambda)$	$(kT_3 - B_{(3)})/(t\lambda)$	
re-estimated mean	$\overline{G} + \tau_1$	$\overline{G} + \tau_2$	$\overline{G} + \tau_3$	

x denotes an observation, $\overline{G}$ is the overall mean of the data set

4.7.4 Other Incomplete Designs

BIBDs do not exist for all combinations of the design parameters that might be employed since the constraint that λ be an integer may force the number of blocks or the block size to be excessively large. This can be overcome by using a **partially balanced incomplete block design** in which some pairs of treatments appear together a different number of times compared to other treatment pairings, resulting in the design having **associate classes**. Other incomplete designs include **Youden squares**, after Jack Youden who developed such structures for use in greenhouse experiments, and **Lattice designs**.

4.8 LATIN SQUARE DESIGN

RBD structures do not eliminate all the possible variation due to block-to-block operations of the treatments. To illustrate this, consider a study to compare three methods of measuring the toxicity of soil samples from a land-fill site using a RBD structure based on analysts as the blocking factor to account for a potentially extraneous source of measurement variability. Suppose also that each recording is time consuming with only one capable of being completed by each analyst in a day so introducing a further source of measurement variability. A modification of the suggested RBD structure would be necessary to account for this second blocking factor to provide a more efficient design structure for the planned study. A **Latin Square (LS)** design structure in which one treatment factor and two blocking factors are accounted for can be used.

4.8.1 Design Structure

Table 4.9 provides an illustration of a LS design structure for an experiment to assess four treatments A to D taking account of two blocking factors. Within the illustrated structure, one blocking factor is assigned to the rows, the second to the columns, and the treatments in a way that ensures each treatment occurs once only in each row and each column. The structure is square with number of rows equal to number of columns equal to number of treatments with both row and column factors orthogonal to the treatments. The illustrated structure is based on simple cyclic rotation of the treatment order across the levels of blocking factor 1. Cyclic rotation across the levels of blocking factor 2 is also possible. Other structures based on randomising the order of the treatment combinations are also possible where randomisation can be based on either or both of the blocking factors.

Table 4.9 — Possible Latin Square design layout for testing four treatments A to D

Blocking factor 2		1	2	3	4
	1	A	B	C	D
Blocking factor 1	2	B	C	D	A
	3	C	D	A	B
	4	D	A	B	C

A LS design blocks in two directions, rows and columns, with each providing a restriction on the allocation of the t treatments. Thus, a LS design for t treatments is a $t \times t$ **Latin Square** and can be represented by a square containing t rows and t columns with a total of t^2 observations. Latin Squares can be further extended to **Graeco-Latin Squares (GLS)** by inclusion of a third blocking factor though, again, assignment of treatments must ensure they appear once only with respect to all blocking factor levels. Replication in LS designs is also possible.

Example 4.3 Five different brands of flash bulbs, A to E, were tested for their effect on the photographic density of film. Film density is measured as the degree of blackness on the film, higher values indicating unacceptable levels. All five brands were compared over five different film types and five different cameras types using these two factors as the blocking factors within a 5 × 5 LS design. The data are shown in Table 4.10.

Table 4.10 — Response data for photographic density study of Example 4.3

Camera	1	2	3	4	5
1	A: 0.64	B: 0.72	C: 0.69	D: 0.66	E: 0.74
Film 2	B: 0.62	C: 0.71	D: 0.69	E: 0.72	A: 0.66
3	C: 0.65	D: 0.62	E: 0.68	A: 0.64	B: 0.74
type 4	D: 0.63	E: 0.73	A: 0.68	B: 0.74	C: 0.72
5	E: 0.74	A: 0.69	B: 0.67	C: 0.75	D: 0.68

4.8.2 Model for the Measured Response

For a recorded observation X_{ij} within a LS experiment, the suggested **response model** structure involves extending the RBD response model (4.1) to include the second blocking factor. This results in the model expression

$$X_{ijk} = \mu_{ijk} + \varepsilon_{ijk} = \mu + \alpha_i + \beta_j + \tau_k + \varepsilon_{ijk} \tag{4.12}$$

where α_i corresponds to the effect of row i, β_j is the effect of column j, τ_k is the effect due to treatment k, and ε_{ijk} is, as usual, the random error. The first part of expression (4.12) again defines the **means model** with the second part corresponding to the **effects model**. This additive response model again specifies that response variation is explained by the effect of the controlled factors, two blocking and one treatment, plus the effect of uncontrolled factors, i.e. the error. Again, the expectation is that controlled variation through, in particular, the treatment effect, explains considerably more of the variability in the measured response than does error.

The inherent **assumptions** remain as stated previously for the CRD and RBD structures namely, equal response variability across all treatments and normality of response, i.e. $\varepsilon_{ijk} \sim \text{NIID}(0, \sigma^2)$. As previously, some **exploratory analysis** of the design data is advised to provide an overall picture of the response data.

Exercise 4.5 The first stage in the analysis of LS study data concerns response model specification and exploratory analysis of the collected data. In this Exercise, we will outline these analysis aspects for the photographic density study of Example 4.3.

Response model: For the LS design structure used, the response model (4.12) will be

$$\text{photographic density} = \mu + \alpha_i + \beta_j + \tau_k + \varepsilon_{ijk}$$

where $i = 1, 2,..., 5$ (five films), $j = 1, 2,..., 5$ (five cameras), and $k =$ A, B, C, D, E (five flash bulb brands). The term α_i defines the film type effect, β_j is the camera effect, τ_k is the flash bulb brand effect, and ε_{ijk} the random error whereby $\varepsilon_{ijk} \sim \text{NIID}(0, \sigma^2)$. This means, as previously, that we are assuming the photographic density data to be normally distributed and that variability in these data is similar for all five bands of flash bulb tested.

Hypotheses: The null and alternative hypotheses for the bulb band effect within this LS based experiment are H_0: no difference in mean density with flash bulb brand against H_1: difference in mean density with flash bulb brand used.

EDA: Exploratory analysis, based on simple SAS plots and summaries of the density by flash bulb brand tested (not shown) highlighted that performance appeared to split into two groups with A and D giving low densities and B, C, and E giving high densities generally. Summary statistics emphasised the same conclusion though brands A and E appeared most consistent while brand B was most variable. □

4.8.3 ANOVA Principle

As with the previous One Factor Designs, application of the **ANOVA principle** in LS designs involves partitioning response variation into the sources of variation specified in the response model (4.12). Based on this response model, this means that *SS*Total can be expressed as

$$SS\text{Rows}\ (SSR) + SS\text{Column}\ (SSC) + SS\text{Treatment}\ (SSTr) + SS\text{Error}\ (SSE) \quad (4.13)$$

The calculations are summarised in the **ANOVA table** displayed in Table 4.11 and are an extension to those described for the CRD and RBD structures.

Determination of a treatment effect is based again on testing a null hypothesis of H_0: no treatment difference using the ratio of treatment variance to error variance. Thus, the **treatment effect test statistic** is expressed in a similar way to that for the CRD and RBD structures as

$$F = MSTr/MSE \quad (4.14)$$

with degrees of freedom $df_1 = t - 1$ and $df_2 = t^2 - 3t + 2$. To decide whether the evidence from the data points to acceptance or rejection of treatment effect, either of the two standard **decision rules** can be used provided software producing p values has been used for ANOVA table determination, i.e. test statistic < critical value ⇒ accept H_0 or p value > significance level ⇒ accept H_0. The relevant tests for row and column effects, the two blocking factors, can similarly be constructed though as with the RBD, testing such effects statistically is not strictly necessary because assessing significance of treatment effect is the primary objective.

Table 4.11 — ANOVA table for a Latin Square design

Source	*df*	*SS*	*MS*
Rows	$t-1$	$SSR = \frac{\sum_{i=1}^{t} R_i^2}{t} - \left(\sum_{i=1}^{t}\sum_{j=1}^{t}\sum_{k=1}^{t} X_{ijk}\right)^2 \Big/ t^2$	$MSR = SSR/df$
Columns	$t-1$	$SSC = \frac{\sum_{j=1}^{t} C_j^2}{t} - \left(\sum_{i=1}^{t}\sum_{j=1}^{t}\sum_{k=1}^{t} X_{ijk}\right)^2 \Big/ t^2$	$MSC = SSC/df$
Treatments	$t-1$	$SSTr = \frac{\sum_{k=1}^{t} T_k^2}{t} - \left(\sum_{i=1}^{t}\sum_{j=1}^{t}\sum_{k=1}^{t} X_{ijk}\right)^2 \Big/ t^2$	$MSTr = SSTr/df$
Error	t^2-3t+2	$SSE\ (= SST - SSR - SSC - SSTr)$	$MSE = SSE/df$
Total	t^2-1	$SST = \sum_{i=1}^{t}\sum_{j=1}^{t}\sum_{k=1}^{t} X_{ijk}^2 - \left(\sum_{i=1}^{t}\sum_{j=1}^{t}\sum_{k=1}^{t} X_{ijk}\right)^2 \Big/ t^2$	

t is the number of treatments, R_i is the sum of responses for row i, X_{ijk} is the experimental response for the combination (row i, column j, treatment k), C_j is the sum of responses for column j, T_k is the sum of responses for treatment k, *df* defines degrees of freedom, *SS* defines sum of squares, *MS* defines mean square

Exercise 4.6 Exercise 4.5 summarised the exploratory analysis findings in respect of the photographic density study of Example 4.3. Some evidence of flash bulb differences was detected. We now need to validate the statistical significance of these detected differences.

Hypotheses: The treatment null hypothesis assessed through this LS design structure is H_0: no difference in mean photographic density with flash bulb brand.

ANOVA table: The SAS output presented in Display 4.7 contains the ANOVA information for the density data and conforms to the general pattern shown in Table 4.11 with response variation split into film type, camera, bulb brand, and error effects. The information conforms to the sum of squares expression *SS*Total = *SS*Filmtype + *SS*Camera + *SS*Bulbbrnd + *SS*Error with the mean square terms derived as the ratio *SS*/*df* for each of the variation effects.

Display 4.7

Source	DF	Sum of Squares	Mean Square	F Value	Pr > F
Model	12	0.02960800	0.00246733	2.66	0.0516
Error	12	0.01112800	0.00092733		
Corrected Total	24	0.04073600			

R-Square	C.V.	Root MSE	Density Mean
0.726826	4.423612	0.03045215	0.68840000

Source	DF	Type III SS	Mean Square	F Value	Pr > F
Filmtype	4	0.00509600	0.00127400	1.37	0.3004
Camera	4	0.00845600	0.00211400	2.28	0.1209
Bulbbrnd	4	0.01605600	0.00401400	4.33	0.0214

Test statistic: Equation (4.14) specifies the bulb brand F test statistic as the ratio of *MS*Bulbbrnd to *MSE*. The figure of 4.33 in the 'F Value' column for the bulb brand effect in Display 4.7 has been evaluated as this ratio, i.e. 0.004014/0.000927. Again, both test statistic and p value approaches are available for testing treatment differences statistically, though only the latter will be demonstrated.

The p value for the bulb band effect F test is quoted as 0.0214. This is below the significance level of 5% ($\alpha = 0.05$) implying rejection of H_0 and the conclusion that it appears photographic density varies according to flash bulb brands tested ($p < 0.05$).

Block effects: Though the objectives of this experiment do not require analysis of the blocking effects, it is instructive to assess their effectiveness in order to confirm whether blocking has been beneficial. From Display 4.7, the film type F test statistic is 1.37, i.e. ratio 0.001274/0.000927, with approximate p value quoted as 0.3004 while the camera type F test statistic is 2.28, i.e. ratio 0.002114/0.000927, with approximate p value listed as 0.1209. Both results imply that inclusion of both as blocking factors has not been as useful as first envisaged. However, the camera effect with a p value of 0.1209 looks to have had some influence and so blocking for this factor may have enhanced the sensitivity of the design to detect a bulb brand effect. □

4.8.4 Follow-up Analysis Procedures

The treatment F test within a LS study is nothing more than a general test of treatment differences. Further analysis through **multiple comparisons** (see Section 3.4.3), **treatment effect estimation** (see Section 3.4.6), or **confidence intervals for treatment means** should be included as necessary to identify specific treatment differences if they exist. Pre-planned comparisons, such as **linear contrasts** (see Section 3.4.4), should be checked through irrespective of significance of treatment effects. The final check, as usual, must be the **diagnostic checking** of the residuals using the graphical methods explained in Section 3.5.1.

Exercise 4.7 For the photographic density study of Example 4.3, we know that bulb brand effects differ statistically. We must finalise the analysis by carrying out multiple comparisons to elaborate on the statistical evidence of bulb brand differences, and also undertake the diagnostic checking of the response model using the residuals.

Multiple comparison: To locate the detected bulb brand differences, we must validate these differences statistically. The SNK multiple comparison procedure, at the α = 0.05 (5%) significance level, will again fulfill this role. The design structure provides us with $n = 5$ and $k = 5$, and from the ANOVA table in Display 4.7, we have $errordf$ = 12 and MSE = 0.000927. The null hypothesis to be tested in this follow-up is H_0: no difference in mean photographic density between compared bulb brands. The mean densities and the table of differences in means, the test statistics for the multiple comparisons, are presented in Table 4.12.

The associated 5% critical values W_r, stemming from evaluation of equation (3.5), are shown in Table 4.13 with the q values read from Table A.5 for the combination of terms df = 12 and r = 2, 3, 4, 5. Using the principles outlined in Exercise 3.4, we can assess the brand differences for any which show statistical difference.

Table 4.12 — Test statistics (differences in means) for multiple comparison of Exercise 4.7

Brand		D	A	B	C	E
	Mean	0.656	0.662	0.698	0.704	0.722
D	0.656	-	0.006	0.042	0.048	0.066 *
A	0.662		-	0.036	0.042	0.060 *
B	0.698			-	0.006	0.024
C	0.704				-	0.018
E	0.722					-

* represents difference statistically significant at the 5% significance level

Table 4.13 — 5% critical values for SNK illustration in Exercise 4.7

r	2	3	4	5
$q_{0.05,r,12}$	3.08	3.77	4.20	4.51
W_r	0.042	0.051	0.057	0.061

For the comparison of bulb brands D and A, the corresponding test statistic from Table 4.12 is 0.006. These bulbs are in positions $i = 1$ (1st) and $j = 2$ (2nd) in the ordered list with associated critical value corresponding to $r = 2 + 1 - 1 = 2$ (two steps apart), i.e. use $W_2 = 0.042$ from Table 4.13. As the test statistic is less than the critical value, we accept H_0 and indicate that there appears insufficient evidence of a statistical difference in photographic density between these flash bulb brands ($p > 0.05$).

For the D and E comparison ($i = 1, j = 5, r = 5$), the test statistic is 0.066 with associated critical value of $W_5 = 0.061$. As the test statistic exceeds the critical value, we conclude significant difference between these bulb brands ($p < 0.05$). Based on all the possible brand comparisons, the results of which are summarised in Table 4.12, we can deduce that the only differences of significance are those between brands D and E, and brands A and E. No other comparisons appear statistically significant. We can conclude therefore that brand groupings (D, A, B, C) and (B, C, E) appear to have statistically similar influences on photographic density. For completeness, the SAS generated SNK comparison information is provided in Display 4.8. The results agree with those discussed.

Diagnostic checking: The final part of the data analysis deals with the analysis of response model residuals through residuals plots of which only the model fits plot is shown. The factor residual plots all hinted at column length differences and possible outliers. The model fits plot in Display 4.9 appears reasonably random. The normal plot (not shown) exhibited some non-linear trend but not sufficient to suggest non-normality. Confirmation of normality for the photographic density data was forthcoming from the Ryan-Joiner test ($R = 0.97226, p > 0.1$). □

Summarising the analyses of the film density data, we can conclude that the flash bulb brands chosen differ ($p < 0.05$) and appear to be producing differing density figures. It appears also that brand E looks to be producing density measurements most different from the other brands. Only the differences between this brand and brands A and D can be statistically verified ($p < 0.05$). The outcome of the diagnostic checking appears to be that the assumptions of equal variability and normality may not be satisfied by the density data obtained within the experiment and future comparable experiments may require such response data to be transformed before analysis.

Display 4.8

```
 Student-Newman-Keuls test for variable: Density

NOTE: This test controls the type I experimentwise error rate under
the complete null hypothesis but not under partial null hypotheses.

          Alpha= 0.05  df= 12  MSE= 0.000927

 Number of Means          2          3         4         5
 Critical Range  0.0419628   0.05138 0.0571785 0.0613878

Means with the same letter are not significantly different.

      SNK Grouping              Mean       N  Bulbbrnd

                 A            0.72200      5  E
                 A
         B       A            0.70400      5  C
         B       A
         B       A            0.69800      5  B
         B
         B                    0.66200      5  A
         B
         B                    0.65600      5  D
```

Display 4.9

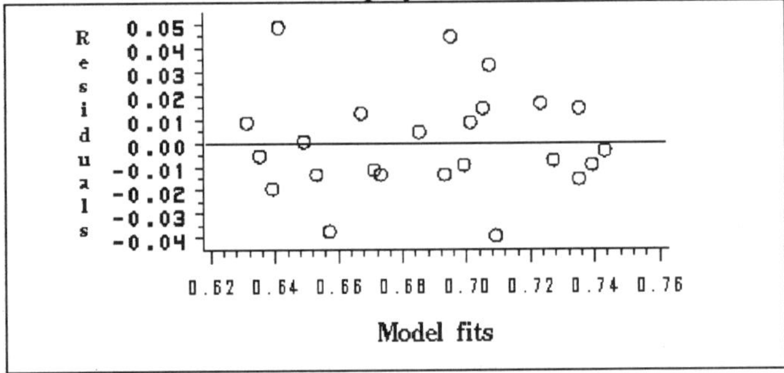

4.8.5 Missing Observations

As with RBD experiments, it is possible that observations may be lost due to unforeseen circumstances. Again, estimation of the missing observation or use of the GLM approach are available. However, missing data should be avoided as it weakens the value of inferences.

When a single missing observation occurs within a LS based experiment, we can still analyse the remaining data by estimating the missing value so as to minimise the error mean square. For this case, we can manually estimate the missing value as

$$X' = \frac{t(R'+C'+T') - 2G'}{(t-1)(t-2)} \tag{4.15}$$

where the primes refer to the row, column, and treatment totals corresponding to the positioning of the missing value, and G' is the overall total of all available observations. In the ANOVA table, the error degrees of freedom and total degrees of freedom are each reduced by one to reflect the loss of information the estimation procedure gives rise to. Manual estimation of missing values is possible when more than one observation is missing though, in such a case, use of the **General Linear Model** (GLM) estimation procedure is more appropriate. Follow-up analysis based on modified multiple comparisons and linear contrasts are available if appropriate to the data analysis.

4.8.6 Efficiency

The efficiency of a LS structure with two blocking elements relative to a RBD structure with only one blocking element can be determined using a similar procedure to that shown in Section 4.5.3 for comparison of a RBD with a CRD. Essentially, efficiency estimation involves evaluating the expression

$$E = \left[\frac{MSEffect + (t-1)MSE}{(t)MSE}\right]100\% \tag{4.16}$$

and comparing its value to 100%. E exceeding 100% would indicate the LS design to be more efficient in comparing treatment means, illustrating that the double blocking has provided a more sensitive test of treatment effects. E less than 100% would suggest that there has been a loss in sensitivity through use of the LS structure and the precision of the treatment comparisons has not been improved. For the photographic density study of Example 4.3, E is 107.5% for the film effect and 125.6% for the camera effect. Both efficiency estimates exceed 100% though it appears inclusion of the camera effect has had a greater influence on design efficiency as confirmed earlier when the block effect p values were assessed (see Exercise 4.6).

PROBLEMS

4.1 The yield obtained from a potato crop is known to be influenced by the intensity of potato root eelworm in the soil. In an attempt to control the parasite and improve yield, plots containing three different intensities of viable eelworm eggs, the blocking factor, were exposed to different chemical sterilants. At the end of the trial period, plot yield, in kgs, was measured. Investigate these data to determine whether sterilants have different effects.

		Sterilant			
		Control	Methamsodium	Chloropicrin	Carbon disulphide
	low	6.3	11.3	12.3	8.7
Eelworm intensity	medium	5.9	10.2	12.6	7.2
	high	3.4	10.1	11.4	3.6

4.2 A proficiency study was conducted to assess potential work plans for the assembly of electrical components. Five plans were investigated. Four assemblers, acting as the blocking factors, were also used with each assembler receiving full training in the new work practices before commencement of the study. Each assembler followed each plan for a full working day and the number of components assembled was recorded. Analyse these results carrying out relevant inferential analysis at the 5% significance level.

		Work plan				
		A	B	C	D	E
	1	10	13	9	14	11
Assembler	2	5	10	5	10	6
	3	6	12	5	10	6
	4	4	8	4	11	6

4.3 In a study of microsomal epoxide hydrolase (mEH) activity in fish livers, specimens of fish livers were treated with three aromatic compounds to assess whether mEH activity differed with compound. Five fish of the same species, age, and sex were randomly selected to act as the blocking factor in the experimental structure. Three equal-sized liver tissue specimens were then taken from each fish and randomly assigned to the three compounds. The mEH activity data in each specimen, in nmol min^{-1} per mg protein, was then recorded as shown below. Carry out a full analysis of these data. Does average activity of fluoranthene differ significantly from average activity associated with the other two compounds?

Fish	Fluoranthene	Naphthalene	Phenanthrene
1	13.8	13.3	14.7
2	12.2	14.3	14.1
3	12.8	13.6	14.5
4	13.4	14.3	13.5
5	13.5	12.9	13.9

4.4 A chemist conducted an experiment to assess if a standard analytical procedure (A) and two variants (B and C) differed in their determination of the concentration of benzocaine in antiseptic throat lozenges. Seven test solutions were prepared and each was divided into three equal amounts which were then randomly assigned to each procedure for analysis. The concentration data collected, as mg of benzocaine, are shown below. Of particular interest is the comparison of method A, the standard method, with the two variants tested. Carry out an appropriate analysis of these data.

		Procedure		
		A	B	C
	1	4.5	4.5	4.9
	2	4.3	5.1	5.3
	3	4.7	4.7	5.1
Test solutions	4	5.2	4.9	5.0
	5	5.3	5.4	5.4
	6	4.7	4.8	5.0
	7	5.2	4.8	5.5

4.5 A plastics manufacturer is concerned with the number of breaks per 100 kg of material during a particular phase of production. Four methods of treatment designed to reduce breakage are to be applied to raw material before this phase of production. Ten lines are used for the production process with five machines to be used as levels of the blocking factor. For treatment differences to be of practical benefit, a minimum detectable difference in number of breakages of at least 6.5 is required. Based on testing treatment differences at the 5% significance level, estimate the power of the proposed significance test assuming an error mean square estimate of 2.25. Is the power acceptable?

4.6 A chemical analyst is planning a comparison experiment to compare three methods of measuring pH in gastric juices from randomly selected patients. The methods to be compared are a glass electrode, an antimony electrode, and an ion-selective field-effect transistor (ISFET). The patient specimens, referring to the blocking elements, are to be split into three equal parts for the analysis with each part randomly assigned to a particular analytical method. A minimum difference in pH of 0.85 is believed to be sufficient to indicate method differences. How many patient specimens are necessary if the power of the F test for method is to be at least 80%? Assume testing at the 5% significance level and an error mean square estimate of 0.347.

4.7 Four forms of insect traps are to be compared in respect of geometric mean number of mosquitoes that each can trap over particular study nights. This response is planned to be transformed using the logarithm transformation for analysis purposes. The entomologist planning the study is of the opinion that a minimum detectable difference of 0.32 in the transformed data would be sufficient to provide evidence of differences in the efficiency of the tested traps. Five study nights are planned. Assuming an error mean square estimate of 0.0163 from past comparable insect studies, assess whether five study nights is sufficient, if the power of associated trap test is to be at least 80% based on testing trap differences at the 5% significance level.

4.8 Six pairs of twin lambs are to be used as blocking factor levels in an investigation into the weight gains achievable from three protein additives added to their normal diet. A control group where no additive is included in the diet is also to be assessed. Using twins means that only two diet variations can be tested at each blocking level so the design structure must be based on a BIBD structure. Construct this design.

4.9 A paper company wished to design an experiment to compare the paper strength from six coatings, A to F, to discern which coating was most beneficial and to improve the appearance of packaging paper. Because the uncoated paper strength could vary along a roll, the experiment is to be conducted using a RBD structure blocking on the position factor with fifteen positions selected as measuring points. Unfortunately due to roll size and machine constraints, it is only possible to obtain strength measurements for four of the six coatings at each selected position on the roll. It is therefore necessary to construct a BIBD for this experiment. Construct this design.

4.10 A potato grower wished to investigate the effect of 16 different varieties of potato on weight harvested. A field was selected for the trial and divided into 20 blocks. However, each block is such that only four potato varieties can be accommodated within it. Construct a possible design for this experiment.

4.11 An experiment was conducted at a manufacturing site which produces four different kinds of petroleum spirit, A to D, used in industrial power tools. The spirit was tested for octane number. A random sample was drawn from the produced stock each day and submitted to one of four tester groups for determination of octane number. The experiment was designed as a 4×4 LS design blocking on the day of production and the tester group producing the figures. The octane data below were obtained. Analyse these data and draw relevant conclusions.

		Tester group			
		1	2	3	4
Day	1	C: 76	A: 89	D: 74	B: 64
	2	D: 75	B: 79	C: 75	A: 83
	3	B: 72	D: 78	A: 88	C: 77
	4	A: 90	C: 90	B: 74	D: 77

4.12 An experiment was carried out into the effects of five different glue formulations, A to E, on the adhesive qualities of a manufactured product. Each formulation consists of different combinations of raw materials and are such that only five formulations can be made from each batch of raw material. In addition, different operators generally prepare the formulations. As the batch of raw material used and operator experience could influence the measured response of pull-off force when two pieces of material are glued together, it was decided to base the experiment on a 5 × 5 LS design using the batch and operator effects as the blocking factors. The recorded pull-off force data are in the table below. Analyse these data carrying out all relevant inference at the 5% significance level.

		Operator				
		1	2	3	4	5
Batch	1	A: 22	B: 19	C: 21	D: 23	E: 27
	2	B: 28	C: 26	D: 30	E: 32	A: 36
	3	C: 19	D: 24	E: 26	A: 27	B: 31
	4	D: 26	E: 28	A: 23	B: 26	C: 21
	5	E: 22	A: 30	B: 27	C: 23	D: 31

4.13 An experiment was set up to assess the effect of five types of fertiliser on the yield of a commercially produced fruit. To account for different soil types and different varieties of the fruit, the experiment was based on a 5 × 5 LS structure using soil type and fruit variety as the blocking factors. The yield data, in coded units, are presented below. Does fertiliser influence yield? Do the blocking factors significantly influence yield?

		Fruit variety				
		1	2	3	4	5
Soil type	1	D: 90	C: 98	E:89	B: 94	A: 96
	2	A: 93	D: 90	B: 92	C: 98	E: 88
	3	E: 83	B: 89	A: 90	D: 87	C: 97
	4	B: 89	E: 87	C: 97	A: 91	D: 90
	5	C: 96	A: 61	D: 88	E: 88	B: 94

4A.1 Appendix: Software Information For Randomised Block Design

Data entry
Data entry for RBDs follows the same principles as CRDs except that there is an additional column of codes corresponding to the blocking factor.

SAS: Data entry in SAS is again achieved through use of the DATA step with PROC PRINT providing the data check. For the sales study of Example 4.1, the following statements were used with statement explanation provided in Appendices 1A and 2A:

```
DATA CH4.EX41;
    INPUT SALES SHOPTYPE $ METHOD $ @@;
    LINES;
    22 I A   14 I B   15 I C   14 I D   20 II A   10 II B   20 II C
    19 II D   11 III A   11 III B   14 III C   10 III D   10 IV A
    8 IV B   8 IV C   10 IV D   16 V A   17 V B   14 V C   9 V D
;
PROC PRINT DATA = CH4.EX41 NOOBS;
    VAR SALES;
    BY SHOPTYPE METHOD;
RUN;
```

Minitab: Entry of response data into column C1 is generally by block-treatment combination. For Example 4.1, this means starting with shop type I and method A then shop type I and method B and so on. Codes entry into column C2 (shop type factor) and column C3 (method factor) then conforms to the data entry pattern enabling the **Calc ➤ Make Patterned Data** and the **Manip ➤ Code ➤ Numeric to Text** options to be used to produce the required factor codes. The menu procedures would be as shown in Appendix 3A for the CRD.

For Example 4.1, either of these forms of data entry would result in a data structure as follows:

Sales	Shoptype	Method	Sales	Shoptype	Method	Sales	Shoptype	Method
22	I	A	19	II	D	8	IV	C
14	I	B	11	III	A	10	IV	D
15	I	C	11	III	B	16	V	A
14	I	D	14	III	C	17	V	B
20	II	A	10	III	D	14	V	C
10	II	B	10	IV	A	9	V	D
20	II	C	8	IV	B			

Exploratory analysis
Data plots and summaries are generated using PROC GPLOT and PROC MEANS in SAS (see Displays 4.1, 4.4, 4.5, and 4.6) and appropriate menu procedures in Minitab.

ANOVA information
As RBDs are balanced, ANOVA procedures associated with balanced structures can be used to generate the necessary ANOVA summaries.

SAS: The PROC ANOVA and PROC GLM procedures, listed below, perform the necessary ANOVA calculations for the sales data of Example 4.1 (see Displays 4.2 and 4.3). In the CONTRAST statement, the specification of contrast coefficients as 3 –1 –1 –1 still retains the necessary contrast property of $\sum c_j = 0$ and is necessary to ensure the contrast is estimable. Output from this statement is shown in Display 4.3. If the contrast was to be written as displayed in Exercise 4.4, then the fraction part would need to specified to at least five decimal place accuracy before estimability of the contrast occurs.

```
PROC ANOVA DATA = CH4.EX41;
    CLASS SHOPTYPE METHOD;
```

```
        MODEL SALES = SHOPTYPE METHOD;
PROC GLM DATA = CH4.EX41;
        CLASS SHOPTYPE METHOD;
        MODEL SALES = SHOPTYPE METHOD;
        CONTRAST 'A versus others' METHOD 3 –1 –1 –1;
        OUTPUT OUT = CH4.EX41A P = FITS R = RESIDS;
RUN;
```

Minitab: The Balanced ANOVA option again provides for ANOVA generation, except the contrast. The necessary menu commands are as follows:

> Select **Stat** ➢ **ANOVA** ➢ **Balanced ANOVA** ➢ for *Responses*, select **Sales** and click **Select** ➢ select the **Model** box and enter **Shoptype Method**.
> Select **Storage** ➢ **Fits** ➢ **Residuals** ➢ click **OK** ➢ click **OK**.

Diagnostic checking
Residual plotting follows the same procedures as described in Appendix 3A for the CRD structure.

4A.2 Appendix: Software Information For Balanced Incomplete Block Design

Data entry
Data entry for BIBDs follows the same reasoning as that adopted for RBDs though certain block-treatment combinations will be missing. Patterned codes entry facilities, such as those within Minitab, cannot therefore be used.

Exploratory analysis
Data plots and summaries can be produced in the usual manner in both software packages.

ANOVA information
With the occurrence of missing treatment combinations in a BIBD (unequal cell counts), then general linear model (GLM) estimation methods must be used to produce the required ANOVA information as balanced ANOVA procedures are no longer appropriate.

SAS: The PROC GLM procedure, listed below, performs analysis of variance for BIBDs.

```
PROC GLM DATA = CH4.EX;
        CLASS BLOCK TREAT;
        MODEL RESPONSE = BLOCK TREAT;
        LSMEANS TREAT / TDIFF ADJUST = BON;
        OUTPUT OUT = CH4.EXA P = FITS R = RESIDS;
RUN;
```

The LSMEANS statement enables adjusted treatment means to be calculated and is appropriate for treatment mean estimation in designs which do not contain all possible treatment combinations. The options appearing after the slash calculate pairwise t tests (TDIFF) with ADJUST = BON providing a multiple comparison adjustment for these tests according to Bonferroni. Other forms of adjustment are also available.

Minitab: Minitab's GLM estimation option provides for ANOVA generation. The necessary menu commands are as follows:

Select **Stat** ➤ **ANOVA** ➤ **General Linear Model** ➤ for *Responses*, select the response and click **Select** ➤ select the **Model** box and enter the factors.
Select **Storage** ➤ **Fits** ➤ **Residuals** ➤ click **OK** ➤ click **OK**.

4A.3 Appendix: Software Information for Latin Square Design

Data entry
Data entry for LS designs follows the same patterning as for other One Factor Designs. For a LS design, there will be one response column and three codes columns. Similar steps to those defined in previous Appendices apply though care must be taken with entry of the treatment codes as not all treatments are tested at all row-column combinations.

For Example 4.3, data entry results in a data structure shown in the table below.

Density	Filmtype	Camera	Bulbbrnd	Density	Filmtype	Camera	Bulbbrnd
0.64	1	1	A	0.64	3	4	A
0.72	1	2	B	0.74	3	5	B
0.69	1	3	C	0.63	4	1	D
0.66	1	4	D	0.73	4	2	E
0.74	1	5	E	0.68	4	3	A
0.62	2	1	B	0.74	4	4	B
0.71	2	2	C	0.72	4	5	C
0.69	2	3	D	0.74	5	1	E
0.72	2	4	E	0.69	5	2	A
0.66	2	5	A	0.67	5	3	B
0.65	3	1	C	0.75	5	4	C
0.62	3	2	D	0.68	5	5	D
0.68	3	3	E				

Exploratory analysis
Generation of data plots and summaries is as described in earlier Appendices.

ANOVA information
Because all treatments are not tested at all row-column combinations in a LS design, the design essentially has unequal cell counts and therefore balanced ANOVA procedures cannot be used. As with BIBDs, general linear model (GLM) estimation procedures require to be used.

SAS: The PROC GLM procedure, listed below, performs the necessary analysis of variance for a LS design such as that described in the photographic density study of Example 4.3 (see Displays 4.7 and 4.8).

```
PROC GLM DATA = CH4.EX43;
    CLASS FILMTYPE CAMERA BULBBRND;
    MODEL DENSITY = FILMTYPE CAMERA BULBBRND;
    MEANS BULBBRND / SNK;
```

```
    OUTPUT OUT = CH4.EX43A P = FITS R = RESIDS;
RUN;
```

Minitab: The general linear model estimation procedure provides the same ANOVA information as PROC GLM. The necessary menu commands are as follows:

Select **Stat** ➢ **ANOVA** ➢ **General Linear Model** ➢ for *Responses*, select **Density** and click **Select** ➢ select the **Model** box and enter **Filmtype Camera Bulbbrnd**.
Select **Storage** ➢ **Fits** ➢ **Residuals** ➢ click **OK** ➢ click **OK**.

Diagnostic checking

Residual plot generation will be as described in Appendix 3A for the CRD structure.

5

Factorial Experimental Designs

5.1 INTRODUCTION

The design structure and ANOVA principles of One Factor Designs can be extended to investigations where the objective is to assess the effect on a response of two or more controlled factors. Such multi-factor investigations occurs often within industrial and scientific experimentation when an outcome may depend on many factors such as temperature of solder, quantity of raw material, pressure setting, concentration of ingredients, and machine operator experience. The factors chosen for examination would be tested at a number of "levels" reflecting likely operating levels for such factors. Experiments of this type come under the domain of **Factorial Designs** (**FDs**) where both **main effects** and **interaction effects** can be investigated to determine the level of influence, if any, such factors have and how this may be occurring in the observed response measurements.

For example, chemical process yield may depend on temperature setting (factor *A*) and production time (factor *B*). By testing combinations of these factors, we could understand how varying factor levels influence yield and assess whether these factors work separately or in combination in their effect on yield. As a result of this assessment, we would be able to conclude on how best to optimise the process in respect of the tested factors. Testing different nutrients (factor *A*) on soils with different pHs (factor *B*) may help provide a picture of how fruit yield is affected by these factors and whether there appears evidence that the effect is influenced by the treatment combination chosen. In a drug trial concerning the effect of two drugs A and B on a pharmacological effect, interest may lie in determining if the combined effect of the two drugs is additive, synergistic, or antagonistic. In developing a new soft drink, interest may lie in testing different sweetness levels (factor *A*), different acidity levels (factor *B*), and different colours (factor *C*) in a three factor experiment to assess how different combinations affect taste of the new product and whether one combination stands out as potentially best. In all these illustrations, a FD structure would be best as it would enable both the effect of each controllable factor and the effect of interaction between treatment factors to be assessed and estimated.

Experimental factors in FDs are generally specified at a number of different levels. For example, temperature: 40°C, 50°C, 60°C, and 70°C; cooling time: 10 minutes, 15 minutes, and 20 minutes; weight of catalyst: 5 g, 7.5 g, 10 g, 12.5 g; angle of tool: 15°, 25°, and 35°; raw material flow rate: slow and fast; and operators: A, B, and C. These **quantitative** and **qualitative** "levels" reflect the different factor settings to be controlled in the experiment. It is the effect of these levels on a measured response, separately and in combination, which FD structures are designed to assess and estimate. Analysis of the data collected from such experiments is comparable to that outlined for One Factor Designs though follow-up analysis is generally more graphical than statistical in practice.

In such experiments, interest lies in assessing how much influence, if any, each controllable factor has on the response on an independent or a dependent, in combination, basis. Independent factor effects reflect the concept of **main effects** measuring the change in response resulting from changing the levels of one factor. Factors working in combination, dependent factor effects, reflect **interaction effects** and measure the differing effect on the response of different treatment combinations. Essentially, interaction measures the joint influence of treatment combinations on the response measurement recorded which cannot be explained by the sum of the individual factor effects. It is this concept which FDs are primarily geared to assessing and estimating.

By way of an illustration, consider an industrial experiment to assess the effect of two factors on the yield of pulp recovered from a cellulosic raw material. The factors to be tested are temperature, in °C, and time of reaction, in hours. Temperature is to be investigated at three levels and time at four levels meaning that the design structure would be that of a 3×4 factorial. By including replication of treatment combinations, we could generate sufficient data to be able to provide estimation of the interdependency of the temperature and time factors, i.e. factor interaction, and how this affects, or does not affect, pulp yield.

Simplicity of practical implementation, more effective use of experimental responses, and provision of more relevant response information are the three fundamental features of factorial structures which make them more appropriate bases for experimentation than **One Factor Designs** (CRD, RBD) or **one-factor-at-a-time (OFAT) experimentation**. The drawback of One Factor Designs is that they are only capable of assessing the effect of a single factor on a measured outcome. In most industrial and scientific experimentation, however, it is highly probable that more than one factor will influence a response either independently or, more likely, in combination. In addition, one factor structures essentially only represent one aspect of a factorial structure with the conclusions reached depending heavily on the aspect chosen to be investigated. In other words, if the assessment of interaction effects is omitted and not given due consideration, then incomplete inferences may be obtained and valuable information lost.

OFAT experimentation is often used by scientists to investigate multi-factor problems and is a sequential optimisation procedure where each factor is optimised in turn (Gardiner 1997, Miller & Miller 1993). Specification of an optimal combination, however, is likely to be due more to chance than design because determination depends on the combinations tested and in OFAT experiments, only a non-structured and limited selection of potential treatment combinations are ever tested. In addition, OFAT approaches are based on the assumption that each factor works independently of every other and so it is unable to provide any measure of the effect of factor interactions.

Such independent influence of factors on a measurable response is highly unlikely in many studies as factors are likely to interact in their effect on a response. As such, the difference in mean response for each level of a factor will vary across the levels of another factor showing that the factors work together, not separately, in their influence on the measured response. Factorial experimentation addresses this point though OFAT experimentation is often adopted in multi-factor experiments when, through adequate planning, simple factorial structures would be more advantageous as they can provide more relevant practical information more efficiently.

The concept of interaction between factors is best explained graphically as the plots in Fig. 5.1 illustrate for a response of absorbance in respect of two factors, temperature at three levels and chemical concentration also at three levels. Absorbance is plotted against concentration for each of the three temperature levels. The response lines in both plots are not parallel indicating non-uniformity of response across the treatment combinations. Both plots highlight that the factors are not acting independently but are inter-dependent. In Fig. 5.1A, we can see that absorbance varies markedly across concentration as temperature changes. For low temperature, the trend is downward while for high temperature, it is upward. Mid temperature indicates a more quadratic effect. This form of interaction plot illustrates the case where interaction is occurring as a difference in the direction of the responses, i.e. non-uniformity of response order. Fig. 5.1B, on the other hand, highlights a magnitude difference form of interaction showing uniformity of response order but magnitude differences. The plot shows that the temperature trends are linear with the same ordering of temperature at each concentration though the difference in mean absorbance varies as concentration changes. Absence of interaction would be shown by simple parallel lines indicative of a simple additive effect when changing factor levels unlike convergent or divergent lines when interaction is present.

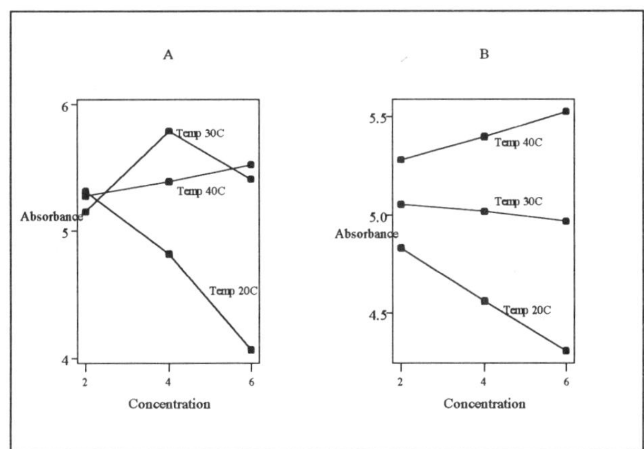

Fig. 5.1 — Illustration of interaction plots arising from a Factorial Design

In certain applications of FFDs, e.g., in a multi-centre pharmaceutical trial, it is often useful to specify the interaction effect according to its nature. The pattern exhibited in Fig. 5.1B is often referred to as a **quantitative interaction** defining similar directional differences but difference in magnitude of the treatment effects, with interpretation using the average treatment difference. Fig. 5.1A characterises a **qualitative interaction** signifying both directional and magnitude differences in respect of the treatment effects, with interpretation based simply on assessing or describing the nature of the interaction.

5.2 TWO FACTOR FACTORIAL DESIGN WITH *n* REPLICATIONS PER CELL

Consider two factors A and B which can be tested at a and b levels respectively providing a total of ab treatment combinations to which the experimental units require to be randomly assigned. Generally, each treatment combination is replicated n times in order to provide sufficient data for efficient effect estimation. It is this replication which enables the interaction effect to be estimated. The two factor design structure will be used as the basis for the description of the underlying concepts and analysis procedures appropriate to factorial based experiments. Blocking in such designs can also occur but will not be discussed (Montgomery 1997).

5.2.1 Design Structure

The experimental set up for a two factor factorial experiment requires each treatment combination to be observed *n* times. Consider the pulp yield experiment introduced earlier. We have three levels for temperature (factor *A*), four levels for the time factor (factor *B*), and suppose three replications are planned for each treatment combination. The set up for such an experiment is illustrated in Table 5.1 and would require 3×4×3 = 36 separate observations to be made. The displayed structure is typical of that associated with FDs in respect of two factors though the number of factor levels and replicates will vary from study to study.

Table 5.1 — Possible Factorial Design layout for pulp yield study

		Replications		
Temperature	Time	1	2	3
140 °C	1 hour	x	x	x
	2 hours	x	x	x
	3 hours	x	x	x
	4 hours	x	x	x
150 °C	1 hour	x	x	x
	2 hours	x	x	x
	3 hours	x	x	x
	4 hours	x	x	x
160 °C	1 hour	x	x	x
	2 hours	x	x	x
	3 hours	x	x	x
	4 hours	x	x	x

x denotes an observation

Example 5.1 The female flower (*A*) and the seed (*B*) of two African plants are widely used in traditional medicine for the treatment of people with certain diseases. It has been claimed, however, that use of these traditional remedies in combination can temporarily impair vision by reducing the patient's field of vision. The measurements provided in Table 5.2 were collected from 18 subjects randomly assigned in groups of two to different combinations of flower and seed concentrations within a 3 × 3 factorial structure. How do the plants appear to affect field of vision?

Table 5.2 — Field of vision data for Example 5.1

Flower *A*	Seed *B*	Field of vision	
0.25 g	0.001 g	67	66
	0.005 g	65	61
	0.009 g	62	64
2.5 g	0.001 g	68	65
	0.005 g	68	61
	0.009 g	55	53

(continued)

(continued)			
5.0 g	0.001 g	65	64
	0.005 g	62	63
	0.009 g	49	47

For two factor experimentation, it may appear that the associated design structure is similar to that of a RBD. This is not the case. A RBD is used to investigate the effect of one treatment factor using blocking as a means of excluding extraneous variation from the response to provide a more sensitive test of treatment effect. A FD, on the other hand, does not sacrifice information on any factor enabling the effect of all factors and their interactions to be fully assessed. Normally, the experimental units are randomly assigned to the treatment combinations in a FD whereas in a RBD, the experimental units are randomly assigned to treatments within the blocks.

5.2.2 Model for the Measured Response

As with One Factor Designs, the first step in the analysis is the specification of a response model for the observations. Denoting X_{ijk} to be the measured response on the kth replicate of the n experimental units receiving factor A level i and factor B level j and assuming additivity of effects, we have

$$X_{ijk} = \mu_{ij} + \varepsilon_{ijk} = \mu + \alpha_i + \beta_j + \alpha\beta_{ij} + \varepsilon_{ijk} \quad (5.1)$$

where $i = 1, 2,..., a$, $j = 1, 2,..., b$, and $k = 1, 2,..., n$. The expression $\mu_{ij} + \varepsilon_{ijk}$ defines the **means model** while $\mu + \alpha_i + \beta_j + \alpha\beta_{ij} + \varepsilon_{ijk}$ refers to the **effects model**. We will primarily discuss the effects model for a **fixed effects**, or **model I**, experiment where the factor levels chosen for study represent only those levels of interest.

The μ_{ij} term defines the population mean for the treatment combination of factor A level i and factor B level j while ε_{ijk} again represents the uncontrolled variation, the within treatment combination effect. The μ term refers to the grand mean which is assumed constant for all treatment combinations tested. The term α_i (*alpha*) represents the effect on the measured response of factor A at level i and is assumed constant for all units receiving factor A at level i but may differ for units receiving a different treatment combination. Similarly, the β_j (*beta*) term defines the effect on the response of factor B at level j and is assumed constant for all units receiving factor B at level j but may differ for units receiving a different treatment combination. The interaction term, denoted by $\alpha\beta_{ij}$, measures the combined effect on the response of factor A level i with factor B level j. As previously, this form of response specification rests on describing the response data in terms of the controlled and uncontrolled components of the experimental structure proposed.

The **assumptions** underpinning FDs are the same as those for One Factor Designs, i.e. equal response variability across the factors, normality of response, and additivity of effects. Additionally, we assume for fixed effects experiments that the model effects conform to the constraints $\Sigma\alpha_i = 0$, $\Sigma\beta_j = 0$, and $\Sigma\Sigma\alpha\beta_{ij} = 0$.

When $n = 1$ in a two factor structure, there is only one observation (single replicate) for each treatment combination. In this case, the response model (5.1) does not contain an interaction term because interaction and error are confounded as there is insufficient data with which to independently estimate these two effects. The response model therefore resembles that of a RBD though interpretation of the terms may be different. The purpose of the factorial experiment is to test for differences in each main effect unlike the RBD which is constructed to test for treatment effects using blocking to provide a more sensitive test. For single replication, interaction plotting can be considered through plotting combination responses against either factor levels or factor effect estimates. In either case, presence of non-parallel lines would provide evidence of interaction while formal testing of interaction is available through **Tukey's test for nonadditivity** (Milliken & Johnson 1992).

5.2.3 Exploratory Data Analysis

As in all previous design illustrations, simple exploratory analysis through plots and summaries should be considered to help gain initial insight into what the tested factor levels specify concerning the experimental objectives.

Exercise 5.1 The first stage of analysis of the field of vision study outlined in Example 5.1 concerns response model specification and exploratory analysis.

Response model: Based on the data collection, the effects model for the field of vision response will be

$$\text{field of vision} = \mu + \alpha_i + \beta_j + \alpha\beta_{ij} + \varepsilon_{ijk}$$

with $i = 1, 2, 3$ (three flower A concentrations), $j = 1, 2, 3$ (three seed B concentrations), and $k = 1, 2$ (two subjects per treatment combination). The term α_i measures the effect of the ith level of concentration of flower A on field of vision, β_j the effect of the jth level of concentration of seed B on field of vision, $\alpha\beta_{ij}$ the interaction effect of the flower and seed concentrations, and ε_{ijk} the uncontrolled variation. Again, model specification indicates the belief that the response is affected by both controlled (flower and seed concentrations) and uncontrolled variation. We assume that the field of vision measurements have equal variability for both sets of concentrations tested and that they are normally distributed.

EDA: Initial analysis of the results corresponding to flower A showed some difference in field of vision with amount tested. Field of vision appeared to vary inversely with concentration showing that high concentration appeared to reduce field of vision most. Variability in results differed also according to concentration with 0.25 g least variable (CV 3.6%) and 5 g most variable (CV 13.9%).

For seed B, a similar trend occurred as concentration increased. High concentration (0.009 g) stood out more as it provided a markedly lower mean (55.00) and higher variability (CV 12.4%) than the other two tested concentrations. Low and medium concentrations produced comparable mean results but marginally different variability summaries.

Initially, it appears that high concentrations have greatest detrimental effect on field of vision, a not unexpected result. □

5.3 ANOVA PRINCIPLE FOR THE TWO FACTOR FACTORIAL DESIGN

The **ANOVA principle**, as for One Factor Designs, provides the basis of the statistical analysis aspects of FDs. It uses statistical theory to help derive the mechanisms for test statistic construction and derivation based on response model specification. The procedures involved mirror those of the CRD and RBD structures (see Sections 3.3 and 4.3).

5.3.1 Hypotheses

Hypotheses associated with factorial designed experiments define a test of absence of effect versus presence of effect. There are three possible effects to investigate in a two factor structure: main effects of factor A and of factor B, and the interaction effect $A \times B$. The associated null hypothesis for each of these cases is as follows:

main effect of factor A: H_0: no factor A effect occurring, i.e. $\alpha_1 = \alpha_2 = ... = \alpha_a = 0$,
main effect of factor B: H_0: no factor B effect occurring , i.e. $\beta_1 = \beta_2 = ... = \beta_b = 0$,
interaction effect A × B: H_0: no interaction effect occurring, i.e. $\alpha\beta_{11} = \alpha\beta_{12} = .. = \alpha\beta_{ab} = 0$.

The corresponding alternative hypotheses simply define presence of the effect being assessed.

5.3.2 ANOVA Table

Application of the **ANOVA principle** within a two factor FD, as with One Factor Designs, involves partitioning response variation into the four sources of variation associated with the specified response model (5.1). Again, this partitioning process rests on the determination of sum of squares (*SS*) terms for the controlled and uncontrolled elements within the response model. From the response model (5.1), this means that *SS*Total can be expressed as

$$SS\text{Total } (SST) = SSA + SSB + SS\text{Interaction } (SSAB) + SS\text{Error } (SSE) \qquad (5.2)$$

The calculation components and general form of **ANOVA table** for a two factor FD are presented in Table 5.3. The obvious difference from the corresponding ANOVA table for a RBD experiment (see Table 4.4) lies in the replacement of the block and treatment effects by the separate factor and interaction effects.

5.3.3 Test Statistics

The test statistics for the three effects associated with a fixed effects two factor FD are again specified, from the associated statistical theory, as a ratio of two mean squares, the mean square term for effect being tested and the mean square term for the error component. For the **factor *A* effect**, the ratio of the factor A mean square to the error mean square provides the test statistic, i.e.

$$F = MSA/MSE \qquad (5.3)$$

with degrees of freedom $df_1 = a - 1$ and $df_2 = ab(n - 1)$. For the **factor *B* effect**, the test statistic is constructed as the ratio of the factor B mean square to the error mean square, i.e.

$$F = MSB/MSE \qquad (5.4)$$

with degrees of freedom $df_1 = b - 1$ and $df_2 = ab(n - 1)$. For the **interaction effect**, the test statistic is based on the ratio of the interaction mean square to the error mean square, i.e.

$$F = MSAB/MSE \qquad (5.5)$$

with degrees of freedom $df_1 = (a-1)(b-1)$ and $df_2 = ab(n-1)$.

Table 5.3 — ANOVA table for a two factor Factorial Design with n replicates of each treatment combination

Source	*df*	*SS*	*MS*
A	$a-1$	$SSA = \dfrac{\sum_{i=1}^{a} A_i^2}{bn} - \dfrac{\left(\sum_{i=1}^{a}\sum_{j=1}^{b}\sum_{k=1}^{n} X_{ijk}\right)^2}{abn}$	$MSA = SSA/df$
B	$b-1$	$SSB = \dfrac{\sum_{j=1}^{b} B_j^2}{an} - \dfrac{\left(\sum_{i=1}^{a}\sum_{j=1}^{b}\sum_{k=1}^{n} X_{ijk}\right)^2}{abn}$	$MSB = SSB/df$
$A \times B$	$(a-1)(b-1)$	$SSAB = \dfrac{\sum_{i=1}^{a}\sum_{j=1}^{b} AB_{ij}^2}{n} - SSA - SSB + \dfrac{\left(\sum_{i=1}^{a}\sum_{j=1}^{b}\sum_{k=1}^{n} X_{ijk}\right)^2}{abn}$	$MSAB = SSAB/df$
Error	$ab(n-1)$	$SSE\ (= SST - SSA - SSB - SSAB)$	$MSE = SSE/df$
Total	$abn-1$	$SST = \sum_{i=1}^{a}\sum_{j=1}^{b}\sum_{k=1}^{n} X_{ijk}^2 - \dfrac{\left(\sum_{i=1}^{a}\sum_{j=1}^{b}\sum_{k=1}^{n} X_{ijk}\right)^2}{abn}$	

a is the number of levels of factor A, A_i is the sum of the responses for level i of factor A, b is the number of levels of factor B, n is the number of replications of each treatment combination, X_{ijk} is the response measurement for experimental unit k receiving factor A level i and factor B level j, B_j is the sum of the responses for level j of factor B, AB_{ij} is the sum of the responses for the treatment combination of factor A level i and factor B level j, *df* defines degrees of freedom, *SS* defines sum of squares, *MS* defines mean square

Essentially, the F ratios (5.3) to (5.5) measure the importance of each factorial effect. Small values will indicate that the tested effect explains little of the variation in the response while large values will be indicative of significant effect and that the factor, or interaction, effect being tested has a significant influence on the response. It is this influence we are interested in detecting if it is occurring within the data. Use of either the **test statistic approach** or the ***p* value approach** follows.

As three statistical tests can be carried out in a two factor FD, it is necessary to decide which should be carried out first. In general, the interaction significance test should be carried out **first** as it is important to detect factor interaction if it exists since this means that the two controlled factors influence the response in combination and not independently. Presence of a significant interaction effect can make testing of main effects (test of factors A and B) unnecessary unless the interaction is orderly (similar pattern of results across all treatment combinations) when a test of main effects can be meaningful. This can often occur with qualitative factors when the interaction plot comes out similar to that displayed in Fig. 5.1B.

Exercise 5.2 Having carried out initial analysis on the field of vision data of Example 5.1, the next step in the analysis is again the statistical assessment of factor effects with interaction tested first.

Hypotheses: The null hypothesis for each possible factorial effect is as follows:

interaction: H_0: no flower $A \times$ seed B interaction effect on field of vision,
flower A: H_0: no flower A concentration effect on field of vision,
seed B: H_0: no seed B concentration effect on field of vision.

ANOVA table: Display 5.1 contains the ANOVA information generated by SAS's PROC ANOVA procedure for the field of vision data, the details of which are presented in Appendices 3A and 5A. The ANOVA details presented mirrors the general form in Table 5.3 with the controlled components ('FlowerA', 'SeedB', 'FlowerA*SeedB') and uncontrolled components ('Error') adding to explain the Total variability in the response data. The mean square terms are again derived as the ratio *SS/df*. The *SS* term for seed *B* ('SeedB') is very much larger than that for flower *A* implying possibly greater influence of concentration of seed *B* on field of vision.

Display 5.1

Source	DF	Sum of Squares	Mean Square	F Value	Pr > F
Model	8	623.77777778	77.97222222	15.77	0.0002
Error	9	44.50000000	4.94444444		
Corrected Total	17	668.27777778			

R-Square	C.V.	Root MSE	VisField Mean
0.933411	3.622171	2.22361068	61.38888889

Source	DF	Anova SS	Mean Square	F Value	Pr > F
FlowerA	2	102.77777778	51.38888889	10.39	0.0046
SeedB	2	386.11111111	193.05555556	39.04	0.0001
FlowerA*SeedB	4	134.88888889	33.72222222	6.82	0.0083

Interaction test: Test statistic (5.5) specifies that the interaction F test statistic is the ratio of the interaction and error mean squares. From the information in Output 5.1, this results in $F = 6.82$ as specified in row 'FlowerA*SeedB' and column 'F Value'. With both test statistic and p value available, we can use either decision rule mechanism for the interaction test.

For the **test statistic approach**, we know $F = 6.82$ with degrees of freedom $df_1 = 4$ (interaction) and $df_2 = 9$ (error). The associated 5% critical value is $F_{0.05,4,9} = 3.63$ (Table A.3). For the ***p* value approach**, the interaction p value is estimated as 0.0083 (row 'FlowerA*SeedB', column 'Pr > F'). This is a low value indicating rejection of the null hypothesis and the conclusion that the evidence implies that concentrations of both flower A and seed B interact in their effect on field of vision ($p < 0.05$). As interaction is significant, it is not appropriate to test the separate flower A and seed B effects on a statistical basis though follow-up analysis may suggest otherwise.

Model fit: As described in Section 3.4.7, we can estimate the error ratio (*SSE*/*SST*)% to help assess adequacy of model fit. For this experiment, we have this ratio as (44.5/668.278)% = 6.7%, a low value, signifying that the specified field of vision response model is a reasonable explanation of how different concentration combinations of the two factors affect a subject's field of vision. □

5.4 FOLLOW-UP ANALYSIS PROCEDURES FOR FACTORIAL DESIGNS OF MODEL I TYPE

Unfortunately, the formal tests (5.3) to (5.5) only point to general evidence within the data of presence of a particular effect. They provide no information on how the detected effect is occurring within the response data so we must consider further analysis to pinpoint and interpret its influence on the measured response. This section details some of the ways in which further analysis can be carried out depending on the type of effect found to be significant.

5.4.1 Significant Interaction

A **significant interaction**, such as found in Exercise 5.2, indicates that the levels of one factor result in a difference in the response across the range of levels of the second factor. This implies that different treatment combinations do not give rise to similar response measurements. Unfortunately, we do not know how this is occurring in respect of the treatment combinations tested. We need, therefore, to analyse the interaction effect further to pinpoint how it is manifesting itself within the set of measured responses.

Normally, for fixed levels of the factors (model I experiment), it is sufficient to plot the interaction means for each treatment combination in an **interaction** or **profile plot** such as those illustrated in Fig. 5.1. By inspection of this summary plot, we can identify the trends and patterns within the data which may explain why significant interaction was detected and so improve our understanding of how interaction is manifested within the response data. In such cases, reporting of interaction summaries including mean and range is generally sufficient.

Exercise 5.3 In Exercise 5.2, the interaction effect of the two factors on field of vision was found to be statistically significant. To assess this further, we need to produce an interaction plot to investigate the interaction effect in more detail. Display 5.2 illustrates the Minitab generated interaction plot using the **Stat ➤ ANOVA ➤ Interactions Plot** options with the annotation, produced using Minitab's graph editing facility, designed to help aid interpretation. The figures inside the brackets refer to the range of the results for each treatment combination.

Display 5.2

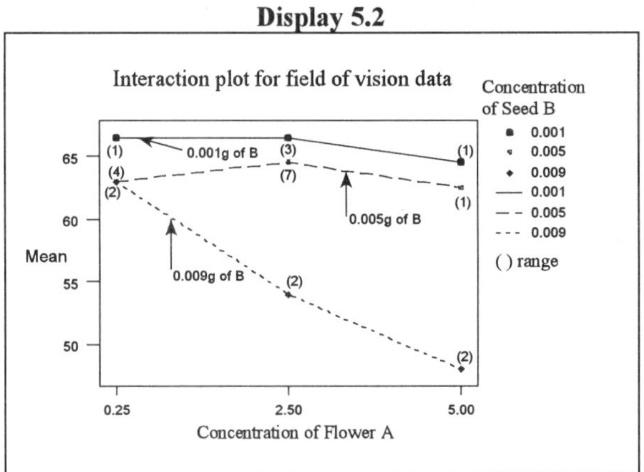

Interaction plot: The presence of non-parallel lines in the interaction plot is indicative of the presence of an interaction effect while the uniform order of presentation shows this to be a magnitude difference form of interaction as illustrated in Fig. 5.1B. The patterns shown indicate that lower concentrations of seed *B* do not seem to affect field of vision greatly as flower *A* concentration increases. In contrast, the high concentration of seed *B* shows a marked decline in field of vision as flower *A* concentration increases. The marked difference apparent in respect of seed *B* concentrations concurs with the comment made on the effect sums of squares suggesting that seed *B* is possibly the major contributing factor.

The orderly nature of the interaction plot suggests that tests of the separate main effects could be useful. This aspect of analysis will not be considered here as it is similar to that already described under multiple comparisons for main effects in Section 3.4.3.

Results variation also changes as flower A concentration increases with least variable effect appearing at 0.25 g. The ranges, specified by the figures in the brackets in Display 5.2, suggest that the combinations (0.25 g A, 0.005 g B) and (2.5 g A, 0.005 g B) give rise to the most variable measurements. Inclusion of ranges in the interaction plot can therefore be particularly useful especially if consistency of results is an important issue. □

In addition, assessment for optimal treatment combination should also be carried out to aid the decision making process. This is particularly appropriate if the goal of the experiment is to generate an optimal treatment combination from those tested. Which combination to choose as best depends on whether attainment of an optimal response and/or good precision in response is the more important. Ideally, we generally want both aspects of the optimal satisfied though in many cases, we may have to play one off against the other, the concept of **trade-off**, when concluding on a best combination of factors.

Optimality assessment is based on picking out the best results from a summary of combination results. Such a summary, based on SAS's PROC MEANS procedure (see Appendix 5A), is provided in Display 5.3 and corresponds to the mean and range results of the interaction plot in Display 5.2. From the summaries, we can see the similarity in most results in respect of both means and ranges except combination (2.5 g A, 0.005 g B) which has wide variation. Also combinations (2.5 g A, 0.009 g B) and (5 g A, 0.005 g B) which have greatly reduced mean values indicating the detrimental effect of these combinations in particular.

Display 5.3

FlowerA	SeedB	N Obs	Mean	Range
0.25	0.001	2	66.50	1.00
	0.005	2	63.00	4.00
	0.009	2	63.00	2.00
2.5	0.001	2	66.50	3.00
	0.005	2	64.50	7.00
	0.009	2	54.00	2.00
5	0.001	2	64.50	1.00
	0.005	2	62.50	1.00
	0.009	2	48.00	2.00

We could consider a formal multiple comparison of the means of each treatment combination. Based on a simple $a \times b$ FD, this would produce a total of $ab(ab - 1)/2$ possible pairings to assess. Even for small a and b, this can be too numerous to consider. For example, for the field of vision study illustrated in Example 5.1 ($a = 3$ and $b = 3$), such an approach would mean 36 possible combination pairings requiring testing which, for the benefits it could provide, is not worth pursuing. Adoption of simple graphical approaches in the assessment of interaction effects is therefore often sufficient and has the added advantage of presenting the information visually rather than numerically. Alternatively, if one factor level seems to have the most diverse behaviour, it could be omitted and the analysis repeated. In this way, the significant interaction effect could perhaps be removed making it possible to focus on main effects tests. Another possible follow-up approach concerns setting one factor at a specific level and apply a multiple comparison to the means of the other factor at that level (Montgomery 1997).

5.4.2 Non-significant Interaction but Significant Factor Effect

If interaction is not significant statistically (accept H_0 in test (5.5)), the evidence appears to point to the factors acting independently on the response. On this basis, we must test the separate main effects for evidence of effect. If such effects are significant, follow-up checks based on multiple comparisons can be carried out for each factor separately to complete the statistical analysis of the response data. Main effect, or standard error, plots could also be considered but only if the associated factor explains most of the inherent response variation.

5.4.3 Linear Contrasts

Planned comparisons of interaction effects and main effects through linear contrasts must still be assessed irrespective of whether any, some, or all of the factorial effects are shown to be statistically significant. The methods associated are as those described in Section 3.4.4.

Suppose an additional aspect of assessment in the field of vision study had been comparison of the lowest (0.25 g) and highest (5 g) concentrations of flower *A*. The linear contrast would be specified as

$$L = 0.25\text{ g} - 5\text{ g} = 64.17 - 58.33 = 5.84.$$

Using the confidence interval (3.7) based on *errordf* = 9 and $n = 6$, we have

$$5.84 \pm (2.262)\sqrt{\frac{(1)^2 + (-1)^2}{6}(4.9444)}$$

which gives a 95% confidence interval of (2.94, 8.74). These limits do not contain zero and so provide evidence that low and high concentrations of flower *A* have statistically different effects ($p < 0.05$) with high concentration appearing to impair field of vision considerably more than low concentration.

5.4.4 Orthogonal Polynomials

Assessment of trend effects through orthogonal polynomials can also be considered in FDs if appropriate to the analysis of the response data and provided factor levels are set at equally spaced intervals over a specified quantitative scale. For the factors, we would use the same procedures as those explained in Section 3.4.5 where k now represents the number of factor levels for the factor being split into different trends. For the interaction effect, application of orthogonal polynomials generally considers splitting the interaction variation into trend effects associated with the interaction of one factor with the trend effects of the second factor. As indicated in Section 3.4.5, orthogonal polynomials can be based on use of treatment means or totals. Box 5.1 summarises the use of orthogonal polynomials for the assessment of trend in interaction effects. It should be noted that adding together the results for expression (5.6) for all trend effects in respect of a particular factor provides the *SS*Interaction result.

Box 5.1 — Orthogonal Polynomials for Interaction Effects

Estimation of trend effect This involves calculating an interaction effect estimate as $\sum c_i \overline{X}_{ij}$ based on $\overline{X}_{ij}$ referring to the mean response for the treatment combination of factor A level i and factor B level j. The c_i values represent the tabulated coefficients of the associated orthogonal polynomials (linear, quadratic, cubic, etc.) as presented in Table A.7.

Test statistic Based on a levels for the factor being investigated for trend effect, b levels for the second factor, and n replications of each combination, the appropriate sum of squares of the trend effect is given by

$$SS(Trend\ effect) = \frac{n\sum_{j=1}^{b}\left(\sum_{i=1}^{a} c_i \overline{X}_{ij}\right)^2}{\sum_{i=1}^{a} c_i^2} - \frac{n\left(\sum_{j=1}^{b}\left(\sum_{i=1}^{a} c_i \overline{X}_{ij}\right)\right)^2}{b\sum_{i=1}^{a} c_i^2} \qquad (5.6)$$

where values for c_i and $\sum c_i^2$ are tabulated for different combinations of trend type and number of factor levels (see Table A.7). Using expression (5.6), we derive the associated trend test statistic as

$$F = SS(Trend\ effect)/MSE \qquad (5.7)$$

based on degrees of freedom (b – 1, *errordf*). The usual decision rule mechanisms for F tests apply.

5.4.5 Estimation

For the **model I** type experiment (all factors classified as fixed effects), use of least squares estimation enables unbiased estimators of the parameters in the response model to be derived. Table 5.4 summarises the form and variance expression for the estimators of the parameters μ_{ij}, μ, α_i, β_j, and $\alpha\beta_{ij}$ of the two factor response model (5.1). With data, these estimators can be used to provide appropriate effect estimates.

Table 5.4 — Least squares estimators for the model components in the two factor Factorial Design

Parameter	Point estimator	Variance of estimator
μ_{ij}	$\overline{AB}_{ij}$	σ^2/n
α_i	$\overline{A}_i - \overline{G}$	$\sigma^2(a-1)/(abn)$
β_j	$\overline{B}_j - \overline{G}$	$\sigma^2(b-1)/(abn)$
$\alpha\beta_{ij}$	$\overline{AB}_{ij} - \overline{A}_i - \overline{B}_j + \overline{G}$	$\sigma^2(a-1)(b-1)/(abn)$

Using the *MSE* term from the ANOVA table as an estimate of the error variance σ^2 and its associated degrees of freedom (*errordf*), it is straightforward to calculate confidence intervals for these parameter estimates using the t distribution and the corresponding variance estimate. The form of confidence interval for each estimated parameter is given by

$$\text{estimate} \pm t_{\alpha/2,errordf}\sqrt{(\text{variance of estimator})} \tag{5.8}$$

If the confidence interval for any factor effect or interaction effect contains 0, then it suggests that the effect being considered has no statistically significant influence on the recorded response. Large confidence intervals for model parameters suggest possible design weakness and in particular, insufficient data.

Based on summary means for main effects and interactions and using $errordf$ = 9, $t_{0.025,9} = 2.262$, $\sigma^2 = 4.9444$ (MSE), $a = 3$, $b = 3$, and $n = 2$, we can generate 95% confidence intervals for the plant interaction effects in the field of vision study as

$$\alpha\beta_{ij} \pm (2.262)\times\sqrt{[(4.9444)\times(3-1)\times(3-1)/(3\times3\times2)]} \quad .$$

The calculated intervals are

Flower A	*Seed B*	*95% confidence interval*
0.25 g	0.001 g	(−4.48, 0.26)
	0.005 g	(−5.48, −0.74)
	0.009 g	(2.85, 7.59)
2.5 g	0.001 g	(−1.98, 2.76)
	0.005 g	(−1.48, 3.26)
	0.009 g	(−3.65, 1.09)
5 g	0.001 g	(−0.64, 4.10)
	0.001 g	(−0.14, 4.60)
	0.009 g	(−6.31, −1.57)

The treatment combinations with greatest influence on field of vision appear to be (0.25 g A, 0.005 g B), (0.25 g A, 0.009 g B), and (5 g A, 0.009 g B) as these are the only intervals which do not straddle zero. The interval (0.25 g A, 0.009 g B) covers positive values only suggesting that this combination could increase field of vision. The other two important intervals both cover solely negative values indicative of detrimental effect of associated treatment combination on field of vision. The intervals shown also highlight that 2.5g of flower A in combination with any concentration of seed B has minimal effect whereas 0.25 g of flower A shows greater effect in combination with any concentration of seed B.

Each interaction effect confidence interval is essentially a linear contrast and strictly speaking, we should have adjusted the critical t value to account for multiple inferences. The above is simply an illustration of how such can be used to aid data interpretation. The adjustment for calculating multiple intervals could be implemented by either using Bonferroni's t statistic or simply adjusting the 0.025 level of the critical t value to suit the number of intervals being calculated.

5.4.6 Diagnostic Checking

As FD structures are based on comparable assumptions to One Factor Designs, inclusion of diagnostic checking must be an integral part of the data analysis. Plotting residuals, measurements – fitted values, against the factors tested and against model fits check for equality of response variability and adequacy of model. Normality of the measured response is again checked by a normal plot of residuals with a straight line considered ideal. Inappropriateness of these assumptions can be overcome through either data re-scaling (see Section 3.8) or a non-parametric routine (see Section 5.7).

Exercise 5.4 The final aspect of the analysis of the field vision data concerns diagnostic checking of the residuals.

Factor plots: The residual plot for flower *A* in Display 5.4 shows distinctly different column lengths where variance of residuals for concentration 2.5 g appears greater that that at the other two concentrations. Unequal variability across the factor levels appears to be occurring.

Display 5.4

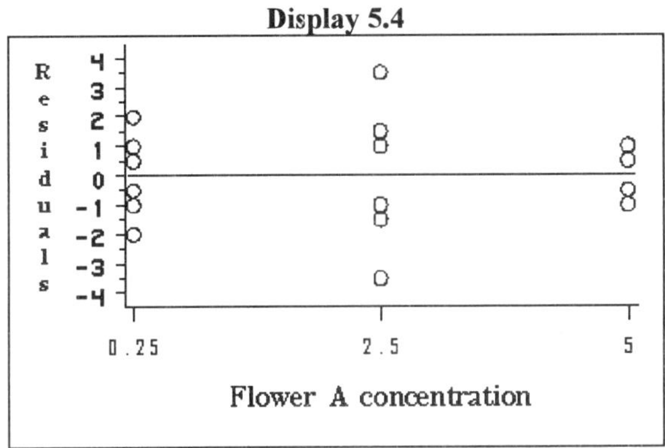

The seed *B* residual plot in Display 5.5 shows comparable characteristics to the flower *A* residual plot leading to comparable interpretation.

Display 5.5

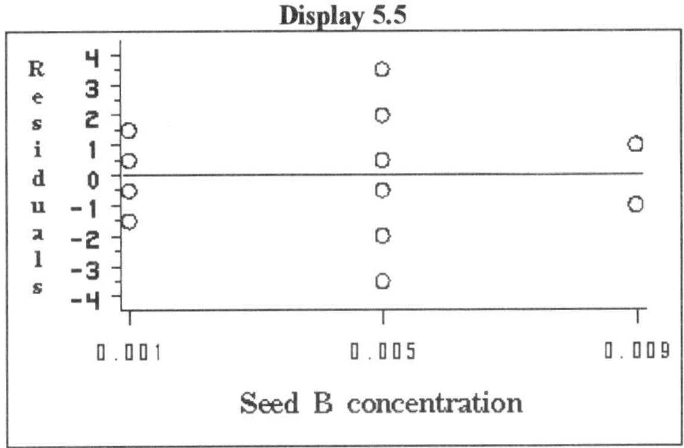

Fits and normal plots: The fits and normal plots, though not shown, were both acceptable. A slight curved trend in the normal plot was apparent but not sufficient to suggest normality of field vision data to be inappropriate. The Ryan-Joiner test provided a suitable back-up ($R = 0.98335, p > 0.1$). □

In summary, the findings from the field of vision study of Example 5.1 suggest that concentration combinations of flower *A* and seed *B* have different influences on field of vision. Statistically, interaction is significant ($p < 0.01$) and the interaction plot shows the non-uniform nature of this effect. High concentrations of seed *B* appear to have greatest influence on field of vision with much lower figures recorded especially at the higher concentrations of flower A. Diagnostic checking suggests that equal variability of response data may not be acceptable and that, in future experiments, some form of data transformation to stabilise variability effects may need to be considered prior to data analysis.

In practical investigations, we would generally carry out as much analyses as appropriate in a coherent manner to provide a method of fully analysing the data. Conclusions, as ever, may influence other experiments. This could mean conducting further experiments around the "best" treatment combination to either home in on a truly optimal treatment combination or to provide further understanding of the phenomenon being investigated.

5.5 OVERVIEW OF DATA ANALYSIS FOR TWO FACTOR FACTORIAL DESIGNS

Exploratory analysis, inferential data analysis, and diagnostic checking should all be part of the data analysis of factorial experiments. The remaining analysis aspects are dependent on which effects are significant and require full analysis and assessment. The field of vision example provides an outline of what can be done when interaction is statistically significant. If interaction is not statistically significant, then each of the two factors may affect the response independently and we should therefore assess each factor separately through the formal F test and associated follow-up.

5.6 POWER ANALYSIS IN TWO FACTOR FACTORIAL DESIGNS

In FDs, it is often necessary to ask "How many replications will be required?". Estimation of the amount of replication is important in the planning phase of a factorial experiment since replication can improve detection of a statistically significant treatment effect. As with the outlined principles of Sections 2.5 and 3.6, information on a minimum experimental outcome is necessary to enable such estimation to take place. Again, this entails the investigator specifying the **minimum detectable difference** δ that they expect to detect if differences exist within the data. For interaction analysis, this corresponds to the minimum difference expected to be detected between any two treatment combinations. For main effects analysis, δ defines the minimum difference expected to be detectable between two factor levels.

Example 5.2 As part of a quality assurance (QA) study, four independent laboratories are to be provided with three similar samples of a food product, each of which is to be analysed a number of times. Past QA studies suggest that a minimum difference of 7.31 ng/g would be indicative of a sample × laboratory interaction. It is planned to test the interaction effect at the 5% significance level with power of at least 80%. An error mean square estimate of 1.22 is available from past QA studies.

To apply **power analysis** in FDs, we implement the same procedure as shown in Sections 3.6.1 and 3.6.2 with ϕ re-defined as

$$\phi = \sqrt{\frac{n' ES^2}{2k'}} \tag{5.9}$$

where n' is used to define the number of observations to be recorded at each tested "level", $ES = \delta/\sigma$ refers to **effect size** based on σ^2 defining the error mean square estimate, and k' refers to the number of "levels" of the effect being examined. This expression is a general one appropriate for all three testable effects in an $a \times b$ factorial experiment. Further information on how k' and n' are defined for each effect and on the related degrees of freedom f_1 and f_2 for each effect is provided in Table 5.5. Replication estimation, as for sample size estimation in a CRD (see Section 3.6.2), is found by estimating n' iteratively until a satisfactory power estimate is obtained. For the interaction effect, the initial number of replications should generally be small so values of $n = 2$, 3, and 4 are appropriate starting points for replication estimation in FD planning.

Table 5.5 — Summary information for power analysis in a two factor Factorial Design

Effect	k'	n'	f_1	f_2
factor A	a	bn	$a-1$	$ab(n-1)$
factor B	b	an	$b-1$	$ab(n-1)$
interaction $A \times B$	ab	n	$(a-1)(b-1)$	$ab(n-1)$

a is the number of levels of factor A, b is the number of levels of factor B, n is the number of replications of each treatment combination

Exercise 5.5 For the planned factorial study of Example 5.2, we need to estimate the number of replicate measurements of each sample that each laboratory should perform to satisfy the specified experimental constraints.

Information: We have $a = 4$ laboratories, $b = 3$ samples, minimum detectable difference $\delta = 7.31$ ng/g for the interaction effect, significance level of 5%, error mean square estimate of $\sigma^2 = 1.22$, and power of at least 80%. From Table 5.5, we can specify that $n' = n$, $k' = 4\times3 = 12$, $f_1 = (4-1)\times(3-1) = 6$, and $f_2 = 4\times3\times(n-1) = 12(n-1)$. From the information provided, we also have $ES = 6.5/\sqrt{(1.22)} = 5.8848$.

Estimation of n: Initially, consider two replicates, i.e. $n' = n = 2$. For this case, $f_2 = 12\times(2-1) = 12$ and expression (5.9) becomes

$$\phi = \sqrt{\frac{2 \times (5.8848)^2}{2 \times 12}} = 1.70 \quad .$$

Consulting Table A.9 for $f_1 = 6$ and $f_2 = 12$, power is estimated as approximately 75.25% (average of entries), marginally too low for adequate experimentation. Increasing the number of replications to three provides $n' = n = 3$, $f_2 = 24$, and $\phi = 2.08$ resulting in a revised power estimate of approximately 97%. Based on this, the advice would be that at least three replicate measurements are necessary. □

Operating characteristic curves (Lipsey 1990), specification of all interaction and factor effects, and simple confidence interval principles (Zar 1996) could also be used for power analysis in FDs.

5.7 NON-PARAMETRIC INFERENCE FOR A TWO FACTOR FACTORIAL DESIGN

As with One Factor Designs, assuming normality for data collected within a FD structure may not be appropriate mitigating against the use of the ANOVA based F

tests (5.3) to (5.5) for assessing the statistical importance of each factorial effect. Access to a non-parametric alternative would therefore be useful to enable such data to be objectively assessed. Such a procedure exists based on an extension of the Kruskal-Wallis procedure (see Section 3.7.1) using ranks of the cell measurements for each treatment combination. The other analysis components illustrated, with the exception of diagnostic checking, can still be used as the basis of the data analysis, the only change being the form of inferential procedure used to assess general factorial effect. Again, such tests are based on testing for difference in medians.

For a two factor FD with a levels of factor A, b levels of factor B, and n replications of each treatment combination, we again have three possible tests: factor A, factor B, and the interaction $A \times B$. The non-parametric alternatives to test statistics (5.3), (5.4), and (5.5) are as follows, with order of testing as recommended in Section 5.3.3, i.e. interaction test **first**:

$$\text{factor } A\text{:}\quad H_A = \frac{12}{N(N+1)} \sum_{i=1}^{a} \frac{R_{i.}^2}{bn} - 3(N+1) \tag{5.10}$$

$$\text{factor } B\text{:}\quad H_B = \frac{12}{N(N+1)} \sum_{j=1}^{b} \frac{R_{.j}^2}{an} - 3(N+1) \tag{5.11}$$

$$\text{interaction } A \times B\text{:}\quad H_{AxB} = \frac{12}{N(N+1)} \sum_{i=1}^{a} \sum_{j=1}^{b} \frac{R_{ij}^2}{n} - 3(N+1) - H_A - H_B \tag{5.12}$$

where $N = abn$ is the total number of observations, $R_{i.} = \Sigma_i R_{ij}$ is the sum of the ranks for the observations for factor A level i, R_{ij} is the sum of the ranks for the combination of factor A level i and factor B level j, $R_{.j} = \Sigma_j R_{ij}$ is the sum of the ranks for the observations for factor B level j. All tests follow approximate χ^2 distributions with degrees of freedom $a - 1$ for factor A, $b - 1$ for factor B, and $(a - 1)(b - 1)$ for the interaction $A \times B$. Again, these tests represent only general tests of factor and interaction effects with follow-up necessary to enhance the conclusion in respect of interaction or factor level differences, whichever is appropriate.

5.8 RANDOM EFFECTS - MODEL II/MIXED MODEL EXPERIMENTS

The description of the two factor FD has been based on assuming each factor is a fixed effect (**model I experiment**). Practical uses of this design may be such that each factor could be classified as a random effect (**model II experiment**) or one factor as random and the other as fixed (**mixed model experiment**). A **random effect** occurs when the levels of a factor are randomly selected from a population of possible levels. For example, in a study involving production line operators as a factor, a large pool of operators may be available. If only three operators are to be tested, then we would require to select a sample of three operators to act as the factor levels. The operator effect would therefore be classified as a random one. When factors are classified as random, inferences concerning the factor are valid for all possible levels of the factor, not just those tested as occurs when fixed effects is the classification.

For model II and mixed model experiments, additional assumptions concerning the effects within the response model occur necessitating modifications to the interpretation of the analysis of the collected data from that illustrated for model I experiments. It is important, in these cases, to have some knowledge of the background statistical theory as many software packages either do not cater for random effects or produce a particular type of mixed model output which may not suit the investigator's feeling about the interaction effects in particular.

The form of the three statistical tests for model II (both factors random) and mixed model experiments can be constructed from the corresponding expected mean square (*EMS*) terms generated from the statistical theory underpinning experimental designs. Construction of appropriate *F* test statistics is straightforward for model I experiments as all test statistics are based on the ratio of appropriate effect mean square and the error mean square (*MSE*). For model II and mixed model experiments, it is important to check **test statistic construction** before analysing the data as the denominator mean square for effect test statistics can vary from that occurring in model I experiments.

5.8.1 Additional Effect Assumptions

The effects model for the response from experiments involving at least one random effect is the same as that shown in expression (5.1), i.e. $X_{ijk} = \mu + \alpha_i + \beta_j + \alpha\beta_{ij} + \varepsilon_{ijk}$. In addition to the usual error assumptions, it is also be necessary to include assumptions about the other model effects. Table 5.6 summarises these additional assumptions for the model II and mixed models within a two factor factorial structure. The terms σ_α^2, σ_β^2, $\sigma_{\alpha\beta}^2$, and σ^2 represent the **variance components** for factor *A*, factor *B*, the interaction $A \times B$, and the error, respectively. If a main effect is classified as a fixed effect, however, the associated variance expression is generally presented as a sum of squared effects and is not strictly a variance term. In mixed models, the interaction effect $\alpha\beta_{ij}$ containing a fixed and a random effect becomes random, i.e. mixing fixed and random effects produces a random effect.

The additional assumptions necessary are not greatly different from those underpinning most formal experimental designs. For the mixed model case (*A* fixed and *B* random for illustration purposes), two model types can be considered: the **restricted** or "standard" model and the **unrestricted** model. The difference between them lies in the assumption surrounding the interaction effect $\alpha\beta_{ij}$. For a **restricted** model, the stated assumption in Table 5.6 means that certain interaction components corresponding to different levels of the fixed factor are assumed not independent, i.e. are assumed correlated, and that interaction effects sum to zero over the levels of the fixed effect. In contrast, the interaction assumption for the **unrestricted** approach has no such restrictions. Choice of model to use depends on the experimental investigation with the restricted model best if a relationship between the effects of a fixed factor across the levels of the random factor is expected.

Most statisticians prefer the restricted model as it is considered more general than the unrestricted model though when the correlation structure of the random factors is not large, there is little to choose between them. It should be noted, however, that software packages such as SAS and Minitab fit the unrestricted mixed model by default when fixed and random effects occur in the same experiment with default modifications necessary if the restricted model is to be fitted. Hocking (1973) provides a summary of mixed models for two factor designs.

Table 5.6 — Additional model assumptions for random effects in a two factor Factorial Design: model II based on A and B random and mixed model based on A fixed and B random

	Model II	Mixed model *Restricted*	Mixed model *Unrestricted*
α_i	NIID(0, σ^2_α)	$\Sigma\alpha_i = 0$	as Restricted
β_j	NIID(0, σ^2_β)	as Model II	as Model II
$\alpha\beta_{ij}$	NIID(0, $\sigma^2_{\alpha\beta}$)	N(0, $(a-1)\ \sigma^2_{\alpha\beta}/a$), $\Sigma_i\alpha\beta_{ij} = 0$	as Model II

Mixed model based on A fixed and B random

5.8.2 EMS Expressions and Test Statistic Derivation

Theoretical underpinning for FD models is provided through the assumptions associated with the model parameters. Based on this, expected mean square (*EMS*) expressions for the four sources of variation associated with a two factor design model can be constructed. These are summarised in Table 5.7 where the σ^2 terms refer to variance components. When an effect is fixed, they correspond to the sum of squared effects and are denoted by Q[effect], e.g., $Q[\alpha] = \Sigma\alpha_i^2/(a-1)$. The expressions shown are similar in form for all model types, the only exception being the expression associated with factor B in the restricted mixed model (non-independence of certain interaction effects). In this case, the *EMS* for the random factor B includes the interaction component whereas in the restricted case, this component is absent. It is believed by some that if interaction is present, then the *EMS* expressions of the main effects should account for the presence of interactions. Rules for writing down *EMS* expressions for balanced designs with random effects can also be considered (Montgomery 1997).

Table 5.7 — EMS expressions for Model II and Mixed models in a two factor Factorial Design

Effect	Model II	Mixed model *Restricted*	Mixed model *Unrestricted*
A	$\sigma^2 + n\sigma^2_{\alpha\beta} + bn\sigma^2_\alpha$	$\sigma^2 + n\sigma^2_{\alpha\beta} + bnQ[\alpha]$	$\sigma^2 + n\sigma^2_{\alpha\beta} + bnQ[\alpha]$
B	$\sigma^2 + n\sigma^2_{\alpha\beta} + an\sigma^2_\beta$	$\sigma^2 + an\sigma^2_\beta$	$\sigma^2 + n\sigma^2_{\alpha\beta} + an\sigma^2_\beta$
$A \times B$	$\sigma^2 + n\sigma^2_{\alpha\beta}$	$\sigma^2 + n\sigma^2_{\alpha\beta}$	$\sigma^2 + n\sigma^2_{\alpha\beta}$
Error	σ^2	σ^2	σ^2

Mixed model based on A fixed and B random

Construction of effect test statistics involves taking a ratio of two effect variances provided by the *EMS* expressions in Table 5.7. The numerator of this ratio corresponds to the *EMS* expression for the effect to be tested while the denominator *EMS* corresponds either to an individual effect *EMS* expression or to a linear combination of *EMS* expressions. For the latter case, the denominator requires to contain common elements (variance terms) to that of the numerator with the only difference being the presence in the numerator of the variance component of the factor effect to be tested. This means the denominator requires to be of a form whereby the ratio of the numerator and denominator *EMS*s becomes equal to **1 + "R"** where the numerator of the term "R" contains solely the variance component corresponding to the factor to be tested. For a two factor FD structure, this means that the numerator of "R" requires to contain the σ^2_α term for factor A, the σ^2_β term for factor B, and the $\sigma^2_{\alpha\beta}$ term for the interaction $A \times B$.

To illustrate this approach, consider a two factor FD experiment to assess the effect of two factors, filters (A) and operators (B), on the amount of material lost at the filtration phase of a production process. Three filters ($a = 3$) and four operators ($b = 4$) are to be used in the study with both factors assumed randomly selected (model II experiment). Consider construction of the test statistic for the factor A effect, the filter effect. From Table 5.7, the numerator *EMS* is shown to be

$$\sigma^2 + n\sigma^2_{\alpha\beta} + 4n\sigma^2_\alpha \quad .$$

To generate the required ratio, the denominator *EMS* must be that of the interaction $A \times B$, i.e.

$$\sigma^2 + n\sigma^2_{\alpha\beta},$$

as it contains the necessary common elements for ratio construction. Expressing this as a ratio, we have

$$(\sigma^2 + n\sigma^2_{\alpha\beta} + 4n\sigma^2_\alpha)/(\sigma^2 + n\sigma^2_{\alpha\beta}) = 1 + (4n\sigma^2_\alpha)/(\sigma^2 + n\sigma^2_{\alpha\beta}) = 1 + \text{“R”}$$

so providing the necessary "R" expression. By similar reasoning, the test statistics for the operator (B) and filter × operator ($A \times B$) interaction effects can be constructed, as summarised in Table 5.8. It can be clearly seen in all cases where random effects occur that the interaction test statistic is always constructed similarly to that for a model I experiment. Test statistic formats for the two types of mixed model are included in Table 5.8 for completeness.

Table 5.8 — Test statistics for a two factor Factorial Design: model I based on both factors fixed, model II based on both factors random, and mixed model based on filter fixed and operator random

Effect	Model I	Model II	Mixed model *Restricted*	Mixed model *Unrestricted*
Filter (A)	*MSA/MSE*	*MSA/MSAB*	*MSA/MSAB*	*MSA/MSAB*
Operator (B)	*MSB/MSE*	*MSB/MSAB*	*MSB/MSE*	*MSB/MSAB*
Interaction ($A \times B$)	*MSAB/MSE*	*MSAB/MSE*	*MSAB/MSE*	*MSAB/MSE*

Hypotheses formation also changes to suit the change in nature of factor effects. Instead of being expressed in terms of effects, they are now expressed in terms of variance components, i.e. H_0: $\sigma^2_{\text{effect}} = 0$ and H_1: $\sigma^2_{\text{effect}} > 0$. Essentially, these specifications mean the same as occurred for fixed effects (see Section 5.3.1). They reflect that the treatments tested represent a random sample of levels and so interest lies more in assessing the variation in response due to these levels than in estimating the magnitude of their influence.

5.8.3 Analysis Components for Models with Random Effects

If significant factor and/or interaction effects are found for any of the requisite random effects, interest lies in the estimation and interpretation of the appropriate main effect, interaction, and error variance components, i.e. the σ^2 terms. Solution for these elements involves solving the *EMS* expressions which essentially correspond to a sequence of simultaneous equations in a number of unknowns. Often, however, a negative variance estimate can result. In such cases, it is assumed that the effect in question has negligible effect on response variability which is generally backed-up by an insignificant F test result.

The variance of any observation in a model II experiment is expressed as

$$\text{Var}(X_{ijk}) = \sigma^2_\alpha + \sigma^2_\beta + \sigma^2_{\alpha\beta} + \sigma^2 \tag{5.13}$$

while for a mixed model experiment with A as a fixed effect, it is expressed as

$$\text{Var}(X_{ijk}) = \sigma^2_\beta + \sigma^2_{\alpha\beta} + \sigma^2 \tag{5.14}$$

Based on these variance estimates, the contribution of each effect to the variability of the response can be gauged using the following simple steps. First, total the corresponding variance component estimates according to expression (5.13) or (5.14). Secondly, express each variance estimate as a percentage of the total, given by $\text{Var}(X_{ijk})$, to provide a measure of the relative importance of each random factor within the explanation of what affects the variability in the experimental data. The resultant percentages are then interpreted to see where the significant influences on response data variability are coming from.

It is also possible to construct confidence intervals for these estimates using similar principles to that for construction of a confidence interval for the variance in a one sample study. For a significant fixed factor within a mixed model experiment, we can still apply multiple comparisons to distinguish any differences between factor levels. The mean square term to be used in the derivation of the critical values of the multiple comparison procedure requires to change compared to that used in the illustrations provided in Chapters 3 and 4. For a fixed effects experiment, this term corresponds to the mean square for error (*MSE*) whereas in a mixed model experiment, this term must be the divisor mean square involved in the construction of the associated factor test statistic.

5.8.4 Power Analysis for Random Effects

To determine the power of a test for a factor classified as a **random effect** in model II or mixed model experiments, we require to estimate power through the probability expression $P\left(F > [F_{\alpha,f_1,f_2}]/c\right)$ using the F distribution. The term α refers to the significance level of the proposed statistical test, f_1 defines the degrees of freedom of the numerator mean square of the associated test statistic, f_2 is the denominator degrees of freedom, and c is as summarised in Table 5.9. Power approaches for random effects requires specification by the investigator of likely variability levels for the random effects, the σ^2 terms, unlike the fixed effects approach of specification of minimum detectable difference. Reversal of this approach can be considered if sample size estimation is necessary. For more details on this concept, refer to Montgomery (1997) and Winer *et al.* (1991).

Table 5.9 — Formulae for c terms in power calculation for random effects in two factor Factorial Designs

Effect	f_1	c
A	$a-1$	$1+(bn\sigma^2_\alpha)/(\sigma^2+n\sigma^2_{\alpha\beta})$
B	$b-1$	$1+(an\sigma^2_\beta)/(\sigma^2+n\sigma^2_{\alpha\beta})$
$A\times B$	$(a-1)(b-1)$	$1+(n\sigma^2_{\alpha\beta})/\sigma^2$

5.9 UNBALANCED TWO FACTOR FACTORIAL DESIGN

In some FDs, it may not be possible to have equal amounts of replication at each treatment combination due to experimental constraints or due to loss of some measurements at certain combinations. Table 5.10 illustrates just such a case where there are unequal numbers of replicates at certain treatment combinations with the design now referred to as **unbalanced**. The response model for an unbalanced design structure still complies with the two factor FD response model (5.1).

Table 5.10 — Illustration of an unbalanced Factorial Design

		Replications		
Factor A	Factor B	1	2	3
Level 1	Level 1	x	x	x
	Level 2	x	x	
	Level 3	x	x	x
Level 2	Level 1	x	x	x
	Level 2	x	x	x
	Level 3	x	x	
Level 3	Level 1	x	x	x
	Level 2	x	x	x
	Level 3	x	x	x

x denotes a recorded observation

Unbalanced FDs have to be handled in a different way to their balanced counterparts. Instead of using balanced ANOVA procedures, the **General Linear Model** (GLM) estimation procedure is required to be used to provide the mechanism for derivation of the ANOVA table and related test statistics. GLM is a regression based estimation procedure for unbalanced and non-hierarchical designs. Additionally, ANOVA principles for unbalanced FDs utilise special sum of squares terms called **Type I**, **Type II**, **Type III**, and **Type IV**. Depending on the experimental objectives, one of these *SS* types should be used as the basis of the statistical test component of the analysis. Choice of sum of squares type to use is important as each, with the exception of Type III, examines a marginally different model hypotheses structure from that associated with a comparable balanced design. Yates (1934) was first to extensively discuss unbalanced factorial structures.

5.9.1 Type I to IV Sums of Squares

In unbalanced FDs, the order in which the terms are fitted to the response model generally leads to different sums of squares for the main effects in particular. Table 5.11 contains a summary of the mechanisms for sum of squares (*SS*) determination for each of the four cases. Explanations of these procedures are provided in the following sections.

Table 5.11 — Type I to III sums of squares determination for an unbalanced two factor Factorial Design

Effect	Type I	Type II	Type III
A	A	$A \mid B$	$A \mid B, AB$
B	$B \mid A$	$B \mid A$	$B \mid A, AB$
$A \times B$	$AB \mid A, B$	$AB \mid A, B$	$AB \mid A, B$

where | denotes conditional on

5.9.1.1 Type I SSs

Type I *SS*s are obtained by sequential fit of the model terms and so each effect is adjusted for effects specified earlier in the model. This technique can be referred to as either the **method for proportional cell sizes**, the **method of sequential sum of squares**, or the **method of weighted marginal means**. Typically, A is fitted first followed by B followed by $A \times B$. In effect, this means that the sum of squares for factor B is conditional on factor A (B given A fitted to model, i.e. $B \mid A$) and the sum of squares of the interaction $A \times B$ is conditional on factors A and B (AB given A and B fitted to model, i.e. $AB \mid A, B$). Type I *SS*s add to the total sum of squares (*SS*Total) for the specified response model. Type I *SS*s are useful in model building and can provide information on the influence of imbalance on the significance of model effects.

5.9.1.2 Type II SSs

Type II *SS*s, sometimes referred to as the **method of fitting constants**, are based on fitting the requisite effect after all other model terms **not** including this effect have been fitted to the model producing sum of square estimates that depend on the number of observations for each treatment combination. This approach provides sum of squares terms corrected for every other effect in the model which does not involve the effect itself so these terms are invariant to the ordering of effects in the model. For factor A, B is fitted followed by A and so the sum of squares for factor A is conditional on factor B, i.e. $A \mid B$. For factor B, Type II *SS* equals the sum of squares for factor B conditional on factor A, i.e. $B \mid A$, which is equivalent to the Type I *SS* for factor B. For the interaction $A \times B$, A and B are fitted first and so the sum of squares of the interaction $A \times B$ is again conditional on factors A and B as for balanced and Type I analysis, i.e. $AB \mid A, B$. Type II *SS*s are mainly appropriate for situations where no interaction is thought to be present and they are such that they do not sum to give *SS*Total.

5.9.1.3 Type III SSs

Type III analysis uses the principle of partial sums of squares and is sometimes called **Yates weighted squares of means technique** or **complete-squares analysis**. The sum of squares terms for all model effects are based on fitting that term into the model last, i.e. all terms in the model are adjusted for all other possible terms in the model. For effect A, for example, B and $A \times B$ are fitted followed by A and so the sum of squares for factor A is conditional on factor B and the interaction $A \times B$, i.e. $A \mid B, AB$. For the interaction $A \times B$, A and B are fitted first and so the sum of squares of the interaction $A \times B$ is again conditional on factors A and B as for balanced, Type I, and Type II analysis, i.e. $AB \mid A, B$. Type III *SS* differ from Type II as adjustment is made for the interaction term which may be inappropriate for main effects particularly if the interaction contains the said main effect. Type III *SS*s take no account of cell sizes making such analysis independent of design structure producing an orthogonality property though they do not add to *SS*Total. This approach is often considered the most desirable of the unbalanced approaches because it can allow for the comparison of main effects even in the presence of interaction and produces hypotheses comparable to those of a balanced design structure (see Section 5.9.2).

5.9.1.4 Type IV SSs

Type IV analysis only differs from Type III when some treatment combinations are not possible. This absence of combination information means that the response model parameters have to be restricted though Type IV sum of squares do possess a balancing property. Illustration of Type IV analysis will not be provided here as when no missing treatment combinations occur, Type III is not different to Type IV.

To illustrate these approaches in practice, consider a two factor design set up to investigate the effect on limb pressure when animals suffering from lameness in an injured limb use the injured limb after receiving, for a given time, one of three treatments at one of three intensity levels. Table 5.12 illustrates the collected data. Unfortunately, one of the animals developed complications during the study and had to be withdrawn. This loss has resulted in unequal replication in one treatment combination so the design is unbalanced.

Table 5.12 — Pressure measurements for animal treatment study with one missing value

Treatment	Intensity of treatment	Pressure measurements		
A	low	28	28	36
	medium	31	36	33
	high	32	31	34
B	low	26	31	
	medium	27	31	32
	high	28	33	35
C	low	24	31	36
	medium	33	35	37
	high	32	29	36

Table 5.13 contains the calculated sums of squares and test statistic values for the unbalanced animal study data presented in Table 5.12. These figures confirm that determination of the sum of squares of the interaction $A \times B$ is identical for all three procedures and that only determination of main effect *SS*s change with approach. All test statistics are based on the ratio *MS*Effect/*MSE* where *MSE* refers to the error mean square for the fully fitted model. Only the intensity of treatment effect appears significant as may be expected though Type III analysis has results closer to non-significance, $p = 0.0448$ compared to $p = 0.0364$ for Type II. Therefore, analysis type can influence statistical conclusions reached, as can the nature of the imbalance.

Table 5.13 — Calculated *SS* and *F* test for unbalanced animal treatment study illustration

		Type I		Type II		Type III	
Effect	*df*	*SS*	*F*	*SS*	*F*	*SS*	*F*
Treatment	2	21.787	1.04	27.298	1.30	24.906	1.19
Intensity	2	84.784	4.05	84.784	4.05	78.456	3.75
Interaction	4	25.480	0.61	25.480	0.61	25.480	0.61

where *errordf* = 17 and *MSE* = 10.461

In software, SAS routinely provides Type I and Type III analysis through PROC GLM with Type II and Type IV available by modifying the MODEL statement. Minitab provides Type I and III analysis through its GLM procedure under the columns 'Seq SS' and 'Adj SS', respectively. With GLIM, only Type I is routinely available using the glm command. For Minitab and GLIM, full analysis for all approaches can be obtained by modifying model order and using the resultant sums of squares values. In the literature, it is unclear which approach is best. Most SAS users in industry favour Type III analysis because it is preferred by regulatory agencies, whereas many researchers who seek a predictive model often adopt Type I. There has also been considerable debate in support of Type II.

5.9.2 Type I to III Hypotheses

In unbalanced FDs, it is important to realise that the Type I to III approaches are generally testing different hypotheses which may reflect different objectives from those specified prior to experimentation. Understanding of how these hypotheses are derived and how they can be interpreted is vital if an investigator is to formulate appropriate hypotheses and choose a corresponding *SS* type for their analysis. Aspects of this part of unbalanced analysis will now be explained beginning first with the hypotheses for a balanced case.

In the two factor factorial, assessment of both row and column factors is of equal interest. Specifically, interest lies in testing hypotheses whether factor effects interact or operate independently. Consider the unbalanced limb pressure study, the data for which are presented in Table 5.12, and suppose that all cells have equal numbers of observations, i.e. the design is balanced. Null hypotheses, in means and contrast forms, for this balanced case are shown in Table 5.14. They show that, in means form, each essentially represents equality of means while in contrast form, the specification indicates that several contrasts are possible with each representing a different combination of treatment means.

Table 5.14 — Null hypotheses for a 3 × 3 balanced Factorial Design

Effect	Null hypothesis
Treatment	$\mu_{1.} = \mu_{2.} = \mu_{3.} \Leftrightarrow \mu_{11} + \mu_{12} + \mu_{13} - \mu_{31} - \mu_{32} - \mu_{33} = \mu_{1.} - \mu_{3.} = 0$ and $\mu_{21} + \mu_{22} + \mu_{23} - \mu_{31} - \mu_{32} - \mu_{33} = \mu_{2.} - \mu_{3.} = 0$
Intensity	$\mu_{.1} = \mu_{.2} = \mu_{.3} \Leftrightarrow \mu_{11} + \mu_{21} + \mu_{31} - \mu_{13} - \mu_{23} - \mu_{33} = \mu_{.1} - \mu_{.3} = 0$ and $\mu_{12} + \mu_{22} + \mu_{32} - \mu_{13} - \mu_{23} - \mu_{33} = \mu_{.2} - \mu_{.3} = 0$
Interaction	$\mu_{11} = \mu_{12} = \ldots = \mu_{32} = \mu_{33} \Leftrightarrow \mu_{11} - \mu_{13} - \mu_{31} + \mu_{33} = 0$, $\mu_{12} - \mu_{13} - \mu_{32} + \mu_{33} = 0$, $\mu_{21} - \mu_{23} - \mu_{31} + \mu_{33} = 0$, and $\mu_{22} - \mu_{23} - \mu_{32} + \mu_{33} = 0$

The presented hypotheses in Table 5.14 are orthogonal, i.e. each hypothesis is uncontaminated by other factor effects, which is an appealing feature since the conclusions reached for each hypothesis test will be free of ambiguity. In unbalanced designs, this property of orthogonality is lost and drawing relevant conclusions becomes a more difficult task. The Type I to III hypotheses associated with the illustrated unbalanced animal pressure experiment which follow were derived in SAS using the estimable function approach.

5.9.2.1 Type I Hypotheses

Hypotheses associated with Type I analysis are functions of cell count and depend on the order of fitting effects to the response model (5.1) resulting in them differing from the balanced equivalents. Type I hypotheses are uncontaminated by preceding model effects but are contaminated by succeeding model effects.

For the treatment factor A, the two contrasts specifying the null hypothesis are

$$\mu_{1.} - \mu_{3.} = 0$$
$$6\mu_{21} + 9\mu_{22} + 9\mu_{23} - 8\mu_{31} - 8\mu_{32} - 8\mu_{33} = 0\,.$$

The former occurred in the balanced case (see Table 5.14) and can be justified for the unbalanced experiment because the contrast of treatment A (level 1) versus treatment C (level 3) is balanced as each respective level contains the same number of observations. The second contrast is less easy to interpret but still measures the treatment B (level 2) versus treatment C (level 3) comparison as in the balanced case. For this contrast, the coefficients, though appearing difficult to interpret, are essentially in cell count ratio. For the $\mu_{2.}$ terms, the ratio is 2:3:3 reflecting the cell counts for treatment B (level 2). The coefficients for the $\mu_{2.}$ terms add to 24 and so the coefficients of the $\mu_{3.}$ terms must add to –24 to ensure the contrast is orthogonal. Given that the counts for treatment C (level 3) are balanced, the coefficients of the $\mu_{3.}$ terms must appear in ratio 1:1:1 and so to ensure addition to –24, these coefficients must be of value –8. Hence, the form of the second contrast shown above.

For the intensity factor B, the associated contrasts are

$$24\mu_{11} - \mu_{12} - 23\mu_{13} + 18\mu_{21} + 2\mu_{22} - 20\mu_{23} + 24\mu_{31} - \mu_{32} - 23\mu_{33} = 0$$
$$\mu_{.2} - \mu_{.3} = 0\,.$$

The latter reflects the medium (level 2) and high (level 3) comparison just as occurred for the balanced case and occurs here because the cells being compared contain equal cell counts. The first contrast, though looking non-interpretable, is still a measure of the low (level 1) and high (level 3) comparison. In this contrast, greater weighting is given to the underpinning components of the contrast, the terms μ_{11}, μ_{13}, μ_{21}, μ_{23}, μ_{31}, and μ_{33}, while considerably less weighting is given to the terms μ_{12}, μ_{22}, and μ_{32} which are unimportant for the specified contrast.

For the interaction $A \times B$, the hypothesis contrasts mirror those shown for the balanced case in Table 5.14 as is expected since the sum of squares for the interaction $A \times B$ in the balanced case is identical to each of the Type I, Type II, and Type III sums of squares for this effect as highlighted in Section 5.9.1.

5.9.2.2 Type II Hypotheses

Type II hypotheses are invariant to effect order but are again different from the balanced case. They are functions of the cell counts and are uncontaminated by succeeding effects unlike Type I hypotheses.

For the treatment factor A, the contrasts specifying the null hypothesis are

$$\mu_{1.} - \mu_{3.} = 0$$
$$2\mu_{11} - \mu_{12} - \mu_{13} + 18\mu_{21} + 24\mu_{22} + 24\mu_{23} - 20\mu_{31} - 23\mu_{32} - 23\mu_{33} = 0\,.$$

For the first case, simplicity of contrast format occurs for the same reasons as explained for Type I hypotheses. For the second specified contrast, greater weighting is given to the underpinning components of the contrast, the terms μ_{21}, μ_{22}, μ_{23}, μ_{31}, μ_{32}, and μ_{33}, while considerably less weighting is given to the terms μ_{11}, μ_{12}, and μ_{13} which are unimportant for the specified contrast. As for factor B Type I, this contrast, though based on various coefficients, still measures the treatment B (level 2) versus treatment C (level 3) comparison as occurred in the balanced case.

For the intensity factor B, the contrasts within the null hypothesis mirror those presented in Section 5.9.2.1 for Type I as Type I *SS* equals Type II *SS* for this factor. For the interaction $A \times B$, the hypothesis contrasts again mirror those shown for the balanced case in Table 5.14 through the relationship that Type I *SS* equals Type II *SS* for this effect.

5.9.2.3 Type III Hypotheses

These hypotheses do not depend on cell counts, only on the cells in which observations occur. They are of the same form as those for the balanced case and are also invariant to effect ordering. This similarity of hypotheses with the balanced case make this approach appear most desirable. Using the estimable function approach for the illustrated unbalanced design produces the same contrast hypotheses as specified for the 3 × 3 balanced design (see Table 5.14).

Contrasts for effect null hypotheses in unbalanced designs can vary considerably in their format and associated interpretation depending, as they do, on the amount and position of the loss within the design structure. Even with only one missing observation, we can see that the hypotheses expressions need not appear in a readily interpretable form. Obviously, the greater the loss the more complex the hypotheses structure and the harder this becomes to relate to the practical aspect of an investigation. Fuller details of unbalanced designs can be found in Milliken & Johnson (1992) and Shaw & Mitchell-Olds (1993).

5.10 THREE FACTOR FACTORIAL DESIGN WITH *n* REPLICATIONS PER CELL

Two factor FDs can be extended to investigate three or more factors. For example, the strength of a nickel-titanium alloy may be influenced by three factors: the raw material used as the basis of alloy production (factor A), the nickel composition (factor B), and the titanium composition (factor C). Each factor could be tested at different levels to enable a full overview of how the factors and their interactions affect the strength of the manufactured alloy in order to assess if different factor combinations result in different strength measurements (interaction effects present). With this knowledge, the manufacturer could improve their understanding of the manufacturing process in order to produce a better quality product.

For such an experiment involving three factors A, B, and C, we would set each factor at levels a, b, and c respectively providing abc treatment combinations, and carry out n replicate determinations of the experimental response at each combination. Seven effects can be estimated and assessed within such an experimental structure. These are the main effects A, B, and C, the second order interactions $A \times B$, $A \times C$, and $B \times C$, and the third order interaction $A \times B \times C$, i.e. seven possible effects.

Example 5.3 A marketing analyst wished to gauge public response to a product, based on daily sales, prior to the start of a new national advertising campaign. The analyst decided to use a factorial structure to compare the effects of three different factors on sales. The factors chosen were week of purchase (three levels), store (three levels), and price (two levels). Monitoring of sales is scheduled to take place over three randomly selected weeks within a three month period. The stores, in which the sales are to be monitored, are to be randomly selected from all the stores taking part in the test marketing. The price factor is to be set at two levels based on the prices charged by rival, comparable products. The volume of sales is to be recorded on three occasions each week in each store at each price category.

Example 5.4 The results from two replicates of a complete factorial experiment to determine the effects of three factors on the length of a tablet pack made from aluminium foil are displayed in Table 5.15. The measurements refer to the excess length, in 10^{-3} cm, of a pack. The factors considered were operators (O) at levels O1, O2, and O3; machines (M) at levels M1, M2, and M3; and foil suppliers (S) at levels S1, S2, and S3. The operators were randomly selected from the production site's pool of operators. For this production facility, three machines can be used and three suppliers provide the foil material.

Table 5.15 — Excess length measurements for tested foil packs

		Supplier					
Operator	Machine	S1		S2		S3	
O1	M1	3	2	4	6	6	8
	M2	1	3	7	5	3	5
	M3	4	3	8	7	10	12
02	M1	2	4	6	4	10	8

(continued)

(continued)

	M2	1	6	7	6	9	3
	M3	2	3	3	5	8	11
O3	M1	5	4	4	6	7	5
	M2	2	3	6	3	3	5
	M3	3	5	3	5	7	6

5.10.1 Model for the Measured Response

The model for an experimental observation within a three factor factorial is given by

$$X_{ijkl} = \mu_{ijk} + \varepsilon_{ijkl} = \mu + \alpha_i + \beta_j + \alpha\beta_{ij} + \gamma_k + \alpha\gamma_{ik} + \beta\gamma_{jk} + \alpha\beta\gamma_{ijk} + \varepsilon_{ijkl} \qquad (5.15)$$

where $i = 1, 2,..., a$, $j = 1, 2,..., b$, $k = 1, 2,..., c$, and $l = 1, 2,..., n$. The first part $\mu_{ijk} + \varepsilon_{ijkl}$ defines the **means model** and the second part the **effects model**. α_i defines the effect of factor A level i on the response, β_j the factor B level j effect, $\alpha\beta_{ij}$ the $A_i \times B_j$ interaction effect, γ_k the factor C level k effect, $\alpha\gamma_{ik}$ the $A_i \times C_k$ interaction effect, $\beta\gamma_{jk}$ the $B_j \times C_k$ interaction effect, $\alpha\beta\gamma_{ijk}$ the $A_i \times B_j \times C_k$ interaction effect, and ε_{ijkl} the error effect. As previously, we can see that expression (5.15) provides a response model specification consisting of both controlled elements (factors and their interactions) and uncontrolled elements (error, unexplained variation). The **assumptions** associated with this response model are as defined for all previous experimental designs, i.e. equality of response variability across all factors and normality of the measured response. Additional effect assumptions may be necessary depending on the nature of the effects to be tested, i.e. fixed or random.

5.10.2 ANOVA Principle and Test Statistics

Again, the **ANOVA principle** underpins the statistical analysis aspects of a three factor FD through its provision for test statistic construction and derivation based on response model specification and the underpinning statistical theory. The associated procedures mirror and extend those shown previously for other formal design structures.

5.10.2.1 Hypotheses

The **hypotheses** for the seven testable effects all specify a test of absence of effect (H_0) versus presence of effect (H_1). In other words, all tests assess whether the effect has or has not an influence on the response measured.

5.10.2.2 ANOVA Table

The underlying ANOVA principle provides the partitioning of response variation into its controlled and uncontrolled elements from which the general sum of squares expression of

$$SST = SSA + SSB + SSAB + SSC + SSAC + SSBC + SSABC + SSE \qquad (5.16)$$

emerges. This expression underpins the general **ANOVA table** illustrated in Table 5.16 where all aspects of the table are as previously described.

Table 5.16 — ANOVA table for a three factor Factorial Design with n replicates of each treatment combination

Source	df	SS	MS
A	$a - 1$	SSA	$MSA = SSA/df$
B	$b - 1$	SSB	$MSB = SSB/df$
$A \times B$	$(a - 1)(b - 1)$	$SSAB$	$MSAB = SSAB/df$
C	$c - 1$	SSC	$MSC = SSC/df$
$A \times C$	$(a - 1)(c - 1)$	$SSAC$	$MSAC = SSAC/df$
$B \times C$	$(b - 1)(c - 1)$	$SSBC$	$MSBC = SSBC/df$
$A \times B \times C$	$(a - 1)(b - 1)(c - 1)$	$SSABC$	$MSABC = SSABC/df$
Error	$abc(n - 1)$	SSE	$MSE = SSE/df$
Total	$abcn - 1$	SST	

5.10.2.3 Test Statistics

The test statistics for the seven possible effects are constructed, as before, as the ratio of two mean square (*MS*), or variance, terms. When all factors are classified as fixed effects (model I experiment), this corresponds to all test statistics being derived as the ratio *MS*Effect/*MSE* as shown in Table 5.17.

Table 5.17 — Test statistics for a fixed effects three factor Factorial Design

Source	MS	F
A	MSA	MSA/MSE
B	MSB	MSB/MSE
$A \times B$	$MSAB$	$MSAB/MSE$
C	MSC	MSC/MSE
$A \times C$	$MSAC$	$MSAC/MSE$
$B \times C$	$MSBC$	$MSBC/MSE$
$A \times B \times C$	$MSABC$	$MSABC/MSE$

When random effects occur in this design structure, test statistic construction for certain effects can differ from that for the model I experiment. Table 5.18 summarises the *EMS* expressions for the eight effects in a three factor FD where the σ^2 terms refer to variance components. When an effect is fixed, they correspond to the sum of squared effects and are denoted by Q[effect]. With knowledge of the nature of factor effects, the *EMS* expressions in Table 5.18 can be used to generate effect mean squares which, together with the ratio derivation principle described in Section 5.8.2, can enable relevant effect test statistics to be constructed. As previously, two forms of mixed model can be considered, the **restricted** model (interaction effects assumed correlated, sum of fixed effects over the levels of random effects zero) and the **unrestricted** model (interaction effects assumed uncorrelated).

For a **model II** (all factors random) or **mixed model** (at least one factor random and one fixed) factorial experiment, an exact F test for certain main effects may not be available by direct means through the ratio of two specific factor mean squares. This lack of available test statistic can be overcome by producing a **pseudo-*F***, or **approximate-*F***, test. The construction of this form of F test uses the *EMS* ratio principle (see Section 5.8.2) of numerator *EMS* over denominator *EMS*, the former again referring to the effect to be tested. The denominator *EMS*, however, requires to be expressed as a linear combination of *EMS* expressions associated with the design effects.

Table 5.18 — *EMS expressions for a three factor Factorial Design*

Effect	*EMS*
A	$\sigma^2 + nD_bD_c\sigma^2_{\alpha\beta\gamma} + bnD_c\sigma^2_{\alpha\gamma} + cnD_b\sigma^2_{\alpha\beta} + bcn\sigma^2_{\alpha}$
B	$\sigma^2 + nD_aD_c\sigma^2_{\alpha\beta\gamma} + anD_c\sigma^2_{\beta\gamma} + cnD_a\sigma^2_{\alpha\beta} + acn\sigma^2_{\beta}$
$A \times B$	$\sigma^2 + nD_c\sigma^2_{\alpha\beta\gamma} + cn\sigma^2_{\alpha\beta}$
C	$\sigma^2 + nD_aD_b\sigma^2_{\alpha\beta\gamma} + anD_b\sigma^2_{\beta\gamma} + bnD_a\sigma^2_{\alpha\gamma} + abn\sigma^2_{\gamma}$
$A \times C$	$\sigma^2 + nD_b\sigma^2_{\alpha\beta\gamma} + bn\sigma^2_{\alpha\gamma}$
$B \times C$	$\sigma^2 + nD_a\sigma^2_{\alpha\beta\gamma} + an\sigma^2_{\beta\gamma}$
$A \times B \times C$	$\sigma^2 + n\sigma^2_{\alpha\beta\gamma}$
Error	σ^2

D is 0 for a fixed effect and 1 for a random effect for the restricted mixed model, D is 1 for the unrestricted mixed model

For example, if A is fixed and B and C are random (restricted mixed model experiment, $D_a = 0$, $D_b = 1$, $D_c = 1$), then

$$F' = MSA/(MSAB + MSAC - MSABC) \tag{5.17}$$

provides a pseudo-F test for testing the statistical significance of the fixed factor A. It is therefore important in three factor FDs, if factor levels are chosen randomly, to use the readily accessible *EMS* expressions for model effects to derive the format of effect test statistics to ensure that derivation of each is properly understood and software output complies with the necessary format. Software can also be used to generate *EMS* expressions, as the information in Appendix 5A.2 demonstrates.

Exercise 5.6 Referring to the marketing study of Example 5.3, we want to show how to generate the effect test statistics for the planned study.

Information: The proposed structure is based on a 3 × 3 × 2 factorial with three replicates. Thus, $a = 3$, $b = 3$, $c = 2$, and $n = 3$. We are told that the week of monitoring (W) and store for the monitoring (S) are both to be randomly selected and so both of these factors can be assumed to be random effects. As price (P) is only to be considered at two levels, we can assume this to be a fixed effect. As both fixed and random effects are occurring, we can say that the study will be of mixed model type. To fully illustrate the connection between *EMS*s and test statistic construction, information will be provided for both the restricted and unrestricted models.

EMS expressions: Based on the information on nature of effects and the definition of D in Table 5.18, we can state that, for the restricted model, $D_a = 1$, $D_b = 1$, and $D_c = 0$ while $D_a = 1 = D_b = D_c$ must be used for the unrestricted model. From the general *EMS* expressions of Table 5.18, the following *EMS* expressions can be derived:

Effect	*Restricted mixed model*	*Unrestricted mixed model*
W (α)	$\sigma^2 + 6\sigma^2_{\alpha\beta} + 18\sigma^2_{\alpha}$	$\sigma^2 + 3\sigma^2_{\alpha\beta\gamma} + 9\sigma^2_{\alpha\gamma} + 6\sigma^2_{\alpha\beta} + 18\sigma^2_{\alpha}$
S (β)	$\sigma^2 + 6\sigma^2_{\alpha\beta} + 18\sigma^2_{\beta}$	$\sigma^2 + 3\sigma^2_{\alpha\beta\gamma} + 9\sigma^2_{\beta\gamma} + 6\sigma^2_{\alpha\beta} + 18\sigma^2_{\beta}$
W×S	$\sigma^2 + 6\sigma^2_{\alpha\beta}$	$\sigma^2 + 3\sigma^2_{\alpha\beta\gamma} + 6\sigma^2_{\alpha\beta}$
P (γ)	$\sigma^2 + 3\sigma^2_{\alpha\beta\gamma} + 9\sigma^2_{\beta\gamma} + 9\sigma^2_{\alpha\gamma} + 27Q[\gamma]$	$\sigma^2 + 3\sigma^2_{\alpha\beta\gamma} + 9\sigma^2_{\beta\gamma} + 9\sigma^2_{\alpha\gamma} + 27Q[\gamma]$
W×P	$\sigma^2 + 3\sigma^2_{\alpha\beta\gamma} + 9\sigma^2_{\alpha\gamma}$	$\sigma^2 + 3\sigma^2_{\alpha\beta\gamma} + 9\sigma^2_{\alpha\gamma}$
S×P	$\sigma^2 + 3\sigma^2_{\alpha\beta\gamma} + 9\sigma^2_{\beta\gamma}$	$\sigma^2 + 3\sigma^2_{\alpha\beta\gamma} + 9\sigma^2_{\beta\gamma}$
W×S×P	$\sigma^2 + 3\sigma^2_{\alpha\beta\gamma}$	$\sigma^2 + 3\sigma^2_{\alpha\beta\gamma}$
Error	σ^2	σ^2

where $Q[\gamma]$ refers to the price effect being fixed and that its associated variance term is expressed as the sum of squared effects.

Test statistics: To generate the necessary test statistics, we must use the 1 + "R" principle explained in Section 5.8.2. Based on this, we can show that the test statistics for each type of mixed model are as follows:

Effect	*Restricted mixed model*	*Unrestricted mixed model*
Week	*MS*W/*MS*WS	*MS*W/[*MS*WS + *MS*WP – *MS*WSP]
Store	*MS*S/*MS*WS	*MS*S/[*MS*WS + *MS*SP – *MS*WSP]
W×S	*MS*WS/*MSE*	*MS*WS/*MS*WSP
Price	*MS*P/[*MS*WP + *MS*SP – *MS*WSP]	as restricted model
W×P	*MS*WP/*MS*WSP	as restricted model
S×P	*MS*SP/*MS*WSP	as restricted model
W×S×P	*MS*WSP/*MSE*	as restricted model

The Week test statistic for the restricted model is derived as the *EMS* ratio

$$MSW/MSWS = (\sigma^2 + 6\sigma^2_{\alpha\beta} + 18\sigma^2_{\alpha})/(\sigma^2 + 6\sigma^2_{\alpha\beta}) = 1 + (18\sigma^2_{\alpha})/(\sigma^2 + 6\sigma^2_{\alpha\beta}) = 1 + \text{"R"}.$$

By similar reasoning, the other effect test statistics for both mixed model types can be constructed. The test statistic formats shown illustrate the differing consequences the choice of model can have when both fixed and random effects occur causing, in particular, modification to test statistics for random effects. Greater use of pseudo-*F* tests are required for the unrestricted model. □

Again, we must order the inferential tests to be carried out. Essentially, we test the three factor interaction $A \times B \times C$ **first** with the other effects, two factor interactions and main effects, tested as appropriate depending on the outcome to the tests of higher order effects. For example, if the $A \times B \times C$ interaction is not significant, we would test all the two factor interactions. Suppose interactions $A \times B$ and $A \times C$ were found to be significant. We would not then carry out the main effect tests for A, B, and C as the significance of all these factors has been adequately shown through the significance of the specified two factor interactions. However, should only the $A \times B$ interaction be significant, we should carry out the main effect test for C. At the end of any such analysis, however, we must be sure that all effects have been tested statistically and those found significant, analysed fully. Follow-up analysis procedures are as those outlined for the two factor case with variability estimation included if random effects are specified for any or all of the factors.

5.10.3 Overview of Data Analysis for Three Factor Factorial Designs

Analysis of data from three factor FDs is based on extending the methods outlined previously (model, exploratory analysis, inferential analysis, follow-up analysis, diagnostic checking). As with the two factor FD, the amount of data analysis depends on the nature of the factors and which effects are significant. If the three factor interaction is significant in a model I experiment, it is advisable to plot this as a two factor interaction at each level of the third factor and assess each interaction plot in itself and in comparison with the plots for the other factor levels. If the three factor interaction is not significant, we would then assess the two factor interactions using the same principles developed for the two factor design. The basic rule in these designs is to ensure that conclusions on the influence of all factors and their interactions are forthcoming through appropriate techniques (formal tests, interaction plots, multiple comparisons, variability analysis).

Exercise 5.7 For the length of tablet pack study outlined in Example 5.4, we want to analyse the length data to assess what influences the three tested factors have.

Response model: Based on the factors information, we can deduce that one factor is random (operator) and two are classified as fixed (machines and suppliers). This experiment is therefore a mixed model. The investigator was prepared to assume that the interaction effects are correlated and that they sum to zero over the machines and suppliers tested. From these assumptions, we can see that a restricted mixed model is to be specified.

As a FD structure is being utilised, we specify the response model as

$$\text{excess length} = \mu + \alpha_i + \beta_j + \alpha\beta_{ij} + \gamma_k + \alpha\gamma_{ik} + \beta\gamma_{jk} + \alpha\beta\gamma_{ijk} + \varepsilon_{ijkl}\,.$$

α_i is the contribution of *i*th operator to excess length, β_j the contribution of the *j*th machine effect, $\alpha\beta_{ij}$ the interaction effect of *i*th operator and *j*th machine, and γ_k the contribution of *k*th supplier with the interaction terms involving γ similarly defined. We additionally assume that the excess length response has equal variability across all factor levels and is normally distributed.

EMS expressions and test statistics: Because one factor is classified as random, before we consider any data analysis, we must first check the construction and format of the test statistics. From the factor information and the knowledge that a restricted mixed model is specified, we have that $D_a = 1$, $D_b = 0$, $D_c = 0$, $a = 3 = b = c$, and $n = 2$. Therefore, from the *EMS* expressions shown in Table 5.18, we have

Effect	*EMS expressions*	*Test statistics*
Operator (α)	$\sigma^2 + 18\sigma^2_\alpha$	*MSO*/*MSE*
Machine (β)	$\sigma^2 + 6\sigma^2_{\alpha\beta} + 18Q[\beta]$	*MSM*/*MSOM*
O×M	$\sigma^2 + 6\sigma^2_{\alpha\beta}$	*MSOM*/*MSE*
Supplier (γ)	$\sigma^2 + 6\sigma^2_{\alpha\gamma} + 18Q[\gamma]$	*MSS*/*MSOS*
O×S	$\sigma^2 + 6\sigma^2_{\alpha\gamma}$	*MSOS*/*MSE*
M×S	$\sigma^2 + 2\sigma^2_{\alpha\beta\gamma} + 6Q[\beta\gamma]$	*MSMS*/*MSOMS*
O×M×S	$\sigma^2 + 2\sigma^2_{\alpha\beta\gamma}$	*MSOMS*/*MSE*
Error	σ^2	

where $Q[.]$ refers to the sum of squared effects.

Hypotheses: The hypotheses for the seven effects associated with this experiment are all based on the test of no effect on excess length response (H_0) against effect on excess length response (H_1) either in terms of effects or variance depending on the nature of the corresponding factors.

EDA: Initial plots and summaries of the excess length data with respect to each factor, though not shown, provided some interesting results. The excess length data were widely variable for all factors and all levels. Operator O3 provided the lowest measurements while machine effect appeared quadratic with minimum for machine M2. Suppliers differed most with supplier S1 being associated with the shortest length of tablet pack and supplier S3 very much higher. Initially, it would appear that the major effect on excess length of tablet pack appears to be the raw material provided by the company's suppliers.

Statistical tests: The Minitab generated ANOVA table using the menu **Stat ➢ ANOVA ➢ Balanced ANOVA** option is presented in Display 5.6 based on fitting a three factor factorial mixed model of restricted type to the excess length response data. The *EMS* expressions printed in Display 5.6 have been generated using the restricted model and agree with those presented.

Display 5.6

```
Source                 DF        SS        MS       F       P
Oper                    2     8.926     4.463    1.64   0.213
Mach                    2    20.481    10.241    1.74   0.286
Suppl                   2   136.704    68.352   10.95   0.024
Oper*Mach               4    23.519     5.880    2.16   0.101
Oper*Suppl              4    24.963     6.241    2.29   0.085
Mach*Suppl              4    40.074    10.019   10.11   0.003
Oper*Mach*Suppl         8     7.926     0.991    0.36   0.930
Error                  27    73.500     2.722
Total                  53   336.093

Source               Variance Error Expected Mean Square
                    component  term (using restricted model)
 1 Oper               0.09671   8   (8) + 18(1)
 2 Mach                         4   (8) + 6(4) + 18Q[2]
 3 Suppl                        5   (8) + 6(5) + 18Q[3]
 4 Oper*Mach          0.52623   8   (8) + 6(4)
 5 Oper*Suppl         0.58642   8   (8) + 6(5)
 6 Mach*Suppl                   7   (8) + 2(7) + 6Q[6]
 7 Oper*Mach*Suppl   -0.86574   8   (8) + 2(7)
 8 Error              2.72222       (8)
```

Looking at the p values, we can see that the three factor interaction is statistically insignificant ($p = 0.930$). The only significant interaction at the 5% level appears to be the one between machines and suppliers with a p value of approximately 0.003 quoted. The supplier effect ($p = 0.024$) is also statistically significant but none of the random effects appear significant. The other two factor interactions, while not significant at the 5% level, do provide p values near to 10% which suggest some statistical importance for the effects in question and that the associated interactions could play important roles in respect of tablet pack length. The *SS* for the supplier effect is by far the largest suggesting it is the most important of the three factors in its influence on variation in the excess length response. We can conclude, therefore, that pack excess length appears to be primarily influenced by the machines × suppliers interaction though other interaction effects may also have some influence though to a lesser extent.

Follow-up analysis: According to the statistical evidence, only the machine × supplier interaction needs further assessment. The associated interaction plot is provided in Display 5.7 and once again the figures in brackets denote the range of the data. The interaction means are shown below the plot. The plot highlights the presence of interaction with the non-parallel lines and the non-uniform order of results for the suppliers across the machines. Suppliers S1 and S2 appear most consistent across all machines though supplier S1 provides lower excess length measurements for all machines. Supplier S3 results in more variable measurements across all machines and would appear worst in terms of quality of raw material. In summary, we can see that interaction is primarily caused by supplier S3 and that supplier S1 looks best.

Display 5.7

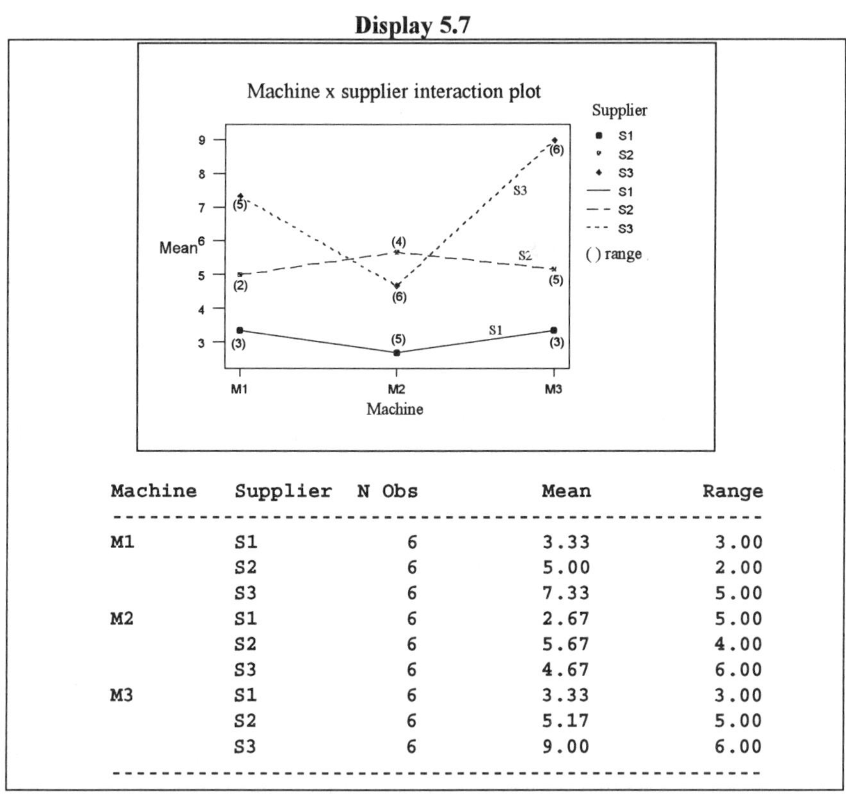

Machine	Supplier	N Obs	Mean	Range
M1	S1	6	3.33	3.00
	S2	6	5.00	2.00
	S3	6	7.33	5.00
M2	S1	6	2.67	5.00
	S2	6	5.67	4.00
	S3	6	4.67	6.00
M3	S1	6	3.33	3.00
	S2	6	5.17	5.00
	S3	6	9.00	6.00

Even though no random effects were declared significant at the 5% level, it would be useful to illustrate how to assess random effects. This involves solving the linear mean square equations in the variance components, the solutions for which are presented in Display 5.6 under the column 'Variance component', and using the variance expression

$$\mathrm{Var}(X_{ijkl}) = \sigma^2 + \sigma_\alpha^2 + \sigma_{\alpha\beta}^2 + \sigma_{\alpha\gamma}^2 + \sigma_{\alpha\beta\gamma}^2$$

to assess the relative contribution of each random effect to response variability. From Display 5.6, we have the variance components $\sigma_\alpha^2 = 0.09671$, $\sigma_{\alpha\beta}^2 = 0.52623$, $\sigma_{\alpha\gamma}^2 =$

0.58642, $\sigma^2_{\alpha\beta\gamma}$ = −0.86574, and σ^2 = 2.72222. The variance of the three factor interaction is negative so the influence of this effect on response variability can be considered negligible and so

$$\text{Var}(X_{ijkl}) = 0.09761 + 0.52623 + 0.58642 + 2.72222 = 3.93248 .$$

Expressing each random effect as a percentage of this total provides

O	O×M	O×S	O×M×S	Error
2.5%	13.4%	14.9%	0	69.2%

showing that error appears to be the greatest contributor to excess length variability and that the operator interactions appear to have similar influences. The high error percentage possibly reflects inadequate measurement accuracy or an inadequate model suggesting other factors could be exerting an influence.

Diagnostic checking: The residual plots for diagnostic checking, though not shown, resulted in distinct column length differences in all the factor plots indicative of variability differences. The fits and normal plots showed no evidence of any patterning. In light of these results, perhaps the excess length response requires to be transformed to stabilise variability if being used as a measurement of product quality in a subsequent experiment. □

Summarising the analysis of Example 5.4, we can say that suppliers appear to have the greatest influence on the level of wastage in the tablet pack production process. The only interaction to be statistically significant was that between machine and supplier ($p < 0.01$). Assessment of this interaction showed supplier S1 appeared best across all machines. Diagnostic checking suggested that response re-scaling might be worth considering for any future experiments on this production facility. In assessment of the random effects, a high error was suggested indicating that the model proposed for the excess length data may not be describing all the influences affecting response variation.

5.10.4 Pooling of Factor Effects

In some FDs, particularly when more than two factors are being investigated, certain factor interactions may be known to be unimportant or can be assumed negligible in their effect. These "negligible" effects can be combined with the error to form a new estimator for the error sum of squares (*SSE*') with modified degrees of freedom (*errordf*'). For example, if in a three factor factorial experiment the three factor interaction $A \times B \times C$ can be assumed negligible, we could combine it with the error sum of squares to form a new error sum of squares *SSE*' = *SSABC* + *SSE* with degrees of freedom *errordf*' = *ABCdf* + *errordf*. Appropriate F test statistics for the remaining model components can then be derived based on the new error mean square estimate of *MSE*' = *SSE*'/*errordf*'. This can be done by re-running the analysis excluding the interaction term. Indeed, some practitioners support the principle of combining non-significant sum of squares with the error sum of squares while others resolutely object to pooling.

5.10.5 Unbalanced Approaches

As with two factor FDs, it is possible that a three factor factorial could be based on unequal numbers of replicates though, preferably, not by design. When this occurs, we again must use one of the Type I to III approaches to produce a relevant and practical analysis of the collected data. The definitions provided in Sections 5.9.1 and 5.9.2 can be extended to the three factor case though the increase from two to three factors can greatly complicate hypotheses formulation and sum of squares evaluation. Table 5.19 provides a summary of one way of generating the Type I to III *SS* terms for a three factor unbalanced design. Again, we can see that each approach is based on a different model formulation though for the highest interaction, $A \times B \times C$, we again have similarity of result irrespective of approach. An additional similarity occurs for effect C and the interaction $B \times C$ where Type I equals Type II.

Table 5.19 — Type I to III sum of squares for an unbalanced three factor Factorial Design

Effect	Type I	Type II	Type III
A	A	$A \mid B, C, BC$	$A \mid B, AB, C, AC, BC, ABC$
B	$B \mid A$	$B \mid A, C, AC$	$B \mid A, AB, C, AC, BC, ABC$
$A \times B$	$AB \mid B, A$	$AB \mid A, B, C, AC, BC$	$AB \mid A, B, C, AC, BC, ABC$
C	$C \mid A, B, AB$	as Type I	$C \mid A, B, AB, AC, BC, ABC$
$A \times C$	$AC \mid A, B, AB, C$	$AC \mid A, B, AB, C, BC$	$AC \mid A, B, AB, C, BC, ABC$
$B \times C$	$BC \mid A, B, AB, C, AC$	as Type I	$BC \mid A, B, AB, C, AC, ABC$
$A \times B \times C$	$ABC \mid A, B, AB, C, AC, BC$	as Type I	as Type I

where | denotes conditional on

5.11 ANALYSIS OF COVARIANCE

Analysis of covariance (ANCOVA) combines both ANOVA and regression techniques. It is appropriate to consider when it may be necessary to control at least one extraneous, or confounding, variable (covariate) that may influence a measurable response, e.g. completely uniform material may not be available at the start of an experiment. Covariates can be included in a design structure at little extra cost in terms of increased experimentation and enable adjustments to be made to the response for chance differences between the groups due to the random assignment of non-uniform experimental units to the treatment groupings. ANCOVA can help to increase the power of an experiment provided a high correlation exists between the selected covariate(s) and the response enabling the final analysis to more precisely reflect the influence of the effects being tested.

PROBLEMS

5.1 A supermarket chain wished to assess how varying product position within a store affected sales of a common household product. Two factors were identified for studying: position of the product on the aisle shelving and position of the product within the aisle. Shelving is normally provided in three sections: bottom, middle, and top. Each of these levels was tested. For the aisle position factor, the product was placed either at the end nearest the checkouts or at the end furthest from the checkouts. Eighteen stores within the chain were selected for the study with each combination assessed in three randomly chosen stores to provide the replicate sales data. The number of product sales after a two week trial period were recorded as shown. Is there an interaction between shelf and position in respect of effect on sales?

Shelf position	Aisle position	Sales		
bottom	nearest	45	39	37
	furthest	40	36	47
middle	nearest	66	76	64
	furthest	37	40	28
top	nearest	53	48	58
	furthest	34	45	38

5.2 An industrial experiment was conducted to investigate the effect of machine insulation and machine speed on the noise produced (measured in decibels). The experiment was based on two levels of insulation factor and three levels of speed factor using a FD structure. Three replicate noise measurements were made for each treatment combination. These data are shown below. Carry out a full and relevant analysis of these data.

Insulation	Speed	Noise measurements		
Low	Low	5.2	4.8	4.7
	Medium	5.8	6.2	6.1
	High	7.6	7.9	7.7
High	Low	3.8	4.2	3.9
	Medium	4.6	4.9	4.8
	High	6.0	6.2	6.0

5.3 Ozonization as a secondary treatment for effluent, following absorption by ferrous chloride, was studied for three reaction times and three pH levels. The study yielded the following results for effluent decline. Do the factors work independently or dependently in their effect on effluent decline? Compare the effect of low and high pH on effluent decline. Carry out all appropriate significance tests at the 5% significance level.

Reaction time (mins)	pH	Effluent decline		
20	7.0	23	21	22
	9.0	16	18	15
	10.5	14	13	16
40	7.0	20	22	19
	9.0	14	13	12
	10.5	12	11	10
60	7.0	21	20	19
	9.0	13	12	12
	10.5	11	13	12

Reprinted from Milton, J. S. & Arnold, J. C. (1990) *Introduction to probability and statistics*. 2nd ed. McGraw-Hill, New York with the permission of The McGraw-Hill Companies.

5.4 An environmental study was conducted to assess how pH adjustment of an extraction procedure affected recovery of the triazine herbicides atrazine, propazine, and simazine from water samples. Samples containing each herbicide spiked to a concentration of 100 ng/l were processed using C_{18} cartridges with varying levels of pH adjustment. Replicate measurements of the recovery percentages for each herbicide at each level of pH adjustment were made and are summarised below. Fully analyse these data and deduce appropriate conclusions.

pH adjustment	Herbicide	Recovery			
None	Atrazine	96.3	92.8	104.4	104.7
	Propazine	84.9	82.8	94.7	94.9
	Simazine	68.4	80.2	72.2	79.5
to 5	Atrazine	84.3	88.5	84.1	80.4
	Propazine	87.9	90.3	84.3	83.5
	Simazine	86.0	86.5	81.2	77.2
to > 8	Atrazine	84.4	76.8	89.0	89.7
	Propazine	80.2	87.1	82.1	80.5
	Simazine	36.8	12.3	27.7	41.0

5.5 A manufacturer of synthetic shoe soles wished to compare the durabilities of three types of sole they manufacture together with the effect that objects of different weight have on the durability of the manufactured item. Durability tests were carried out on three combinations of sole type and weight in simulated tests, durability being measured by the time taken until the sole tested reaches a specified state of wear. The design and data collected are provided in the following table. Is durability of sole type affected by weight? Compare sole types I (lowest manufacturing cost) and III (most expensive manufacturing cost).

Sole type	Weight (kg)	Durability			
I	60	5.4	5.1	5.2	5.5
	80	4.6	4.8	4.9	5.1
	100	4.3	4.1	4.2	4.4
II	60	5.4	5.6	5.4	5.8
	80	5.0	5.4	5.2	5.1
	100	4.7	4.9	4.6	4.9
III	60	5.5	5.8	5.9	5.6
	80	5.4	5.2	5.1	5.3
	100	4.9	5.1	5.0	5.2

5.6 Primer paints are applied to surfaces of metals by different methods. In addition, the type of primer added could affect the adhesion of the primer paint. A study is to be conducted into the effect of three different primers and two application methods on the adhesion force of the manufactured paint. The engineer investigating this problem is of the opinion that a minimum detectable difference of at least 0.86 in the adhesion force would be reflective of a significant interaction effect. It is planned to carry out the associated interaction test of significance at the 5% level with power of at least 80%. It is thought six test specimens at each treatment combination would be sufficient to detect such an interaction effect. Assuming an error mean square estimate of 0.0622, assess if six test specimens are sufficient.

5.7 In industry, fermentable carbohydrates have a number of potentially important uses. Generation of these carbohydrates is often by steaming peat and collecting the residual content. A study of this process is to be initiated to determine how two factors, temperature and time, affect the level of fermentable carbohydrates released by the steaming process.

Three temperature settings and four time periods are available for use. Based on past production information, it is thought that a minimum detectable difference in the level of fermentable carbohydrates of 7.28 would be acceptable for evidence of a significant temperature $\times$ time interaction effect. The associated interaction test is planned for the 5% significance level with power at least 80%. It has been suggested that the experiment will need at least two replications per treatment combination. Using an error mean square estimate of 2.4, estimate the number of treatment combinations necessary for the planned experiment. Suppose also it was of interest to be able to test for minimum detectable difference in level of fermentable carbohydrates of 1.8 for the temperature effect. Using the replication estimate obtained for the interaction effect, estimate the power associated with this test. Is the initial advice requiring modification?

5.8 The compressive strength of metals used in the construction industry is important given the need to ensure that buildings satisfy building control regulations. A metals manufacturer wishes to examine how the four types of metal they produce and the sintering time of the manufacturing process across three standard levels affect the compressive strength of the produced materials. A minimum detectable difference of 8.6 kg cm^{-2} in the measured compressive strength is considered to be indicative of a significant interaction effect between metal type and sintering time, the test for which is planned for the 5% significance level with power at least 80%. An error mean square estimate of 3.75 is available from past production records. How many replications of each treatment combination are necessary to satisfy the design and test constraints? Comment.

5.9 An investigation is to be carried out into the effect on young rabbits of adding differing amounts of protein and carbohydrates to their basic diet which can be provided in three ways: liquid, solid, and a mixture of each. The amounts of protein and carbohydrates to be added to the diets are to be varied between three randomly chosen levels. Five rabbits are to be randomly assigned to each treatment combination. Thus, a $3 \times 3 \times 3$ FD with five replications is to be the experimental base. The response to be measured is the total weight gain, in gms, over the study period of two months. Assuming a mixed model of restricted type is to be fitted, determine the *EMS* components for this design. Construct the corresponding test statistics for each effect.

5.10 Rosy Cheeks Distillery manufacture alcohol by the action, during fermentation, of yeast on maize. The Distillery production manager believes that three specific factors have a strong influence on the amount of alcohol produced within the fermentation process, these being the temperature at which fermentation occurs, the type of yeast used, and the type of maize used. A controlled factorial experiment is to be performed to assess the effects of these three factors on the alcohol yield from the fermentation process to ascertain the best combination of factors to maximise alcohol yield. Three settings of the temperature factor are to be used based on the usual operational temperatures associated with the fermentation process. The three yeast types correspond to those used routinely in fermentation. Four maize varieties from those used within alcohol fermentation are to be randomly chosen. Two replications are to be performed at each treatment combination and the alcohol yield of each, in grams alcohol per 250 grams of liquid produced, measured. Determine the appropriate expected mean square, *EMS*, terms for the model effects associated with this design assuming an unrestricted mixed model. Show, using the *EMS* terms, how the test statistic for each effect is constructed.

5.11 The safe operation of an electronic device used in a special lamp system is contingent upon the successful performance of a thermal fuse which deactivates the unit in case of thermal overload. A three factor FD was set up to view the behaviour of the fuse under various operating conditions. The three factors tested in the experiment were start condition, ambient temperature, and line voltage. The response collected was the operating temperature of a thermocouple placed in the fuses after 10 minutes operating time. The collected temperature data, based on ten randomly selected fuses at each treatment combination, are presented below. Carry out a full analysis of these data assuming all effects are fixed.

Start condn	Ambient temp	Line voltage	Thermocouple operating temperature (°F)
Hot	75 °F	110	250.0 232.2 248.2 236.7 256.6 250.2 241.9 242.2 235.0 243.9
		120	281.8 265.3 264.6 256.1 264.2 258.6 264.9 269.2 262.0 257.0
		126	284.4 261.0 276.8 289.8 251.4 275.9 266.0 283.1 266.4 280.9
	110 °F	110	259.5 235.9 255.4 253.6 245.5 246.4 243.7 245.7 247.1 240.0
		120	285.4 251.6 268.3 277.5 249.6 268.9 251.1 269.8 262.4 273.6
		126	264.7 252.1 278.6 265.5 282.0 275.9 287.2 257.3 263.1 271.9
Cold	75 °F	110	204.5 188.2 190.2 203.1 190.5 200.5 204.1 208.6 203.7 211.6
		120	223.3 202.9 211.2 212.3 196.4 213.6 210.0 212.3 207.2 219.4
		126	233.8 226.2 227.4 227.1 225.5 227.9 226.2 225.0 226.1 218.8
	110 °F	110	226.6 219.8 234.3 232.3 223.9 231.5 235.3 240.0 226.6 228.0
		120	255.3 243.0 249.1 252.7 240.2 249.7 242.8 253.4 241.7 254.1
		126	253.9 244.2 250.4 256.4 251.1 252.8 248.7 258.3 249.2 260.0

5.12 Chromium stress and type of medium can affect the growth of aquatic micro-organisms. To assess this, a simple three factor factorial experiment was conducted using carbon broths, chromium concentration, and organism type as the three factors. The broths used were glucose, arabinose, lactose, and mannitol which are those commonly used in growth experiments involving aquatic micro-organisms. Chromium concentration, in μg ml^{-1}, was set at levels 0, 30, 60, and 90 with each broth and chromium combination tested on identical specimens from each of three randomly selected organism types. The response measured was the percentage change in conductance over a 48-hour period at a temperature of 25 °C. Change in conductance is used in impedance microbiology testing to assess growth of organisms through its relation to changes in growth levels, positive values implying organism growth and negative values growth inhibition. The data obtained were as follows based on two replicate measurements at each treatment combination:

Broth	Chromium concn	Organism type A		B		C	
Glucose	0	10.0	10.2	2.0	2.2	2.0	1.8
	30	1.5	1.7	0.5	0.4	1.5	1.7
	60	2.5	2.3	1.5	1.7	1.5	1.5
	90	4.5	4.4	2.5	2.3	2.0	2.1
Arabinose	0	−19.0	−21.0	−2.5	−2.3	−1.0	−1.4
	30	−3.0	−2.5	−2.0	−2.2	−2.0	−2.3
	60	−3.0	−2.7	−2.5	−2.3	−1.5	−1.4
	90	−4.0	−4.2	−2.0	−1.9	−0.5	−0.7

(continued)

(continued)

Lactose	0	−1.0 −0.8	5.5 5.6	−1.0 −0.8
	30	3.5 3.2	4.5 4.4	5.0 5.4
	60	2.5 2.7	4.0 3.9	4.0 3.8
	90	3.5 3.4	3.5 3.7	2.5 2.7
Mannitol	0	−3.0 −3.2	3.5 3.4	−0.5 −0.8
	30	15.0 14.7	3.2 3.5	7.0 7.3
	60	4.0 3.7	4.0 4.1	4.0 4.2
	90	2.5 2.6	4.0 4.3	1.5 1.7

Fully analyse these data to assess for presence or absence of factor interactions. Contrast the average effect of high chromium concentration (90 μg ml^{-1}) with that of low concentration (30 μg ml^{-1}).

5.13 Rosy Cheeks Distillery manufacture alcohol by the action, during fermentation, of yeast on maize. The Distillery production manager believes that the temperature at which fermentation occurs (Temp), the type of yeast used (Yeasts), and the type of maize used (Maizes) have an influence on the amount of alcohol produced within the fermentation process. A controlled experiment was performed to assess the effects of these factors on alcohol yield. The purpose of the experiment was to ascertain if a best combination of these factors for maximisation of alcohol yield existed. The temperature factor was set at three levels over the range 21°C to 31°C. The yeast types Y1, Y2 and Y3 chosen for the experiment are those generally used by the Distillery for the fermentation process and each was tested using the same quantity of yeast. Four standard types of maize referred to as MUS1, MUS2, MC1, and MC2 were used. MUS1 and MUS2 are two varieties of maize obtained from sources in the USA and MC1 and MC2 are the same two varieties obtained from sources in Canada. As with the yeast factor, the same quantity of maize was used. The design was therefore based on 36 temperature-yeast-maize combinations. Two replications were performed at each treatment combination and the alcohol yield, in grams alcohol per 250 grams of liquid produced, of each recorded. The alcohol yield data are presented below. Carry out a full and relevant analysis of these data using 5% as the significance level. The production manager was also interested in comparing the average alcohol yield for the maize varieties obtained from the USA (MUS1 and MUS2) against the average alcohol yield for the same varieties obtained from Canadian sources (MC1 and MC2).

		Maizes			
Temp	Yeasts	MUS1	MUS2	MC1	MC2
21°C	Y1	46.8 52.2	54.2 53.6	49.2 51.0	54.8 59.6
	Y2	54.4 53.8	55.0 60.8	54.8 55.4	65.2 61.4
	Y3	70.8 71.2	74.6 78.6	74.4 69.6	80.0 80.4
26°C	Y1	57.8 59.4	68.0 65.6	61.2 61.6	74.0 75.6
	Y2	69.4 71.6	77.8 83.0	66.6 71.8	84.8 86.0
	Y3	69.8 70.0	80.4 83.0	63.6 69.4	85.2 84.2
31°C	Y1	64.0 66.2	67.2 67.6	62.6 66.8	77.0 75.0
	Y2	66.6 62.2	69.2 71.6	65.4 64.6	75.8 75.0
	Y3	67.6 68.4	74.6 76.2	70.2 71.4	78.0 83.2

5A.1 Appendix: Software Information for Two Factor Factorial Designs

Data entry

Data entry for two factor FDs follows the same principles as those stated for RBDs with response as one variable, the factors as others, and codes to distinguish the factor levels.

SAS: Data entry in SAS is again achieved through use of the DATA step which is as follows for the field of vision study of Example 5.1:

```
DATA CH5.EX51;
    INPUT FLOWERA SEEDB VISFIELD @@;
    LINES;
    0.25 0.001 67   0.25 0.001 66   0.25 0.005 65   0.25 0.005 61
    0.25 0.009 62   0.25 0.009 64   2.5 0.001 68   2.5 0.001 65
    2.5 0.005 68   2.5 0.005 61   2.5 0.009 55   2.5 0.009 53   5 0.001 65
    5 0.001 64   5 0.005 62   5 0.005 63   5 0.009 49   5 0.009 47
;
PROC PRINT DATA = CH5.EX51;
    VAR VISFIELD;
    BY FLOWERA SEEDB;
RUN;
```

Minitab: Entry of response data into column C1 is generally by treatment combination. For Example 5.1, this means starting with (flower *A* 0.25 g, seed *B* 0.001 g) then (flower *A* 0.25 g, seed *B* 0.005 g) then (flower *A* 0.25 g, seed *B* 0.009 g) and so on. Codes entry into column C2 (flower *A* concentration) and column C3 (seed *B* concentration) can then conform to a pattern enabling the **Calc ➤ Make Patterned Data** option to be used to produce the requisite factor codes. The menu procedures would be similar to those described in Appendix 3A for the CRD.

For Example 5.1, either of these forms of data entry would result in the data structure displayed below.

VisField	FlowerA	SeedB	VisField	FlowerA	SeedB
67	0.25	0.001	61	2.5	0.005
66	0.25	0.001	55	2.5	0.009
65	0.25	0.005	53	2.5	0.009
61	0.25	0.005	65	5	0.001
62	0.25	0.009	64	5	0.001
64	0.25	0.009	62	5	0.005
68	2.5	0.001	63	5	0.005
65	2.5	0.001	49	5	0.009
68	2.5	0.005	47	5	0.009

Exploratory analysis

Data plots and summaries can be generated using PROC GPLOT and PROC MEANS in SAS or appropriate menu procedures in Minitab (see previous Appendices for details).

ANOVA information

As FDs are balanced, ANOVA procedures associated with balanced structures can be used to generate the ANOVA table.

SAS: The PROC ANOVA and PROC GLM procedures, listed below, perform the necessary ANOVA calculations. For the field of vision study of Example 5.1 (see Display 5.1), the statements would be as follows with statement explanation as provided in Appendix 3A:

```
PROC ANOVA DATA = CH5.EX51;
    CLASS FLOWERA SEEDB;
    MODEL VISFIELD = FLOWERA SEEDB FLOWERA*SEEDB;
PROC GLM DATA = CH5.EX51;
    CLASS FLOWERA SEEDB;
    MODEL VISFIELD = FLOWERA SEEDB FLOWERA*SEEDB;
    OUTPUT OUT = CH5.EX51 P = FITS R = RESIDS;
RUN;
```

In the MODEL statement, the entry provided could equally be written as FLOWERA | SEEDB where | is a shorthand notation for the two factor interaction model.

Minitab: The Balanced ANOVA menu again provides for ANOVA calculation. The necessary menu commands to generate ANOVA output comparable to that in Display 5.1 would be as follows:

Select **Stat** ➢ **ANOVA** ➢ **Balanced ANOVA** ➢ for *Responses*, select **VisField** and click **Select** ➢ select the **Model** box and enter **FlowerA SeedB FlowerA*SeedB**.
Select **Storage** ➢ **Fits** ➢ **Residuals** ➢ click **OK** ➢ click **OK**.

The **Model** entry could equally be written as FLOWERA | SEEDB using the | symbol as a shorthand notation for the two factor interaction model.

Interaction plot and summaries

SAS: Interaction plot derivation is not easily achieved in SAS. However, interaction summaries can be easily generated using PROC MEANS as shown in Display 5.3. The following statements were used to generate this information:

```
PROC MEANS DATA = CH5.EX51 MAXDEC = 2 MEAN RANGE;
    CLASS FLOWERA SEEDB;
    VAR VISFIELD;
RUN;
```

Use of both factors in the CLASS statement enable the summaries to be presented in interaction form.

Minitab: Plot and summary production is simple to achieve in Minitab. The necessary menu procedures are as follows:

Select **Stat** ➢ **ANOVA** ➢ **Interactions Plot** ➢ for *Factors*, enter **SeedB FlowerA** ➢ for *Source of response data*, select the **Raw response data in** box, select **VisField**, and click **Select** ➢ select the **Title** box and enter **Interaction plot for field of vision data** ➢ click **OK**.
Select **Stat** ➢ **Tables** ➢ **Cross Tabulation** ➢ for *Classification variables*, highlight **FlowerA** and **SeedB,** and click **Select**.

Select **Summaries** ➢ for *Associated variables*, select **VisField** and click **Select** ➢ for *Display*, select **Means, Minimums**, and **Maximums** ➢ click **OK**.

Select **Options** ➢ for *first*, enter **2** in the box ➢ select the **next** box and enter **0** ➢ click **OK** ➢ click **OK**.

Diagnostic checking

Production of residual plots, such as those shown in Displays 5.4 and 5.5, is as outlined in Appendix 3A for the CRD.

5A.2 Appendix: Software Information for Three Factor Factorial Designs

Data entry

Data entry for three factor FDs is a simple extension of the methods described for two factor FDs though there is now one response variable and three factors. Entry by specific treatment combinations can simplify codes entry for the factors. For Example 5.4, data entry would result in a data structure as follows:

ExLgth	Oper	Mach	Suppl	ExLgth	Oper	Mach	Suppl	ExLgth	Oper	Mach	Suppl
3	O1	M1	S1	2	O2	M1	S1	5	O3	M1	S1
2	O1	M1	S1	4	O2	M1	S1	4	O3	M1	S1
4	O1	M1	S2	6	O2	M1	S2	4	O3	M1	S2
6	O1	M1	S2	4	O2	M1	S2	6	O3	M1	S2
6	O1	M1	S3	10	O2	M1	S3	7	O3	M1	S3
8	O1	M1	S3	8	O2	M1	S3	5	O3	M1	S3
1	O1	M2	S1	1	O2	M2	S1	2	O3	M2	S1
3	O1	M2	S1	6	O2	M2	S1	3	O3	M2	S1
.				.				.			
.				.				.			

Exploratory analysis

As explained in previous Appendices.

ANOVA information

Balancing generally prevails in three factor FDs so similar procedures can be used for ANOVA calculation as were used for two factor FDs.

SAS: The PROC GLM procedure can perform the necessary ANOVA calculations for a three factor study. When both fixed and random factors occur, PROC GLM can only deal with the unrestricted mixed model. Typical code for such a case, based on Example 5.4, would be as follows:

```
PROC GLM DATA = CH5.EX54;
    CLASS OPERATOR MACHINE SUPPLIER;
    MODEL EXLGTH = OPERATOR | MACHINE | SUPPLIER;
    RANDOM OPERATOR OPERATOR*MACHINE OPERATOR*SUPPLIER
        OPERATOR*MACHINE*SUPPLIER / TEST;
RUN;
```

The RANDOM statement specifies the model effects which are of random type stemming from the classification of the OPERATOR factor as random. This statement enables the unrestricted *EMS* expressions to be generated. The TEST option specifies that test results for each model effect be calculated using appropriate error terms as determined by the *EMS* expressions. Without this option, SAS will generate test statistics assuming all effects fixed.

Minitab: Use of the Balanced ANOVA procedure can generate the necessary ANOVA table (see Display 5.6). The associated menu commands are as follows, based on a restricted mixed model:

> Select **Stat** ➢ **ANOVA** ➢ **Balanced ANOVA** ➢ for *Responses*, select **ExLgth** and click **Select** ➢ select the **Model** box and enter **Operator | Machine | Supplier** ➢ select the **Random Factors (optional)** box and enter **Operator**.
>
> Select **Options** ➢ **Use the restricted form of the mixed model** ➢ **Display expected mean squares** ➢ click **OK**.
>
> Select **Storage** ➢ **Fits** ➢ **Residuals** ➢ click **OK** ➢ click **OK**.

Exclusion of the **Options** selection **Use the restricted form of the mixed model** will provide *EMS* information and variance estimation for the unrestricted mixed model.

Diagnostic checking
As previous Appendices.

6

Hierarchical Designs

6.1 INTRODUCTION

Hierarchical Designs typically arise in situations where several observations are collected from the same sampling or experimental unit. The simplest example of such a design is the CRD as different experimental units are restricted to each treatment level. However, hierarchical designs with two and more factors are much more useful. In a two factor experiment, the levels of one factor (factor *B*) are typically obtained from different sampling units within the different levels of the other treatment factor (factor *A*). These differences mean that this design structure cannot be construed to be a factorial one in which all levels of factors must be tested across all levels of factor *A*. For such a design structure, we say factor *B* is **nested** within factor *A* and we have a **two factor Nested Design (ND)**.

To illustrate a ND structure, consider a study investigating the characteristics of a manufactured product produced at each of a company's three manufacturing sites A1, A2, and A3 (factor *A*). Each site has a number of production lines (factor *B*) with three measurements recorded from each production line. The two factors being studied will be manufacturing site and production line within site of manufacture. Clearly, production lines cannot be taken as identical across sites. As such, the production line factor is classified as being nested within sites as the line tested is associated with the specific site of manufacturing. The layout for such a design is shown in Fig. 6.1 and highlights a typical hierarchical design structure in which one factor is of the nested type.

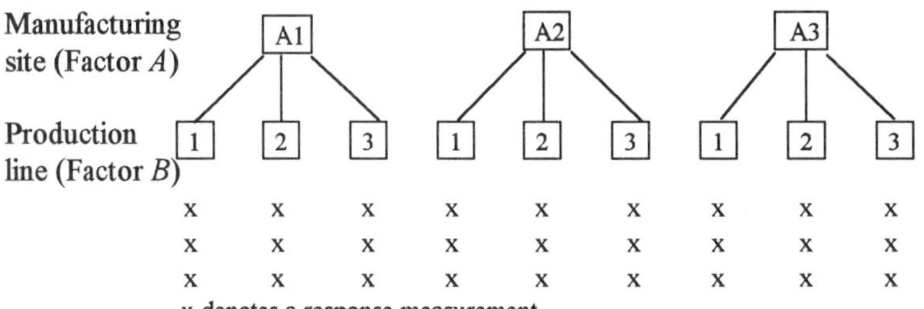

Figure 6.1 — Possible two factor Nested Design layout for manufactured product study

6.2 TWO FACTOR NESTED DESIGN

6.2.1 Design Structure

Suppose we have *a* levels of treatment factor *A* and the experimental units are randomly assigned so that each treatment group receives *b* similarly defined but not identical units. From each unit within each treatment group, a total of *n* observations is collected. The general layout for a two factor ND is shown in Table 6.1. The levels of factor *B*, though similarly labelled, are not identical for each level of factor

A and are therefore nested within factor A. In the absence of replicates, we would simply have a CRD structure. In a ND, variation in the measured response can arise from three possible sources: factor A differences, factor B differences within different levels of factor A (nested factor effect), and experimental error or uncontrolled variation. The ND structure described is often also referred to as a **three-stage nested sampling experiment**.

Table 6.1 — Two factor Nested Design structure

Factor A (Treatment)	1	2	.	.	a
Factor B	1 . . . b	1 . . . b	.	.	1 . . . b
	x . . . x	x . . . x	.	.	x . . . x
	. . . . .	. . . . .	.	.	. . . . .
	. . . . .	. . . . .	.	.	. . . . .
	x . . . x	x . . . x	.	.	x . . . x

x denotes a response measurement

To illustrate a two factor ND structure, consider a study to compare blood glucose levels in two groups of three patients where one group receives human insulin and the other porcine insulin. A blood specimen is to be collected from each randomly selected subject and five replicate blood glucose determinations made on each specimen according to the design structure displayed in Table 6.2. In this study, factor A represents the type of insulin being administered while factor B corresponds to the patient effect, and no patient receives both treatments. The patients are clearly nested within the treatments and we have a two factor ND with $a = 2$ (two insulin types), $b = 3$ (three patients to each insulin type), and $n = 5$ (five replicate measurements of blood glucose for each patient).

Table 6.2 — Possible Nested Design layout for blood glucose study

Treatment	Human insulin			Porcine insulin		
Patient	1	2	3	1	2	3
	x	x	x	x	x	x
	x	x	x	x	x	x
	x	x	x	x	x	x
	x	x	x	x	x	x
	x	x	x	x	x	x

x denotes a blood glucose measurement

Example 6.1 A company maintains three manufacturing sites for the production of photographic film. Some customer complaints have been received in respect of the light sensitivity, or speed, of the film produced. The company decided to investigate this quality problem by conducting a simple study into whether the light sensitivity of the film varied with manufacturing site. The data on light sensitivities, as ISO speed values, are presented in Table 6.3, and were obtained by randomly selecting four production batches from each site and collecting two replicate measurements from each batch. The purpose of this study is to assess if sites differ and if batches produced within the sites affect variation in the light sensitivity of the film.

Table 6.3 — Response data for light sensitivity study of Example 6.1

Site	S1				S2				S3			
Batch	1	2	3	4	1	2	3	4	1	2	3	4
	69	84	78	58	71	54	52	64	57	62	78	56
	71	84	79	61	72	56	49	66	71	64	79	58

6.2.2 Model for the Measured Response

An observation X_{ijk} recorded in such a two factor Nested Design can be described by the **linear model**

$$X_{ijk} = \mu_{ij} + \varepsilon_{k(ij)} = \mu + \alpha_i + \beta_{j(i)} + \varepsilon_{k(ij)} \tag{6.1}$$

where $i = 1, 2,..., a$, $j = 1, 2,..., b$, and $k = 1, 2,..., n$. The first part $\mu_{ij} + \varepsilon_{k(ij)}$ again defines the **means model** and the second part the **effects model**. The terms μ and α_i are as defined for a two factor FD (see Section 5.2.2). The error is expressed as $\varepsilon_{k(ij)}$ where the brackets refer to nested within treatment combination (A_i, B_j) though it could equally be expressed in the standard notation ε_{ijk}. The term $\beta_{j(i)}$ represents the effect of factor B at level j nested within the ith level of factor A, with the brackets surrounding i defining nesting. This term is assumed constant for all of factor B at level j within factor A at level i but may differ for other units within factor A at level i. Essentially, compared to the two factor FD, $\beta_{j(i)} = \beta_j + \alpha\beta_{ij}$ so the nested factor B effect is essentially a measure of the B plus interaction $A \times B$ effects though it is difficult to give a practical interpretation to the $A \times B$ interaction.

In NDs, the usual error assumptions apply, i.e. $\varepsilon_{k(ij)} \sim \text{NIID}(0, \sigma^2)$. In most applications of this structure, factor B generally defines a **random effect**, e.g., randomly selected subjects or units, and so it is generally also assumed that $\beta_{j(i)} \sim \text{NIID}(0, \sigma^2_\beta)$. Factor A may be either a fixed effect with $\sum\alpha_i = 0$ assumed or random effect with $\alpha_i \sim \text{NIID}(0, \sigma^2_\alpha)$ assumed. Therefore, ND based experiments will be mostly of **mixed** or **model II** type.

Exercise 6.1 For the light sensitivity ND study of Example 6.1, we first require to set up the relevant response model.

Response model: Based on the presented information, we can deduce that the site effect is fixed and that the batch effect is random resulting in a mixed model. Additionally, the batches tested refer to that site's production so the batch effect defines a nested factor and the design corresponds to a two factor nested structure. As a ND structure is being used, we specify the response model (6.1) as

$$\text{light sensitivity} = \mu + \alpha_i + \beta_{j(i)} + \varepsilon_{k(ij)}.$$

where α_i is the contribution of ith site to light sensitivity, $\beta_{j(i)}$ is the contribution of the jth batch nested within the ith site, and $\varepsilon_{k(ij)}$ refers to the random error effect. We assume that $\varepsilon_{k(ij)} \sim \text{NIID}(0, \sigma^2)$, $\sum\alpha_i = 0$ as the site effect is fixed, and $\beta_{j(i)} \sim \text{NIID}(0, \sigma^2_\beta)$ as the batch effect is random. □

6.2.3 Exploratory Data Analysis

The initial data analysis to carry out depends very much on how the factors being tested have been defined. Simple plots and summaries generally suffice for fixed effects.

Exercise 6.2 After response model specification for the light sensitivity study of Example 6.1, the next stage of analysis concerns conducting exploratory analysis of the data.

> *EDA*: Initial plots and summaries of the light sensitivity data with respect to the fixed site factor provided some useful information. Site effect showed little difference though site S2 was lowest overall and wide variation in light sensitivity was apparent across all sites. The plot and summaries with respect to the batch effect, shown in Display 6.1, highlight distinctive trends. Site differences are obvious in respect of the second and third batches selected in particular. Additionally, most batches appear to give consistent measurements except batch 1 from site S3 which exhibits inconsistent measurements (wide range). Initial analysis suggests no site effect but possibly an important batch effect. □

Display 6.1

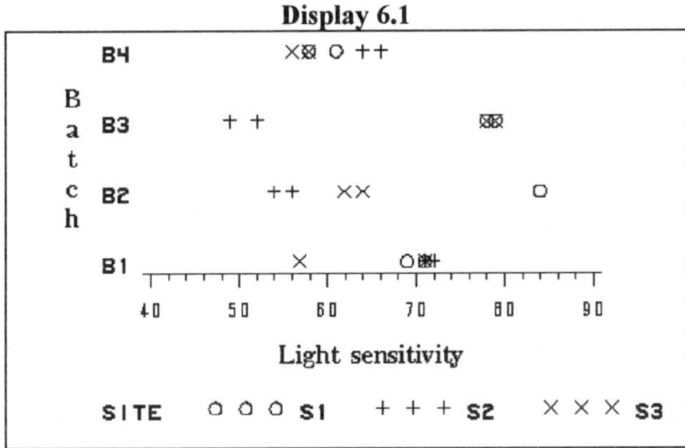

6.3 ANOVA PRINCIPLE FOR THE TWO FACTOR NESTED DESIGN

Partitioning response variation into its component parts using ANOVA procedures provides the base on which statistical analysis of NDs rests. Theoretical underpinning, similar to that briefly mentioned for One Factor Designs and FDs, enables variation partitioning, and test statistic construction and derivation to be developed from the response model (6.1).

6.3.1 Hypotheses

The effect **hypotheses** within ND experiments are similarly structured to those of previous design structures. They again define a test of no effect on response (H_0) against effect on response (H_1). Hypotheses format depends also on the nature of the design effects as follows:

factor	*fixed effect*	*random effect*
A	H_0: all $\alpha_i = 0$ and H_1: not all $\alpha_i = 0$	H_0: $\sigma_\alpha^2 = 0$ and H_1: $\sigma_\alpha^2 > 0$
B	H_0: all $\beta_{j(i)} = 0$ and H_1: not all $\beta_{j(i)} = 0$	H_0: $\sigma_\beta^2 = 0$ and H_1: $\sigma_\beta^2 > 0$

6.3.2 ANOVA Table

The variation in the measured response within a two factor ND can be decomposed into the sources specified within the proposed response model. These sources are specified as factor A, factor B nested within factor A, denoted $B(A)$, and experimental error. The associated total sum of squares (SSTotal) expression underpinning the **ANOVA principle** is therefore

$$SS\text{Total } [SST] = SSA + SSB(A) + SS\text{Error } [SSE] \tag{6.2}$$

The corresponding mean square (MS) terms are $MSA = SSA/(a-1)$, $MSB(A) = SSB(A)/[a(b-1)]$, and $MSE = SSE/[ab(n-1)]$ for each of the response model effects.

The general **ANOVA table** for a two factor ND takes the form illustrated in Table 6.4 where EMS defines the expected mean squares which stem from the theoretical concepts underpinning experimental design structures. The σ^2 terms define the variance components for each of the model effects, except those classified as fixed effects when the σ^2 terms define the sum of squared effects and not proper variance components.

Table 6.4 — ANOVA table for a two factor Nested Design

Source	df	SS	MS	EMS
A	$a-1$	SSA	$MSA = SSA/df$	$\sigma^2 + nD_b\sigma^2_\beta + nb\sigma^2_\alpha$
B within A	$a(b-1)$	$SSB(A)$	$MSB(A) = SSB(A)/df$	$\sigma^2 + n\sigma^2_\beta$
Error	$ab(n-1)$	SSE	$MSE = SSE/df$	σ^2
Total	$abn-1$	SST		

a is the number of levels of treatment factor A, b is the number of levels of nested factor B, n is the number of replicate observations collected, df defines degrees of freedom, SS defines sum of squares, MS defines mean square, EMS defines expected mean square, D is 0 for a fixed effect and 1 for a random effect

6.3.3 Test Statistics

The two testable factor effects in a two factor ND are factor A and factor B nested within factor A. Construction of the associated test statistic, based on the standard mean square ratio definition, depends on the classification, either **fixed** or **random**, of the tested factors. **Test statistic construction** for a **model I** experiment (both factors fixed), a **model II** experiment (both factors random), and a **mixed model** experiment (generally A fixed and B random) can be summarised as follows:

factor	*Model I*	*Model II*	*Mixed model*	
A	MSA/MSE	$MSA/MSB(A)$	$MSA/MSB(A)$	(6.3)
B	$MSB(A)/MSE$	$MSB(A)/MSE$	$MSB(A)/MSE$	(6.4)

The associated degrees of freedom for each effect test statistic depends on the mean square ratio of the associated test statistic.

For model I, $D_b = 0$ in the factor A EMS expression in Table 6.4 while for the model II and mixed model cases, $D_b = 1$ as the nested factor B is classified as random. For a model I experiment, both test statistics are based on use of the MSE

term as divisor. When B is random, the test statistic for factor A alters and is based on a different divisor term. In practice, the most frequent use of two factor NDs is as a mixed model when factor A is fixed and B is a random effect. Order of testing is generally based on A first followed by B, though again such tests can only provide general information on factor effects. Specific information requires use of relevant follow-up methods.

Exercise 6.3 Having carried out the response model specification and exploratory analysis on the light sensitivity data of Example 6.1, the next step in the analysis requires to be the statistical assessment of the factor effects.

Hypotheses: The hypotheses for effects associated with this experiment are based on the simple test of no effect on light sensitivity response (H_0) against effect on light sensitivity response (H_1). For the site effect, the hypotheses are expressed either in terms of model effects or factor means, while for the batch effect, the batch variance term is used.

EMS expressions and test statistics: Because one factor is classified as random and before we consider any data analysis, we must first check the construction and format of the effect test statistics. From the factor information and the fact that a mixed model is being applied, we know that $D_b = 1$ (batch random), $a = 3$, $b = 4$, and $n = 2$. Therefore, from the *EMS* expressions shown in Table 6.4, we have

Effect	*EMS*	*Test statistics*
Site (α)	$\sigma^2 + 2\sigma_\beta^2 + 8Q[\alpha]$	*MSS*/*MSB*(S)
Batch(Site) (β)	$\sigma^2 + 2\sigma_\beta^2$	*MSB*(S)/*MSE*
Error	σ^2	

where $Q[\alpha]$ refers to the sum of squared effects for the fixed site effect. Test statistic construction is therefore based on different denominator *EMS*s for each effect.

ANOVA table: Display 6.2 contains the ANOVA information generated by SAS's PROC NESTED procedure. The *EMS* coefficients table agree with those derived manually with the denominator mean square for test statistic derivation shown in the column 'Error Term'. The ANOVA information as presented mirrors the general form in Table 6.4 with the controlled components (Site, Batch) and uncontrolled components (Error) adding to explain the Total variability in the response data. The sum of squares term for the batch effect is very much larger than that for the site effect, implying possibly strong influence of batch on the light sensitivity data. The final part of Display 6.2 contains the variance component estimation, though it must be noted that PROC NESTED assumes a model II experiment so a variance estimate is produced for the fixed site effect which is unnecessary. An alternative to use of PROC NESTED is described in Appendix 6A.1.

Effect tests: Test statistic (6.3) for a mixed model specifies that the site F test statistic is the ratio of the site and batch mean squares. From the information in Display 6.2, this ratio results in the test statistic $F = 1.646$ as specified in row 'Site' and column 'F Value' with p value estimated as 0.2460. The batch test statistic (6.4) is the ratio of the batch and error mean squares producing a test statistic of $F = 19.436$ with p value estimated at 0.0000, i.e. $p < 0.00005$. With both test statistic and p value available, we can use either decision rule mechanism for both tests. The associated 5% critical values for the site and batch effects are $F_{0.05,2,9} = 4.26$ and $F_{0.05,9,12} = 2.80$ respectively.

The high p value for the site effect indicates acceptance of the null hypothesis and the conclusion that there appears no evidence to suggest that light sensitivity of the film differs with manufacturing site ($p > 0.05$). However, the p value for the batch effect is very low indicating rejection of the null hypothesis and the conclusion that the evidence implies that light sensitivity appears to vary markedly with batch sampled ($p < 0.05$). Therefore, it would appear that production batch is the primary cause of the problems reported by customers. □

Display 6.2

```
Coefficients of Expected Mean Squares
                                  Source          Site         Batch          Error
                                  Site               8             2              1
                                  Batch              0             2              1
                                  Error              0             0              1

Nested Random Effects Analysis of Variance for Variable LghtSens
              Degrees
Variance           of          Sum of                                          Error
Source        Freedom         Squares        F Value          Pr > F          Term
Site                2      631.750000          1.646          0.2460          Batch
Batch               9     1727.375000         19.436          0.0000          Error
Error              12      118.500000
Total              23     2477.625000

     Variance                                 Variance                  Percent
     Source        Mean Square               Component                 of Total
     Site           315.875000               15.493056                  13.3107
     Batch          191.930556               91.027778                  78.2054
     Error            9.875000                9.875000                   8.4840
     Total          107.722826              116.395833                 100.0000

                            Mean                                      66.37500000
                            Standard error of mean                     3.62787243
```

6.4 FOLLOW-UP ANALYSIS

As previously, formal tests of factor effects only provide general information on the importance of different factor levels to the measured response. No information is available on the specific nature of such effects so follow-up analysis must be considered to help enhance the conclusions. However, it should be noted that the nature of the factors strongly influences the approaches to consider.

6.4.1 Analysis of Factor Effects

For fixed effects, the usual effect estimation, multiple comparisons, and linear contrast procedures can be used to understand more how factor effects are influencing the response. In each case, care must be taken with the application of the procedure to ensure the correct "error" variance (*MS* term) and degrees freedom are used based on the nature of model effects and the influence this has on model type and effect test statistic construction. For mean and effect estimation, least squares estimators can be obtained in the usual fashion as shown in Table 6.5 with confidence interval construction and interpretation as outlined in Sections 3.4.6 and 5.4.5.

Table 6.5 — Least squares estimators for fixed effect model components in the two factor Nested Design

Parameter	Point estimator	Variance of estimator
$\mu_{i.}$	$\overline{A}_i$	$\sigma^2/(bn)$ model I, $MSB(A)/(bn)$ mixed model
μ_{ij}	$\overline{AB}_{ij}$	σ^2/n
α_i	$\overline{A}_i - \overline{G}$	$\sigma^2(a-1)/(abn)$
$\beta_{j(i)}$	$\overline{AB}_{ij} - \overline{A}_i$	$\sigma^2(b-1)/(abn)$

Mixed model based on A fixed and B random, $\overline{A}_i$ is the mean response for factor A level i, $\overline{AB}_{ij}$ is mean response for nested combination (A_i, B_j), $\overline{G}$ is the mean of the collected data

6.4.2 Mixed Model - Significant Nested Factor

For a two factor ND with factor A classified as a fixed effect and the nested factor B as a random effect, we could obtain a statistically significant nested factor. This would suggest that the levels of the nested factor within the levels of the main treatment factor are differing in their effect on the response and differences in the nested factor are present for all or only some levels of the treatment factor. In such cases, it would be useful to enhance the conclusion by determining, on a statistical basis, which levels of factor A this nested difference corresponds to. This can be achieved by carrying out a simple follow-up based on partitioning the sum of squares for the nested factor into the contributing components for each level of factor A and testing each separately by means of a simple F test. Box 6.1 summarises the elements of this procedure.

Box 6.1 — Follow-up for Significant Nested Factor in a Mixed Model

Hypotheses The null hypothesis of this procedure is expressed as H_0: no factor B differences within the specified levels of factor A.

Test statistic Sum of squares for nested factor B for level i of factor A, denoted $SSB(A_i)$, can be expressed as

$$SSB(A_i) = n\sum_{j=1}^{b}(\overline{AB}_{ij} - \overline{A}_i)^2 \tag{6.5}$$

where $\overline{AB}_{ij}$ corresponds to the mean of the response data for factor B level j nested within factor A level i, $\overline{A}_i$ is the mean of the response data for factor A level i, and $SSB(A_i)$ is such that $SSB(A) = \sum SSB(A_i)$. From this, we derive the test statistic

$$F = \frac{[SSB(A_i)/(b-1)]}{MSE} \tag{6.6}$$

which follows an F distribution with degrees of freedom $df_1 = b - 1$ and $df_2 = ab(n-1)$.

As with previous applications of the F test, small values of the test statistic (6.6) will signify no detectable factor B differences within the tested level of the treatment factor while large values will provide the opposite interpretation.

Exercise 6.4 The light sensitivity study of Example 6.1 was based on a mixed model ND with random nested effect. As the batch effect was shown to be significant, it would be useful to apply the follow-up procedure in Box 6.1 to investigate if this effect is prevalent across just one, just two, or all three of the manufacturing sites.

Design information: From the ND experiment carried out and the data collected, we have $a = 3$, $b = 4$, $n = 2$, $\bar{A}_1 = 73.00$, $\bar{A}_2 = 60.50$, $\bar{A}_3 = 65.63$, and $MSE = 9.875$. The nested factor means $\overline{AB}_{ij}$ are presented in Table 6.6. From the values of a, b, and n, we determine that $df_1 = 4 - 1 = 3$ and $df_2 = 3\times4\times(2 - 1) = 12$ providing a 5% critical value of $F_{0.05,3,12} = 3.49$ for each comparison.

Table 6.6 — Response summaries for nested levels in light sensitivity study of Example 6.1

Site	Batch	Mean	Range	Site	Batch	Mean	Range	Site	Batch	Mean	Range
S1	B1	70.0	2	S2	B1	71.5	1	S3	B1	64.0	14
	B2	84.0	0		B2	55.0	2		B2	63.0	2
	B3	78.5	1		B3	50.5	3		B3	78.5	1
	B4	59.5	3		B4	65.0	2		B4	57.0	2

Test statistics: Expression (6.5) can be evaluated for site S1 as

$$SSB(\text{S1}) = 2[(70 - 73)^2 + (84 - 73)^2 + (78.5 - 73)^2 + (59.5 - 73)^2] = 685.0$$

providing a test statistic (6.6) of $F = [685/3]/9.875 = 23.12$. As the test statistic greatly exceeds the associated critical value, we can conclude that significant batch differences within site S1 appear to be occurring ($p < 0.05$). Applying similar reasoning to the information pertaining to site S2, we have SSB(S2) as 543 and $F = 18.33$ again signifying significant batch differences ($p < 0.05$). For site S3, SSB(S3) = 499.375 and $F = 16.86$ leading to the same conclusion as for the other two sites ($p < 0.05$).

Application of this follow-up shows that variability in the light sensitivity of the manufactured film differs in all three manufacturing sites though, comparing the associated *SS* terms reveals that this variation is marginally greater in site S1. Note that addition of the *SS* components for each site produces the nested sum of squares of 1727.375 as printed in Display 6.2. □

6.4.3 Mixed Model and Model II Variability Analysis

Both of these model types are such that they provide estimates for the variance of the nested factor variance σ_β^2 (variance between nested experimental units) and the error variance σ^2 (variance within nested experimental units). Using these variance estimates enables $\text{Var}(X_{ijk})$, the variance estimate for any observation, to be derived. For a model II experiment, this corresponds to

$$\text{Var}(X_{ijk}) = \sigma_\alpha^2 + \sigma_\beta^2 + \sigma^2 \quad (6.7)$$

as factor A is random. For a mixed model experiment based on nested factor B being random, this is modified to

$$\text{Var}(X_{ijk}) = \sigma_\beta^2 + \sigma^2 \quad (6.8)$$

Estimation of the variance components is summarised in Table 6.7 based on solving, in turn, the *EMS* expressions in Table 6.4 for each model type.

Table 6.7 — Variability estimation for the two factor Nested Design

	Model II	Mixed model
σ^2	MSE	MSE
σ^2_β	$[MSB(A) - MSE]/n$	$[MSB(A) - MSE]/n$
σ^2_α	$[MSA - MSB(A)]/(nb)$	–

Using these variance estimates, the contribution of each effect to the variability of the response can be gauged using the same simple steps as outlined in Section 5.8.3. This entails totalling the corresponding variance component estimates according to expression (6.7) or (6.8) and then expressing each variance estimate as a percentage of the total to provide a measure of the relative importance of each random factor. The resultant percentages are interpreted accordingly to assess which effects are the significant influences on response data variability.

Exercise 6.5 For the light sensitivity study of Example 6.1, a mixed model experiment was conducted based on the batch effect classified as random. Assessment of the variance estimates of this and the random error is therefore necessary.

Variability estimation: The variance estimates, expressed in the column 'Variance Component' in Display 6.2, provide $\sigma^2_\beta = 91.0278$ and $\sigma^2 = 9.875$. Using expression (6.8), we can determine the percentage contribution to response variability of the batch effect to be approximately 90%. This indicates that most of the differences in the light sensitivity measurements are associated with the batches sampled so confirming the observations obtained in earlier analysis. □

It is also possible to construct confidence intervals for these variability estimates using similar principles to that for construction of a confidence interval for the sample variance in a one sample experiment. For the error variance σ^2, a 100(1 – α) percent confidence interval is expressed as

$$\frac{ab(n-1)MSE}{\chi^2_{1-\alpha/2,ab(n-1)}} < \sigma^2 < \frac{ab(n-1)MSE}{\chi^2_{\alpha/2,ab(n-1)}} \tag{6.9}$$

using the χ^2 distribution (Table A.2) to provide the appropriate critical values. Confidence intervals for the other variance components are more difficult to determine because the associated variance estimates are linear functions of mean squares and thus are linear functions of chi-square random variables which do not conform to known distribution patterns. Information on approximation methods for such confidence intervals can be found in Mendenhall (1968) and Searle (1997).

6.4.4 Diagnostic Checking

As Hierarchical Design structures are based on comparable assumptions to One Factor Designs and FDs, inclusion of **diagnostic checking** as an integral part of the data analysis is important. Plotting residuals (measurements – fitted values) against the factors tested and against model fits check for equality of response variability and adequacy of model. Normality of the measured response can again be checked through the normal plot of residuals. Inappropriateness of these assumptions may be overcome, in future comparable studies, through data transformation (see Section 3.8).

Exercise 6.6 The final aspect of the analysis of the light sensitivity data of Example 6.1 concerns diagnostic checking of the residuals using factor, model, and normal plots.

Factor plots: The residual plot for site in Display 6.3 shows distinctly different column lengths especially in respect of site S3 though this difference is primarily due to the presence of two likely outliers. Unequal variability across the sites could be concluded to be occurring. The batch residual plot showed similar patterning with three columns very similar but the fourth, corresponding to batch 1 from each site, exhibiting greater length and the presence of the same two outliers.

Display 6.3

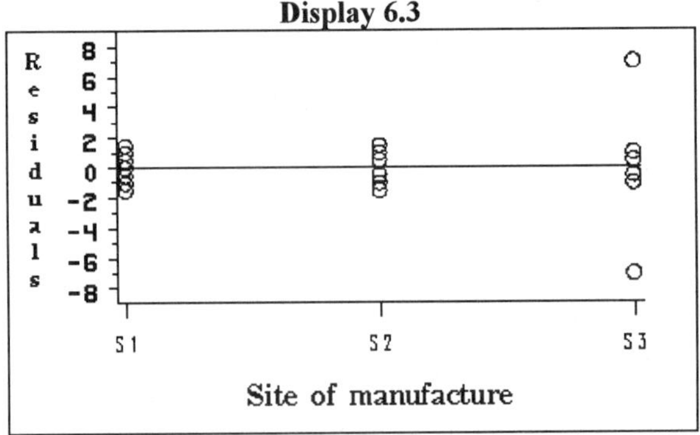

Fits and normal plots: The fits plot exhibited acceptable model fit on balance except for the detected outliers. The normal plot, shown in Display 6.4, exhibits a linear trend through most of the points with the presence of the two outliers causing the full trend to appear non-linear. This suggests that the normality of the light sensitivity data may be inappropriate as confirmed by the Ryan-Joiner test ($R = 0.86995, p < 0.1$). □

Display 6.4

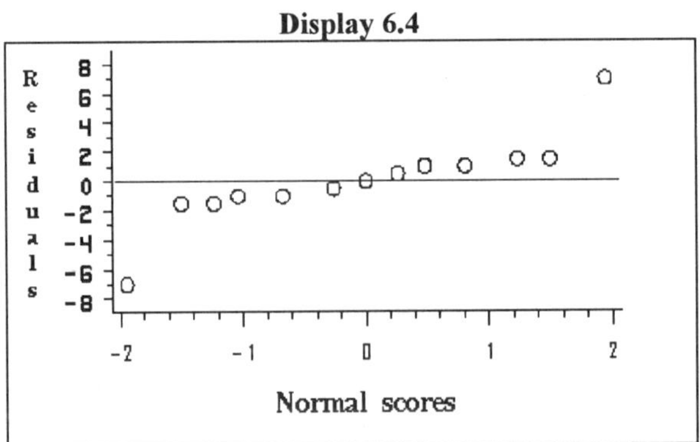

Summarising the findings from the light sensitivity study of Example 6.1, we have found manufacturing sites do not differ (p > 0.05) and the batch effect dominates (p <

0.05). All sites appear to suffer from similar batch variation ($p < 0.05$) so it would appear necessary to investigate this further to understand why batches are so different. Diagnostic checking suggests that equal variability and normality of the light sensitivity response data may not be acceptable and that, in future experiments based on such a response, some form of data transformation to stabilise variability and normality effects may need to be considered prior to data analysis or the protocol should be designed to avoid outliers.

6.5 OTHER FEATURES ASSOCIATED WITH TWO FACTOR NESTED DESIGNS

6.5.1 Relative Efficiency

The relative efficiency of a two factor ND of mixed model type is often compared on the basis of the variance of the treatment means. The mean of treatment i (factor A level i) is provided by $\mu + \alpha_i$ and has variance $(\sigma^2 + n\sigma_\beta^2)/(bn)$. Using estimates of the within nested experimental units variance σ^2 and the between nested experimental units variance σ_β^2, the relative efficiencies of combinations of n (number of replications) and b (number of nested factor levels) can be approximated and compared with the outcome of the initial experiment in respect of the change in precision of the treatment mean estimate. This procedure, therefore, provides a simple method for assessing, for future mixed model ND studies, the number of levels of nested factor B and the number of replicates to consider.

To illustrate, consider the light sensitivity study of Example 6.1. From this study, we have $b = 4$, $n = 2$, $\sigma^2 = 9.875$, and $\sigma_\beta^2 = 91.0278$ producing a variance estimate of 23.99 for treatment i. Increasing the number of replications n to three gives a treatment variance estimate of 23.58 providing only a minor improvement in precision. Changing to $b = 5$ batches and keeping $n = 2$ replicates results in a variance estimate of 19.19 showing that sampling more batches would appear a useful course of action provided it does not increase the cost or difficulty of data collection. Since most of the variability arises from batches, increasing the number of batches sampled makes sense.

6.5.2 Power Analysis

The efficiency comparison principles in Section 6.5.1 represent a form of **power analysis** of a proposed nested structure. The aspects of power and sample size estimation discussed in Sections 2.5, 3.6, and 5.6 can be readily adapted to NDs in respect of the treatment factor. As for FDs (see Section 5.6), k' is the number of levels of the factor, n' defines the number of replicate measurements planned at each level, and $f_1 = k' - 1$. The appropriate source of error mean square estimate differs depending on the classification of the nested factor. For a fixed nested factor, the requisite estimate corresponds to the actual error mean square with f_2 equal to $ab(n - 1)$ while for a random nested factor, the estimate required must refer to the nested factor mean square with f_2 specified as $a(b - 1)$.

6.5.3 Unequal Replicates

As in the two factor FD, use of the **General Linear Model** (GLM) estimation technique can provide the necessary estimation and ANOVA information for data analysis if replication is unequal. Both fixed and random effects can still be considered though the latter results in very complicated analyses. Unbalanced designs, however, should be avoided if possible.

6.6 THREE FACTOR NESTED DESIGN

The simple principle of factor nesting can be extended readily to more than two factors. For three factors *A*, *B*, and *C*, we generally consider *C* nested within *B* and *B* nested within *A*. Again, the nested factors are generally classified as random effects so most applications of such a design structure will be either of **model II** type or **mixed model** type depending on the classification of factor *A*.

By way of an illustration, consider a study designed to investigate the factors influencing the determinations of the level of fat in dried eggs by means of acid hydrolysis. The study has to assess the relative variabilities of the laboratories carrying out the analysis, the analysts undertaking the analysis, and the samples used by the analysts. As analysts are located within different laboratories, we must take the analyst factor to be a nested one. Samples presented to the laboratories may also differ with different analysts in each laboratory analysing different samples giving rise to a second nested factor, the samples. We would therefore have an experiment based on three factors: laboratories, analysts nested within laboratories, and samples nested within analysts. A three factor ND, as Fig. 6.2 illustrates, would therefore be required for such a study.

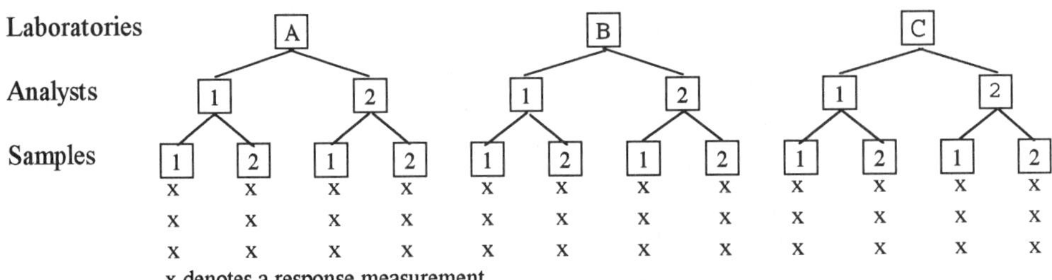

Figure 6.2 — Three factor Nested Design layout for acid hydrolysis study

6.6.1 Model for the Measured Response

The **linear response model** for a three factor ND is expressed as

$$X_{ijkl} = \mu_{ijk} + \varepsilon_{l(ijk)} = \mu + \alpha_i + \beta_{j(i)} + \gamma_{k(ij)} + \varepsilon_{l(ijk)} \tag{6.10}$$

using γ to define the nested factor *C* effect on the response and where $i = 1, 2,..., a$, $j = 1, 2,..., b$, $k = 1, 2,..., c$, and $l = 1, 2,..., n$. The error is denoted $\varepsilon_{l(ijk)}$ with the brackets again being used to emphasise the nested nature of the measured observations. This term could equally be expressed as ε_{ijkl}. This response model rests on comparable **assumptions** to those underpinning the two factor nested structure (see Section 6.2.2).

6.6.2 ANOVA Table and Test Statistics

The associated **ANOVA table** for a three factor ND structure is summarised in Table 6.8 using the usual format. The ANOVA principle of partitioning response variation into its contributing components as per the response model (6.10) is again its base. **Test statistic construction** is based on similar *EMS* ratio principles to those explained previously based on factor effect classification as either fixed or random. Analysis of data collected from such a design follows the principles and ideas outlined for the two factor ND with the amount and type of analysis dependent on the data collected, the nature of the factors being assessed, and the statistical significance of the tested factors.

Table 6.8 — ANOVA table for a three factor Nested Design

Source	*df*	*SS*	*MS*	*EMS*
A	$a-1$	SSA	$MSA = SSA/df$	$\sigma^2 + nD_c\sigma_\gamma^2 + ncD_b\sigma_\beta^2 + nbc\sigma_\alpha^2$
$B(A)$	$a(b-1)$	$SSB(A)$	$MSB(A) = SSB(A)/df$	$\sigma^2 + nD_c\sigma_\gamma^2 + nc\sigma_\beta^2$
$C(B)$	$ab(c-1)$	$SSC(B)$	$MSC(B) = SSC(B)/df$	$\sigma^2 + n\sigma_\gamma^2$
Error	$abc(n-1)$	SSE	$MSE = SSE/df$	σ^2
Total	$abcn-1$	SST		

a is the number of levels of treatment factor *A*, *b* is the number of levels of the first nested factor *B*, *c* is the number of levels of the second nested factor *C*, *n* is the number of replicate observations collected, *df* defines degrees of freedom, *SS* defines sum of squares, *MS* defines mean square, *EMS* defines expected mean square, *D* is 0 for a fixed effect and 1 for a random effect

6.7 REPEATED MEASURES DESIGN

This design arises very frequently in experimental studies where repeated measurements are to be made on one factor, the treatment factor. Typically, the repeated measurements are over time and time trends in the response are an important consideration. **Repeated Measures Design (RMD)** structures frequently occur in clinical trials when patient responses are measured at regular intervals and in inter-laboratory studies when preparing precision statements. The data collected provide information on the effects of time on each treatment and how this may vary across the treatments tested. This represents a more efficient use of resources than if different experimental units were considered at each time point. As such units can be liable to wide variation, use of the same unit at each time point improves the precision of the time effect estimates and making use of repeated measurements is similar in its effect to the inclusion of blocking in One Factor Designs.

In the RMD structure, both factorial and nested factors appear providing a powerful design with applications in a wide range of observational studies. The analysis of the design data involves methods appropriate to both FDs and NDs.

6.7.1 Design Structure

Consider *a* levels of factor *A* (treatment groups) to which *b* subjects (factor *B*) are randomly assigned as in a CRD structure. Now suppose the response variable is measured for each subject at *c* levels of factor *C*, e.g., successive points in time. This gives rise to a factorial experiment with repeated measurements on the time factor. To illustrate such a RMD structure, consider an experiment set up to study the effect of three types of thermometer on the determination of the melting points of a solid by capillary tube technique. Thermometers represent the treatment factor (factor *A*) but different technicians could also affect the measurements. To account for this, four technicians (factor B) are to be assigned to each selected thermometer whereby each technician is only in one group. The technician effect is therefore a nested one as different groups of technicians will be testing each thermometer. Each technician is to make two replicate melting point determinations each day (factor C) for five working days. A possible design layout for such an experiment is displayed in Table 6.9 and represents a typical repeated measures structure for three factors.

Table 6.9 — Possible Repeated Measures Design layout for thermometers experiment

Thermometer (Factor A)	Technicians (Factor B)	Day (Factor C) 1	2	3	4	5
A	1	x x	x x	x x	x x	x x
	2	x x	x x	x x	x x	x x
	3	x x	x x	x x	x x	x x
	4	x x	x x	x x	x x	x x
B	5	x x	x x	x x	x x	x x
	6	x x	x x	x x	x x	x x
	7	x x	x x	x x	x x	x x
	8	x x	x x	x x	x x	x x
C	9	x x	x x	x x	x x	x x
	10	x x	x x	x x	x x	x x
	11	x x	x x	x x	x x	x x
	12	x x	x x	x x	x x	x x

x denotes a melting point measurement

In a RMD, as the illustration shows, each treatment (the thermometers) is tested at each time point though using different technicians introduces the nested factor. As daily replication is suggested, this means we can use this structure to assess five effects: thermometer, technician nested within thermometer, time, thermometer × time interaction, and technician nested within thermometer × time interaction. Thus main, nested, and interaction terms are open to assessment providing a powerful design structure. However, in most practical applications, only one measurement is generally made at each time point as Example 6.2 demonstrates. Information on certain analysis aspects of RMDs when time point replication occurs will also be provided for completeness.

Example 6.2 An antibiotic can be prepared using three different formulations. To study the effect of these formulations, a RMD structure was adopted with six subjects selected and randomly assigned, two per formulation. Each formulation was then administered to the appropriate subjects. Blood samples were taken from each subject 1 hour, 12 hours, and 24 hours after administration and the level of activity of the antibiotic measured as presented in Table 6.10. Do the formulations differ and is time of sampling important?

6.7.2 Model for the Measured Response

Response model specification for a RMD will be explained for n replications at each time point and for single replicate measurement at each time point.

Table 6.10 — Data for antibiotic activity for the RMD study of Example 6.2

Formulation	Subject	Time 1h	12h	24h
A	1	37.5	30.1	5.4
	2	34.8	22.1	1.6
B	1	26.9	11.3	5.5
	2	20.4	8.4	3.9
C	1	40.1	30.3	2.1
	2	36.1	26.3	2.3

6.7.2.1 n Replications at Each Time Point

The **linear effects model** for the measured response X_{ijkl} in a RMD based on n replicate measurements at each time point is

$$X_{ijkl} = \mu + \alpha_i + \beta_{j(i)} + \gamma_k + \alpha\gamma_{ik} + \beta\gamma_{j(i)k} + \varepsilon_{l(ijk)} \tag{6.11}$$

where $i = 1, 2,..., a$, $j = 1, 2,..., b$, $k = 1, 2,..., c$, and $l = 1, 2,..., n$. The model parameters are similarly defined as for FD and ND response models. α_i defines the effect of the ith level of factor A (ith treatment), $\beta_{j(i)}$ is the effect of level j of factor B nested within level i of factor A, γ_k is the effect of level k of factor C (kth time point), $\alpha\gamma_{ik}$ is the treatment × time interaction effect, $\beta\gamma_{j(i)k}$ is the subject nested within treatment × time interaction effect, and $\varepsilon_{l(ijk)}$ defines the random error.

Error **assumptions** conform to those associated with all formal experimental designs namely, $\varepsilon_{l(ijk)} \sim \text{NIID}(0, \sigma^2)$. In general, factors A and C are assumed fixed while factor B, the subject effect, is mostly of random type leading to the additional assumption of $\beta_{j(i)} \sim \text{NIID}(0, \sigma^2_\beta)$. Additionally, it is assumed that the errors within a subject are equally correlated and have equal variances, representing what is called the **compound symmetry property**. A further assumption made is that the within subjects error covariance matrices are identical for all subjects (Huynh-Feldt condition, Huynh & Feldt 1970). Milliken & Johnson (1992) provide further discussion of these assumptions as well as more comprehensive advice on analysis than is provided in this section while Everitt & Der (1996) discuss both univariate and multivariate approaches. As with NDs, most practical applications of RMDs will be of **model II** or **mixed model** type, the latter being the more common.

6.7.2.2 Single Replicate at Each Time Point

The **linear effects model** for the measured response X_{ijk} in a RMD based on a single replicate measurement at each time point is

$$X_{ijk} = \mu + \alpha_i + \beta_{j(i)} + \gamma_k + \alpha\gamma_{ik} + \varepsilon_{ijk} \tag{6.12}$$

where $i = 1, 2,..., a$, $j = 1, 2,..., b$, and $k = 1, 2,..., c$. The model parameters are as defined for the n replicate case with the only difference in models (6.11) and (6.12) being the absence of the $\beta\gamma_{j(i)k}$ term, the subject nested within treatment × time interaction effect, which cannot be estimated when only one observation is measured at each time point. **Assumptions** are as previously defined with again, the nested factor generally classified as random.

Exercise 6.7 The first stage in the analysis of the repeated measures data described in Example 6.2 concerns model specification and exploratory analysis.

Response model: Based on the factors information, we can deduce that the formulation factor is fixed, the subject effect is random, and the time points of measurement are fixed resulting in a mixed model. Additionally, the subjects testing each formulation are independent so the subject effect refers to a nested factor. We therefore have a repeated measures structure.

As a RMD structure with one replicate was used, we can use the model expression (6.12) to specify the response model as

$$\text{antibiotic activity} = \mu + \alpha_i + \beta_{j(i)} + \gamma_k + \alpha\gamma_{ik} + \varepsilon_{ijk}$$

where α_i is the contribution of *i*th formulation to antibiotic activity, $\beta_{j(i)}$ is the contribution of the *j*th subject nested within the *i*th formulation, γ_k is the effect of time period k, $\alpha\gamma_{ik}$ is the effect of the interaction of formulation i and time period k, and ε_{ijk} refers to the random error. We assume that the error $\varepsilon_{ijk} \sim \text{NIID}(0, \sigma^2)$, $\Sigma\alpha_i = 0$ as the formulation effect is fixed, $\Sigma\gamma_k = 0$ as the time effect is fixed, and $\beta_{j(i)} \sim \text{NIID}(0, \sigma_\beta^2)$ as the subject effect is random. It is also assumed that the errors within a subject are equally correlated and have equal variances.

EDA: The formulation effect showed A and C to be similar with B generally resulting in lower antibiotic activity levels though for each, wide variation in measurements occurred due, mostly, with the time span over which measuring took place. Time trends, as shown in Display 6.5, demonstrate a distinct reduction in antibiotic activity with time. For each subject, the trend over time appears almost linear with formulation differences apparent given that subjects 1 and 2 received formulation A and subjects 3 and 4 formulation B. Significant formulation × time interaction appears likely as the trends over time for the groups do not appear to be parallel. □

Display 6.5

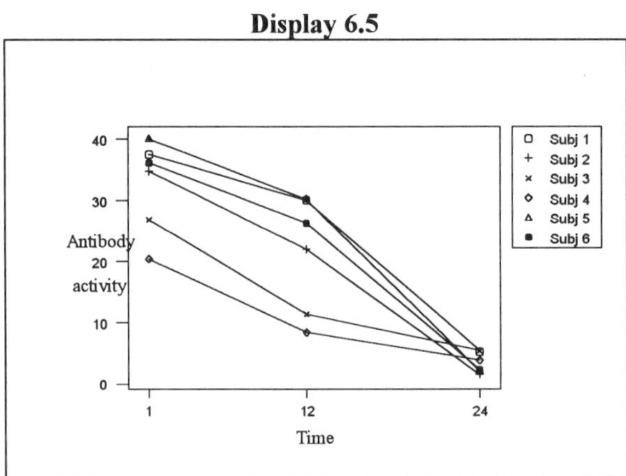

6.7.3 ANOVA Table and Test Statistics

The application of the **ANOVA principle** to repeated measures data again rests on the partitioning of response variation into the component elements defined within the response model (6.11) or (6.12). The cases for n replications at each time point and single replicate measurement at each time point will be presented. In a RMD, there are a number of tests of interest: factor A, factor B nested within A, factor C, the interaction $A \times C$, and the interaction $B(A) \times C$. The usual ANOVA approaches are acceptable provided all the repeated measures assumption, described in Section 6.7.2.1, are satisfied.

6.7.3.1 n Replications at Each Time Point

When n replications are collected, all five tests associated with the RMD structure can be assessed. The associated **ANOVA table** and *EMS* expressions for a typical replicated RMD structure are shown in Table 6.11.

Table 6.11 — ANOVA table for a Repeated Measures Design with one treatment factor and n replications at each time point

Source	*df*	*SS*	*EMS*
Between subjects	$ab - 1$		
A	$a - 1$	SSA	$\sigma^2 + nD_bD_c\sigma^2_{\beta\gamma} + nbD_c\sigma^2_{\alpha\gamma} + ncD_b\sigma^2_{\beta} + nbc\,\sigma^2_{\alpha}$
$B(A)$	$a(b - 1)$	$SSB(A)$	$\sigma^2 + nD_c\sigma^2_{\alpha\gamma} + nc\,\sigma^2_{\beta}$
Within subjects	$ab(c - 1)$		
C	$c - 1$	SSC	$\sigma^2 + nD_b\sigma^2_{\alpha\gamma} + nbD_a\sigma^2_{\alpha\gamma} + nab\,\sigma^2_{\gamma}$
A × C	$(a - 1)(c - 1)$	$SSAC$	$\sigma^2 + nD_b\sigma^2_{\alpha\gamma} + nb\,\sigma^2_{\alpha\gamma}$
$B(A) \times C$	$a(b - 1)(c - 1)$	$SSB(A)C$	$\sigma^2 + n\sigma^2_{\alpha\gamma}$
Error	$abc(n - 1)$	SSE	σ^2
Total	$abcn - 1$	SST	

a is the number of levels of treatment factor A, b is the number of levels of nested factor B, c is the number of time points for measurement, n is the number of replicate observations collected at each time point, *df* defines degrees of freedom, *SS* defines sum of squares, *EMS* defines expected mean square, D is 0 for a fixed effect and 1 for a random effect

The *EMS* expressions for each model effect provide the numerator and denominator elements of the appropriate F test statistics of model effects. Such construction hinges on factor classification as either fixed or random. As most uses of RMDs classify B as **random** ($D_b = 1$), it is relatively straightforward to use the *EMS* expressions in Table 6.11 to provide the **test statistic construction**. Assuming factor C (time) is fixed ($D_c = 0$), this would be as follows:

		Mixed model [A fixed, $D_a = 0$]	*Mixed model* [A random, $D_a = 1$]	
Between subjects	A	$MSA/MSB(A)$	$MSA/MSB(A)$	(6.13)
	$B(A)$	$MSB(A)/MSE$	$MSB(A)/MSE$	(6.14)
Within subjects	C	$MSC/MSB(A)C$	$MSC/MSAC$	(6.15)
	$A \times C$	$MSAC/MSB(A)C$	$MSAC/MSB(A)C$	(6.16)
	$B(A) \times C$	$MSB(A)C/MSE$	$MSB(A)C/MSE$	(6.17)

The only difference occurs in the test statistic for factor C which is based on a different denominator mean square depending on A's classification.

6.7.3.2 *Single Replicate at Each Time Point*

For single replication, of the five tests possible in a RMD, only that associated with the interaction $B(A) \times C$ cannot be considered. The **ANOVA table** and associated *EMS* expressions for a typical single replicate RMD structure are displayed in Table 6.12. The *EMS* expressions for each model effect again provide the numerator and denominator elements of the appropriate F test statistics using the usual ratio principles. As factor B is generally defined as **random** ($D_b = 1$) in RMDs and assuming factor C (time) is fixed ($D_c = 0$), the test statistic derivations for both treatment effect classifications would be as follows:

		Mixed model [A fixed, $D_a = 0$]	*Mixed model* [A random, $D_a = 1$]	
Between subjects	A	$MSA/MSB(A)$	$MSA/MSB(A)$	(6.18)
	$B(A)$	$MSB(A)/MSE$	$MSB(A)/MSE$	(6.19)
Within subjects	C	MSC/MSE	$MSC/MSAC$	(6.20)
	$A \times C$	$MSAC/MSE$	$MSAC/MSE$	(6.21)

Again, only the test for factor C differs depending on the A classification. Compared to the n replications case, we can see that the test for the interaction $A \times C$ uses MSE as divisor instead of $MSB(A)C$ which is confounded with the error in the single replicate case.

Table 6.12 — ANOVA table for a Repeated Measures Design with one treatment factor and single measurements at each time point

Source	*df*	*SS*	*EMS*
Between subjects	$ab - 1$		
A	$a - 1$	SSA	$\sigma^2 + bD_c\sigma^2_{\alpha\gamma} + cD_b\sigma^2_\beta + bc\sigma^2_\alpha$
$B(A)$	$a(b - 1)$	$SSB(A)$	$\sigma^2 + c\sigma^2_\beta$
Within subjects	$ab(c - 1)$		
C	$c - 1$	SSC	$\sigma^2 + bD_a\sigma^2_{\alpha\gamma} + ab\sigma^2_\gamma$
$A \times C$	$(a - 1)(c - 1)$	$SSAC$	$\sigma^2 + b\sigma^2_{\alpha\gamma}$
Error	$a(b - 1)(c - 1)$	SSE	σ^2
Total	$abc - 1$	SST	

a is the number of levels of treatment factor A, b is the number of levels of nested factor B, c is the number of time points for measurement, *df* defines degrees of freedom, *SS* defines sum of squares, *EMS* defines expected mean square, D is 0 for a fixed effect and 1 for a random effect

The general order of statistical testing in RMDs is, as in FDs, interactions **first** then other effects as appropriate. As previously, the amount of analysis will depend on which, if any, effects appear statistically important.

Exercise 6.8 The second stage of the analysis of the antibiotic activity data of Example 6.1 is the statistical assessment of the model effects.

EMS expressions and test statistics: Because the subject factor is classified as random and before we consider the statistical inference aspect, we must first check the construction and format of the effect test statistics. From the factor information and the fact that a mixed model is necessary, we know that $D_a = 0$ (formulation fixed), $D_b = 1$ (subject random), $D_c = 0$ (time fixed), $a = 3$, $b = 2$, and $c = 3$. Therefore, from the *EMS* expressions shown in Table 6.12, we have

Effect	*EMS*	*Test statistics*
Formulation (α)	$\sigma^2 + 3\sigma_\beta^2 + 6Q[\alpha]$	*MSF*/*MSS*(F)
Subject(Formulation) (β)	$\sigma^2 + 3\sigma_\beta^2$	*MSS*(F)/*MSE*
Time (γ)	$\sigma^2 + 6Q[\gamma]$	*MST*/*MSE*
Formulation × Time	$\sigma^2 + 2Q[\alpha\gamma]$	*MSFT*/*MSE*
Error	σ^2	

where $Q[.]$ refers to the sum of squared effects. Only the formulation effect test statistic is based on a different denominator mean square.

ANOVA table: Display 6.6 contains the ANOVA information generated by Minitab's Balanced ANOVA procedure. The ANOVA information presented mirrors the general form in Table 6.12 while the *EMS* information conforms to that presented. The *SS* term for the time effect is largest by far implying that antibiotic activity varies significantly with time of measurement. Variability estimation for the subject and error random factors uses the variance estimates provided in the column 'Variance component'.

Display 6.6

```
Source                DF         SS          MS        F        P
Formn                  2     375.84      187.92     8.63    0.057
Subject(Formn)         3      65.35       21.78     6.49    0.026
Time                   2    2597.42     1298.71   386.78    0.000
Formn*Time             4     282.52       70.63    21.03    0.001
Error                  6      20.15        3.36
Total                 17    3341.28

Source              Variance  Error  Expected Mean Square
                   component  term   (using restricted model)
 1 Formn                       2     (5) + 3(2) + 6Q[1]
 2 Subject(Formn)      6.142   5     (5) + 3(2)
 3 Time                        5     (5) + 6Q[3]
 4 Formn*Time                  5     (5) + 2Q[4]
 5 Error               3.358         (5)
```

Effect tests: Test statistic (6.21) for a mixed model with A fixed specifies that the formulation × time interaction F test statistic is the ratio of the interaction and error mean squares. From the information in Display 6.6, this ratio has provided the test statistic F = 21.03 (row 'Formn*Time', column 'F') with p value presented as 0.001. From these values, we can deduce that the formulation × time interaction is statistically significant suggesting that formulation effects appear to differ with time ($p < 0.05$) as highlighted in the discussion of Display 6.5.

The ratio of formulation and subject mean squares provide the formulation test statistic (6.18). From Display 6.6, this is F = 8.63 with p value 0.057 suggesting that formulation differences appear insignificant ($p > 0.05$) though only just. Based on this result, it could be said that the formulations appear similar in their effect on antibiotic activity but the closeness of the p value to the 5% significance level means that the evidence is not clear cut.

The subject test statistic (6.19) is the ratio of the subject and error mean squares producing a test statistic of F = 6.49 with p value estimated at 0.026 also signifying statistical difference as might be expected ($p < 0.05$). □

Often, the time or within subjects analysis can be influenced by deviations from the compound symmetry assumptions, particularly because responses may be correlated through time so violating the independence assumption. Also, the repeated measurements collected may not satisfy the Huynh-Feldt assumption. In such cases, modifications of the inference procedures for the C and $A \times C$ effects should be considered to account for departures from the assumptions.

One way of doing this involves adjusting the degrees of freedom of the time and error components by a 'correction factor'. Box (1954) proposed that $1/(c - 1)$ could be used. For A fixed and considering the single replicate case (see Table 6.12), use of correction factor would mean the test of factor C being based on degrees of freedom $df_1 = 1$ and $df_2 = a(b - 1)$, while the degrees of freedom for the test of the interaction $A \times C$ would become $df_1 = a - 1$ and $df_2 = a(b - 1)$. Other correction factors include one reflecting the level of deviation from compound symmetry (Milliken 1990), the Greenhouse-Geisser estimate (Kuehl 1994), and the less conservative Huynh-Feldt estimate (Kuehl 1994). The last two adjustments can be performed within SAS (see Appendix 6A.2). An alternative to the correction factor approach is to use a multivariate approach (Everitt & Der 1996). It is because this is such an important and widely used design that a great deal of research and variations on the design assumptions and interpretations exist.

6.7.4 Follow-up Procedures

Follow-up analysis within RMDs, as with FDs and NDs, depends on which effects are significant and the nature of such effects.

6.7.4.1 Significant $A \times C$ Interaction

For A and C fixed, significant $A \times C$ interaction can be investigated using a simple interaction plot of the treatment combination mean responses. Analysis of factor effects in the presence of interaction can be considered but is extremely complex.

Exercise 6.9 For the antibiotic activity study of Example 6.2, we found the interaction between formulation and time to be statistically significant. To understand this effect more fully, we must assess the interaction plot of combination means presented in Display 6.7.

Display 6.7

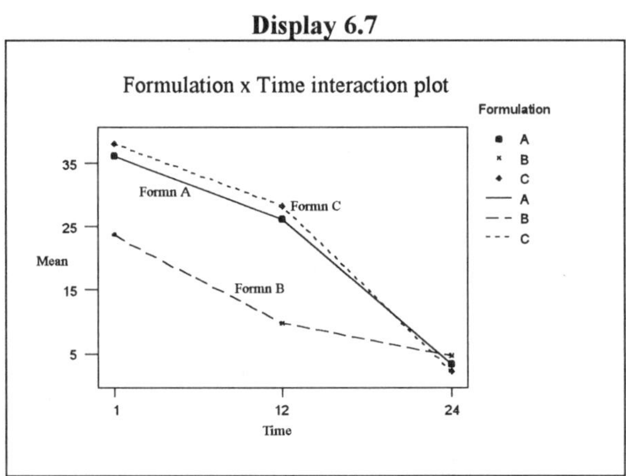

Formulation × time interaction: Presence of interaction is manifested through the non-parallel nature of the lines particularly from the 12 hour to 24 hour measurement. The plot shows the similarity in effect of formulations A and C but difference in trend of formulation B. A and C have higher antibiotic activity initially dropping slowly to 12 hours then more rapidly over the next 12 hour period. Formulation B is lower except at 24 hours when it appears to result in marginally higher mean activity. If high initial activity level is required, it would appear that formulations A and C appear better suited to the task. □

6.7.4.2 Significant Factor Effects but Insignificant $A \times C$ Interaction

When the $A \times C$ interaction is not significant, significant fixed factor effects can be assessed using multiple comparisons where the error variance term and error degrees of freedom must correspond to the relevant mean square denominator for the associated factor effect test statistic. When A is classified as a fixed effect, multiple comparison for factor A would use $MSB(A)$ as the error variance term and $a(b-1)$ as the error degrees of freedom while multiple comparison for factor C would be based on MSE and error degrees of freedom of $a(b-1)(c-1)$. For A classified as a random effect, the error variance for the multiple comparison for factor C is $MSAC$ with $(a-1)(c-1)$ error degrees of freedom. Linear contrasts of treatments can also be considered subject to the same principles. Milliken (1990) provides further details.

6.7.4.3 Variability Estimation

Variability estimation for the random effects can also be considered based on similar principles to those described in Sections 5.8.3 and 6.4.3.

Exercise 6.10 In the antibiotic activity study of Example 6.2, there were two random effects, subject and error. The associated variance estimates are provided in Display 6.6.

Variability estimation: The variance estimates for the subject and error effects are $\sigma^2_\beta = 6.142$ and $\sigma^2 = 3.358$ respectively. Based on $\text{Var}(X_{ijk}) = \sigma^2_\beta + \sigma^2$, we can specify that the subject effects account for 64.7% of the variation in the measured antibiotic activity levels while error accounts for 35.3% of this variation. Most variation appears to stem from the subjects chosen which is to be expected though the error percentage is high suggesting possible model misspecification and occurrence of missing factors. □

6.7.4.4 Assessment of Trend Effects

In a RMD, the response of subjects to treatments is measured over time. It would therefore be useful to be capable of comparing treatment trends over time. This can be achieved by modelling the response over time for each subject within each treatment and using the characteristics of the models to make treatment comparison. Modelling is based on expressing the relationship between response and time as a polynomial for each treatment, assuming time points equally spaced:

$$\text{response} = \alpha_{ij} + \beta_{1ij}T_k + \beta_{2ij}T_k^2 + \beta_{3ij}T_k^3 + \ldots + \varepsilon_{ijk} \qquad (6.22)$$

where $i = 1, 2, \ldots, a$, $j = 1, 2, \ldots, b$, and $k = 1, 2, \ldots, c$. The regression coefficients α and β are defined for the jth subject receiving the ith treatment where α_{ij} represents the intercept coefficient, β_{1ij} the linear coefficient, β_{2ij} the quadratic coefficient, etc., and T_k defines the kth time point of measurement.

Least squares estimation is used to fit an appropriate model to the responses for each subject to obtain coefficient estimates. One-way ANOVA (CRD) is used to compare the coefficient types for each treatment. The intercept test represents a comparison of the treatments at time $t = 0$ while the other coefficient tests essentially enable examination of the trend occurring within a significant interaction. Follow-up of significant trend effects is possible through application of multiple comparison procedures. Orthogonal polynomial regression contrasts can also be considered.

Exercise 6.11 To illustrate trend effect comparison, consider the antibiotic study of Example 6.2. A significant formulation × time interaction was detected and discussed in Exercise 6.9. For illustration of trend effect assessment, a linear trend will be fitted and assessed.

Trend model: The trend model fitted to the three observations corresponding to each subject is

$$\text{antibiotic activity} = \alpha_{ij} + \beta_{1ij}T_k + \varepsilon_{ijk}$$

where $i = 1, 2, 3$, $j = 1, 2$, $c = 1, 2, 3$, α_{ij} defines the intercept coefficient for the jth subject receiving the ith formulation, and β_{1ij} is the associated slope. The least squares estimates for the six subjects tested are presented in Table 6.13.

Table 6.13 — Parameter estimates for the linear model fit of each subject's antibiotic activity data

Formulation	Subject	Intercept (α_{ij})	Slope (β_{1ij})
A	1	41.6699	−1.40567
A	2	37.3523	−1.44748
B	1	25.9587	−0.92368
B	2	19.6840	−0.71222
C	1	44.6736	−1.66272
C	2	39.7902	−1.47758

ANOVA analysis: The one-way ANOVA based treatment comparisons of the estimated linear model parameters are summarised in Table 6.14. The results show significant differences in both intercepts and slopes ($p < 0.05$) suggestive of trend differences of the formulations over time. Applying the SNK multiple comparison procedure resulted in detection of formulation B as different from the other formulations for both trend effects ($p < 0.05$), lowest for intercept and nearest to zero for slope. □

6.7.5 Diagnostic Checking

As the RMD is based on formal model specification, inclusion of diagnostic checking must be an integral part of the data analysis. As previously, plotting of residuals against factors and model fits check for equality of response variability and adequacy of model while a normal plot of residuals enables normality of measured response to be assessed.

Table 6.14 — ANOVA tables for comparison of intercept and slope coefficients of the linear trend models for the antibiotic activity study of Example 6.2

	Source	*df*	*SS*	*MS*	*F*	*p*
Intercept	Formn	2	441.81	220.91	16.19	0.025
	Error	3	40.93	13.64		
	Total	5	482.74			
Slope	Formn	2	0.63790	0.31895	23.70	0.015
	Error	3	0.04037	0.01346		
	Total	5	0.67827			

Exercise 6.12 The final aspect of the analysis of the antibiotic activity data of Example 6.2 concerns diagnostic checking of the residuals using factor, model, and normal plots.

Factor plots: The formulation and time residual plots (not shown) provided acceptable plots and indicated no assumption violations. The time plot when split by formulation and subject, as shown in Display 6.8, highlight distinctive subject trends with time indicative of unequal response variability across the nested subjects.

Display 6.8

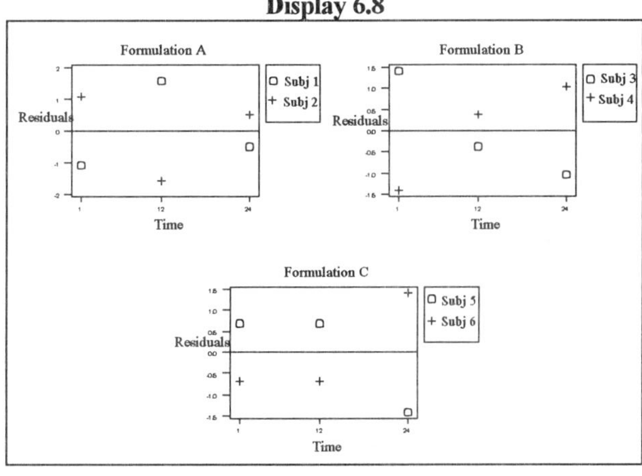

Fits and normal plots: The fits plot (not shown) exhibited two distinct clusters reflecting the range of magnitude of the responses suggesting inadequate model explanation. The normal plot shown in Display 6.9, though hinting at a curved trend, does not suggest any major departure from normality of antibiotic activity data. The Ryan-Joiner test confirmed this observation ($R = 0.96789, p > 0.1$). □

6.7.6 Unbalanced Designs

Unequal numbers of subjects in each treatment group, i.e. an unbalanced design, can also be handled by means of the **General Linear Model** (GLM) estimation procedure. It is necessary, however, to have complete time (factor *C*) measurements for each subject! In SAS, when there are both random and fixed effects, PROC MIXED is often preferred to PROC GLM when the design is unbalanced. Once again, unbalanced designs should be avoided as the determination and interpretation of sums of squares is not always clear cut.

Display 6.9

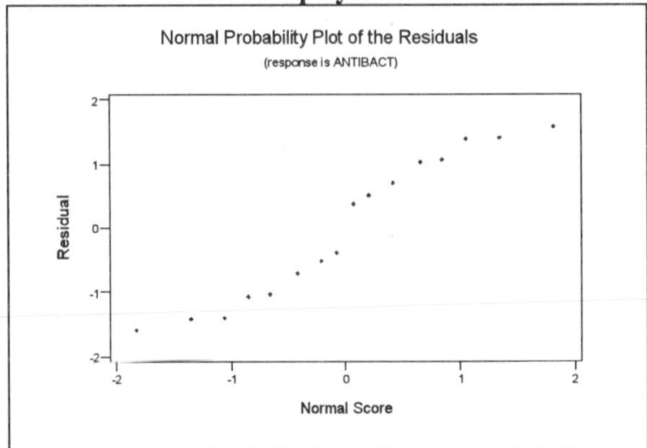

There are several additional points to raise concerning the practical application of RMDs. A key assumption is the independence of the repeated measures. It is possible that time points may be correlated leading to this assumption being invalid and causing difficulties with the inferences generated. Time points may also be unequally spaced so adjustment to the analysis of mean responses may be required, as such assumed time points equally spaced. Depending on the objective of the study, assessment of average response over time may not provide the most meaningful base for data interpretation. In some clinical trial applications of RMDs, the area under the curve (AUC) or the duration of a certain level of effect reduction may be used as the clinical response. In such cases, the RMD may be replaced by the simpler CRD.

6.8 CROSSOVER DESIGN

A **CrossOver Design (COD)** can be considered as a further enhancement of the RMD experimental design which controls for a further possible source of variation. Consider the general RMD structure of a treatments, b subjects in each treatment grouping, and c time measurements. In this structure, different subjects receive different treatments depending on treatment allocation procedure. Not all subjects therefore receive all treatments. This introduces possible bias in the treatment effect which can be easily avoided if each subject can be assigned to each treatment. Typically, treatments are assigned to the subjects according to a pre-determined sequence whereby the subjects are switched to alternative treatments after a 'washout' period which removes the possibility of a treatment carryover effect. This mode of operation ensures that all subjects receive all treatments with response measurement occurring within a particular time period.

The difference between a COD and a RMD therefore lies in the way the subjects receive the treatments. In a COD, all subjects are equally exposed and factor B (the subject effect) is still considered nested but within the treatment sequences rather than the treatments themselves. The COD arises often in clinical trials where it is important that new drugs or new formulations of a drug are tested across a selection of subject types. CODs will be discussed for the case where single replication is made at each time point.

6.8.1 Design Structure

To illustrate a COD structure, consider the antibiotic activity study of Example 6.2. To provide an ideal comparison independent of subject effects, we would require all subjects to be given each formulation. A potential structure satisfying such a requirement is presented in Table 6.15. Each of the formulations would be administered to the randomly assigned subjects in one of three possible sequences with a suitable washout period between administrations. Typically in drug studies, a guideline is to set the washout period at five times the half-life of the drug. Three sequences are necessary as three formulations are to be tested. Notice also that each subject receives each formulation once and each formulation is measured at each time point just as in the repeated measures study of Example 6.2. The structure essentially resembles a 3 × 3 LS design.

Table 6.15 — Possible CrossOver Design layout for revised antibiotic activity study

		Time		
Sequence	Subject	1h	12h	24h
1 (*ABC*)	1	*A*	*B*	*C*
	2	*A*	*B*	*C*
2 (*BCA*)	1	*B*	*C*	*A*
	2	*B*	*C*	*A*
3 (*CAB*)	1	*C*	*A*	*B*
	2	*C*	*A*	*B*

Example 6.3 Consider the COD variation to the antibiotic activity study outlined in Example 6.2. Suppose the antibiotic activity data measured within such a structure was as presented in Table 6.16 based on the testing of three treatment sequences.

Table 6.16 — Data for antibiotic activity for CrossOver Design study of Example 6.3

Formulation		Time		
sequence	Subject	1h	12h	24h
1 (ABC)	1	37.5 (A)	11.3 (B)	2.1 (C)
	2	34.8 (A)	8.4 (B)	2.3 (C)
2 (BCA)	1	26.9 (B)	30.3 (C)	5.4 (A)
	2	30.4 (B)	26.3 (C)	1.6 (A)
3 (CAB)	1	40.1 (C)	30.1 (A)	5.5 (B)
	2	36.1 (C)	22.1 (A)	3.9 (B)

Balancing for carryover effect in CODs occurs when treatment sequencing ensures each sequence of treatments occurs with equal frequency and so period and treatment effects will not be confounded. The COD structure presented in Table 6.15 is not balanced for carryover effect because three of the six possible treatment sequences are not accounted for, these being ACB, BAC, and CBA. The study could be made balanced by increasing the number of sequences to six through the inclusion of a second 3 × 3 Latin Square comprising the omitted treatment sequences. A fuller description of CODs is contained in Kuehl (1994).

A COD is essentially a mixture of a Repeated Measures structure (within subjects) and a Latin Square structure (order combinations of treatments). As such, CODs are often also called **Latin Square change-over designs**. Ideally, the number of sequences should equal the number of periods and the number of treatments, though extensions to number of sequences and periods can be handled. For an even number of treatments, a single LS Design is necessary for treatment assignment while for an odd number of treatments, two LS Designs are generally necessary, one for a first sequence of treatments and one for a second treatment sequence to ensure balancing for carryover effects. Design structures based on two Latin Squares are often referred to as **Williams Designs**. Incomplete designs are also possible.

CODs can increase the precision of treatment comparisons through the treatment of experimental units as blocks to help reduce experimental error and improve design efficiency. In addition, such designs efficiently use the experimental units available and as these often correspond to animal and human subjects, cost of experimentation can be reduced. Treatments and period (time) of measurement can affect a response with CODs enabling the influence of each to be separated and independently estimated. Use of treatment sequences provides a means of estimating and assessing carryover effects of treatments from one period to the next. Ideally, such an effect should be minimal within the subjects but cannot be ignored in design construction in case it could have an effect.

A disadvantage of CODs is that they can take longer as subjects receive treatments in succession with washout periods between treatments. In such cases, it may be necessary to settle for parallel treatment groups and the RMD, but the design will be less powerful than the COD.

6.8.2 Model for the Measured Response

Consider a general COD for a treatments tested across s sequences using b subjects for each sequence over c treatment periods. The **linear model** for the measured response X_{ijk} associated with such a COD is

$$X_{ijk} = \mu + \alpha_i + \beta_{j(i)} + \gamma_k + \tau_{m(i,k)} + \lambda_{m(i,k-1)} + \varepsilon_{ijk} \tag{6.23}$$

where $i = 1, 2,..., s$, $j = 1, 2,..., b$, $k = 1, 2,..., c$, and $m = 1, 2,..., a$. α_i denotes the effect of the ith treatment sequence, $\beta_{j(i)}$ is the effect of subject j nested within the ith treatment sequence, γ_k is the kth period effect, and ε_{ijk} is the random error. In general, we **assume** $\sum\alpha_i = 0$, $\sum\gamma_k = 0$, $\beta_{j(i)} \sim \text{NIID}(0, \sigma_\beta^2)$, and $\varepsilon_{ijk} \sim \text{NIID}(0, \sigma^2)$. $\tau_{m(i,k)}$ defines the **direct** effect of the mth treatment administered in period k of sequence i while $\lambda_{m(i,k-1)}$ is used to define the **carryover** effect of the mth treatment administered in period $(k - 1)$ of sequence i. Direct refers to the actual effect of the treatment and carryover to the residual effect of the treatment in the next treatment period. As no carryover can occur in the first time period, $\lambda_{m(i,0)} = 0$.

Exercise 6.13 Consider the crossover based antibiotic study of Example 6.3. First steps in the analysis concern, as usual, model specification for the antibiotic activity response.

Response model: Given that the design structure is one of crossover type, then the response model (6.23) for the antibiotic activity response will be

$$\text{antibiotic activity} = \mu + \alpha_i + \beta_{j(i)} + \gamma_k + \tau_{m(i,k)} + \lambda_{m(i,k-1)} + \varepsilon_{ijk}$$

where $i = 1, 2, 3$ ($s = 3$ sequences), $j = 1, 2$ ($b = 2$ subjects per sequence), $k = 1, 2, 3$ ($c = 3$ time points for measurement), and $m = 1, 2, 3$ ($a = 3$ formulations). α_i denotes the effect of the ith formulation sequence, $\beta_{j(i)}$ is the effect of subject j nested within the ith formulation sequence, and γ_k is the kth time point of measurement. $\tau_{m(i,k)}$ defines the direct effect of the mth formulation measured at time point k of formulation sequence i while $\lambda_{m(i,k-1)}$ refers to the carryover effect of the mth formulation administered in period $(k - 1)$ of formulation sequence i where we assume that $\lambda_{m(i,0)} = 0$ as no carryover effect can be present at the first measuring point.

ε_{ijk} refers to the random error and is assumed NIID(0, σ^2). For the fixed formulation sequence and time point effects, it is assumed that $\sum\alpha_i = 0$ and $\sum\gamma_k = 0$. As subjects have been randomly selected for the study, we assume also that the nested subject effect $\beta_{j(i)} \sim$ NIID(0, σ_β^2).

Explanation of treatment effects: Treatment effects in this COD can correspond to the direct and carryover effects of the formulations administered. Direct effect of formulation A, denoted τ_1, will be estimated from the formulation sequence and time combinations (1, 1h), (2, 24h), and (3, 12h). By similar reasoning, direct formulation B effects (τ_2) will be estimated from combinations (1, 12h), (2, 1h), and (3, 24h) while direct formulation C effects (τ_3) are estimable from the (1, 24h), (2, 12h), and (3, 1h) combinations.

Source of carryover effect estimation can be similarly determined. Formulation A was administered over time period 1 for sequence 1, time period 3 for sequence 2, and time period 2 for sequence 3. Carryover effects of formulation A, denoted λ_1, will be estimable from the measurements made in the time period following administration which will only apply to the first and last of these combinations. Therefore, only the observations collected from the (1, 12h) and (3, 24h) combinations will contribute to estimation of this carryover effect. Similar reasoning can be applied to source the observations contributing to carryover effects for formulations B (λ_2) and C (λ_3) which are (1, 24h) and (2, 12h), and (2, 24h) and (3, 12h) respectively. □

6.8.3 ANOVA Table and Test Statistics

Application of the **ANOVA principle** in a COD involves partitioning response variation into the between and within subject components defined in the response model (6.23). The associated **ANOVA table**, corresponding to fixed sequence, time, and treatment effects and random subject effect, is displayed in Table 6.17 where treatment effects are split into direct and carryover effects.

The sum of squares for the two types of treatment effects, direct and carryover, are non-orthogonal and so must be adjusted for the presence of the other form of treatment effect. Adjustment is based on computing the sum of squares differences for the experimental error between full and reduced models for the response X_{ijk} as outlined in Table 6.18. Completion of the crossover ANOVA table is then possible.

All factor effects within a COD can be analysed. **Test statistic construction** is similar to that for the single replicate RMD (see Section 6.7.3.2) with *MSE* used as divisor for all except the sequence test which is based on the mean square for the nested subject factor given its random classification. Sequence and period testing can be considered first with the former ideally expected to be insignificant. In respect of treatment effects, carryover effects should be tested first using carryover adjusted for direct effects and direct effects second using, in turn, direct adjusted for carryover effects.

Table 6.17 — ANOVA table for a CrossOver Design with single measurements at each time point assuming subjects random

Source	*df*	*SS*	*F* test
Between subjects	$sb-1$		
Sequence	$s-1$	*SS*Seq	*MS*Seq/*MS*Sub(Seq)
Subject(Sequence)	$s(b-1)$	*SS*Sub(Seq)	*MS*Sub(Seq)/*MSE*
Within subjects	$c-1+2(a-1)$		
Period	$c-1$	*SSP*	*MSP*/*MSE*
Treats (Direct)	$a-1$	*SSD*(unadj)	
Carryover (adjusted for Direct)	$a-1$	*SSC*(adj)	*MSC*(adj)/*MSE*
AND			
Treats (Carryover)	$a-1$	*SSC*(unadj)	
Direct (adjusted for Carryover)	$a-1$	*SSD*(adj)	*MSD*(adj)/*MSE*
Error	by subtraction	*SSE*	
Total	$sbc-1$	*SST*	

s is the number of treatment sequences tested, b is the number of subjects tested in each sequence, c is the number of time points for measurement, a is the number of levels of the treatment factor, *df* defines degrees of freedom, *SS* defines sum of squares, *MS* defines mean square, *F* test defines the associated test statistic

Table 6.18 — Adjusted sum of squares derivations for carryover and direct treatment effects in a CrossOver Design

	Carryover adjusted for Direct	Direct adjusted for Carryover
Reduced model	$\mu+\alpha_i+\beta_{j(i)}+\gamma_k+\tau_D+\varepsilon_{ijk}$	$\mu+\alpha_i+\beta_{j(i)}+\gamma_k+\lambda_C+\varepsilon_{ijk}$
Full model	$\mu+\alpha_i+\beta_{j(i)}+\gamma_k+\tau_D+\lambda_C+\varepsilon_{ijk}$	$\mu+\alpha_i+\beta_{j(i)}+\gamma_k+\lambda_C+\tau_D+\varepsilon_{ijk}$
SS calculation	$SSC(\text{adj}) = SSE_R - SSE_F$	$SSD(\text{adj}) = SSE_R - SSE_F$

D refers to direct effects, C refers to carryover effects, R refers to reduced model, F refers to full model

For software implementation, generally, the General Linear Model (GLM) procedures have to be used. However, special attention must be paid to data coding particularly in respect of inclusion of codes for the carryover effects. Detailed coding strategies for the crossover data associated with Example 6.3 are shown in Appendix 6A.3. An alternative coding strategy based on specification of carryover effects as covariates is also possible.

Exercise 6.14 The ANOVA information pertaining to the antibiotic activity crossover study of Example 6.3 is presented in Display 6.10. The information was computed using PROC GLM in SAS and provide requisite adjusted sums of squares for both types of treatment effect based on use of Type I sequential approaches for *SS* derivation.

Effect tests: The sequence test appears not significant at $p = 0.0973$ as would be hoped. The time effect, however, appears highly significant ($p = 0.0001$) indicating that time of measurement is important to antibiotic activity measurement.

Carryover effects adjusted for the presence of direct effects are tested through the test statistic $F = MS\text{Carry(adj)}/MSE = 10.35$ with p value estimated as 0.0114.

Both of these results point to significant carryover effects of the formulations ($p < 0.05$), a less than ideal result.

The test statistic $F = MS\text{Formn(adj)}/MSE = 14.59$ with p value of 0.0050 provide the means for assessing the presence of direct formulation effects accounting for carryover effects. The results suggest significance of effect indicating that the tested formulations appear to result in different antibiotic activity levels ($p < 0.05$). □

Display 6.10

```
Source            DF   Sum of Squares     Mean Square    F Value    Pr > F
Model             11    3428.54944444    311.68631313      58.94    0.0001
Error              6      31.72666667      5.28777778
Corrected Total   17    3460.27611111

       R-Square               C.V.          Root MSE     AntiBact Mean
       0.990831           11.65624        2.29951686       19.72777778

Source            DF        Type I SS     Mean Square    F Value    Pr > F
Sequence           2     144.43444444     72.21722222
Subj(Seq)          3      38.76833333     12.92277778       2.44    0.1620
Time               2    2877.75444444   1438.87722222     272.11    0.0001

       Formn       2     258.17444444    129.08722222      24.41    0.0013
       Carry(adj)  2     109.41777778     54.70888889      10.35    0.0114

       Carry       2     213.31822222    106.65911111      20.17    0.0022
       Formn(adj)  2     154.27400000     77.13700000      14.59    0.0050

Error              6      31.72666667      5.28777778
Corrected Total   17    3460.27611111

Tests of Hypotheses using the Type I MS for Subj(Seq) as an error term
Source          DF         Type I SS     Mean Square    F Value    Pr > F
Sequence         2      144.43444444     72.21722222       5.59    0.0973
```

Normally, if carryover effects are not significant, treatment inference based on means unadjusted for carryover effects is carried out. There is a school of thought, however, who consider that treatment comparisons should always be based on carryover effect adjustment irrespective of the significance of carryover effects. Such treatment inference can be based on the usual multiple comparison and linear contrast procedures with appropriate modification.

In CODs where the number of periods equals the number of treatments, the direct and carryover effects are not orthogonal. This nonorthogonality effect can be removed by the addition of an **extra period** which repeats the treatment sequence administered in the final sequence. Such a design provides for more efficient estimation of direct and carryover effects. The most widely used COD is the 2×2 structure comprising two treatments, two sequences, and two periods, i.e. one group receives sequence $A \rightarrow B$ and the other the sequence $B \rightarrow A$. Often, this structure is modified to include an extra period to provide better estimation for the direct and carryover effects. Information on these and other features of CODs are described in Kuehl (1994).

6.9 SPLIT-PLOT DESIGNS

Situations arise where neither the usual two factor experiment in randomised blocks nor a design with several observations for each treatment combination is convenient. A **Split-Plot Design** (**SPD**) is a multi-factor design involving randomised blocks. It occurs frequently in agricultural investigations where the experimental unit is the plot and where the levels of one or more factors are kept at a fixed value for a set of experimental units, while the levels of the remaining factors are randomised within each of the experimental units. This final randomisation is not possible in RMDs as the time intervals cannot be randomly assigned to the levels of time. This feature essentially represents the difference between the RMD and SPD structures. SPDs can often be referred to as **confounded Factorial Designs**.

6.9.1 Design Structure

Consider two factors, A at five levels and B at two levels. We could design the related experiment as illustrated in Table 6.19. In this layout, each level of factor A is used in one main plot in each block, with each main plot sub-divided and the two levels of factor B allocated at random to the sub-plots. This arrangement of the treatment combinations within each block is a more systematic one than would have occurred if the combinations had been allocated randomly. If the sub-plot treatment is required to be compared with the greater precision, then it is best to assign that treatment to the sub-plots.

Table 6.19 — Possible two factor Split-Plot Design layout

	Block 1	Block 2	Block 3
Main plots	*A*3 *A*5 *A*2 *A*1 *A*4	*A*4 *A*2 *A*1 *A*3 *A*5	*A*2 *A*5 *A*1 *A*4 *A*3
Sub-plots	*B*2 *B*1 *B*2 *B*2 *B*2	*B*2 *B*2 *B*1 *B*1 *B*2	*B*1 *B*1 *B*2 *B*1 *B*2
	*B*1 *B*2 *B*1 *B*1 *B*1	*B*1 *B*1 *B*2 *B*2 *B*1	*B*2 *B*2 *B*1 *B*2 *B*1

This type of treatment combination arrangement can also occur in industrial experimentation, where one series of treatments may require a large bulk of experimental material while another series can be compared with much smaller amounts. For instance, batches of an alloy from different furnaces might form the main plots and these would then be sub-divided, perhaps to test out different types of mould into which the alloy might be poured. If several replicates are used, the presence of an interaction between the two factors could also be considered.

6.9.2 Model for the Measured Response

Consider the general case of factor A at a levels applied to main plots and factor B at b levels applied to the sub-plots, the whole experiment being carried out over n blocks. The **linear effects model** for the measured response X_{ijk} within such a design may be written as

$$X_{ijk} = \mu + \rho_i + \alpha_j + \rho\alpha_{ij} + \beta_k + \alpha\beta_{jk} + \rho\alpha\beta_{ijk} \quad (6.24)$$

where $i = 1, 2,..., n$, $j = 1, 2,..., a$, and $k = 1, 2,..., b$. The term ρ_i defines the effect of the ith block, α_j is the effect of the jth level of factor A (main treatment), $\rho\alpha_{ij}$ is the main plot error, β_k is the effect of the kth level of factor B (sub-plot treatment), $\alpha\beta_{jk}$ is the interaction effect of the treatment combination (A_j, B_k), and $\rho\alpha\beta_{ijk}$ defines the sub-plot error. This form of model parameter definition means that the main plot error is the block × A interaction and the sub-plot error is the addition of the block × B and block × A × B interactions. Model **assumptions** concern the main plot and sub-plot errors which are generally assumed NIID(0, σ_1^2) and NIID(0, σ_2^2), respectively.

6.9.3 ANOVA Table and Test Statistics

In the statistical analysis, account must be taken of the fact that observations from different sub-units in the same unit may be correlated. In field experiments, this correlation may just be a reflection of the fact that, for example, neighbouring pieces of land tend to be similar in fertility and other agronomic properties. Thus, it is reasonable to assume that errors within a main plot may be correlated, while those of sub-plots in different main plots may be independent. The main effects of factor A are calculated entirely from the main plots, while the main effects of factor B are estimated within main plots from differences between the sub-plots and their error variance.

The **ANOVA table** and test statistic formats for a SPD structure is shown in Table 6.20 based on the mixed model where blocks are assumed random but factors A and B are assumed fixed effects. Error1 represents the main plot error and is equivalent to the block × A interaction. Error2 is the sub-plot error and is equivalent to the addition of the block × B and block × A × B interactions. In practice, MSE_1 should generally be larger than MSE_2. The block effect is usually assumed random and provided interaction between blocks and either A or B can be assumed negligible, **test statistic construction** for effects A, B, and $A \times B$ is the same irrespective of whether the experiment is of **model I**, **model II**, or **mixed model** type. Follow-up analysis using multiple comparisons or pairwise contrasts are available using appropriate error terms based on the "error" component associated with test statistic construction.

Table 6.20 — ANOVA table for a Split-Plot Design

Source	*df*	*SS*	*MS*	*F* test
Main plots	$na - 1$	$SSMp$		
Blocks	$n - 1$	$SSBl$	$MSBl$	
A	$a - 1$	SSA	MSA	MSA/MSE_1
Error1	$(n - 1)(a - 1)$	SSE_1	MSE_1	
Sub-plots	$na(b - 1)$	$SSSp$		
B	$b - 1$	SSB	MSB	MSB/MSE_2
$A \times B$	$(a - 1)(b - 1)$	$SSAB$	$MSAB$	$MSAB/MSE_2$
Error2	$(n - 1)a(b - 1)$	SSE_2	MSE_2	
Total	$nab - 1$	SST		

n is the number of blocks used, a is the number of levels of treatment factor A, b is the number of levels of treatment factor B, *df* defines degrees of freedom, *SS* defines sum of squares, *MS* defines mean square, *F* test defines the associated test statistic

A SPD is advantageous if the factor B and the interaction $A \times B$ effects are of greater interest than the factor A effects, or if the factor A effects cannot be tested on small amounts of material. It is a design in which the main effects are confounded with the blocks. This is in contrast to factorial experiments where it is generally interactions which are selected for confounding. The idea of split-plots can be extended by using a Latin Square design (two randomisation restrictions) instead of randomised blocks for the main treatments and by further splitting of the sub-plots into sub-sub-plots to provide a **Split-Split-Plot Design**.

PROBLEMS

6.1 As part of a pollution study, lead levels at five monitoring sites in an urban area were measured. Grass specimens from three areas within each monitoring site were collected and their lead levels, in μg g^{-1} dry weight, measured in duplicate as presented below. Does lead pollution differ with monitoring site and areas within monitoring site? Would increasing the number of replicate measurements to 3 improve design efficiency?

Site	A			B			C			D			E		
Area	1	2	3	1	2	3	1	2	3	1	2	3	1	2	3
	40	37.5	22.5	31	42.5	32.5	27.5	11	15	25	11	14	12.5	9	10
	40	37.5	20	37.5	43	36.5	30	10	14	25	11	15	10	8	9

6.2 The strain readings of glass cathode supports produced by three machines within a manufacturing plant were studied. Each machine has 10 heads on which glass is formed with a sample of four from each machine selected for assessment. The data below refer to replicate strain readings for each selected machine head. Do machines differ? Does the head used for glass formation cause significant variation in strain readings?

Machine	A				B				C			
Head	1	2	3	4	1	2	3	4	1	2	3	4
	6	13	1	7	10	2	4	0	0	10	8	7
	2	3	10	4	9	1	1	3	0	11	5	2
	0	9	0	4	7	1	7	4	5	6	0	5
	8	8	6	9	12	10	9	1	5	7	7	4

It has been suggested that increasing the number of heads sampled to five and reducing the number of replicate measurements to three could improve design efficiency. Is this the case?

6.3 In a genetic study of an insect population which damages crops, a random sample of four male insects were each mated with two females chosen at random. The weights, in g, of three eggs produced by each of the eight mated females were measured. Analyse the egg weight data fully to ascertain the importance of parents to egg weight.

Males	A		B		C		D	
Females	1	2	1	2	1	2	1	2
	3.8	3.7	5.1	6.1	7.1	4.9	7.7	6.4
	5.6	6.2	4.5	7.7	5.5	4.7	7.9	5.6
	4.1	5.7	7.5	5.5	7.0	4.1	7.9	7.4

The geneticist conducting this study wished to estimate the heritability index given by

$$(4\sigma_m^2)/(\sigma^2 + \sigma_m^2 + \sigma_f^2)$$

which measures the proportion of the weight variation caused by the average effect of genes and where the subscripts m and f of the variance terms refer to male and female respectively. True heritability is restricted to range from 0 to 1. Estimate the heritability and comment on whether egg weight is a heritable trait.

6.4 A study was carried out into how variations in sampling and testing affected a furnace lining material. The furnace lining manufacturer was interested in comparing how variations between production runs, between production batches produced within each run, and between samples from the same production batch affected the quality of the finished product. The quality response measured, as displayed below, was the ratio of two of the prime ingredients making up the lining material where high values correspond to good quality. Carry out a full and relevant assessment of these data to determine which factors, if any, have a strong influence on product quality.

Run	A				B			
Batch	1		2		1		2	
Sample	1	2	1	2	1	2	1	2
	0.46	0.49	0.50	0.52	0.57	0.64	0.59	0.62
	0.48	0.47	0.51	0.54	0.54	0.61	0.59	0.65

6.5 Garden fertiliser is sold in bags. A fertiliser supplier was concerned that their production process was not meeting its quality requirements in respect of the nitrogen content stated on the produced bags of fertiliser. Variations in the two standard ways in which the fertiliser components are mixed, the bags sampled, and the samples taken from the bags were all thought to be having a possible influence on the variation in the nitrogen content. A simple experiment based on a three factor nested structure using two sampled bags, three samples per bag, and two replications per sample was conducted to investigate this problem using percent nitrogen content as the response. Analyse these data fully commenting on the findings reached.

Mixture method	A						B					
Bag sampled	1			2			1			2		
Sample	1	2	3	1	2	3	1	2	3	1	2	3
	7.33	8.13	7.56	7.43	7.34	7.69	7.34	7.87	8.02	7.67	7.78	7.54
	7.45	8.07	7.64	7.51	7.31	7.65	7.38	7.83	7.97	7.71	7.74	7.62

6.6 The voltage regulator fitted to a motor driven lawnmower is required to operate within the range 15.8 to 16.4 volts. As a test of adequacy of voltage regulators bought in from their two regular suppliers, a lawnmower manufacturer subjected each of seven sampled regulators from each supplier to a quality assurance test. The test consisted of simulating lawnmower operation and measuring the operating voltage of the regulator at four time points. The operating voltage data are given below. Compare these test results for each factor carrying out the appropriate statistical tests at the 5% significance level.

Supplier	Regulator	Time of measurement 1	2	3	4
A	1	16.5	16.5	16.6	16.6
	2	15.8	16.7	16.2	16.3
	3	16.2	16.5	15.8	16.1
	4	16.3	16.5	16.3	16.6
	5	16.2	16.1	16.3	16.5
	6	16.9	17.0	17.0	17.0
	7	16.0	16.2	16.0	16.0
B	1	16.1	16.0	16.0	16.1
	2	16.0	15.9	16.2	16.0
	3	15.7	15.8	15.7	15.7
	4	15.6	16.4	16.1	16.2
	5	16.0	16.2	16.1	16.1
	6	15.7	15.7	15.7	15.7
	7	16.1	16.1	16.1	16.0

6.7 Three treatments A, B, and C are generally used to treat lameness in horses. Treatment effectiveness can be measured by the pressure that a horse can exert using the injured limb with high pressure considered ideal. To compare these three treatments, a repeated measures study was conducted on randomly selected horses suffering from lameness in one injured limb. Pressure was measured each day for five days after administration of the treatment. The horses were randomly assigned to the treatments in groups of five so that the horse effect corresponds to a nested factor. The pressure data were as follows:

Treatment	Horse	Time after administration (days) 1	2	3	4	5
A	1	28	30	31	33	36
	2	28	29	30	32	33
	3	26	28	28	31	32
	4	32	32	33	35	35
	5	29	30	31	31	33
B	1	26	27	29	31	32
	2	34	35	35	36	37
	3	31	31	32	34	35
	4	30	31	32	33	35
	5	27	27	28	30	31
C	1	27	28	29	31	33
	2	30	31	32	32	34
	3	31	31	32	34	36
	4	35	35	37	38	38
	5	24	26	27	28	29

Using these data, test for the presence of factor effects and fully analyse any that appear significant. Use 5% as the significance level for all relevant inference procedures.

6.8 A company wished to investigate the effect of three factors on the thickness of a coating substance it manufactures. The coating manufacturing process can be run at any one of three settings. Operator differences were also thought potentially important so 12 operators were randomly selected and assigned in groups of four to each setting level. The operators then ran the process for a fixed time period with thickness measurements taken at four time points over the study period. The study was therefore based on a RMD structure giving rise to the thickness measurements below. How does each factor affect coating thickness? Fully analyse these data for the information they contain.

		Time point of measurement			
Setting	Operator	1	2	3	4
A	1	83	85	82	81
	2	88	89	89	86
	3	81	85	83	80
	4	86	84	82	86
B	1	73	78	82	81
	2	82	77	76	75
	3	75	78	77	75
	4	80	82	81	77
C	1	95	91	88	89
	2	95	96	97	94
	3	86	87	85	84
	4	85	86	85	84

6.9 As part of an investigation of respiratory illnesses, a COD was used to compare four forms of treatment. The four treatments were an inhaler with special decongestant chemicals (A), a drug administered in solution (B), the same drug in tablet form (C), and another drug in solution (D). Each patient received their assigned treatment with an interval of one week between treatments. Twenty-four hours after administration, a biochemical test was conducted to measure the level of breathing abnormality present, low levels being indicative of significant relief. The design and biochemical measurement data are presented below. Carry out an appropriate analysis of these data to assess for factor effects, in particular, direct and carryover treatment effects.

		Period			
Sequence	Patient	1	2	3	4
1 (ADBC)	1	5.3	6.1	5.3	6.2
	2	5.6	6.3	5.5	6.4
2 (BACD)	1	5.4	5.5	6.1	5.6
	2	5.7	5.2	6.7	5.9
3 (CBDA)	1	6.1	5.8	6.7	5.7
	2	5.6	5.1	5.7	6.3
4 (DCAB)	1	6.5	5.3	6.1	5.8
	2	6.1	6.7	6.1	5.7

6.10 A COD was used to compare two homeopathic remedies and a placebo (A) in the initial treatment of dental neuralgia following tooth extraction. The remedies tested were Arnica (B) and Hypericum (C). Eighteen randomly selected patients requiring extraction of three teeth were divided into six groups of three with each group randomly assigned to each of six treatment sequences. Each patient had a tooth extracted at the start of the day. There was a week between extractions. A single dose, comprising four pellets, was administered to each patient three times on the extraction day at four hourly intervals. On the day following extraction, pain intensity was evaluated using a standard scale, low values signifying low pain. The collected data are shown below. Do the treatments differ in their effectiveness? Is there any evidence of carryover effects?

		Period		
Sequence	Patient	1	2	3
1 (ABC)	1	9	8	5
	2	5	4	6
	3	7	6	5
2 (BCA)	1	5	6	7
	2	8	6	7
	3	5	6	9
3 (CAB)	1	7	5	4
	2	8	6	5
	3	6	6	5
4 (ACB)	1	7	8	5
	2	6	5	8
	3	7	5	8
5 (BAC)	1	6	4	6
	2	6	7	4
	3	7	5	5
6 (CBA)	1	8	6	5
	2	7	9	6
	3	6	8	5

6A.1 Appendix: Software Information for Nested Designs

Data entry

Data entry for NDs conforms to the standard form of data coding used for most formal experimental design structures. For the nested factor, similar codes are used for the nested factor levels even if corresponding to different treatments. For Example 6.1, codes entry is necessary for the site (S) and batch (B) effects. The data structure necessary for software implementation would therefore look like

LghtSens	Site	Batch	LghtSens	Site	Batch	LghtSens	Site	Batch
69	S1	B1	71	S2	B1	57	S3	B1
71	S1	B1	72	S2	B1	71	S3	B1
84	S1	B2	54	S2	B2	62	S3	B2
84	S1	B2	56	S2	B2	64	S3	B2
78	S1	B3	52	S2	B3	78	S3	B3
79	S1	B3	49	S2	B3	79	S3	B3
58	S1	B4	64	S2	B4	56	S3	B4
61	S1	B4	66	S2	B4	58	S3	B4

Plots and summaries

Data plots and summaries can be generated in the usual manner.

SAS: PROC GPLOT and PROC MEANS can be used as described previously.

Minitab: The menu selections **Graph ➤ Plot** and **Stat ➤ Basic Statistics ➤ Descriptive Statistics** in Minitab provide access to the required summary presentations. Plotting in Minitab, however, requires numeric not character coding for the factor levels.

ANOVA information

SAS: For NDs within SAS, ANOVA output can be generated using the PROC NESTED procedure though the variance estimates produced refer to a model II experiment and they will not be wholly appropriate if the treatment factor A is a fixed effect. The statements necessary for generation of the information presented in Display 6.2 are as follows:

```
PROC NESTED DATA = CH6.EX61;
    CLASS SITE BATCH;
    VAR LGHTSENS;
PROC GLM DATA = CH6.EX61 NOPRINT;
    CLASS SITE BATCH;
    MODEL LGHTSENS = SITE BATCH(SITE);
    RANDOM BATCH(SITE);
    OUTPUT OUT = CH6.EX61A P = FITS R = RESIDS;
RUN;
```

PROC GLM is included to provide diagnostics information for the diagnostic analysis aspect. The NOPRINT option prevents the ANOVA output from being printed.

An alternative way of generating similar information is to use PROC GLM followed by PROC VARCOMP. PROC GLM can be used to provide the test statistics for each model effect, as well as comparison information for the fixed site effect. PROC VARCOMP generates variance component estimates for random effects in design models.

```
PROC GLM DATA = CH6.EX61 NOPRINT;
    CLASS SITE BATCH;
    MODEL LGHTSENS = SITE BATCH(SITE);
    RANDOM BATCH(SITE) / TEST;
    MEANS SITE / SNK E = BATCH(SITE);
    OUTPUT OUT = CH6.EX61A P = FITS R = RESIDS;
PROC VARCOMP DATA = CH6.EX61;
    CLASS SITE BATCH;
    MODEL LGHTSENS = SITE BATCH(SITE) / FIXED = 1;
RUN;
```

The TEST option specifies that test results for each model effect be calculated using appropriate error terms as determined by the *EMS* expressions. The MEANS statement enables the SNK multiple comparison to be applied to the fixed site effect, with the E option ensuring that the error *MS* used is that for batch nested within site, i.e. the correct denominator (error) *MS*. In the MODEL statement of PROC VARCOMP, the FIXED option specifies that the first (1) factor is classified as a fixed effect and thus, only variance component estimation is only to be produced for the random nested factor and the error.

Minitab: Comparable ANOVA output is generated within Minitab by choosing appropriate menu commands. For Display 6.2, these would be:

Select **Stat** ➤ **ANOVA** ➤ **Balanced ANOVA** ➤ for *Responses*, select **LghtSens** and click **Select** ➤ select the **Model** box and enter **Site Batch(Site)** ➤ select the **Random Factors (optional)** box and enter **Batch(Site)**.

Select **Options** ➤ **Use the restricted form of the mixed model** ➤ **Display expected mean squares** ➤ click **OK**.

Select **Storage** ➤ **Fits** ➤ **Residuals** ➤ click **OK** ➤ click **OK**.

Diagnostic checking

Residual plot derivation follows the same procedures as shown in Appendix 3A for the CRD.

6A.2 Software Information for Repeated Measures Designs

Data entry

Software data entry for RMD based studies follows similar principles to those for Nested Designs. Again, the nested factor codes must be repeated despite the fact that the subjects tested generally differ with treatment. Codes and response data entry for Example 6.2 would be as follows:

AntibAct	Formn	Subject	Time	AntibAct	Formn	Subject	Time
37.5	A	1	1	20.4	B	2	1
30.1	A	1	12	8.4	B	2	12
5.4	A	1	24	3.9	B	2	24
34.8	A	2	1	40.1	C	1	1
22.1	A	2	12	30.3	C	1	12
1.6	A	2	24	2.1	C	1	24
26.9	B	1	1	36.1	C	2	1
11.3	B	1	12	26.3	C	2	12
5.5	B	1	24	2.3	C	2	24

ANOVA information

SAS: ANOVA output for RMDs can be generated in SAS using the PROC GLM procedure though all printed test results must be treated with caution as all are based on the *MSE* term as divisor. Use of the RANDOM option within PROC GLM enables the *EMS* expressions to be produced and corrected tests for appropriate effects to be printed. The statements necessary for generation of the repeated measures ANOVA information for Example 6.2 (see Display 6.6) is as follows:

```
PROC GLM DATA = CH6.EX62;
    CLASS FORMN SUBJECT TIME;
    MODEL ANTIBACT = FORMN SUBJECT(FORMN) TIME FORMN*TIME;
    RANDOM SUBJECT(FORMN) / TEST;
    OUTPUT OUT = CH6.EX62A P = FITS R = RESIDS;
RUN;
```

The TEST option in the RANDOM statement ensures that the nested factor test statistic is calculated using the correct ratio of *EMS* expressions.

SAS can also generate both univariate and multivariate analysis information for RMDs, with the univariate information including both the Greenhouse-Geisser and Huynh-Feldt corrections to the factor and interaction tests. To do this, we require to enter the experimental data for each time point as a variable and include the formulation effect also.

The sequence of SAS statements which would could be used in this case would be as follows:

```
DATA CH6.EX62B;
     INPUT TIME1 TIME2 TIME3 FORMN $ @@;
     37.5 30.1 5.4 A   34.8 22.1 1.6 A   26.9 11.3 5.5 B   20.4 8.4 3.9 B
     40.1 30.3 2.1 C   36.1 26.3 2.3 C
;
PROC GLM DATA = CH6.EX62B;
     CLASS FORMN;
     MODEL TIME1-TIME3 = FORMN / NOUNI SS1;
     REPEATED TIME / PRINTE;
RUN;
```

In the MODEL statement, TIME1-TIME3 specifies that we have multivariate data. The option NOUNI suppresses the printing of univariate statistics while SS1 specifies that Type I *SS*s be used for test statistic derivation. The REPEATED statement requests that univariate and multivariate tests of the time factor and its interaction with formulation be printed. For the univariate tests, both unadjusted and adjusted results are printed, the adjusted results corresponding to both the Greenhouse-Geisser and Huynh-Feldt adjustments. The PRINTE option prints correlation matrices for the measurements as well as a simple test providing for assessment of the Huynh-Feldt condition (SAS/STAT User's Manual).

Minitab: ANOVA output creation, as illustrated in Display 6.6, is based on selection of appropriate menu options as follows:

Select **Stat** ➤ **ANOVA** ➤ **Balanced ANOVA** ➤ for *Responses*, select **AntibAct** and click **Select** ➤ select the **Model** box and enter **Formn Subject(Formn) Time Formn*Time** ➤ select the **Random Factors (optional)** box and enter **Subject(Formn)**.

Select **Options** ➤ **Use the restricted form of the mixed model** ➤ **Display expected mean squares** ➤ click **OK**.

Select **Storage** ➤ **Fits** ➤ **Residuals** ➤ click **OK** ➤ click **OK**.

6A.3 Appendix: Software Information for CrossOver Designs

Data entry

Data coding for a COD study follows similar principles to those described in Appendices 6A.1 and 6A.2. For the COD study of Example 6.3, codes for the sequence (Seq), subject (Sub), time (T), and formulation (F) effects are relatively straightforward to prepare. However, codes are also required for the carryover effects (C) to enable the reduced and full model fitting process to be implemented.

A simple coding scheme is generally adopted based on specifying "0" for the first period observation, "1" for cases where treatment 1 occurred in the previous period, "2" for cases where treatment 2 occurred in the previous period, and so on. Choice of "0" code for the first period observations reflects that there are no carryover effects in the first period. Carryover codes therefore correspond to the treatment code from the previous period. The complete coding for the antibiotic activity data of Example 6.3 is therefore as follows:

AntibAct	Seq	Sub	Time	Formn	Carry	AntibAct	Seq	Sub	Time	Formn	Carry
37.5	ABC	1	1	A	0	30.4	BCA	2	1	B	0
11.3	ABC	1	12	B	1	26.3	BCA	2	12	C	2
2.1	ABC	1	24	C	2	1.6	BCA	2	24	A	3
34.8	ABC	2	1	A	0	40.1	CAB	1	1	C	0
8.4	ABC	2	12	B	1	30.1	CAB	1	12	A	3
2.3	ABC	2	24	C	2	5.5	CAB	1	24	B	1
26.9	BCA	1	1	B	0	36.1	CAB	2	1	C	0
30.3	BCA	1	12	C	2	22.1	CAB	2	12	A	3
5.4	BCA	1	24	A	3	3.9	CAB	2	24	B	1

ANOVA information

SAS: ANOVA output for CODs can be generated using the PROC GLM procedure which produces test results based on the *MSE* term as divisor. As a COD contains an element of imbalance, it is necessary to use the Type I, or sequential, sums of squares as the ANOVA analysis base. Use of the TEST option within PROC GLM enables the corrected sequence test to be printed based on a random subject factor. The statements necessary for generation of the crossover ANOVA information for Example 6.3, as summarised in Display 6.10, are as follows:

```
PROC GLM DATA = CH6.EX63;
    CLASS SEQ SUBJ TIME FORMN CARRY;
    MODEL ANTIBACT = SEQ SUBJ(SEQ) TIME FORMN CARRY;
    TEST H = SEQ E = SUBJ(SEQ) / ETYPE = 1 HTYPE = 1;
PROC GLM DATA = CH6.EX63;
    CLASS SEQ SUBJ TIME FORMN CARRY;
    MODEL ANTIBACT = SEQ SUBJ(SEQ) TIME CARRY FORMN;
RUN;
```

First use of PROC GLM generates all the required ANOVA information including that for carryover adjusted for formulation effect. The TEST statement is included to ensure that the correct Sequence test statistic is printed based on the *MS* ratio required for this test, i.e. *MS*Seq/*MS*Subj(Seq), with H indicating numerator *MS* and E denominator (error) *MS*. The ETYPE and HTYPE elements are included to indicate that the type of *SS* terms to use, 1 signifying that the Type I. The second use of PROC GLM is included to provide the information for the effect of formulation adjusted for carryover by switching the order of the FORMN and CARRY variables.

Minitab: In Minitab, the general linear model (GLM) estimation procedure, outlined below, must be used but it will only provide the sequential sums of squares. Manual derivation of test statistics based on the produced sums of squares is therefore necessary. The first menu selection will generate information for the effect of carryover adjusted for formulation, while the second will provide the information for the formulation adjusted for carryover effect.

Select **Stat** ➢ **ANOVA** ➢ **General Linear Model** ➢ for *Responses*, select **AntibAct** and click **Select** ➢ select the **Model** box and enter **Seq Subject(Seq) Time Formn Carry** ➢ click **OK**.

Select **Stat** ➢ **ANOVA** ➢ **General Linear Model** ➢ for *Responses*, select **AntibAct** and click **Select** ➢ select the **Model** box and enter **Seq Subject(Seq) Time Carry Formn** ➢ click **OK**.

7

Two-level Factorial Designs

7.1 INTRODUCTION

Factorial experimentation is a powerful technique for scientific and technological research. However, when many factors require to be assessed, full FDs can be costly to implement. A **two-level Factorial Design** represents a particular form of factorial structure which can be very useful in multi-factor experimentation. Using only two levels of each factor reduces the number of treatment combinations within the experimental region but does not affect the ability of the design to provide independent effect estimates of all factors and their interactions. Such structures provide good estimation precision and enable the knowledge base of a practical investigation to be extended incrementally. They are extremely useful designs for exploratory screening purposes within product and process development, and they can also underpin process optimisation.

Two-level FDs are generally used to study the effect of k factors on a response, each factor set at two levels generally denoted as "low" and "high", e.g. the absence or presence of an ingredient, a temperature of 40 °C or 50 °C, and a mixing time of 10 or 20 seconds. Such factor level setting essentially assumes linearity of response between levels. Each level of every factor is tested with each level of every other factor resulting in all 2^k treatment combinations being tested. Further reduction in experimental effort is attained by only carrying out one experiment for each treatment combination, i.e. single replication, though replication may be necessary if repeatability and reproducibility elements are also to be assessed.

To illustrate the operation of such a design, consider a two-level FD to investigate the radius of propellant grain based on three possible factors: powder temperature (A), extrusion rate (B), and die temperature (C), i.e. a 2^3 design. The experimental region covered by the eight possible treatment combinations is displayed in Fig. 7.1 and shows that the design points provide full coverage of the experimental region.

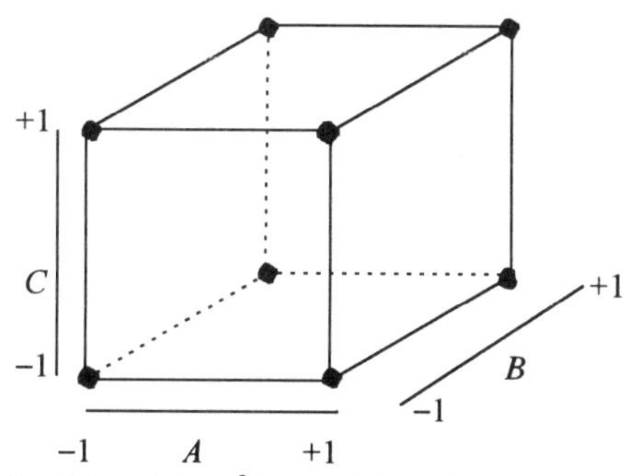

Fig. 7.1 — The 2^3 Factorial Design

Generally, such experimental structures can be summarised by means of a **design matrix**. The design matrix for the propellant grain study is displayed in Table 7.1 in standard Yates order based on single replicate measurement for each treatment combination. The minus sign denotes that the factor is at the "low" level and the plus sign that the factor is at the "high" level. This is known as **geometric notation** though specification of levels can also be numeric in form as 0 and 1 for low and high levels respectively. All combinations of factor levels are included in the design matrix. In addition, each factor level, or combination of levels, is experimented on an equal number of times as shown by the four – and four + in each column, so this unreplicated 2^3 experiment contains hidden replication.

Table 7.1 — 2^3 Factorial Design matrix

Combination	A	B	C	Radius
(1)	–	–	–	x
a	+	–	–	x
b	–	+	–	x
ab	+	+	–	x
c	–	–	+	x
ac	+	–	+	x
bc	–	+	+	x
abc	+	+	+	x

x denotes a response measurement

The 'Combination' column in Table 7.1 specifies the eight treatment combinations tested using a standard form of notation for two-level designs. Such combinations would be tested in a random order if feasible though there may be a necessity to run the experiment on the basis of trend-free run orders which negate the need for randomising run order as pointed out by Cheung (1988) and Wang (1995). The notation is based on denoting the higher levels of the factors A, B, and C by the letters a, b, and c, respectively. Lower levels of each factor are denoted by the absence of the letter referring to the factor. Thus, the treatment combination **(1)** denotes all factors at low level while combination ***a*** denotes factor A high level and factors B and C both at low level. Combination ***b*** denotes factor A low level, B high, and C low while the combination ***ab*** defines factors A and B at high level and C low level. The other four treatment combinations ***c***, ***ac***, ***bc***, and ***abc*** can be similarly referenced.

All full two-level FDs are based on a group of properties which highlight the appropriateness of such design structures. These properties are

- every treatment combination is included and the design points provide full coverage of the experimental region;
- the design is perfectly balanced as every factor, and factor level combination, appear an equal number of times;
- all response measurements can be used to estimate the factor effects providing estimates of high precision;

- all main effects and interaction effects can be estimated independently.

7.2 CONTRASTS AND EFFECT ESTIMATION

Analysis of data from two-level FDs are based on the specification of **contrasts** and the **estimation of factor effects**. Contrasts essentially represent measures of the difference in the level of the response caused by changing factor levels or treatment combination levels. Effect estimation provides a measure of the average effect on the response of changes in the levels of factors or treatment combinations. Effect estimates are essentially the average of the associated contrast values.

Example 7.1 Data were collected on the effects of three two-level factors on fish survival. The factors were temperature (A), cyanide pollution (B), and oxygen concentration (C). Cyanide is used to control fish parasites and other diseases. The survival time of the fish was recorded once for each treatment combination as shown in Table 7.2. Single replication was necessary due to restricted testing facilities. The question arises as to which effects exert strongest influence on fish survival.

Table 7.2 — Design matrix and fish survival time data for Example 7.1

Run	Temperature (A)	Cyanide (B)	Oxygen (C)	Combination	Time response
3	low	low	low	(1)	122
5	high	low	low	a	143
2	low	high	low	b	85
6	high	high	low	ab	64
4	low	low	high	c	131
1	high	low	high	ac	168
8	low	high	high	bc	101
7	high	high	high	abc	79

7.2.1 Contrasts

Contrasts can be constructed for all factor effects. A **main effect contrast** for a controlled factor is constructed as the difference between the responses at the high level and the responses at the low level, i.e. (high level – low level). The **contrast for a two factor interaction** is constructed as the difference between the responses when each factor is set at the same level and the responses when each factor is set at the opposing level, i.e. same – opposite. Such general definitions can be easily extended to higher order interactions. To explain contrasts further, consider the 2^3 study described in Example 7.1 and let (1), a, b, ab, c, ac, bc, and abc represent the survival time observation of the eight possible treatment combinations. The order in which these treatment combinations is listed is known as **standard order**. Using such response definitions, general contrast expressions for all possible effects within a 2^3 design can be easily derived. These are summarised in Table 7.3.

Table 7.3 — Contrasts for a three factor two-level (2^3) Factorial Design

Effect	Contrast
A	$a + ab + ac + abc - (1) - b - c - bc$
B	$b + ab + bc + abc - (1) - a - c - ac$
AB	$(1) + ab + c + abc - a - b - ac - bc$
C	$c + ac + bc + \text{abc} - (1) - a - b - ab$
AC	$(1) + b + ac + abc - a - ab - c - bc$
BC	$(1) + a + bc + abc - b - ab - c - ac$
ABC	$a + b + c + abc - (1) - ab - ac - bc$

Using the simple main effect contrast definition for factor A, a contrast expression of

$$\text{contrast}_A = a + ab + ac + abc - (1) - b - c - bc \tag{7.1}$$

can be determined for the temperature factor A within Example 7.1. The responses a, ab, ac, and abc represent those for temperature set at the high level while (1), b, c, and bc correspond to those for temperature set at the low level. Contrasts for cyanide (B) and oxygen (C) can be similarly constructed. From the data presented in Table 7.2, the contrast for temperature, using expression (7.1), can be evaluated as

$$143 + 64 + 168 + 79 - 122 - 85 - 131 - 101 = 15$$

suggesting that increasing temperature can increase fish survival time by 15 units. The contrast effects for cyanide and oxygen are –235 and 65 respectively.

For the temperature and cyanide interaction ($A \times B$), the contrast can be summarised as the difference between the responses for the two factors tested at the same level and the responses for them at the opposite level. This leads to the $A \times B$ contrast being defined as

$$A_{\text{low}}B_{\text{low}} + A_{\text{high}}B_{\text{high}} - A_{\text{low}}B_{\text{high}} - A_{\text{high}}B_{\text{low}} \quad .$$

Based on the standard notation for the fish survival time observations, we can specify this contrast as

$$\text{contrast}_{A \times B} = (1) + ab + c + abc - a - b - ac - bc \tag{7.2}$$

This contrast is independent of the third factor oxygen (C) as each level of this factor appears at both low and high level in both the positive and negative parts of the contrast. The contrasts for the temperature × oxygen interaction ($A \times C$) and the cyanide × oxygen interaction ($B \times C$) can be similarly defined. The data presented in Table 7.2 results in the expression (7.2) becoming

$$122 + 64 + 131 + 79 - 143 - 85 - 168 - 101 = -101$$

suggestive of a strong interaction effect. The $A \times C$ and $B \times C$ interactions are very much lower by comparison at 15 and –3 respectively.

The contrast for the three factor interaction $A \times B \times C$ is a measure of the differences in a related two factor interaction at the two levels of the third factor. One possible way of finding this contrast is to take the difference in the interaction $A \times B$ at the two levels of C as follows:

$$\text{contrast}_{A \times B \times C} = AB_{\text{high}C} - AB_{\text{low}C} \tag{7.3}$$

From the $A \times B$ contrast definition, we have

$$AB_{\text{high}C} = AB_{\text{same}} \text{ at high } C - AB_{\text{opp}} \text{ at high } C \qquad AB_{\text{low}C} = AB_{\text{same}} \text{ at low } C - AB_{\text{opp}} \text{ at low } C\ .$$

In the usual notation, these become

$$AB_{\text{high}C} = c + abc - ac - bc \qquad AB_{\text{low}C} = (1) + ab - a - b.$$

Combining these, as per expression (7.3), provides the contrast expression

$$\text{contrast}_{A \times B \times C} = a + b + c + abc - (1) - ab - ac - bc \tag{7.4}$$

Thus, for the fish survival study of Example 7.1, the contrast for the three factor interaction becomes

$$143 + 85 + 131 + 79 - 122 - 64 - 168 - 101 = -17.$$

Contrasts can therefore be expressed in terms of the treatment combinations tested with contrast coefficients capable of being summarised in a simple table as illustrated in Table 7.4. The coefficients in each column of this table sum to 0 for each factorial effect, except the mean, and multiplication of any pair of columns provides another column in the table. Therefore, once contrasts are established for the main effects in two-level designs, simple column multiplication, row by row, can establish the contrast patterning for the interaction effects. Contrasts exhibiting these properties are called **orthogonal contrasts** resulting in each factorial effect being independently estimable. On the basis of this property, two-level designs are often referred to as **orthogonal designs** and, as each factor level and combination of levels appear the same number of times in the design structure, they also exhibit a **balancing property**.

Table 7.4 — Contrast matrix for a three factor two-level (2^3) Factorial Design

Treatment combination	Factorial effect Mean	A	B	C	AB	AC	BC	ABC
(1)	+	−	−	−	+	+	+	−
a	+	+	−	−	−	−	+	+
b	+	−	+	−	−	+	−	+
ab	+	+	+	−	+	−	−	−
c	+	−	−	+	+	−	−	+
ac	+	+	−	+	−	+	−	−
bc	+	−	+	+	−	−	+	−
abc	+	+	+	+	+	+	+	+

7.2.2 Effect Estimation

Analysis of results from single replicate two-level designs is based, primarily, on estimating and analysing factor and interaction effects. An effect estimate is essentially an average measurement reflecting how changing levels affect average response. From the effect contrasts specified in Table 7.3, effect estimates are generally expressed as

$$\textit{Effect estimate} = (\textit{Effect contrast})/2^{k-1} \tag{7.5}$$

as each factor level, or combination of factor levels, has been experimented on 2^{k-1} times. Estimation of factor effects is relatively straightforward once data are collected and use is made of the relevant contrast expressions. If replication of treatment combinations has occurred, the divisor in expression (7.5) changes to $n2^{k-1}$ to include n, the number of replications.

Exercise 7.1 Using the fish survival study of Example 7.1, we will now demonstrate effect estimation manually and introduce software generation.

Contrast derivation: Consider the temperature factor. Expression (7.1) provided a temperature contrast estimate of 15. We know $k = 3$ so $2^{k-1} = 4$ represents the number of times each level of temperature was tested. The effect estimate expression (7.5) therefore produces an average effect of temperature on fish survival time of $15/4 = 3.75$. By similar means, the average effect of the cyanide factor is estimated as $-235/4 = -58.75$ while that for the oxygen factor becomes $65/4 = 16.25$.

For the temperature × cyanide interaction ($A \times B$), the effect estimate using Table 7.3, expression (7.2), and expression (7.5) is $-101/4 = -25.25$. For the temperature × oxygen ($A \times C$) and cyanide × oxygen ($B \times C$) interactions, the effect estimates are $15/4 = 3.75$ and $-3/4 = -0.75$ respectively. The effect estimate for the three factor interaction is $-17/4 = -4.25$.

Software generation: Most statistical software can provide effect estimation procedures for two-level designs. Display 7.1 contains the Minitab output of effect estimates for the observed fish survival times based on using the **Stat ➤ DOE ➤ Analyze Custom Model** procedure (see Appendix 7A). The output contains the necessary effect estimates in the column 'Effect' with the values presented agreeing with those derived using the contrast approach. The column 'Coef' provides the coefficient estimates for a regression model of response as a function of the seven possible effects within a 2^3 design. Essentially, these coefficient estimates are half the effect estimates.

From the effect estimates obtained, we can see that three effects stand out from the rest numerically, these being the main effects of cyanide and oxygen and the temperature × cyanide interaction. These represent the likely important effects in respect of fish survival time. □

Display 7.1

Estimated Effects and Coefficients for SurvTime		
Term	Effect	Coef
Constant		111.62
Temp	3.75	1.87
Cyanide	−58.75	−29.37
Oxygen	16.25	8.12
Temp*Cyanide	−25.25	−12.62
Temp*Oxygen	3.75	1.88
Cyanide*Oxygen	−0.75	−0.38
Temp*Cyanide*Oxygen	−4.25	−2.13

Contrast approaches can be feasible for two-level designs based on few factors where specification of contrasts is relatively straightforward. An alternative to this mechanism of effect estimation is to use **Yates' method**. This involves setting up a simple table, the first column of which is the treatment combinations in logical order and the second the response, or total responses for replicate measurement. The remaining columns are determined by adding and subtracting column entries in pairs for as many columns as there are factors, i.e. the table will have k columns, with the final column providing the contrast estimate. Table 7.5 demonstrates application of Yates' method for Example 7.1. Essentially, the procedure involves adding the first two results (122 + 143) and putting the answer to the top of column 1. The second, third, and fourth pairs of results are treated similarly to provide the next three entries of column 1. Then subtract the results in pairs as (second − first), (fourth − third), and so on to fill in the rest of column 1. This procedure is repeated on column 1 to generate column 2 and finally on column 2 to generate the contrast column (column 3). Once contrasts are determined, we can use expression (7.5) to derive the relative effect estimates as provided in the 'Estimate' column, i.e. the estimate of the a effect will be $15/2^{3-1} = 3.75$.

Table 7.5 — Effect estimation by Yates' method for fish survival study of Example 7.1

Combination	Time	1	2	Contrast	Estimate
(1)	122	265	414	893	111.625
a	143	149	479	15	3.75
b	85	299	0	−235	−58.75
ab	64	180	15	−101	−25.25
c	131	21	−116	65	16.25
ac	168	−21	−119	15	3.75
bc	101	37	−42	−3	−0.75
abc	79	−22	−59	−17	−4.25

Effect estimates can be either positive or negative with appropriate interpretation depending on the objective of the experimentation. For maximising a response, positive estimates imply high level best, or same levels for interaction effects, while a negative estimate would suggest low level best, or opposite levels for interaction effects. For response minimisation, the opposite interpretation is required. However, it is generally not feasible to specify a *'target'* estimate for discerning important effects with interpretation based solely on comparing estimates within themselves.

In two-level experimentation, it is hoped that effects will break down into two distinct categories, important (active) and unimportant (inactive). This is defined as **effects sparsity** or **factor sparsity** where it is assumed only a few effects are active and have a real influence on the response with these hopefully being main effects and low order interactions. Effect estimation also depends on the factor levels chosen for experimentation so choice of levels is important as it can influence the practical importance of the controlled factors.

7.2.3 Missing Values

In two-level FDs, an observation could be missing resulting in incomplete data. A simple way of estimating a single missing observation is to set the highest order interaction to zero as its effect is most likely to be negligible. Setting the contrast for this effect equal to zero enables the missing observation to be estimated. For two missing observations, two contrast effects must be set equal to zero providing two equations in the two unknowns which can be readily solved. Box (1990) describes this process in more detail.

7.3 INITIAL ANALYSIS COMPONENTS

7.3.1 Exploratory Analysis

Though effect assessment is the main aspect of two-level design analysis, EDA of the response data, including assessment of run order, should also be considered to gain initial insight into how the factors tested affect the response. Obviously, such an approach cannot provide interaction information but can provide a useful starting point for the data analysis.

Exercise 7.2 Simple plots and summaries of fish survival time against the run order of experiments from Table 7.2 and the factors (not shown) represent the first stage in the data analysis.

EDA: The run order plot appeared reasonably random though a comparable upward trend across runs 3 to 5 and 6 to 8 appeared to occur. In respect of temperature, survival time showed little difference with level tested but much higher variability at high temperature. A downward trend was apparent for cyanide as might be expected. A moderate upward trend occurred with changes to oxygen concentration with both levels providing widely varying results. □

7.3.2 Effect Estimate Plots

Numerical assessment of effect estimates represents the first step in the data analysis for two-level designs. If the number of factors is large, however, this approach becomes impractical. Graphical presentations of the effect estimates then become the most appropriate way of presenting and assessing the estimate information. Several forms of graphical presentation can be adopted for this purpose including **normal plot**, **half-normal plot**, **Pareto chart**, and **active contrast plot**. Often, it is advisable to use at least two of these plots to assess effect estimates though all should produce comparable results in respect of important and unimportant effects.

7.3.2.1 Normal Plot

The first, and most often used, is the **normal plot** corresponding to a plot of the normal scores of each effect estimate against the estimates themselves. The plot structure is similar to that used in diagnostic checking (see Section 3.5.1). An "ideal" normal plot should exhibit a distinct split between unimportant and important effects similar to that displayed in Fig. 7.2. The effectiveness of this plot is best therefore when only a small proportion of effects are active. Unimportant effects should cluster around zero with effect sizes such that a simple, almost vertical, straight line links them together. Important effects, by contrast, should appear at the bottom left (large negative: C) or top right (large positive: D, $A \times C$, and A) of the plot in positions distinctly different from those occupied by the unimportant effects signifying that the important effects are not due to random chance variation. Such a plot would provide appropriate evidence for the existence of effects sparsity.

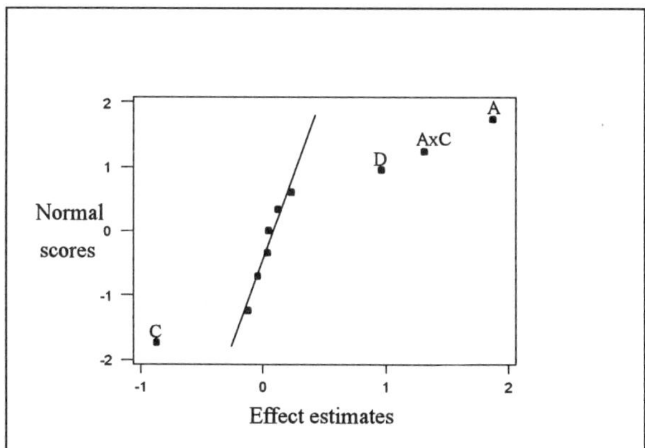

Fig. 7.2 — Illustration of normal plot of effect estimates

Exercise 7.3 Exercise 7.1 illustrated effect estimation for the fish survival experiment described in Example 7.1. A Minitab generated normal plot of these effect estimates is presented in Display 7.2 based on $\alpha = 0.20$ as the significance level for distinguishing important and unimportant effects. See Appendix 7A for menu instructions.

Display 7.2

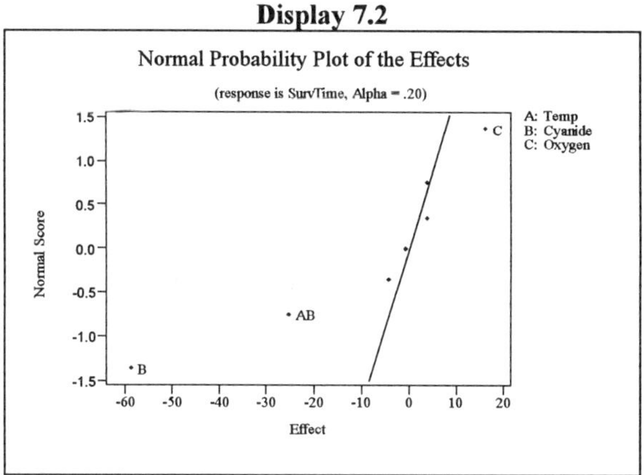

Normal plot of effect estimates: The normal plot in Display 7.2 exhibits an "ideal" pattern with distinct separation of effects shown. The same three effects, as indicated in Exercise 7.1, stand out, these being cyanide (B), the temperature × cyanide interaction (AB), and oxygen (C). The first two of these effects have large negative estimates while oxygen is large positive and least of the three. For cyanide, the negative estimate suggests, based on maximum survival time as optimal, that low cyanide pollution improves survival time as might be expected. The negative estimate for the temperature × cyanide interaction suggests that opposing factor levels may be best for improving survival time. Oxygen is the only important effect providing a positive estimate suggestive that high oxygen concentration appears beneficial. □

7.3.2.2 Half-normal Plot

The **half-normal plot**, first described by Daniel (1959), uses the absolute values of the effect estimates to provide a picture of the relative size of effects. The X axis of the plot refers to the absolute values of the effect estimates while the Y axis can be based on either the position value of the absolute effect estimates or the approximate standard normal probability $p_i = (i - 3/8)(n + 1/4)$ of the position values i of the absolute effect estimates. Fig. 7.3, based on the same data used to generate Fig. 7.2, demonstrates an "ideal" half-normal plot. This shows that a half-normal plot is similar to the normal plot except that the lower left of the graph (negative estimates) is flipped over with important effects ideally showing up as points appearing in the top right corner deviating markedly from a near vertical straight line through the unimportant effects. Guard-rail values for half-normal plots enabling statistical decisions to be made on effect significance also exist though their application can often result in a misleading picture (Bayne & Rubin 1986).

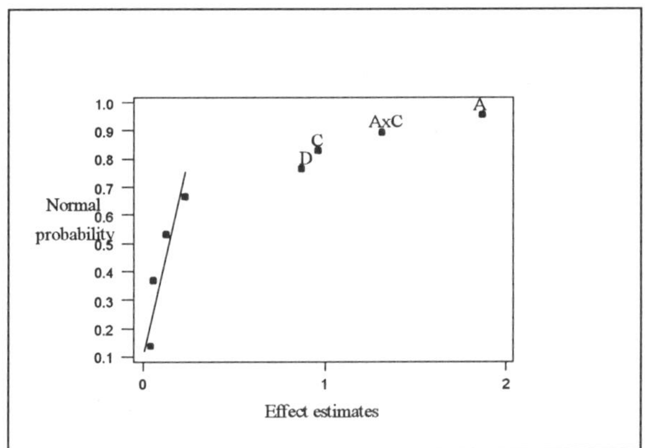

Fig. 7.3 — Illustration of half-normal plot of effect estimates

7.3.2.3 Pareto Chart

A **Pareto chart**, illustrated in Fig. 7.4, is a means of effect estimate plotting similar to a horizontal histogram. As with other effect estimate plots, a Pareto chart is particularly useful when effects sparsity is present. Important effects will lie in the top right corner of the plot as illustrated with a distinct "elbow" effect apparent specifying the jump from unimportant to important effects. A Pareto chart is therefore not dissimilar to the half-normal plot in appearance and interpretation.

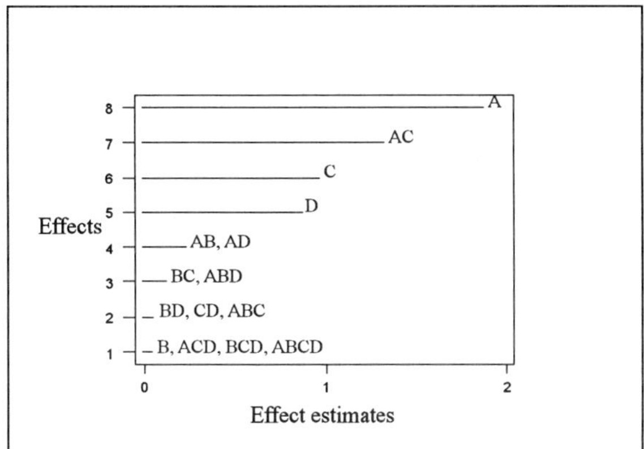

Fig. 7.4 — Illustration of Pareto chart of effect estimates

Exercise 7.4 A Minitab generated Pareto chart for the effects associated with the fish survival time study of Example 7.1 is provided in Display 7.3. The chart, as with the normal plot of Display 7.2, has been based on choice of significance level of 20% ($\alpha = 0.20$).

Display 7.3

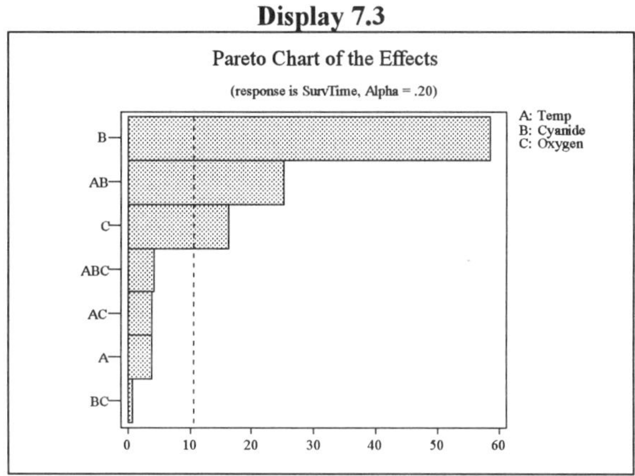

Pareto chart: The chart exhibits the classic characteristics of effect sparsity as would be expected given the normal plot presentation of Display 7.2. A distinct elbow can be seen between effects ABC and C where the divide between unimportant and important effects appears, The three important effects are as those discussed in Exercise 7.3 with the remaining four effects appearing to have little effect on fish survival time. □

7.3.2.4 Active Contrast Plot

The **active contrast plot**, or **Bayes plot**, graphically displays the probability that an effect has an important influence on the response (Haaland 1989). It is similar in style to a Pareto chart with the posterior probability of an effect being active replacing the effect estimate as the X axis base of the plot. Probabilities greater than 0.5 are generally suggestive of important effect. Construction of the plot depends also on the investigator specifying the prior probability of effect importance and a scale factor describing how much larger estimates for "real" effects are likely to be compared to unimportant effects. For screening experiments, a prior probability of 0.4 is considered a reasonable choice. The scale factor can be set at values of 5, 10,

15, and 20. An active contrast plot should be produced for each value of the scale factor and the plots reviewed. The 'best' plot which clearly separates important and unimportant effects is then chosen.

7.3.3 Data Plots and Summaries

Plot and summary analysis of important effects provide the final part of the initial data analysis for two-level designs to help understand the active effects and explain how they affect the measured response. For main effects, we plot the average response, in conjunction with response range, for each level while for two factor interactions, interaction plots should be used. Ideally, we want to use these methods to understand how these effects influence the experimental response and possibly decide on an optimal combination of factors corresponding to the "best" response, maximum or minimum with acceptable consistency. As in all experimentation, however, such a goal may not always be attainable and it may be necessary to **trade-off** to decide "best" results.

Exercise 7.5 Effect estimate analysis for the fish survival experiment in Example 7.1 indicated that cyanide (B), oxygen (C), and the temperature × cyanide interaction ($A \times B$) appeared to be the most important effects. To understand more fully how these affect fish survival time, we should assess plots of them and describe their practical implications. Such summary information is displayed in Display 7.4 for the main effects and Display 7.5 for the highlighted interaction.

Main effects analysis: The main effects plot indicates that the major change in survival time (sharply sloping line) occurs when cyanide levels increase from low to high as might be expected. For cyanide, the point at low level is the average of the treatment combinations (1), *a*, *c*, and *bc* and so equal to 564/4, i.e. 141. At the high level, the point is the average of the combinations *b*, *ab*, *bc*, and *abc* and so equal to 329/4, i.e. 82.25. Evaluating the difference (high − low) for these points provides the cyanide effect estimate, i.e. 82.25 − 141 = −58.75.

A slight increase in survival time is highlighted as oxygen concentration increases. Variability differences, using the range (maximum − minimum), enables dispersion effects to be assessed. The values shown in Display 7.4 indicate that consistency of observation between factor levels is comparable though higher for the oxygen factor. This comparability means that the decision on "best" levels appears unaffected by measurement variability.

Interaction effects analysis: From the plot in Display 7.5, the non-parallel lines provide evidence of the presence of an interaction effect. We can see that, as cyanide levels increase, a different pattern emerges for the two temperature levels. For low temperature, survival time marginally decreases while for high temperature, the decrease is more marked. From the plot, it would appear that low cyanide and high temperature could provide for longest survival time. The original data in Table 7.2 highlight that this combination did provide the best results with oxygen at high concentration. The range values, presented as the figures in brackets, indicate marginally higher variability for this "best" combination with lowest variability occurring at low levels of each factor. Higher levels of cyanide appear to produce more variable results as noted in Exercise 7.1. No other combination comes near to this one in either mean or consistency.

Display 7.4

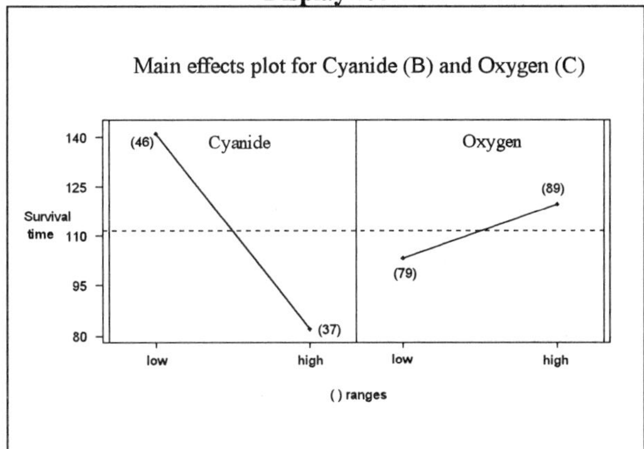

Display 7.5

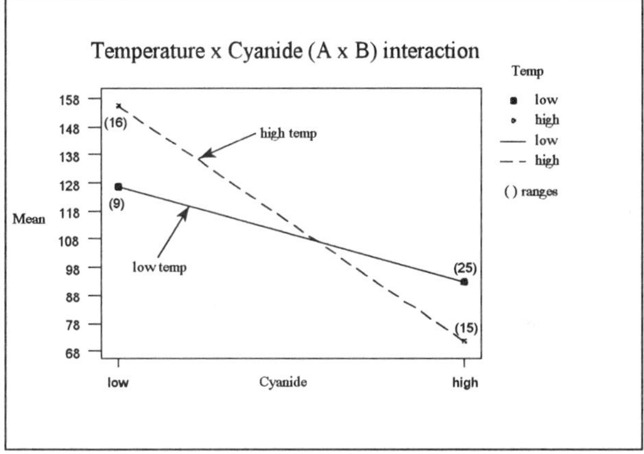

Summary of findings: In two-level experiments, it is generally useful to summarise the findings in order to ascertain if a "best" treatment combination occurs. Table 7.6 provides such a summary for the fish survival study indicating that the optimal combination, based on the experiment undertaken, may be high temperature, low cyanide, and high oxygen. Further experimentation around this combination may prove fruitful in respect of survival time optimisation based on the conclusion that the combination of temperature and cyanide appears important while oxygen has an important main effect also. □

Table 7.6 — Summary of analysis of fish survival study of Example 7.1

	Factors		
Source	Temperature (A)	Cyanide (B)	Oxygen (C)
Main effects		low	high
A x B interaction	high	low	
Possible "best"	high	low	high

Square plots and **cube plots** of treatment combination results represent further data plotting procedures (Montgomery 1997). The vertex values for each plot type correspond to the response, or average response, and/or range of responses at the associated treatment combination. Again, we would be looking to find the vertex which provides the "best" result as regards the experimental objective.

7.4 STATISTICAL COMPONENTS OF ANALYSIS

The data analysis discussed so far has only used some of the analysis and interpretational methods associated with unreplicated two-level designs. Often, inclusion of statistical analysis can provide further back-up on the importance of highlighted effects. For replicated experiments, statistical analysis is more readily available as the replication provides sufficient data for error estimation and thus, statistical assessment. Several statistical approaches can be considered with a few discussed here.

7.4.1 Statistical Assessment of Proposed Model

Unreplicated two-level experiments, unlike more formal experimental designs, are not built round specification of a response model at the outset. Screening of the effects through effect estimate analysis can enable a response model in terms of the active effects to be constructed. By so doing, the statistical significance of such effects can be assessed as well as enabling some diagnostic checking to take place. Generally, it is advisable to express the model in terms of importance of effects, most important first.

As with One Factor and Factorial Designs, analysis of the proposed model involves applying the **ANOVA principle** to split response variation into the sources of variation deemed to be affecting the response. The sum of squares (*SS*s) for active effects are expressed as

$$SS\text{Effect} = (\textit{Effect estimate})^2 2^{k-2} \tag{7.6}$$

with each *SS* based on one degree of freedom. The total sum of squares, *SS*Total, is given by

$$SS\text{Total} = \Sigma X_j^2 - (\Sigma X_j)^2/2^k \tag{7.7}$$

with associated degrees of freedom ($2^k - 1$) where X_j defines the response measurement for the *j*th treatment combination. The error sum of squares (*SSE*) is obtained by subtraction in the usual manner.

As each effect is based on only one degree of freedom, then *SS*Effect = *MS*Effect for all active effects included. Test statistic construction is as before, i.e.

$$F = MS\text{Effect}/MSE \tag{7.8}$$

enabling the statistical significance of the specified active effects to be tested. In two-level designs, all factors are generally regarded as fixed effects. Construction of the **ANOVA table** can be carried out manually or by use of statistical software though software such as Minitab require the proposed model to be hierarchical, i.e. factors contributing to interaction terms must be included in the model even if unimportant. General Linear Model estimation and regression methods can also be used through fitting a model of the response as a function of active main and interaction effects.

Exercise 7.6 For the fish survival study of Example 7.1, effects analysis suggested three effects to be important: cyanide (B), temperature × cyanide interaction ($A \times B$), and oxygen (C). From this, we could propose a model of the form

$$\text{survival time} = \text{cyanide } (B) + \text{temperature} \times \text{cyanide } (A \times B) + \text{oxygen } (C) + \text{error}$$

to explain the data collected. The error term comprises those effects deemed unimportant from the effects analysis, i.e. temperature (A), temperature × oxygen ($A \times C$), cyanide × oxygen ($B \times C$), and the three factor interaction ($A \times B \times C$). This pooling of terms assumes that these effects are negligible in their influence on the survival time response. Display 7.6 contains the ANOVA information generated by the general linear model (GLM) procedure in Minitab for the proposed model. For expression (7.6), effect estimates can be obtained from Display 7.1. The summation components in the *SS*Total expression (7.7) are $\Sigma X_j = 893$ and $\Sigma X_j^2 = 108481$. Using these values together with $k = 3$ generates a *SS*Total result of 8799.9, as provided in the last line of the 'Seq SS' column in Display 7.6.

Display 7.6

Source	DF	Seq SS	Adj SS	Adj MS	F	P
Cyanide	1	6903.1	6903.1	6903.1	295.32	0.000
TempxCyanide	1	1275.1	1275.1	1275.1	54.55	0.002
Oxygen	1	528.1	528.1	528.1	22.59	0.009
Error	4	93.5	93.5	23.4		
Total	7	8799.9				

Analysis of proposed model: As $k = 3$ and the effect estimate for cyanide (B) is given as −58.75, expression (7.6) provides an *SS* value of

$$SS\text{Cyanide} = (-58.75)^2 \times 2^{3-2} = 6903.125$$

as given in Display 7.6. Re-application of expression (7.6) for each of the other model effects will provide the *SS* given in the Minitab output.

The critical values for test statistic comparison from Table A.3 are $F_{0.05,1,4} = 7.71$ and $F_{0.01,1,4} = 21.20$ (the 0.1% critical value is 74.14). Based on these, or use of the printed p values, we can deduce that the cyanide effect appears significant at the 0.1% significance level ($p < 0.001$) while the temperature × cyanide interaction and oxygen effects are significant at a lower level ($p < 0.01$). We can conclude, however, that the three active effects specified are all highly significant, much as expected since these effects were the ones highlighted as having important influence on survival time. □

7.4.2 Prediction

Once a potentially optimal solution has been suggested by the data analysis (see Exercise 7.5), it may be useful to predict the response for this combination and compare it to the measured response at the same, or nearly equivalent, treatment combination. The prediction obtained represents an estimate of what is believed will be the response if the treatment combination suggested is run. If the prediction gives a more favourable result than the recorded measurement with respect to the type of response required, then the suggested optimal combination may well be "best" for the problem being investigated. Prediction is determined as

$$\overline{X} + \tfrac{1}{2}[\Sigma\{(\text{factor optimal level})\times(\textit{Effect estimate})\}] \tag{7.9}$$

where $\overline{X}$ is the overall mean of the response data. The summation component in expression (7.9) simply means sum the product of factor level and effect estimate across the significant effects. For interaction effects, the level to use refers to a simple product of the corresponding main effect levels.

Exercise 7.7 Exercise 7.6 indicated that all three active effects were statistically significant and Exercise 7.5 pointed to the combination high temperature, low cyanide, and high oxygen as being possibly best. We want to predict fish survival time for this combination and compare the result against the measured experimental response of 168 for that combination (see Table 7.2).

Prediction of optimal: The steps in the prediction calculation are shown in Table 7.7. The optimal of '−' for the interaction effect corresponds to the fact that opposing levels appeared best for this effect. The predicted survival time of 161.75 is marginally lower than the corresponding experimental measurement of 168. Despite this discrepancy, we can say that the prediction is of comparable magnitude suggesting that the specified combination may well be "best" based on the factor levels tested. □

Table 7.7 — Fish survival prediction for potential "best" combination for Exercise 7.7

Significant effects	Cyanide	Temp × Cyanide	Oxygen
Optimal levels	−	−	+
Effect estimate	−58.75	−25.25	16.25

Prediction = 111.625 + ½[(−)(−58.75) + (−)(−25.25) + (+)(16.25)] = 161.75

This concept of prediction is often extended to assess predictions for all treatment combinations to fully investigate model adequacy. Essentially, this corresponds to carrying out a practical validity check of the fitted model. Ideally, if the model is to be a good explanation of the response, measurements and predicted responses should match well. Inadequate matching would be indicative of inappropriate model suggesting that missing explanatory factors may be a likely cause. Fig. 7.5 shows a plot of observed and predicted survival time measurements for the model proposed in Exercise 7.6. Predictions appear close to the experimental responses throughout indicating the model proposed appears reasonably adequate.

7.4.3 Diagnostic Checking

Through prediction, we can obtain estimates of the error, or noise, in the proposed response model (the residuals) enabling **diagnostic checking** to be considered. Such analysis is useful, in two-level designs, for assessing factor influences on response variability for both active and inactive factors. Often, inactive factors can have a strong influence on response variability even though no detectable effect on response average is forthcoming. In factor residual plots, this type of effect generally corresponds to different column lengths. Additionally, a plot of residuals against run order can provide information on the randomness of the response data in a similar way to the procedures associated with **Statistical Process Control** (SPC). A formal statistical test of factor effects on response variability also exists (Montgomery 1997).

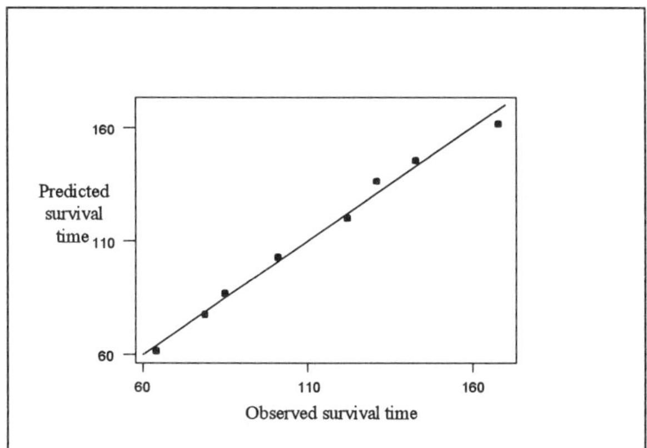

Fig. 7.5 — Observed and predicted fish survival times

Exercise 7.8 The final aspect of analysis of the fish survival study of Example 7.1 concerns the diagnostic checking of the residuals associated with the model proposed in Exercise 7.6.

Diagnostic checking: From the factor plots, those for cyanide and oxygen exhibited distinctive trends. The plot for oxygen is shown in Display 7.7 and shows that low oxygen concentration appears to result in lower variability in the average survival time.

Display 7.7

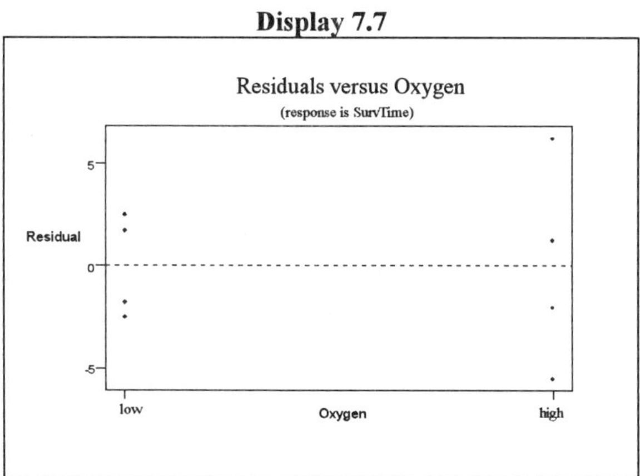

The plot of residuals versus predicted survival time, in Display 7.8, shows a possible quadratic trend and suggests problems with equality of variance. The trends effect suggests that, though the proposed model appears adequate, it could be that factor effects are more curved than linear. A normal plot of the residuals showed no deviations from linearity. □

Display 7.8

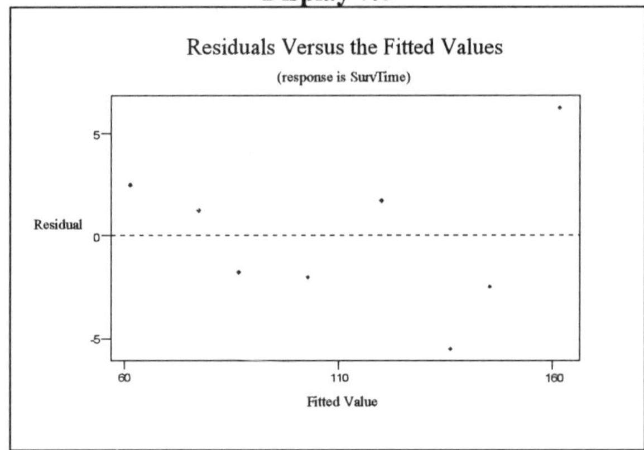

7.5 GENERAL UNREPLICATED 2^k DESIGNS

Discussion so far has centred on the analysis of a 2^3 experiment. In two-level experiments, the number of factors can be increased easily though the number of experiments will obviously double for each new factor introduced. A 2^4 experiment requires 16 experiments, double that of a 2^3, while a 2^5 experiment requires 32 experiments, double that of a 2^4 and four times that of a 2^3. Though there is an increase in experimentation, the increase in number of factors can enable more wide ranging conclusions on factor effects to be reached.

Example 7.2 A laboratory experiment was undertaken to investigate the determination of aluminium through complexation with Solochrome Violet RS (SVRS). Four two-level factors were identified for experimentation: SVRS volume (ml), pH, heating time (HT, seconds), and delay time (DT, seconds). Heating time refers to the heating of the aluminium solution in a microwave oven at 360W for the specified time period. The delay time factor corresponds to the waiting time before measuring the fluorescence intensity at 590 nm on a Perkin Elmer LS50 Luminescence Spectrometer. The design matrix and collected fluorescence intensity (FI) measurements are presented in Table 7.8.

Table 7.8 — Design matrix and fluorescence response data for Example 7.2

Run	SVRS	pH	HT	DT	FI response	Run	SVRS	pH	HT	DT	FI response
1	1	2	15	120	60	9	5	2	15	240	90
2	1	6	5	240	162	10	5	6	5	120	71
3	1	6	5	120	219	11	1	2	5	240	182
4	1	2	15	240	127	12	5	2	5	120	117
5	1	2	5	120	159	13	1	6	15	240	211
6	5	6	15	240	49	14	1	6	15	120	172
7	5	6	15	120	45	15	5	6	5	240	71
8	5	2	15	120	86	16	5	2	5	240	123

The structural and analysis components of 2^k designs mirrors those discussed for the 2^3 design. Contrasts for factor effects in 2^k designs can be easily constructed using the general principles of Section 7.2.1. For a 2^4 design structure, this results in the contrast matrix displayed in Table 7.9 where, again, each effect column adds to 0 and each column can be derived from a pair of columns just as occurred for the 2^3

design structure. Data analysis follows that outlined: EDA, effect estimation, formulation of model, statistical assessment of model, analysis of residuals, and interpretation of results.

Table 7.9 — Contrast matrix for a four factor two-level (2^4) Factorial Design

Treatment combn	Factorial effect Mean	*A*	*B*	*C*	*D*	*AB*	*AC*	*AD*	*BC*	*BD*	*CD*	*ABC*	*ABD*	*ACD*	*BCD*	*ABCD*
(1)	+	−	−	−	−	+	+	+	+	+	+	−	−	−	−	+
a	+	+	−	−	−	−	−	−	+	+	+	+	+	+	−	−
b	+	−	+	−	−	−	+	+	−	−	+	+	+	−	+	−
ab	+	+	+	−	−	+	−	−	−	−	+	−	−	+	+	+
c	+	−	−	+	−	+	−	+	−	+	−	+	−	+	+	−
ac	+	+	−	+	−	−	+	−	−	+	−	−	+	−	+	+
bc	+	−	+	+	−	−	−	+	+	−	−	−	+	+	−	+
abc	+	+	+	+	−	+	+	−	+	−	−	+	−	−	−	−
d	+	−	−	−	+	+	+	−	+	−	−	−	+	+	+	−
ad	+	+	−	−	+	−	−	+	+	−	−	+	−	−	+	+
bd	+	−	+	−	+	−	+	−	−	+	−	+	−	+	−	+
abd	+	+	+	−	+	+	−	+	−	+	−	−	+	−	−	−
cd	+	−	−	+	+	+	−	−	−	−	+	+	+	−	−	+
acd	+	+	−	+	+	−	+	+	−	−	+	−	−	+	−	−
bcd	+	−	+	+	+	−	−	−	+	+	+	−	−	−	+	−
abcd	+	+	+	+	+	+	+	+	+	+	+	+	+	+	+	+

Exercise 7.9 Consider the fluorescence experiment described in Example 7.2. We want to assess the results for importance of the tested factors.

EDA: Plots and summaries of each factor (not shown) indicated that all factors had some influence with SVRS by far the most important. Variability in response was also quite large.

Effect estimates: The effect estimation results in Display 7.9 show that SVRS, the SVRS × pH interaction, and heating time (HT) appear to be the most important effects. The remaining effects split into two groups, one based on numerical estimates of the order 10 to 20 and the other on numerical estimates below 10. Given the closeness and lack of distinct separation of these estimates, it is difficult to state which are unimportant and so can be ignored within the subsequent analysis. This means that the principle of effects sparsity does not appear to be prevalent in this study.

Display 7.9

Estimated Effects and Coefficients for Fluoresecence

Term	Effect	Coef	Term	Effect	Coef
Constant		121.50	pH*HT	21.50	10.75
SVRS	−80.00	−40.00	pH*DT	−14.25	−7.13
pH	7.00	3.50	HT*DT	17.75	8.88
HT	−33.00	−16.50	SVRS*pH*HT	−17.50	−8.75
DT	10.75	5.37	SVRS*pH*DT	12.75	6.38
SVRS*pH	−52.00	−26.00	SVRS*HT*DT	−17.25	−8.62
SVRS*HT	5.00	2.50	pH*HT*DT	7.25	3.63
SVRS*DT	−7.25	−3.62	SVRS*pH*HT*DT	−5.75	−2.87

Normal plot of effect estimates: The normal plot in Display 7.10, based on effect significance level of 20% (α = 0.20), does not conform to the ideal pattern of two distinct groups of estimates, important and unimportant effects, due to the numerical similarity of the effect estimates and the lack of effect dispersion across the scale of estimates. SVRS (A), the SVRS × pH interaction (AB), and heating time (C) do, however, stand out as large negative estimates. These effects would obviously require further analysis. Additionally, though the output suggests unimportant effects (effects close to straight line), it would be ill advised to ignore all such effects because the line plotted is not as vertical as would be ideal for the occurrence of effects sparsity as all such effects are clustered close together with none obviously different.

Display 7.10

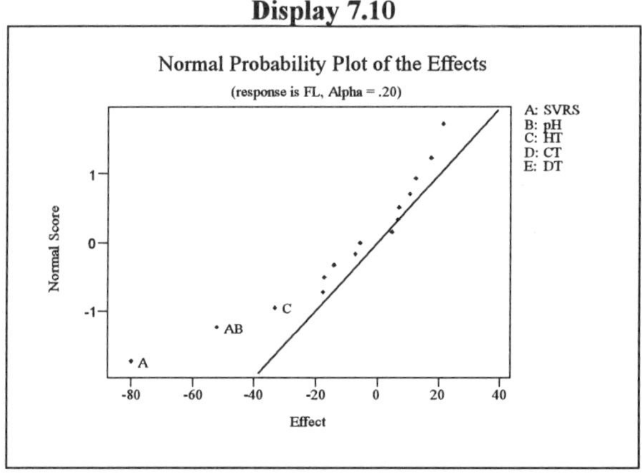

Data analysis: The advice, in this case, would be to analyse, at least, the main effects and all two factor interactions. Consideration of certain three factor interactions could also be appropriate given the similarity of effect estimates. By analysing all such effects, it may be feasible to obtain a good overall picture of how the tested factors affect fluorescence intensity of the aluminium complex. From such an analysis, an optimal combination of factors could result from which it may be possible to specify a model for the fluorescence response measurement and suggest avenues for future experimentation. □

7.6 USE OF REPLICATION

Applications of two-level designs in screening experiments, in particular, tend to involve only single replication of each treatment combination. It is possible to run two-level designs such that replicate observations are obtained for each treatment combination. Replication provides data from which the experimental error can be estimated and enables more formal statistical analysis to be considered though effect estimation and analysis should still be used.

Two-level designs are generally concerned with assessing how tested factors affect the mean response. Inclusion of replication can also be beneficial if it is of interest to assess how the experimental factors affect response variability, the treatment combination replication providing replicate data for variability estimation for each combination. It is this summary measure which will be the response to be analysed by the methods explained in this chapter.

Two-level designs are based on the assumption of linearity of factor effects between the two levels though the designs are generally robust to violations of this assumption. Replicating **centre points** in such a structure enables curvature of effects to be estimated provided all factors are **quantitative**. Assessment of such an effect is useful since optimal factor levels may occur in the interior of the experimental region rather than at the extremes. Centre points, consisting of replications of the treatment combination corresponding to mid-level (0) of the factors, can be added to the design structure easily as Fig. 7.6 illustrates as well as allowing for error estimation. The basic experimental region of eight treatment combinations is still covered with the additional experimentation provided by the replication at the design centre. This form of additional experimentation does not affect the process of effect estimation which remains as described in this chapter.

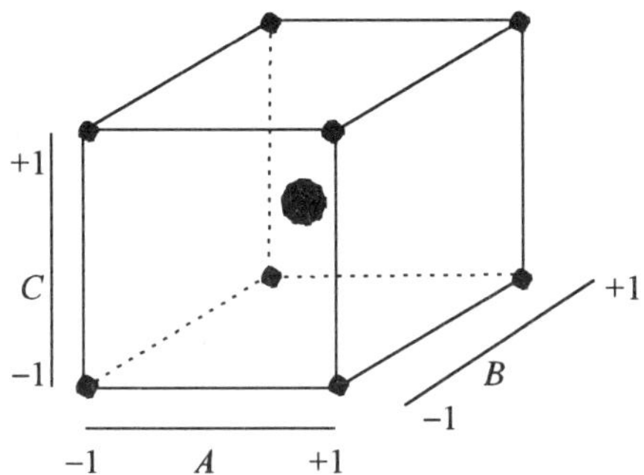

Fig. 7.6 — The 2^3 Factorial Design with centre point replication

7.7 THREE-LEVEL DESIGNS

Increasing the levels of each factor from two to three levels through use of **three-level Factorial Designs** is a logical extension to two-level designs. Such designs can be particularly helpful in the assessment of curvature effects though this inevitably will result in increased experimentation and modified analysis procedures. This ability to assess for curvature means that any main effect can be partitioned into single degree of freedom linear and quadratic components enabling trend types associated with such effects to be readily estimated and assessed. Interaction effects can be similarly partitioned to enable combinations of linear and quadratic trends in the main effects comprising an interaction to be assessed. Experiments based on **mixed factor levels** can also be considered if the situation demands. Fuller details of this and other aspects of three-level FDs are available in Montgomery (1997).

PROBLEMS

7.1 In an agroforestry trial, three factors were varied at two levels with the timber yield, in coded units, from eight plantations summarised below. The factors were variety, spacing between trees, and dates of sowing. Identify those effects which influence yield.

Variety	Spacing	Date	Yield response	Variety	Spacing	Date	Yield response
X	narrow	early	50.3	X	narrow	late	62.8
Y	narrow	early	43.4	Y	narrow	late	53.6
X	wide	early	36.8	X	wide	late	45.0
Y	wide	early	41.5	Y	wide	late	59.2

7.2 A plastics manufacturer wished to develop a new product. It was believed that the yield from the new process may be affected by three factors: time of processing, in hours, temperature, in °C, and quantity of catalyst, in g, added towards the end of the manufacturing process. Manufacture of the new product was only possible in small batches, each sufficient to test only eight combinations so a 2^3 FD was chosen as the design base. The design matrix and process yield results, in kg, are shown below. Carry out a relevant analysis of these data and draw conclusions.

Run	Time	Temperature	Catalyst	Yield response
5	1	110	10	46
4	4	110	10	34
2	1	150	10	55
3	4	150	10	41
6	1	110	20	47
7	4	110	20	68
1	1	150	20	53
8	4	150	20	63

7.3 An experiment was carried out to investigate the effects of three factors: tool type (A), angle of tool edge bevel (B), and type of cut (C), on the power requirements for cutting metals with ceramic tools. A two-level FD based on the three specified factors was undertaken using the levels of A of I and II, B 15° and 30°, and C continuous and interrupted. A single replicate measure of power requirement was made. The results data were as follows:

Run	Combination	Response	Run	Combination	Response
2	(1)	−1	8	c	−3
1	a	−1.25	7	ac	−4.25
5	b	3.75	4	bc	−0.5
3	ab	3.25	6	abc	−1.75

Analyse these data to assess for factor and interaction effects.

7.4 A two-level experiment was set up to investigate the dependence of the yield from a chemical reactor when four factors are varied. The four factors were temperature (A) at 40 °C and 65 °C, pressure (B) at 1 atmosphere and 1.5 atmosphere, time (C) at 20 minutes and 30 minutes, and stirring (D) at not stirred and stirred. Sixteen single replicate runs were conducted resulting in the yield data below. Which effects have most influence on reactor yield? Can reactor yield be modelled by a combination of these factors?

Run	Yield	Run	Yield	Run	Yield	Run	Yield
9	(1) = 77.2	3	c = 51.1	4	d = 48.1	15	cd = 17.1
1	a = 32.7	16	ac = 104.5	10	ad = 5.0	5	acd = 66.7
14	b = 145.7	7	bc = 61.8	13	bd = 110.1	11	bcd = 36.5
8	ab = 93.6	2	abc = 125.1	6	abd = 68.4	12	$abcd$ = 96.4

7.5 A semiconductor manufacturer conducted a 2^4 FD and obtained the following set of data:

(1) = 18.3 a = 32.8 b = 7.8 ab = 21.5 c = 6.3 ac = 27.4 bc = 8.3
abc = 17.0 d = 15.4 ad = 27.5 bd = 5.5 abd = 22.1 cd = 6.9 acd = 25.7
bcd = 7.0 $abcd$ = 16.0

What are the important effects? Which interactions, if any, are important?

7.6 An experiment was conducted into the insulation properties of a new product. Four factors were selected for investigation. These were density of material (A), addition of a specific ingredient (B), moisture content (C), and age (D). Each factor was set at two levels for the initial experiment. For factors A, C, and D, the levels were simply low (−1) and high (1) while for factor B, the levels corresponded to absence (−1) or presence (1) of the ingredient. The insulation performance (IP) data collected are presented below and are in coded units whereby high values suggest better insulation performance. Which effects appear active? Which appear inactive? Carry out a full and relevant analysis of these data.

Run	A	B	C	D	IP response	Run	A	B	C	D	IP response
6	−1	−1	−1	−1	2.47	9	−1	−1	−1	1	4.97
15	1	−1	−1	−1	4.53	12	1	−1	−1	1	6.53
3	−1	1	−1	−1	2.47	5	−1	1	−1	1	8.97
13	1	1	−1	−1	7.53	16	1	1	−1	1	7.53
7	−1	−1	1	−1	8.53	11	−1	−1	1	1	7.03
1	1	−1	1	−1	3.47	8	1	−1	1	1	1.47
14	−1	1	1	−1	10.03	4	−1	1	1	1	10.03
10	1	1	1	−1	8.47	2	1	1	1	1	4.47

7.7 A 2^5 experiment was conducted into gas production from an underground source. Coded gas flow rate data are given below. Analyse these data for active and inactive effects.

(1) = 7.3 a = 15.2 b = 15.5 ab = 35 c = 3.8 ac = 13.7 bc = 14 abc = 30
d = 8 ad = 16 bd = 14.8 abd = 35 cd = 7.3 acd = 15.5 bcd = 16.8 $abcd$ = 35
e = 4 ae = 8 be = 9 abe = 18.3 ce = 4.2 ace = 9 bce = 9.8
$abce$ = 20.2 de = 2.8 ade = 5.7 bde = 6.5 $abde$ = 13.3 cde = 3.2 $acde$ = 7
$bcde$ = 7.5 $abcde$ = 15.8

7A Appendix: Software Information For Two-Level Factorial Designs

Data entry

Data entry for two-level FDs is relatively straightforward and is based on entry of run order number, factor codes, and response data. Generally, the actual levels of the factors can be used, e.g. low and high, though the −1/+1 system may also be necessary as software cannot always handle character code data for its plotting, summary, and ANOVA facilities.

SAS: Data entry for two-level FDs is as shown in previous Appendices. The following DATA step program was used for the fish survival data from Example 7.1:

```
DATA CH7.EX71;
    INPUT RUN TEMP CYANIDE OXYGEN SURVTIME T C O @@;
    LINES;
    3 LOW LOW LOW 122 -1 -1 -1    5 HIGH LOW LOW 143 1 -1 -1
    2 LOW HIGH LOW 185 -1 1 -1    6 HIGH HIGH LOW 64 1 1 -1
    4 LOW LOW HIGH 131 -1 -1 1    1 HIGH LOW HIGH 168 1 -1 1
    8 LOW HIGH HIGH 101 -1 1 1    7 HIGH HIGH HIGH 790 1 1 1
;
RUN;
```

It is necessary, however, after entering the data, to include additional data generation to create variables that will be necessary if SAS is to generate effect estimates for such designs. This entails using the DATA step with the SET option to create the necessary variables by simple multiplication of the numerical factor codes as shown below. Essentially, the created variables represent the signs of the contrasts for the specified interaction.

```
DATA CH7.EX71;
    SET CH7.EX71;
    TC = T*C;
    TO = T*O;
    CO = C*O;
    TCO = T*C*O;
RUN;
```

Minitab: In Minitab, data entry is more straightforward following the data presentation shown in Table 7.2 but also with additional columns of numeric factor codes.

Data entry for Example 7.1 would therefore look exactly as that shown in Table 7.2 and below.

Run	Temp	Cyanide	Oxygen	SurvTime	T	C	O
3	low	low	low	122	−1	−1	−1
5	high	low	low	143	1	−1	−1
2	low	high	low	85	−1	1	−1
6	high	high	low	64	1	1	−1
4	low	low	high	131	−1	−1	1
1	high	low	high	168	1	−1	1
8	low	high	high	101	−1	1	1
7	high	high	high	79	1	1	1

Exploratory analysis

Plotting and summarising two-level design data for exploratory analysis can be produced using similar procedures to those described in earlier Appendices.

Effect estimation and plotting
Production of effect estimates and associated data plots for two-level FDs can be easily achieved within most software packages.

SAS: In SAS, there is no default facility for direct evaluation of effect estimates in unreplicated two-level FDs. It is necessary to develop a program based on regression modelling of a full response model through PROC REG and normal score plotting, as in Appendix 3A, to provide a comparable form of analysis to that available by default within Minitab. The sequence of statements below show how this can be achieved making use of the additional variables created after data entry.

```
PROC REG DATA = CH7.EX71 OUTEST = CH7.EX71A NOPRINT;
     MODEL SURVTIME = T C O TC TO CO TCO;
DATA CH7.EX71A;
     SET CH7.EX71A;
     KEEP T C O TC TO CO TCO EFFEST;
     EFFEST = 'EFFEST';
PROC TRANSPOSE DATA = CH7.EX71A OUT = CH7.EX71A NAME = EFFECT;
     ID EFFEST;
DATA CH7.EX71A;
     SET CH7.EX71A;
     EFFEST = EFFEST + EFFEST
PROC RANK DATA = CH7.EX71A NORMAL = BLOM OUT = CH7.EX71A;
     VAR EFFEST;
     RANKS NSCOREST;
PROC PRINT DATA = CH7.EX71A NOOBS SPLIT = '*';
     LABEL EFFECT = 'Factor';
     LABEL EFFEST = 'Estimate';
     LABEL NSCOREST = 'Normal scores*of estimates';
PROC GPLOT DATA = CH7.EX71A;
     PLOT NSCOREST*EFFEST / HAXIS = AXIS1 VAXIS = AXIS2 HREF = 0;
     AXIS1 LABEL = (F = CENTB J = C 'Effect estimates') LENGTH = 40;
     AXIS2 LABEL = (F = CENTB J = C H = 0.8 'Normal scores') LENGTH = 15;
RUN;
```

The statements first use regression modelling (PROC REG) to generate and store (OUTEST) regression coefficients for a full response model. Estimates of these coefficients are half the effect estimates. The DATA step with the SET option is used to tidy up the file storing these coefficients so that only coefficient estimates and labels are retained with PROC TRANSPOSE modifying the presentational form into two columns, one for labels and one for coefficient estimates. As the generated coefficients correspond to half the correct effect estimates, the second DATA step is used to ensure that correct effect estimates are produced. PROC RANK is used to generate the normal scores as per Appendix 3A and PROC GPLOT will produce the associated normal plot with a vertical reference line at zero on the X axis (HREF = 0).

Minitab: The **Stat** menu provides access to Minitab's presentation and analysis procedures for two-level designs. Displays 7.1, 7.2, and 7.3 illustrate the generated presentations for the two-level fish survival study of Example 7.1. The information contained in these plots was produced using the following menu selections:

Select **Stat** ➤ **DOE** ➤ **Analyze Custom Design** ➤ for *Design*, select **2-Level factorial** ➤ **Fit Model** ➤ for *Responses*, select **SurvTime** and click **Select** ➤ for *Factors*, select the box and enter **Temp Cyanide Oxygen**.

Select **Terms** ➤ ensure that **Include terms in the model up through order** is set at **3** ➤ ensure that the **Selected Terms** box contains all the possible effects ➤ click **OK**.

Select **Graphs** ➤ for *Effects Plots*, select **Normal** and **Pareto** ➤ for *Alpha*, enter **0.2** ➤ click **OK**.

Select **Storage** ➤ for *Model Information*, select **Effects** ➤ click **OK** ➤ click **OK**.

Data plots

Minitab: Main effect plots, illustrated in Display 7.4, can be produced by the following menu selections though the illustrated plot has been edited using Minitab's graph edit facility before production in order to provide the range data and some labelling.

Select **Stat** ➤ **ANOVA** ➤ **Main Effects Plot** ➤ for *Factors*, enter **Cyanide Oxygen** ➤ for *Raw response data in*, click the box, select **SurvTime**, and click **Select** ➤ for *Title*, click the box and enter **Main effects plot for Cyanide (B) and Oxygen (C)** ➤ click **OK**.

The interaction plot macro is accessed by the menu selection below and was used to produce the interaction plot provided in Display 7.5 though again plot editing has been carried out to incorporate the ranges.

Select **Stat** ➤ **ANOVA** ➤ **Interactions Plot** ➤ for *Factors*, enter **Temp Cyanide** ➤ for *Raw response data in*, click the box, select **SurvTime** and click **Select** ➤ for *Title*, click the box and enter **Temperature x Cyanide (A × B) interaction** ➤ click **OK**.

Main effect and interaction plots can also be produced through the **Stat** ➤ **DOE** ➤ **Analyze Custom Design** menu after running the 2^k FD analysis routine. The initial menu selection for this mode of data plot generation is

Select **Stat** ➤ **DOE** ➤ **Analyze Custom Design** ➤ ensure **Design** is set at **2-Level factorial** ➤ **Plot Response** ➤ choose plot required ➤ **Setup** ➤ then follow menu to generate the required data plots.

ANOVA information

Derivation of ANOVA information for a proposed response model in two-level designs can be handled using software.

SAS: The PROC GLM procedure can provide the route to the ANOVA information required for a proposed two-level design model after screening. The program below would produce comparable ANOVA information to that shown in Display 7.6 for the fish survival study of Example 7.1. The TC variable corresponds to the interaction between T and C obtained by multiplying together the T and C variables as shown in ***Data entry***.

```
PROC GLM DATA = CH7.EX71;
    CLASS C O TC;
    MODEL SURVTIME = C TC O;
    OUTPUT OUT = CH7.EX71A P = FITS R = RESIDS;
```

```
RUN;
```

Residual plots for diagnostic checking can be produced using the procedures described in Appendix 3A.

Minitab: The ANOVA and general linear model (GLM) facilities in Minitab cannot handle a proposed response model if it is not hierarchical. This difficulty can be overcome, however, by creating a column of codes for interaction terms by multiplying together numerical code columns for the said factors. For Exercise 7.6, this would entail multiplying together the numerical T and C columns to generate a TC column for their interaction. The calculation facility within the **Calc** ➢ **Calculator** menu selection can be used. The ANOVA output in Display 7.6 was obtained from the menu selection as follows:

Select **Stat** ➢ **ANOVA** ➢ **General Linear Model** ➢ for *Responses*, select **SurvTime** and click **Select** ➢ for *Model*, select the box and enter **C TC O**.
Select **Storage** ➢ **Fits** ➢ **Residuals** ➢ click **OK**.
Select **Graphs** ➢ select the residual plots necessary ➢ click **OK** ➢ click **OK**.

8

Two-level Fractional Factorial Designs

8.1 INTRODUCTION

The number of experimental runs required for two-level FDs can become impractical when k, the number of factors, is large. For example, a 2^6 design requires 64 experimental runs and a 2^8 design 256 experimental runs. In these designs, all possible factor interactions can be estimated though some may be known to be negligible in their effect or may not be readily interpretable. In addition, most of the degrees of freedom are associated with third and higher order interactions and, since these can be used as estimates of the error, then such designs can have substantial degrees of freedom associated with error estimation. Such design structures represent wasted experimental effort. A more appropriate way of structuring such experiments to reduce the level of experimentation would be to discount higher order interactions. **Fractional Factorial Designs** (FFDs) cater for such approaches allowing one-half, one-quarter, etc. of the treatment combinations to be tested. Constraint concerns such as time, cost, and availability of resources may also dictate that reduced experimentation is necessary.

A major use of FFDs occurs in **screening experiments** in which many controllable factors are considered in order to identify those factors, if any, that exert most influence on the response. Such experiments are usually performed in the early stages of a project with the factors identified as important then investigated more fully in subsequent experiments. They provide for small, efficient designs though there is no "best" design as choice depends on study requirements and experimental constraints. In fact, it is often prudent to carry out a sequence of FFDs, each succeeding experiment being influenced by the results of the preceding experiment. In such designs, however, it may not be possible to estimate all potential effects as estimation of certain ones is sacrificed and independent estimation of others is not possible. Before discussing FFDs, we will first discuss the principles of **confounding** which underpin FFD structures.

8.2 CONFOUNDING

In two-level FDs, it may not be possible to make simultaneous measurements on all treatment combinations and the experiment may have to be run using **incomplete blocks**. For a k factor experiment, it is often useful to use a design in 2^p blocks ($p < k$) when the entire 2^k treatment combinations cannot be applied under homogeneous conditions due to lack of resources such as manpower, raw material, time, and cost, or because of physical constraints such as size of the pilot plant. The advantage of this is that certain effects, generally high order interactions, are sacrificed as a result of the blocking, the degree of sacrifice depending on the number of blocks required. This is called **confounding**.

Example 8.1 A study is to be set up to examine the effect of three factors on the weight gains of young pigs. The three factors and their levels are

Factors	*Levels*
Lysine (A)	0% and 1%
Soybean meal (B)	type 1 and type 2
Sex (C)	male and female.

Factors A and B correspond to supplements added to the feed. The soybean meal levels reflect the amount to be added to supply 10% and 15% protein. Unfortunately, only eight animals are available for the study and pen constraints dictate that only four animals per pen can be used. The study must therefore be run using two separate pens, i.e. using a two block structure.

To illustrate the concept of confounding, consider the 2^3 design outlined in Example 8.1 based on three factors each at two levels where the treatment combinations are defined in **standard order** as (1), a, b, ab, c, ac, bc, and abc using the notation introduced in Sections 7.1 and 7.2. Simultaneous observations on all treatment combinations is not possible because of the pen constraints and so the experiment will have to be implemented in two incomplete blocks (pens) of four treatment combinations each instead of one pen of eight treatment combinations. How should the four treatment combinations be assigned to each pen? Which treatment combinations are to be tested in which pen?

Seven possible options exist as shown in Table 8.1. The lysine (factor A) contrast is

$$a + ab + ac + abc - (1) - b - c - bc$$

from expression (7.1). Based on this expression, option 1 would lead to this contrast providing a measure of both the lysine effect (high – low) and the block effect (block 2 – block 1) as the combinations with positive coefficients all refer to lysine high level and are also tested only in block 2. In such a case, the lysine and block effects would be said to be **confounded**. In option 3, the lysine × soybean interaction ($A \times B$) and the block effect are confounded based on the contrast expression

$$(1) + ab + c + abc - a - b - ac - bc$$

which corresponds to (same – opposite) for the $A \times B$ effect (see expression (7.2)) and also to (block 1 – block 2) for the block effect. Option 7 results in the three factor interaction and the block effect being confounded (see $A \times B \times C$ contrast expression (7.4)). Confounded effects are unable to be estimated independently and share the same sum of squares and test statistic.

In practice, the more factors there are the more design options that are available. The preferred option is generally the design structure that confounds the highest order interaction with block as it can be difficult to interpret this interaction practically. It is also hoped that such an interaction has no significant effect on the response, a generally appropriate assumption in most multi-factor experimentation. For the 2^3 weight gain study of Example 8.1, option 7 in Table 8.1 is preferred as it confounds the highest order interaction with block. Other economic and practical constraints could also influence this choice.

Table 8.1 — Design options for blocking in the weight gain weight gain study of Example 8.1

Option	Block 1	Block 2	Effect confounded with blocks
1	(1) *b c abc*	*a ab ac abc*	*A*
2	(1) *a c ac*	*b ab bc abc*	*B*
3	(1) *ab c abc*	*a b ac bc*	*AB*
4	(1) *a b ab*	*c ac bc abc*	*C*
5	(1) *b ac abc*	*a ab c abc*	*AC*
6	(1) *a bc abc*	*b ab c ac*	*BC*
7	(1) *ab ac bc*	*a b c abc*	*ABC*

For b blocks, we require to specify c confounded effects where $b = 2^c$. In addition to the specified effects, there will also be $(2^c - c - 1)$ additional confounded effects generated from multiples of the assigned effects providing a total of $b - 1$ confounded effects. For $b = 2$ blocks, the block effect degree of freedom must be found from the treatment combinations through one effect ($c = 1$) being confounded. For $b > 2$ blocks, $b - 1$ degrees of freedom must be obtained from the treatment combinations through $b - 1$ effects being confounded. For example, in a 2^4 design with four blocks, then three degrees of freedom must be found from the factor effects. In such a case, two confounded effects must be specified with the third emerging from multiplication of the two effects modulus 2 (even powers equal the identity I).

8.3 BLOCK CONSTRUCTION PROCEDURES

Determining all design options for confounded two-level experiments is not always easy. A procedure for treatment combination assignment is therefore necessary. This can be achieved by deciding an effect to confound with the blocks, the **defining contrast**. Once this decision is reached, the next step requires allocation of the treatment combinations to the blocks. Several methods exist, all of which result in the same block structure.

8.3.1 Even/Odd Method

Consider each treatment combination separately. For a **two block structure** (one defining contrast), assign to block 1 those combinations that have an even number of letters in common with the defining contrast and assign to block 2 those combinations with an odd number of letters in common with the defining contrast. For the weight gain study described in Example 8.1, the ideal defining contrast would be the three factor interaction lysine × soybean × sex, i.e. the $A \times B \times C$ interaction. Thus, the *ABC* effect will be confounded with the blocks. Table 8.2 illustrates the block assignment resulting from this even/odd allocation method.

Table 8.2 — Block construction for weight gain study of Example 8.1 using even/odd assignment method and defining contrast *ABC*

Treatment combination	(1)	*a*	*b*	*ab*	*c*	*ac*	*bc*	*abc*
Common letters	0	1	1	2	1	2	2	3
Block	1	2	2	1	2	1	1	2

For treatment combination (1), i.e. all factors low level, there are no letters in common with the defining contrast, hence the entry of 0 for "Common letters" in Table 8.2. This value is considered even and so the combination (1) is assigned to block 1, referred to as the **principal block**. Treatment combination *a*, i.e. lysine high, has one letter in common with *ABC* and as 1 is odd, the combination is assigned to block 2. Similar reasoning can be applied to the allocation of the other six treatment combinations. Thus, block 1 would contain the treatment combinations (1) (lysine 0%, soybean 1, sex male), *ab* (lysine 1%, soybean 2, sex male), *ac* (lysine 1%, soybean 1, sex female), and *bc* (lysine 0%, soybean 2, sex female). It should be noted that through this allocation process, the **balancing property** of two-level designs is retained as each factor level will still appear an even number of times within each proposed block.

For a **four block structure**, we would require to specify two defining contrasts ($c = 2$). One ($2^2 - 2 - 1$) additional generalised contrast will emerge from multiplication of the specified contrasts modulus 2. For example, suppose *ABC* and *BCD* are selected for confounding in a four factor study. The additional generalised contrast will therefore be *AD* as

$$ABC(BCD) = AB^2C^2D = AIID = AD,$$

recalling that even powers equal the identity *I*. The process of treatment combination allocation involves selecting any two of these contrasts to generate pairs of commonality elements with each pair of values attached to a particular block. For this form of structure, these components will be (even, even) for block 1, (even, odd) for block 2, (odd, even) for block 3, and (odd, odd) for block 4. The treatment combinations are then assigned to appropriate blocks depending on the resulting pair of commonality elements for that combination.

To illustrate this approach, suppose in the growth study of Example 8.1 pen size was such that only two animals per pen was possible. The experimental structure would have to be modified to be run using four separate pens, i.e. using a four block structure. Suppose that interactions *AB* and *AC* were selected for confounding. The additional generator would therefore be *BC* since $AB(AC) = A^2BC$. Choosing the contrasts *AB* and *AC* as the basis of block construction would provide the four block structure shown in Table 8.3.

Table 8.3 — Block construction for modified weight gain study using even/odd assignment method and defining contrasts *AB* and *AC*

Combination	Common letters *AB*	*AC*	Block
(1)	0 (even)	0 (even)	1
a	1 (odd)	1 (odd)	4
b	1 (odd)	0 (even)	3
ab	2 (even)	1 (odd)	2
c	0 (even)	1 (odd)	2
ac	1 (odd)	2 (even)	3
bc	1 (odd)	1 (odd)	4
abc	2 (even)	2 (even)	1

8.3.2 Linear Combination Method

A second method of assigning treatment combinations to the blocks uses the linear combination

$$L = \alpha_1 X_1 + \alpha_2 X_2 + \ldots + \alpha_k X_k \tag{8.1}$$

where α_i is the index (exponent) of the ith factor in each defining contrast, X_i is the level of the ith factor (low or high) appearing in a particular treatment combination, and k is the number of factors. Each specified defining contrast has a unique linear combination expression. For two-level designs, we define $\alpha_i = 0$ if the exponent of the ith factor is 0 in the defining contrast or $\alpha_i = 1$ if the exponent of the ith factor is 1 in the defining contrast. For defining contrast ABC in a 2^3 design, i.e. $A^1B^1C^1$, we would have $\alpha_1 = \alpha_2 = \alpha_3 = 1$ and expression (8.1) would become

$$L = X_1 + X_2 + X_3 \quad .$$

For a defining contrast of AB, i.e. $A^1B^1C^0$, the α terms would be $\alpha_1 = \alpha_2 = 1$ and $\alpha_3 = 0$ giving rise to

$$L = X_1 + X_2 \quad .$$

The X terms in expression (8.1) are such that $X_i = 0$ if the ith factor effect is not present in the specific treatment combination ("low" level) and $X_i = 1$ if the ith factor effect is present ("high" level). For the combination (1) in a 2^3 design, all three factors are absent and so $X_1 = X_2 = X_3 = 0$ whereas the treatment combination ac would result in $X_1 = 1$, $X_2 = 0$, and $X_3 = 1$.

For a **two block structure** (one defining contrast), the treatment combinations are assigned to blocks based on addition modulus 2 in which the only possible values of L (mod 2) are 0 and 1. Using these values, the 2^k treatment combinations can therefore be assigned to the two blocks based on 0 for block 1 and 1 for block 2. Again, treatment allocation will result in the block effect being confounded with the defining contrast.

For Example 8.1, the ideal defining contrast is the three factor lysine $\times$ soybean $\times$ sex ($A \times B \times C$) interaction which provides $\alpha_i = 1$ for all factors and a linear combination (8.1) of

$$L = X_1 + X_2 + X_3 \quad .$$

The resultant treatment combination assignment is shown in Table 8.4. For treatment combination (1), i.e. all factors low level, we have $X_1 = X_2 = X_3 = 0$ resulting in $L = 0$ (mod 2) and allocation of this combination to block 1. Treatment combination a, i.e. lysine high, provides $X_1 = 1$, $X_2 = 0$, and $X_3 = 0$ and so $L = 1$ (mod 2) which is odd resulting in allocation of this combination to block 2. The combination abc, i.e. all factors high level, provides $X_1 = 1$, $X_2 = 1$, and $X_3 = 1$ and so $L = 3 = 1$ (mod 2) providing block 2 assignment. Allocation of the other five treatment combinations can be explained similarly. It can be seen that this method of allocation provides the same block structure as the even/odd method (see Table 8.2).

Table 8.4 — Block construction for weight gain study of Example 8.1 using linear combination assignment method and defining contrast *ABC*

Combination	*X* values	*L* value	Block
(1)	$X_1 = 0, X_2 = 0, X_3 = 0$	$L = 0 = 0 \text{ (mod 2)}$	1
a	$X_1 = 1, X_2 = 0, X_3 = 0$	$L = 1 = 1 \text{ (mod 2)}$	2
b	$X_1 = 0, X_2 = 1, X_3 = 0$	$L = 1 = 1 \text{ (mod 2)}$	2
ab	$X_1 = 1, X_2 = 1, X_3 = 0$	$L = 2 = 0 \text{ (mod 2)}$	1
c	$X_1 = 0, X_2 = 0, X_3 = 1$	$L = 1 = 1 \text{ (mod 2)}$	2
ac	$X_1 = 1, X_2 = 0, X_3 = 1$	$L = 2 = 0 \text{ (mod 2)}$	1
bc	$X_1 = 0, X_2 = 1, X_3 = 1$	$L = 2 = 0 \text{ (mod 2)}$	1
abc	$X_1 = 1, X_2 = 1, X_3 = 1$	$L = 3 = 1 \text{ (mod 2)}$	2

When more than one defining contrast occurs, a unique linear combination can be constructed for each defining contrast. For a **four block structure** (two defining contrasts), an additional generalised contrast can be generated from the two defined resulting in the specification of three possible linear combinations. Treatment allocation involves selecting any two of these linear combinations to generate a pair of *L* values. Each pair of values is attached to a particular block generally according to (0, 0) for block 1, (0, 1) for block 2, (1, 0) for block 3, and (1, 1) for block 4. Treatment combination assignment to blocks therefore depends on the resulting pair of *L* values, in a similar way to the even/odd assignment method illustrated in Section 8.3.1.

Signs of contrasts and **group properties** are two other assignment methods. Both would provide the same block structure as shown in Tables 8.2, 8.3, and 8.4 if applied to the weight gain study of Example 8.1. Montgomery (1997) provides a fuller explanation of each of these methods. If replication of confounded designs is feasible, it can be useful to confound a different effect in each replicate so that some information on all factor effects can be obtained. This is called **partial confounding**. Such an approach can enable all the interactions being confounded to be tested statistically for their effect on the response but at the expense that such interactions can only be estimated from a subset of the blocks used. Further details on the concepts and principles relating to confounding can be found in Montgomery (1997) and Winer *et al.* (1991).

8.4 FRACTIONAL FACTORIAL DESIGNS

Reduction of experimental effort in two-level designs can also be achieved by carrying out a fraction of the experimental runs of a full 2^k design. Designs of this type, known as **Fractional Factorial Designs** (FFDs), can be used to assess large numbers of factors in designs based on few experimental runs. The basic premise is that only one-half, one-quarter, one-eighth, etc. of the total factorial plan is actually carried out resulting in only a fraction of the possible treatment combinations being tested. FFDs are widely used in **screening experiments** within product and process development and quality improvement to provide quick and easy procedures for the identification of important factors.

FFDs adhere to the same basic concepts as full two-level designs. They enable large numbers of factors to be assessed in designs which retain the **orthogonality** and **balancing properties** of two-level designs. The **effects sparsity** principle of few important effects, primarily main effects and low-order interactions, and many unimportant effects, mostly higher order interactions, is again a key feature.

Consider a 2^3 experiment based on three factors each at two levels where the treatment combinations are specified as (1), *a*, *b*, *ab*, *c*, *ac*, *bc*, and *abc* as before. Such an experiment can be split into two fractions each containing four treatment combinations. This half-fraction of a 2^3 experiment is generally referred to as a $\mathbf{2^{3-1}}$ **design**. In the exponent, 3 refers to the number of factors being assessed and −1 reflects that the design is a half-fraction based on $1/2^1$ expressed as 2^{-1}. Negative powers of 2 are therefore used to specify the design fractionation. A general notation of $\mathbf{2^{k-p}}$ is used to denote FFDs where k is the number of factors and p is the level of fractionation such that $2^{-p} = 1/2^p$.

Consider the weight gain study of Example 8.1 and suppose that the experimental constraints require that only one pen of four animals can be used. This means that only four treatment combinations can be tested and so a half-fraction 2^{3-1} design must be considered. Table 8.5 shows the signs of contrasts for the 2^3 design split into two fractions through the assignment of treatment combinations according to their sign in the *ABC* contrast (8.1), fraction 1 the plus sign and fraction 2 the minus sign. This commonality of sign leads to *ABC* being defined as the **generator** of the fractions and $I = ABC$ being called the **defining relation**, or **defining contrast**, for the design. Each fraction satisfies the orthogonality property of two-level designs while each effect, except the design generator, satisfies the balancing property.

Table 8.5 — Contrast matrix for a 2^{3-1} Fractional Factorial Design

Fraction	Treatment combination	Factorial effect *A*	*B*	*AB*	*C*	*AC*	*BC*	*ABC*
1	*a*	+	−	−	−	−	+	+
	b	−	+	−	−	+	−	+
	c	−	−	+	+	−	−	+
	abc	+	+	+	+	+	+	+
2	(1)	−	−	+	−	+	+	−
	ab	+	+	+	−	−	−	−
	ac	+	−	−	+	+	−	−
	bc	−	+	−	+	−	+	−

Example 8.2 The time taken by an industrial plant to produce filtered water is to be studied with respect to its dependence on five factors each set at two levels: origin of water supply (*A*), temperature (*B*), recycle device (*C*), rate of addition of caustic soda (*D*), and filter cloth (*E*). Only 16 experimental runs are possible, so it has been decided to base the study on a half-fraction of a 2^5 design with the defining relation $I = ABCDE$.

To avoid referring to contrast expressions each time, a **basic design approach** can be used for treatment combination assignment in FFDs. The method is based on using a full 2^{k-p} design in the first $k - p$ factors and adding to this design the columns for the additional factors. This involves determining the plus and minus levels of these factors from the combination of the plus and minus signs associated with each design generator, for each row of the basic design. Illustrations of this approach will be provided in Exercises 8.1 and 8.3.

Exercise 8.1 Refer to the proposed 2^{5-1} design for the water filtration study of Example 8.2. We want to show how to specify the treatment combinations for the two fractions associated with this design.

Information: The design is to be a 2^{5-1} FFD ($k = 5$) with defining relation $I = ABCDE$. The study will require 16 runs with the -1 exponent signifying that the design will be a half-fraction ($p = 1$).

Treatment combination assignment: Assignment of the treatment combinations to the fractions can be carried out using the basic design approach. This entails using the full $2^{5-1} = 2^4$ design in the first four factors A, B, C, and D, as shown in Table 8.6, which allows for 16 experimental runs. The design has been written in standard order for convenience though, normally, design implementation requires treatment combination order to be randomised. To find the levels of the fifth factor E, we solve the defining relation $I = ABCDE$ for the additional factor E, i.e. $E = ABCD$, which we refer to as the **design generator**. The levels of E for each treatment combination will correspond to the product of the plus and minus signs of the A, B, C, and D columns. The outcome to this is displayed in column E in Table 8.6. Those treatment combinations associated with a 2^5 structure which are not specified would appear in fraction 2, which could be obtained similarly using the design generator $E = -ABCD$.

Table 8.6 — Fraction 1 for a 2^{5-1} Fractional Factorial Design with defining contrast $I = ABCDE$

Basic design					
A	B	C	D	$E = ABCD$	Combination
−	−	−	−	+	e
+	−	−	−	−	a
−	+	−	−	−	b
+	+	−	−	+	abe
−	−	+	−	−	c
+	−	+	−	+	ace
−	+	+	−	+	bce
+	+	+	−	−	abc
−	−	−	+	−	d
+	−	−	+	+	ade
−	+	−	+	+	bde
+	+	−	+	−	abd
−	−	+	+	+	cde
+	−	+	+	−	acd
−	+	+	+	−	bcd
+	+	+	+	+	$abcde$

Software generation: Statistical software generally have design generation facilities for two-level designs. Display 8.1 illustrates the Minitab generated design for this study obtained using the menu procedures described in Appendix 8A. It can be seen that the design matrix displayed is identical to that shown in Table 8.6. SAS's design build facility, also explained in Appendix 8A, generated the same result. The concepts of 'Alias Structure' and 'Resolution' will be explored later. □

Display 8.1

```
Factors:     5    Base Design:        4, 16      Resolution: V
Runs:       16    Replicates:            1       Fraction: 1/2

Design Generators:  E = ABCD      Defining Relation:   I = ABCDE

         Alias Structure
         A + BCDE  B + ACDE  C + ABDE  D + ABCE  E + ABCD

         AB + CDE  AC + BDE  AD + BCE  AE + BCD  BC + ADE
         BD + ACE  BE + ACD  CD + ABE  CE + ABD  DE + ABC

         Data Matrix
           Run  A  B  C  D  E           Run  A  B  C  D  E
             1  -  -  -  -  +             9  -  -  -  +  -
             2  +  -  -  -  -            10  +  -  -  +  +
             3  -  +  -  -  -            11  -  +  -  +  +
             4  +  +  -  -  +            12  +  +  -  +  -
             5  -  -  +  -  -            13  -  -  +  +  +
             6  +  -  +  -  +            14  +  -  +  +  -
             7  -  +  +  -  +            15  -  +  +  +  -
             8  +  +  +  -  -            16  +  +  +  +  +
```

Treatment combination assignment to the fractions can also be based on the methods of Section 8.3. For half-fraction designs with defining relation as the highest order interaction, i.e. the interaction of all the k factors, fraction 1 will consist of those combinations with even commonality or with $L = 0 \pmod 2$ when k is even. The opposite will be the case for k odd.

8.4.1 Alias Structure

From the contrast matrix in Table 8.5, we can generate the contrasts associated with the two fractions of a 2^{3-1} design. These are shown in Table 8.7.

Table 8.7 — Contrasts for the 2^{3-1} Fractional Factorial Design

Fraction	Effect	Contrast	Effect	Contrast
1	A	$a - b - c + abc$	BC	$a - b - c + abc$
	B	$-a + b - c + abc$	AC	$-a + b - c + abc$
	C	$-a - b + c + abc$	AB	$-a - b + c + abc$
2	A	$-(1) + ab + ac - bc$	BC	$(1) - ab - ac + bc$
	B	$-(1) + ab - ac + bc$	AC	$(1) - ab + ac - bc$
	C	$-(1) - ab + ac + bc$	AB	$(1) + ab - ac - bc$

Consider fraction 1 first. The contrasts for the main effects are the same as those for the two factor interactions. As a result of this, differentiation between the influence of main effects and two factor interactions is not possible. Thus, when main effects are estimated, we are in fact estimating A and BC, B and AC, and C and AB. Effects which exhibit this property of sharing contrasts are called **aliases**. The occurrence of this effect can be denoted by $A \equiv BC$, $B \equiv AC$, and $C \equiv AB$ where $\equiv$ defines aliased with. For fraction 2, the main effect contrasts are negative of those of the two factor interactions and so main effect estimation really is providing estimates of A and $-BC$, B and $-AC$, and C and $-AB$ suggesting an alias structure of $A \equiv -BC$, $B \equiv -AC$, and $C \equiv -AB$.

We can see, therefore, that each fraction of a FFD gives rise to the same alias structure and we say both fractions belong to the same **family**. It does not matter which is selected though ease of experimentation and experimental constraints may have a bearing on this choice. Often after running one fraction, the other fraction may be run to produce a full 2^3 design to help **de-alias** effects to provide independent effect estimates. Without prior information or assumptions about effects, it may not be possible to discern which aliased effect is actually important. This is a disadvantage of FFDs, though only a minor one. Such designs are most appropriate when there is a large number of factors and there is some prior knowledge of the negligible effect of higher order interactions, or where information on higher order interaction effects is not a priority.

The rule for determining aliases is straightforward. Multiply any factorial effect by the defining contrast(s) modulus 2 (even powers equal the identity I). For fraction 1 of a 2^{3-1} design, the aliases are determined by multiplying the three main effects by ABC as shown in Table 8.8.

Table 8.8 — Aliases for 2^{3-1} Fractional Factorial Design

Fraction	Main effect	Alias determination
1	A	$A(ABC) = A^2BC = BC \rightarrow A \equiv BC$
	B	$B(ABC) = AB^2C = AC \rightarrow B \equiv AC$
	C	$C(ABC) = ABC^2 = AB \rightarrow C \equiv AB$
2	A	$A(-ABC) = -A^2BC = -BC \rightarrow A \equiv -BC$
	B	$B(-ABC) = -AB^2C = -AC \rightarrow B \equiv -AC$
	C	$C(-ABC) = -ABC^2 = -AB \rightarrow C \equiv -AB$

$\equiv$ defines aliased with

For A, the product becomes A^2BC and since $A^2 = I$, we have $A \equiv BC$. The aliasing of B and C can be determined similarly. This fraction, with defining relation $I = +ABC$ based on its association with the plus coefficients in the ABC contrast (8.1), is called the **principal fraction**. Fraction 2, called the **complementary fraction**, is based on the minus coefficients in the ABC contrast and has defining relation of the form $I = -ABC$.

Exercise 8.2 Refer to the proposed 2^{5-1} design for the water filtration study of Example 8.2. We want to derive the associated alias structure and discuss its implications for the planned study.

Alias structure: Multiplying each effect by the defining contrast $ABCDE$ will provide information on the alias structure of the planned study. This is shown as follows:

Effect	*Alias derivation*
A	$A(ABCDE) = A^2BCDE = BCDE \rightarrow A \equiv BCDE$
B	$B(ABCDE) = AB^2CDE = ACDE \rightarrow B \equiv ACDE$
C	$C(ABCDE) = ABC^2DE = ABDE \rightarrow C \equiv ABDE$
D	$D(ABCDE) = ABCD^2E = ABCE \rightarrow D \equiv ABCE$
E	$E(ABCDE) = ABCDE^2 = ABCD \rightarrow E \equiv ABCD$
AB	$AB(ABCDE) = A^2B^2CDE = CDE \rightarrow AB \equiv CDE$
AC	$AC(ABCDE) = A^2BC^2DE = BDE \rightarrow AC \equiv BDE$
AD	$AD(ABCDE) = A^2BCD^2E = BCE \rightarrow AD \equiv BCE$

Continuing in this way, we can derive the remaining aliases as $AE \equiv BCD$, $BC \equiv ADE$, $BD \equiv ACE$, $BE \equiv ACD$, $CD \equiv ABE$, $CE \equiv ABD$, and $DE \equiv ABC$. Thus, main effects are aliased with four factor interactions and two factor interactions with three factor interactions. Main effects and two factor interactions are therefore estimable independently. This alias structure corresponds exactly to that shown in the software generated output of Display 8.1 produced using the Minitab menu procedures shown in Appendix 8A. Provided the three and four factor interactions can be assumed negligible, this 2^{5-1} FFD looks feasible for the planned study. □

In FFDs, it is also useful to know how many aliases are associated with each effect. All factorial effects have at least one alias with the total number dependent on the number of defining contrasts specified and the level of fractionation. The number can be specified as

$$\text{number of aliases} = p + (2^p - p - 1) \tag{8.2}$$

for a 2^{k-p} FFD, i.e. a $1/2^p$ fraction of a 2^k design. p is the level of fractionation and also the number of specified defining contrasts while $(2^p - p - 1)$ corresponds to the number of additional defining contrasts attainable by multiplying together the specified contrasts modulus 2. In the 2^{5-1} design discussed in Exercise 8.2, there is only **one defining contrast** ($p = 1$) so each factorial effect has only one alias, i.e. $1 + (2^1 - 1 - 1)$. When **two defining contrasts** require to be specified, we have $p = 2$ and one $(2^2 - 2 - 1)$ additional contrast. Each factorial effect will thus have $2 + 1 = 3$ aliases from expression (8.2), one from each of the defining contrasts and one from the additional contrast. Further degrees of fractionation will obviously lead to a greater number of defining contrasts and consequently, more complex aliasing.

Exercise 8.3 Suppose in the water filtration study of Example 8.2, it had been necessary to include a sixth factor F specifying the length of time spent on a particular portion of the filtering operation. To run the experiment within 16 runs will require that a quarter-fraction design be constructed, i.e. a 2^{6-2} design. Suppose $I = ABCE$ and $I = BCDF$ are specified to be the defining contrasts. Construct the principal fraction and determine the aliases of the six main effects.

Defining relation: Two defining contrasts have been specified so $p = 2$. As a quarter-fraction design is planned, this means that one $(2^2 - 2 - 1)$ additional defining contrast will exist. This additional contrast is derived as

$$ABCE(BCDF) = AB^2C^2DEF = ADEF.$$

Thus, the full defining relation for the planned 2^{6-2} FFD is

$$I = ABCE = BCDF = ADEF.$$

Fraction structure: Illustration of treatment combination assignment will be based on the basic design approach using the basic 2^4 structure for the first four factors (see Table 8.9) and solving the defining contrasts $I = ABCE$ and $I = BCDF$ for the factors E and F. This leads to the design generators $E = ABC$ and $F = BCD$. Appropriate multiplication of the plus and minus signs in the first four columns will then provide the signs for the two additional factors and thereby the associated treatment combinations. The results of this procedure are shown in Table 8.9 with the design matrix presented in standard order for convenience. The three other fractions associated with this 2^{6-2} FFD can be similarly derived using the generator pairings ($E = ABC$, $F = -BCD$), ($E = -ABC$, $F = BCD$), and ($E = -ABC$, $F = -BCD$).

Table 8.9 — Principal fraction for a 2^{6-2} Fractional Factorial Design with defining contrasts $I = ABCE$ and $I = BCDF$

Basic design						
A	B	C	D	$E = ABC$	$F = BCD$	Combination
−	−	−	−	−	−	(1)
+	−	−	−	+	−	*ae*
−	+	−	−	+	+	*bef*
+	+	−	−	−	+	*abf*
−	−	+	−	+	+	*cef*
+	−	+	−	−	+	*acf*
−	+	+	−	−	−	*bc*
+	+	+	−	+	−	*abce*
−	−	−	+	−	+	*df*
+	−	−	+	+	+	*adef*
−	+	−	+	+	−	*bde*
+	+	−	+	−	−	*abd*
−	−	+	+	+	−	*cde*
+	−	+	+	−	−	*acd*
−	+	+	+	−	+	*bcdf*
+	+	+	+	+	+	*abcdef*

Alias structure: As $p = 2$ defining contrasts were initially specified, expression (8.2) indicates that each effect will have three aliases. Using the technique illustrated in Exercise 8.2, we can show the main effect aliasing to be

$$A \equiv BCE \equiv ABCDF \equiv DEF \quad B \equiv ACE \equiv CDF \equiv ABDEF \quad C \equiv ABE \equiv BDF \equiv ACDEF$$
$$D \equiv ABCDE \equiv BCF \equiv AEF \quad E \equiv ABC \equiv BCDEF \equiv ADF \quad F \equiv ABCEF \equiv BCD \equiv ADE$$

where the aliases are derived from the defining contrasts $ABCE$, $BCDF$, and $ADEF$ in turn.

Software generation: The output in Display 8.2 shows the Minitab generated design and alias structure for the planned 2^{6-2} design (see Appendix 8A for menu procedures). As before, the design matrix is as derived by the basic design approach. The alias structure for the main effects agrees with that presented above. The additional alias information provided indicates that two factor interactions are aliased primarily with other two factor interactions and four factor interactions while three factor interactions are aliased with main effects, three factor interactions, and five factor interactions. □

8.4.2 Design Resolution

Though we have shown how to construct the fractions and how to determine effect aliases, it is often easier to consider aliasing in a more general way through **resolution**. Resolution essentially defines the ability of the design to provide acceptable and independent effect estimates of the main components of interest, the main effects and lower order interactions, i.e. the best possible alias structure for recovery of effect estimates. The resolution of two-level FFDs can be defined as the length of the smallest (least complex) interaction within the defining relation (set of defining contrasts) where length corresponds to number of letters comprising the interaction. A Roman numeral is generally used to denote design resolution.

Display 8.2

```
Factors:    6     Base Design:        4, 16      Resolution: IV
Runs:      16     Replicates:             1      Fraction: 1/4

Design Generators: E = ABC F = BCD

Defining Relation: I = ABCE = BCDF = ADEF

Alias Structure
A + BCE + DEF + ABCDF    B + ACE + CDF + ABDEF    C + ABE + BDF + ACDEF
D + AEF + BCF + ABCDE    E + ABC + ADF + BCDEF    F + ADE + BCD + ABCEF

AB + CE + ACDF + BDEF    AC + BE + ABDF + CDEF    AD + EF + ABCF + BCDE
AE + BC + DF + ABCDEF    AF + DE + ABCD + BCEF    BD + CF + ABEF + ACDE
BF + CD + ABDE + ACEF

ABD + ACF + BEF + CDE    ABF + ACD + BDE + CEF
```

Data Matrix

Run	A	B	C	D	E	F	Run	A	B	C	D	E	F
1	-	-	-	-	-	-	9	-	-	-	+	-	+
2	+	-	-	-	+	-	10	+	-	-	+	+	+
3	-	+	-	-	+	+	11	-	+	-	+	+	-
4	+	+	-	-	-	+	12	+	+	-	+	-	-
5	-	-	+	-	+	+	13	-	-	+	+	+	-
6	+	-	+	-	-	+	14	+	-	+	+	-	-
7	-	+	+	-	-	-	15	-	+	+	+	-	+
8	+	+	+	-	+	-	16	+	+	+	+	+	+

8.4.2.1 Resolution III

For the weight gain study of Example 8.1, the illustrated 2^{3-1} design with defining contrast $I = ABC$ is a 2^{3-1}_{III} FFD as *ABC* contains three letters. For designs of this resolution, main effects are clear of each other but are aliased with two factor interactions, resulting in main effects and two factor interactions not being estimable independently. In addition, some two factor interactions may be aliased with each other.

8.4.2.2 Resolution IV

A 2^{4-1} design with defining contrast $I = ABCD$ would be a resolution IV design as *ABCD* contains four letters, i.e. a 2^{4-1}_{IV} FFD. The 2^{6-2} design for the water filtration study of Exercise 8.3 is also an illustration of a resolution IV design, a 2^{6-2}_{IV} design, as each contrast in the defining relation $I = ABCE = BCDF = ADEF$ is of length 4. This resolution level specifies that no main effects are aliased with any other main effect or with any two factor interaction though some two factor interactions may be aliased with each other. This means main effects can be estimated separately but two factor interactions may be confounded with each other. Resolution IV designs can be obtained from resolution III designs by a process of **foldover** which entails re-running the resolution III experiment using the original design matrix format but with the plus and minus signs switched. Foldover enables missing treatment combinations to be tested allowing effect estimation to be de-aliased.

8.4.2.3 Resolution V

The 2^{5-1} design for the water filtration study of Example 8.2 is an illustration of a resolution V design as the contrast within the defining relation $I = ABCDE$ is of length 5, i.e. a 2_{V}^{5-1} FFD. Designs of this type have no main effect aliased with any other main effect, two factor interaction, or three factor interaction. In addition, no two factor interaction is aliased with any other two factor interaction though some may be aliased with three factor interactions. This type of design allows for unique estimation of all main effects and two factor interactions, provided all higher order interactions can be assumed negligible.

When deciding on the most appropriate FFD to use, it is best to choose one with high resolution consistent with the degree of fractionation required. For half-fraction designs, this generally corresponds to choosing the defining contrast as the highest order interaction. Higher resolution reduces the influence of aliasing on effect estimation and interpretation, particularly in respect of main effects and lower order interactions. It also provides for less restrictive assumptions in respect of which interactions require to be assumed negligible in the assessment of the associated effect estimates. The maximum resolution of half-fraction designs is k, the number of factors. Full two-level designs in up to k factors ($k \leq 7$) have resolution ($2k + 1$).

The real benefits of fractional replication are not seen until k is large with the higher order interactions used for error estimation. A 2^{6-1} design with highest order interaction as defining contrast, for example, results in an **ANOVA table** of the form illustrated in Table 8.10. In such a design, the three factor interaction effects can be used for error estimation provided their effect can be assumed negligible. This enables statistical assessment of main and two factor interaction effects to occur on the assumption that the four and five factor interactions are of little importance. FFD structures form the basis of **Orthogonal Arrays** (see Chapter 9), **Taguchi Methods** (see Chapter 10), and **Response Surface Methods** (see Chapter 11). Further details on fractional replication can be found in Kuehl (1994), Montgomery (1997), and Winer *et al.* (1991) with Montgomery (1997) also providing a listing of selected FFDs for up to 15 factors.

Table 8.10 — ANOVA table for a 2^{6-1} Fractional Factorial Design with defining contrast $I = ABCDEF$

Source	*df*
main effects A to F	1 each for 6 effects
(*aliased with five factor interactions*)	
two factor interactions $A \times B$, $A \times C$, etc.	1 each for 15 effects
(*aliased with four factor interactions*)	
three factor interactions $A \times B \times C$, $A \times B \times D$, etc.	1 each for 20 effects
(*aliased with three factor interactions*)	(*use as error*)

8.5 ANALYSIS COMPONENTS FOR FRACTIONAL FACTORIAL DESIGNS

The analysis procedures associated with FFDs are similar to those discussed in Chapter 7 for full two-level FDs. The methods of analysis can therefore be exploratory (effect estimates, effect plots, data plots) and statistical (proposed response model, ANOVA modelling, diagnostic checking) as necessary depending on the information the data provide in respect of effect separation into distinctly different influence groupings. The presence of aliasing, however, does mean that account must be taken of any assumptions necessary concerning aliased effects.

Example 8.3 An experiment was conducted to investigate which variables were affecting the colour of a product produced by a chemical process. The objective of the experimentation was to develop a process control procedure to aid reduction in product variation. Five two-level factors were identified:

Factor	*Levels*
Solvent/reactant ratio (A)	low and high
Catalyst/reactant ratio (B)	0.025 and 0.035
Temperature (C)	150 °C and 160 °C
Reactant purity (D)	92% and 96%
pH of reactant (E)	8.0 and 8.7.

The process variables A and B represent ratios of solvent to reactant and catalyst to reactant within the production process. The experiment was based on a 2^{5-1} FFD with defining contrast $I = ABCDE$. The collected data are presented in Table 8.11 where the response is the colour of the product in coded units.

Table 8.11 — Product colour data from 2^{5-1} Fractional Factorial Design with defining contrast $I = ABCDE$ for Example 8.3

Run	A	B	C	D	E	Combination	Colour
14	–	–	–	–	+	*e*	−0.63
9	+	–	–	–	–	*a*	2.51
16	–	+	–	–	–	*b*	−2.68
3	+	+	–	–	+	*abe*	−1.66
2	–	–	+	–	–	*c*	2.06
10	+	–	+	–	+	*ace*	1.22
13	–	+	+	–	+	*bce*	−2.09
7	+	+	+	–	–	*abc*	1.93
12	–	–	–	+	–	*d*	6.79
4	+	–	–	+	+	*ade*	6.47
1	–	+	–	+	+	*bde*	3.45
5	+	+	–	+	–	*abd*	5.68
11	–	–	+	+	+	*cde*	5.22
15	+	–	+	+	–	*acd*	9.38
6	–	+	+	+	–	*bcd*	4.30
8	+	+	+	+	+	*abcde*	4.05

Reprinted from Snee, R. D. (1985). Experimenting with a large number of variables. In: Snee, R. D., Hare, L. B. & Trout, J. R. (eds.), *Experiments in industry: Design, analysis, and interpretation of results*, American Society for Quality Control, Milwaukee, pp. 25–35 with permission of the American Society for Quality Control.

8.5.1 Exploratory Analysis

Effect assessment is still the main aspect of the analysis of two-level FFDs though inclusion of EDA should also be considered to gain initial insight into the effect the factors have on the response. As previously, such an approach cannot provide interaction information but can be a useful starting point for the analysis.

Exercise 8.4 Simple plots and summaries of the colour data with respect to the run order of Table 8.10 and each factor (not shown) represent the first stage in the data analysis.

Alias information: The design run was a 2^{5-1} FFD with defining contrast $I = ABCDE$ meaning the design is of resolution V. For such a design, it can be shown that main effects are aliased with four factor interactions and two factor interactions are aliased with three factor interactions. Assuming three and four factor interactions are negligible, we can see that the design will provide independent estimation of the main effects and two factor interactions, representing efficient use of experimental effort.

EDA: The run order plot suggested acceptably random responses. A definite difference in colour was obvious with reactant purity (D) with higher values occurring at its high level. The catalyst/reactant ratio (B) also showed some difference with high colour occurring at the low ratio setting. Variability in response did not appear to differ with factor levels for each factor. The most consistent colour responses were for reactant purity. □

8.5.2 Effect Estimates Analysis

Effect estimation in two-level FFDs is based on the same principles as illustrated for two-level FDs in Section 7.2.2. Effect estimates in single replicate FFDs are specified similarly to expression (7.5) as

$$Effect\ estimate = (Effect\ contrast)/2^{k-p-1} \tag{8.3}$$

as each factor level, or combination of factor levels, is experimented on 2^{k-p-1} times where p defines the level of fractionation. If each treatment combination has been replicated n times, the divisor in expression (8.3) becomes $n2^{k-p-1}$.

When the number of factors is large, however, numerical assessment of effect estimates can become impractical. Graphical presentations of the effect estimates again represent the most advantageous way of presenting and assessing the information contained within the effect estimates. As with two-level FDs, several forms of graphical presentation can be considered namely, **normal plot**, **half-normal plot**, **Pareto chart**, and **active contrast plot**. It is generally advisable, as before, to use at least two of these plots to assess effect estimates though all should produce comparable results in respect of important and unimportant effects.

Exercise 8.5 Exploratory analysis of the colour data of Example 8.3 was discussed in Exercise 8.4. The next step in the data analysis concerns effect estimation and assessment. When carrying out this assessment, the associated alias structure must be taken into account.

Effect estimation: Using appropriate contrasts for the FFD used, it is possible to generate the effect estimates for the main effects and two factor interactions. These are presented in Display 8.3 based on use of Minitab's **Stat ➤ DOE ➤ Analyze Custom Model** procedure, the effect estimates being provided in the column 'Effect' with the 'Coef' column providing the regression coefficients for the response as a model of all possible effects where 'Effect' = 2×'Coef'. From the presented values, it certainly appears that only the main effects of reactant purity (D), catalyst/reactant ratio (B), reactant pH (E), and solvent/reactant ratio (A) are the largest effects.

Display 8.3

Term	Effect	Coef	Term	Effect	Coef
Constant		2.875	A	1.645	0.822
B	−2.505	−1.252	C	0.768	0.384
D	5.585	2.792	E	−1.742	−0.871
A*B	0.110	0.055	A*C	0.128	0.064
A*D	−0.190	−0.095	A*E	−0.612	−0.306
B*C	0.083	0.041	B*D	−0.090	−0.045
B*E	0.373	0.186	C*D	−0.627	−0.314
C*E	−0.575	−0.288	D*E	0.002	0.001

Normal plot of effect estimates: The normal plot of the colour effect estimates is provided in Display 8.4 based on $\alpha = 0.20$ as the significance level for effect separation. The four highlighted effects stand out again by virtue of their difference from the line through the unimportant effects. Reactant purity (D) and solvent/reactant ratio (A) provide positive effect estimates suggesting high colour with high level. The negative estimates of the catalyst/reactant ratio (B) and reactant pH (E) suggest the opposite interpretation. The corresponding Pareto chart, though not shown, generated the same conclusion though the closeness of temperature (C) being classified as important was more apparent. It would appear, therefore, that the main effects only have an additive effect. As main effects are aliased with four factor interactions, it is safe to assume that the calculated estimates correspond to the associated main effect only. □

Display 8.4

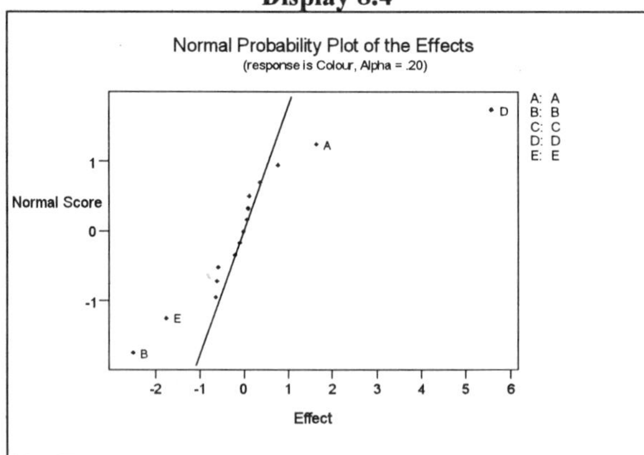

8.5.3 Data Plots and Summaries

Once screening has indicated which are the active effects, it is important to investigate these effects more fully to understand how they influence the measured response. **Main effect** and **interaction plots** again provide the mechanisms for such assessment. **Trade-off** between the two natures of response, mean and variability, may again be necessary to decide "best" results.

Exercise 8.6 Analysis of effect estimates for the product colour study of Example 8.3 showed that only four of the main effects appeared important, provided the aliased four factor interactions could be assumed negligible. To interpret these effects further, we should assess the associated main effect plots for the information they contain.

Main effects analysis: The main effect plots for the four highlighted effects are presented in Display 8.5. The plots show that reactant purity (D) provides greatest colour change when changing levels, the change being positive. Changing the level of solvent/reactant ratio (A) also results in an increase in the colour measurement. Lowering of colour occurs to a similar extent with both catalysts/reactant ratio (B) and reactant pH (E). The effect of variability, through assessment of range, shows comparability for each factor at each level with reactant purity providing the most consistent results. Choice of "best" levels therefore appears unaffected by measurement variability.

From this analysis, depending on the requirements for product colour, we can suggest a "best" treatment combination. Table 8.12 provides such a summary. As only main effects appear important, it is obvious why different levels are associated with different types of outcome. Only temperature (C) does not vary with level because this was the one factor deemed unimportant in its influence. It is interesting to note that the combinations *ad* and *be*, suggested as possible optimal treatment combinations, were not tested in the original experiment. This is often a feature of FFDs where the optimal treatment combination has not been tested within the experiment conducted. □

Display 8.5

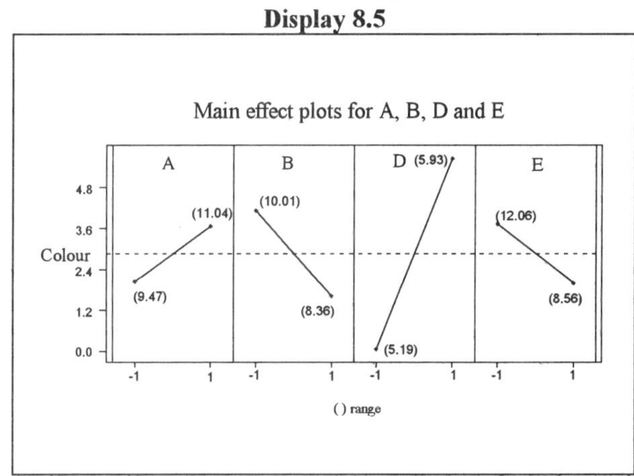

Table 8.12 — Summary of analysis of product colour study of Example 8.3

Factors	A	B	C	D	E	Combinations
Max colour	+	–	–/+	+	–	*ad*, *acd*
Min colour	–	+	–/+	–	+	*be*, *bce*

8.5.4 Statistical Components

Unreplicated two-level FFDs, as full two-level FDs, are not built around specification of a response model at the outset. Screening of the effects through effects analysis highlights the important effects appropriate to include in a **response model**. Statistical assessment of such a model enables significance of active effects to be assessed as well as allowing for diagnostic checking of factors to take place. It is generally advisable to express the response model in order of importance of effects, most important first.

Statistical assessment of a proposed response model within FFDs rests on use of the **ANOVA principle** of partitioning of response variation and on construction of an associated **ANOVA table**. Effect sum of squares are calculated using the expression

$$SS\text{Effect} = (\textit{Effect estimate})^2 2^{k-p-2} \tag{8.4}$$

where k is the number of factors and p is the level of fractionation of the FFD. For a half-fraction, $p = 1$ while a quarter-fraction would provide $p = 2$. In Minitab, correctly calculated effect estimates are produced but in SAS, the ADX macros generate half the correct estimate. Thus, when using SAS, the produced effect estimate should be doubled before being used in expression (8.4). The total sum of squares, SSTotal, is calculated as

$$SS\text{Total} = \sum X_j^2 - (\sum X_j)^2/n \tag{8.5}$$

where X_j is the response of the jth experimental run and n refers to the number of experimental runs of the fractional experiment conducted. For instance, the 2^{4-1} half-fraction design has $n = 8$ runs while the 2^{5-1} FFD would provide $n = 16$ and the 2^{6-2} FFD $n = 16$ runs also. Construction of the **ANOVA table** can be carried out manually or by use of statistical software using the GLM procedure (see Appendix 8A). The general ANOVA procedures within software can also be used though in Minitab, the proposed model requires to be hierarchical, i.e. effects contributing to interaction terms must be included in the model even if unimportant. Regression modelling methods provide alternative procedures for model fitting.

Exercise 8.7 For the product colour study of Example 8.3, effects analysis suggested four effects to be important. From this, we could propose a model of the form

colour = reactant purity (D) + catalyst/reactant ratio (B) + reactant pH (E)
+ solvent/reactant ratio (A) + error

to explain the colour measurements. The error term comprises those effects deemed unimportant from the exploratory analysis, i.e. temperature and all the two factor interactions. Display 8.6 contains the ANOVA information generated by Minitab's GLM procedure. For expression (8.4), effect estimates can be obtained from Display 8.3. The summation components for the SSTotal expression (8.5) are $\sum X_j = 46$ and $\sum X_j^2 = 312.7152$. Using these values together with $n = 16$ generates a SSTotal result of 180.465, as presented in the last line of the 'Seq SS' column in Display 8.6.

Display 8.6

Source	DF	Seq SS	Adj SS	Adj MS	F	P
D	1	124.769	124.769	124.769	179.95	0.000
B	1	25.100	25.100	25.100	36.20	0.000
E	1	12.145	12.145	12.145	17.52	0.002
A	1	10.824	10.824	10.824	15.61	0.002
Error	11	7.627	7.627	0.693		
Total	15	180.465				

Analysis of proposed model: As $k = 5$, $p = 1$, and the effect estimate for reactant purity (D) is 5.585, expression (8.4) provides an SS value for factor D of

$$SSD = (5.585)^2 \times 2^{5-1-2} = 124.7689$$

as given in Display 8.6. Applying expression (8.4) for each of the other model effects will provide the SS given in the Minitab output.

The critical values for test statistic comparison from Table A.3 are $F_{0.05,1,11} = 4.84$ and $F_{0.01,1,11} = 9.65$ (the 0.1% critical value is 19.69). Based on these, or use of the printed p values, it can be seen that reactant purity (D) and catalyst/reactant ratio (B) appear significant at the 0.1% significance level ($p < 0.001$). Reactant pH (E) and solvent/reactant ratio (A) are significant at a lower level ($p < 0.01$). In conclusion, therefore, we can state that the four active effects specified are all highly significant as expected given the marked difference in their effect estimates from those of the other effects. □

Once statistical assessment is complete, the final two aspects of **prediction of optimal solution** and **diagnostic checking** should be considered to round off the analysis. For the product colour study of Example 8.3, prediction is not appropriate as the study objective of minimisation or maximisation is not known. Assessment of model predictions for the proposed response model of Exercise 8.7 produced more unreasonable than reasonable predictions indicating that the proposed model may not be fully explaining what affects product colour.

Exercise 8.8 The final aspect of analysis for the product colour study of Example 8.3 is the diagnostic checking of the residuals generated by fitting the colour model of Exercise 8.7.

Diagnostic checking: From the factor plots, those for catalyst/reactant ratio (B), shown in Display 8.7, reactant purity (D), and reactant pH (E) exhibited distinctive trends. Display 8.7 indicates that low ratio appears to suggest lower variability in average colour. The plots for reactant purity and reactant pH were similar though the column patterning was reversed with high level appearing to suggest lower variability.

Display 8.7

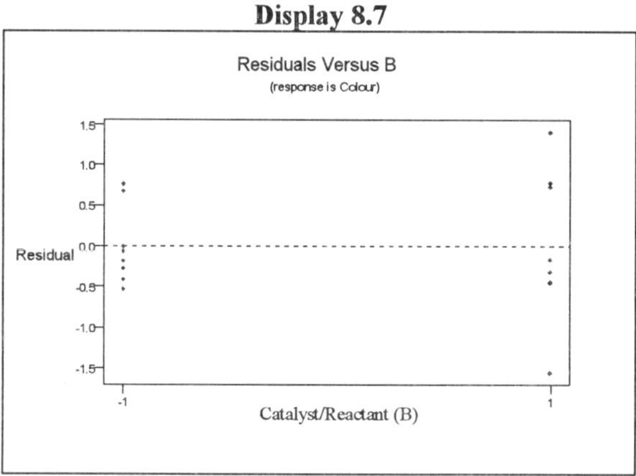

The plot of residuals against run order showed no obvious trend. The plot with respect to fitted values, shown in Display 8.8, does hint at a model trend with predictions better for high colour than low. This again emphasises that, though a picture of what affects product colour has emerged, it is by no means the full picture. A normal plot of the residuals showed no deviations from linearity. □

Display 8.8

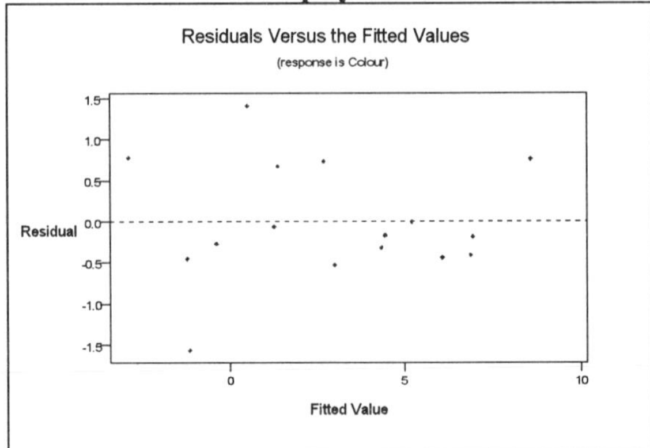

Example 8.3 and the subsequent analysis has provided an insight into how to handle the analysis of a half-fraction design in five or more factors. This is made simpler by main effects being aliased with at least four three factor interactions which, in general, can be safely assumed negligible. In FFDs, interpretation of two factor interactions can be made difficult as often they may be aliased with another of the same order, e.g. $AB \equiv CE$, so we cannot tell whether the estimated effect is due to either interaction separately or some combination of them. If a factor has a large effect, it is possible that it will have an appreciable interaction with other factors. Thus, by looking at which main effects are important, we could determine which of the aliased interactions may be important. Other ways of establishing the importance of aliased interactions involve the running of further experiments as another fraction of the fractional structure, or as a design generated by the foldover technique, or as small factorials in the factors of the aliased interactions (Haaland 1989).

Example 8.4 A tube manufacturer was concerned about the quality of the plastic coating of manufactured cabling tubes. The tubes are used to carry cables in buildings. Six two-level factors were identified for investigation. These are temperature to which the plastic is heated (A), the force at which the hot plastic is pushed down a coating tube (B), the type of plastic (C), the thickness applied (D), the pressure of the coating chamber (E), and the cooling time (F). The investigation was based on a 16-run 2^{6-2} FFD with defining contrasts $I = ABCE$ and $I = BCDF$. The design, which corresponds to that provided in Exercise 8.3, and quality data are displayed in Table 8.13. The data refer to the coating quality of the finished product based on a scale from 0 to 40, high values signifying good quality. Columns A to D are assigned – and + values as in a 2^4 design. Column E is the product of columns A, B, and C since the defining contrast $I = ABCE$ provides the design generator $E = ABC$. Column F is similarly found using the defining contrast $I = BCDF$ and associated design generator $F = BCD$.

Table 8.13 — Plastic coating quality data from 2^{6-2} Fractional Factorial Design with defining contrasts $I = ABCE$ and $I = BCDF$ for Example 8.4

Run	A	B	C	D	E	F	Combination	Coating quality
11 −	−	−	−	−	−	−	(1)	5.670
7 +	−	−	−	+	−	−	*ae*	4.995
9 −	+	−	−	+	+	+	*bef*	15.795
1 +	+	−	−	−	+	+	*abf*	29.700
13 −	−	+	−	+	+	+	*cef*	3.375
15 +	−	+	−	−	+	+	*acf*	7.245
2 −	+	+	−	−	−	−	*bc*	12.825
10 +	+	+	−	+	−	−	*abce*	29.700
6 −	−	−	+	−	+	+	*df*	12.015
8 +	−	−	+	+	+	+	*adef*	5.940
5 −	+	−	+	+	−	−	*bde*	16.875
4 +	+	−	+	−	−	−	*abd*	29.700
16 −	−	+	+	+	−	−	*cde*	7.965
12 +	−	+	+	−	−	−	*acd*	2.430
14 −	+	+	+	−	+	+	*bcdf*	18.360
3 +	+	+	+	+	+	+	*abcdef*	25.785

Exercise 8.9 We want to illustrate aspects of the analysis of the data associated with the product quality problem described in Example 8.4.

Alias structure: The 2^{6-2} FFD of Example 8.4 is a resolution IV design with defining relation

$$I = ABCE = BCDF = ADEF.$$

This means that an alias structure identical to that in Exercise 8.3 prevails (see Table 8.14) with main effects aliased with three and five factor interactions. Assuming such interactions are negligible, then main effect estimation appears acceptable. Two factor interactions, however, are aliased with interactions of the same order making practical effect interpretation difficult but not impossible.

Table 8.14 — Complete alias structure for coating quality study of Example 8.4

$A \equiv BCE \equiv DEF \equiv ABCDF$	$B \equiv ACE \equiv CDF \equiv ABDEF$	$C \equiv ABE \equiv BDF \equiv ACDEF$
$D \equiv BCF \equiv AEF \equiv ABCDE$	$E \equiv ABC \equiv ADF \equiv BCDEF$	$F \equiv BCD \equiv ADE \equiv ABCEF$
$AB \equiv CE \equiv ACDF \equiv BDEF$	$AC \equiv BE \equiv ABDF \equiv CDEF$	$AD \equiv EF \equiv ABCF \equiv BCDE$
$AE \equiv BC \equiv DF \equiv ABCDEF$	$AF \equiv DE \equiv ABCD \equiv BCEF$	$BD \equiv CF \equiv ABEF \equiv ACDE$
	$BF \equiv CD \equiv ABDE \equiv ACEF$	
$ABD \equiv ACF \equiv BEF \equiv CDE$		$ABF \equiv ACD \equiv BDE \equiv CEF$

EDA: The plot of coating quality with respect to the run order number of each experiment showed no trend. Only temperature (A) and force (B) showed up as having an influence. The other four factors exhibited comparable coating quality irrespective of level.

Effect estimation: Through contrasts for the 2^{6-2} FFD run, it is again possible to generate effect estimates. Display 8.9 provides the Minitab generated estimates for this investigation.
Normal plot of effect estimates: Display 8.10 provides the normal plot of effect estimates based on a significance level for effect separation of $\alpha = 0.20$. Four effects stand out: force (B), temperature × force interaction (AB), temperature (A), and temperature × thickness interaction (AD). The AD effect is smallest and also the only important one which is negative. The related Pareto chart, though not shown, provided the same conclusion.

Display 8.9

Estimated Effects and Coefficients for CoatQual

Term	Effect	Coef	Term	Effect	Coef
Constant		14.285	A	5.349	2.675
B	16.116	8.058	C	−1.603	−0.802
D	1.198	0.599	E	−0.962	−0.481
F	1.029	0.515	A*B	7.408	3.704
A*C	0.354	0.177	A*D	−3.189	−1.595
A*E	0.253	0.127	A*F	−0.523	−0.262
B*D	−0.523	−0.262	B*F	−0.894	−0.447
A*B*D	0.557	0.278	A*B*F	−1.569	−0.785

Display 8.10

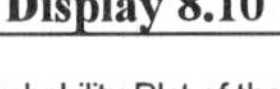

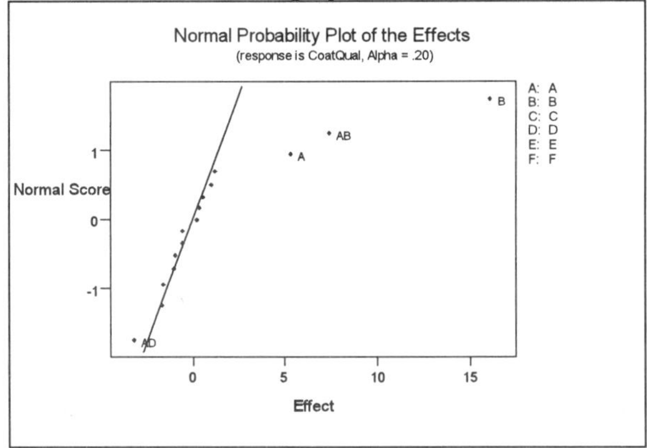

Unfortunately, interaction effects $A \times B$ and $C \times E$ are aliased while effects $A \times D$ and $E \times F$ are also aliased. How can we distinguish between these aliased effects in respect of which underpins the associated effect estimate? We can do this by looking at the effect estimates of the contributing main effects presented in Display 8.9. Consider the interaction $A \times B$. Both main effects appear important (largest main effects) so it is highly likely that the associated interaction will also be important. Effects C and E are low signifying unimportant effects so it could be argued that interaction $C \times E$ can be assumed unimportant. By similar reasoning, we could suggest that interaction $A \times D$ is the effect underpinning its associated effect estimate.
Main effects analysis: Main effect plots for the temperature (A) and force (B) factors are illustrated in Display 8.11. They show that force is more important (steeper slope) with high level possibly best for each. However, variability in coating quality, as

measured by the range, is higher at this level for both factors indicating a lack of consistency in coating quality.

Display 8.11

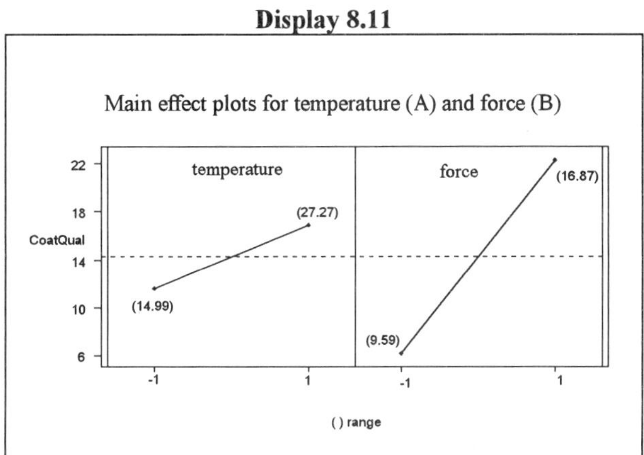

Interaction analysis: The interaction plot for temperature (A) and force (B) is shown in Display 8.12, the non-parallel lines confirming the presence of interaction. Greater change in coating quality is apparent at high temperature as the force factor is increased. Coating quality appears best at high levels of both factors.

Display 8.12

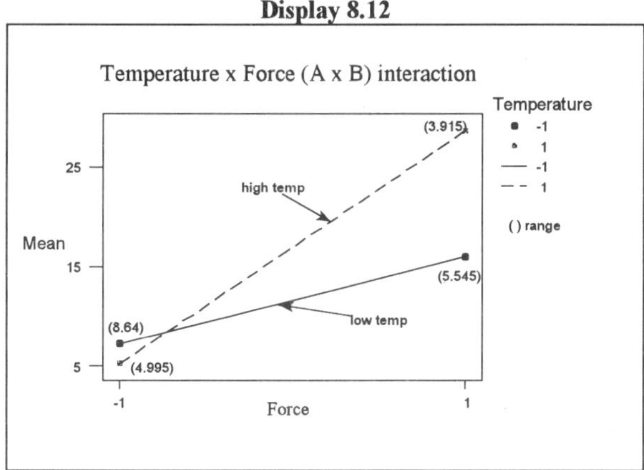

Display 8.13 presents the interaction plot for the temperature × thickness ($A \times D$) interaction. The plot shows that as coating thickness increases, coating quality improves at low temperature but diminishes at high temperature. Variability is also high at high temperature.

Summary: Table 8.15 contains a summary of the findings. It would appear to suggest that treatment combinations involving the combination *ab* may be optimal for maximising coating quality. This combination was not tested in the experiment conducted and provides another illustration of how FFDs can generate an optimal solution, despite the apparent lack of coverage of all possible treatment combinations. □

Display 8.13

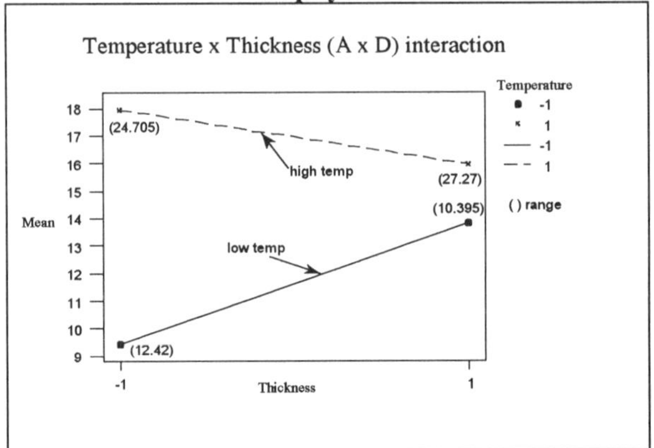

Table 8.15 — Summary of analysis of plastic coating quality study of Example 8.4

Factors	*A*	*B*	*C*	*D*	*E*	*F*
Main effects		+	+			
A x *B* interaction	+	+				
A x *D* interaction	+			−		
Possible "best"	+	+	−/+	−	−/+	−/+

Further analysis of the coating quality problem of Example 8.4 based on a proposed model of

$$\text{coating quality} = \text{force } (B) + \text{temperature} \times \text{force } (A \times B) + \text{temperature } (A) + \text{temperature} \times \text{thickness } (A \times D) + \text{error}$$

could be carried out. The analysis points to temperature, force, and coating thickness as being the major factors influencing the quality of the tube coating. It is these factors that would be advised to be assessed further to understand their influence more fully.

8.6 THREE-LEVEL FRACTIONAL FACTORIAL DESIGNS

Fractional replication can be extended to three-level FDs to provide **3^{k-p} FFDs**, p again specifying the level of fractionation. It is possible to construct a one-third fraction (3^{k-1}), a one-ninth fraction (3^{k-2}), a one-twenty seventh fraction (3^{k-3}), etc. Fractionation is based on similar concepts to two-level FFDs in respect of design construction with p components of interaction selected and used to generate the 3^p fractions of treatment combinations. Alias determination is based on effect multiplication as before though for three-level FFDs, this consists of multiplication modulus 3 of each effect by each defining contrast and also by the square of each defining contrast. An additional constraint is that the only exponent allowed for the first letter is 1 so if this is not 1, the expression must be squared through multiplication modulus 3.

For example, a 3^{3-1} FFD with defining contrast $I = AB^2C$ would have the alias structure for main effects of

$$A = A(AB^2C) = A^2B^2C = (A^2B^2C)^2 = A^4B^4C^2 = ABC^2 \rightarrow A \equiv ABC^2$$
$$A = A(AB^2C)^2 = A^3B^4C^2 = BC^2 \rightarrow A \equiv BC^2$$
$$B = B(AB^2C) = AB^3C = AC \rightarrow B \equiv AC$$
$$B = B(AB^2C)^2 = A^2B^5C^2 = (A^2B^2C^2)^2 = A^4B^4C^4 = ABC \rightarrow B \equiv ABC$$
$$C = C(AB^2C) = AB^2C^2 \rightarrow C \equiv AB^2C^2$$
$$C = C(AB^2C)^2 = A^2B^4C^3 = (A^2B)^2 = A^4B^2 = AB^2 \rightarrow C \equiv AB^2$$

where the terms involving factor powers define the components of interaction. Thus, three-level FFDs often have complex alias structures involving interaction effects and components of interactions. In turn, this can lead to difficulties with alias interpretation unless all interactions can be assumed negligible. Such designs can also require many experimental runs which can make fractional replication of three-level FDs unattractive. Fuller details on three-level FFDs can be found in Montgomery (1997) while Connor & Zelen (1959) list 3^{k-p} FFD plans for four to ten factors.

PROBLEMS

8.1 An experiment is to be run within a 2^4 design structure but not all treatment combinations can be run in the same session. It is possible to run eight treatment combinations in one session with two sessions available for experimentation. A 2^4 design confounded in two blocks is deemed appropriate where the blocks represent the session effect. Construct a design in two blocks using *ABCD* as the defining relation.

8.2 An experiment is to be run within a 2^5 design but not all treatment combinations can be run at any one time. It is possible to run eight treatment combinations at a time so a 2^5 design confounded in four blocks is considered an appropriate design structure. Construct a design in four blocks using *ACDE* and *BCD* as the defining contrasts. What is the other defining contrast?

8.3 List the aliases for the main and two factor interactions effects in a 2^{5-1} FFD with defining relation $I = ACDE$. What is the design resolution?

8.4 A quarter-fraction of a 2^6 FD is to be constructed using *BCDE* and *ABCF* as the defining contrasts. The experiment is to be used by an electrical engineer to assess the effects of six two-level factors on the power output of an electrical device. Determine what effects are aliased with the main effects and the two factor interactions. Is the design resolution acceptable?

8.5 The manager of a postal sorting office wished to analyse the factors affecting the number of items of mail damaged by the automatic sorting system. The factors tested were belt angle (A) at 10° and 15°, belt speed (B) at slow and fast, belt material (C) at type 1 and type 2, and number of letters sorted per hour (D) at low and high. In order to check the system in a single day, only eight runs of the system were possible. A 2^{4-1} design with defining relation $I = ABCD$ was used for the study. The average number of damaged items of mail per hour were recorded over a 10 hour period of operation as given below. Are any effects important? What combination of factors, if any, minimise mail damage?

Run	Response	Run	Response	Run	Response	Run	Response
5	(1) = 8.5	7	*ad* = 17.0	1	*bd* = 13.0	4	*ab* = 22.5
8	*cd* = 7.0	2	*ac* = 16.25	3	*bc* = 13.75	6	*abcd* = 23.25

8.6 A 2^{5-1} experiment using the five factor interaction as defining contrast was set up to assess how five different factors and their interactions affect ion-interaction HPLC retention time (min) of the triazine herbicide atrazine. The factors considered were pH of the mobile phase, ion-interaction reagent alkyl chainlength N, organic modifier concentration CM (% methanol), interaction reagent concentration CR, and flow rate F. The design and responses are listed below. Which effects appear active? Is a response model for atrazine possible?

pH	N	CM	CR	F	Atrazine response	pH	N	CM	CR	F	Atrazine response
5	9	45	2	1.5	16.34	5	9	35	2	1	67.69
5	5	35	2	1.5	57.50	5	5	45	2	1	22.12
8	5	35	10	1.5	39.49	8	5	45	2	1.5	16.67
8	9	45	2	1	22.84	5	5	35	10	1	71.12
8	9	35	10	1	27.53	8	9	45	10	1.5	9.68
5	5	45	10	1.5	16.81	5	9	35	10	1.5	31.73
8	9	35	2	1.5	32.85	8	5	45	10	1	23.69
8	5	35	2	1	65.33	5	9	45	10	1	21.42

Reprinted from Marengo, E., Gennaro, M. C. & Abrigo, C. (1996) Investigation by experimental design and regression models of the effect of five experimental factors on ion-interaction high-performance liquid chromatographic retention. *Analytica Chimica Acta* **321** 225–236 with kind permission of Elsevier Science – NL, Sara Burgerhartstraat 25, 1055 KV Amsterdam, The Netherlands.

8.7 An instant soup product manufacturer produces dry soup mix packages for supermarkets. Weight variations between the packages have been causing customer complaints. In order to find out more about the process, the Quality Manager decided to conduct a screening experiment to try to determine which factors, if any, affected the weight variation between the packages with a view to minimising this variation. Five factors were identified for testing as follows: number of mixer ports (*A*) at 1 port and 3 ports, temperature (*B*) at cooled (Co) and ambient (Am), mixing time (*C*) at 60 seconds and 80 seconds, batch weight (*D*) at 1500 lbs and 2000 lbs, and days delay (*E*) at 1 day and 7 days. Factor *A* refers to the number of mixer ports through which the vegetable oil is added, *B* is the temperature surrounding the mixer, and *E* corresponds to the number of days delay between mixing and packaging. Owing to raw material constraints, it was only possible to test 16 treatment combinations. To account for this, the screening experiment was based on a 2^{5-1} FFD with defining contrast $I = ABCDE$. Each batch, corresponding to a particular treatment combination, consisted of a group of 25 sets of five samples of soup mix taken at regular intervals from the production process. The performance standard deviation, measuring the combined effects of all sources of variation on package weight, was computed as the response measure. The design and standard deviation responses are shown below. Analyse these data and draw appropriate conclusions.

Batch	*A*	*B*	*C*	*D*	*E*	Standard deviation	Batch	*A*	*B*	*C*	*D*	*E*	Standard deviation
1	1	Co	60	2000	7	0.78	9	1	Am	80	2000	7	0.76
2	3	Co	80	2000	7	1.10	10	3	Am	60	2000	7	0.62
3	3	Am	60	1500	1	1.70	11	1	Co	80	2000	1	1.09
4	3	Co	80	1500	1	1.28	12	1	Co	60	1500	1	1.13

(continued)

(continued)

5	1	Am	60	1500	7	0.97		13	3	Co	60	1500	7	1.25
6	1	Co	80	1500	7	1.47		14	3	Am	80	1500	7	0.98
7	1	Am	60	2000	1	1.85		15	3	Co	60	2000	1	1.36
8	3	Am	80	2000	1	2.10		16	1	Am	80	1500	1	1.18

Reprinted from Hare, L. B. (1988) In the soup: A case study to identify contributors to filling variability. *Journal of Quality Technology* **20** 36–43 © 1997 American Society for Quality Control Reprinted with Permission.

8.8 A 2^{5-2} experiment with defining contrasts $I = ABD$ and $I = ACE$ was set up to assess how five different factors and their interactions affect ion-interaction HPLC retention time (min) of the triazine herbicide simazine. The factors considered were pH of the mobile phase, ion-interaction reagent alkyl chainlength N, organic modifier concentration CM (% methanol), interaction reagent concentration CR, and flow rate F. The responses recorded are listed below. Carry out a relevant analysis of these data.

pH	N	CM	CR	F	Simazine response	pH	N	CM	CR	F	Simazine response
5.8	6	35	7.75	1.3	22.29	5.8	6	45	7.75	0.9	11.12
7.2	6	35	2.75	0.9	31.91	7.2	6	45	2.75	1.3	16.08
5.8	8	35	2.75	1.3	28.50	5.8	8	45	2.75	0.9	16.00
7.2	8	35	7.75	0.9	20.28	7.2	8	45	7.75	1.3	9.39

Reprinted from Marengo, E., Gennaro, M. C. and Abrigo, C. (1996) Investigation by experimental design and regression models of the effect of five experimental factors on ion-interaction high-performance liquid chromatographic retention. *Analytica Chimica Acta* **321** 225–236 with kind permission of Elsevier Science – NL, Sara Burgerhartstraat 25, 1055 KV Amsterdam, The Netherlands.

8.9 Parts manufactured in an injection moulding process are showing excessive shrinkage. This is causing problems in the assembly operation upstream from the injection moulding area. A quality improvement team has decided to use a designed experiment to study the injection moulding process so that shrinkage can be reduced. The objective of the experiment was to learn how each factor affects shrinkage and also, something about how the factors interact. The team decided to investigate seven two-level factors thought to influence shrinkage. The factors and their levels were mould temperature (A) at low and high, screw speed (B) at slow and fast, holding time (C) at min and max, cycle time (D) at short and long, gate size (E) at narrow and wide, moisture content (F) at low and high, and holding pressure (G) at low and high. The design used was a 2^{7-3} FFD with defining contrasts $I = ABCE$, $I = BCDF$, and $I = ACDG$. The observed shrinkage (x 10) for the test part produced at each of the 16 experimental runs is given in the table below. Using the data presented, carry out a full and relevant analysis. Is there evidence for a best combination of factors? Comment on the findings.

Run	Shrinkage	Run	Shrinkage	Run	Shrinkage
1	$abcdefg = 52$	7	$dfg = 8$	12	$adef = 12$
2	$abce = 60$	8	$bcg = 26$	13	$aeg = 10$
3	$cefg = 4$	9	$bcdf = 37$	14	$abfg = 60$
4	$acf = 15$	10	$bdeg = 34$	15	$bef = 32$
5	$abd = 60$	11	$cde = 16$	16	$acdg = 5$
6	$(1) = 6$				

Reprinted from Montgomery, D. C. (1990) Using fractional factorial designs for robust product development. *Quality Engineering* **3** 193–205 by courtesy of Marcel Dekker, Inc.

8.10 Ruggedness studies are used in analytical laboratories to assess the resistance of an analytical procedure to slight deviations in procedure factors. FFDs are often used for this purpose. One such study concerned a procedure for measuring the amount of analyte in a solution. Seven two-level factors were considered for the study, these being sample weight (A), concentration of first reagent (B), concentration of second reagent (C), total volume (D), time of heating (E), reaction temperature (F), and pH of solution (G). Each factor was varied at two levels, low level representing the normal setting of the factor and high level corresponding to the perturbed level. This perturbed level represents the maximum likely excusion from the normal setting. A 2^{7-3} design with defining contrasts $I = ABCE$, $I = BCDF$, and $I = ACDG$ was used to run the study within 16 experimental runs. The treatment combinations and data collected are shown below. Analyse these data for factor effects.

Run	Analyte found	Run	Analyte found	Run	Analyte found
1	$abcdefg = 68$	7	$abce = 67$	12	$adef = 71$
2	$abfg = 59$	8	$bcdf = 61$	13	$aeg = 61$
3	$(1) = 70$	9	$bef = 70$	14	$cde = 59$
4	$bdeg = 60$	10	$bcg = 67$	15	$dfg = 64$
5	$abd = 58$	11	$acdg = 66$	16	$cefg = 57$
6	$acf = 64$				

8A Appendix: Software Information for Two-level Fractional Factorial Designs

Design construction facilities

Most statistical software have default design build facilities to enable construction of FFDs together with creation of a sheet for data recording.

SAS: In SAS, FFD construction facilities are accessible through the PROC FACTEX procedure. The statements necessary to produce an output comparable to that in Display 8.1 are

```
PROC FACTEX;
    FACTORS A B C D E;
    SIZE DESIGN = 16;
    MODEL RESOLUTION = MAXIMUM;
    EXAMINE ALIASING (5) DESIGN;
RUN;
```

The SIZE statement specifies that a design of 16 runs is required while the MODEL statement specifies to search for the 16-run design that provides for best resolution. The ALIASING option within the EXAMINE statement requests a print of the alias structure of the constructed design with 5 specifying the highest order of aliased effects to be printed. DESIGN specifies that the design matrix is to be listed in the printed output.

PROC FACTEX does not allow, however, for re-specification of the design generators from SAS's defaults. For Exercise 8.3, SAS would generate a design with the defining contrasts $I = BCDE$ and $I = ACDF$ which do not correspond to those specified for the planned study. SAS also has an Automated Design of Experiments (ADX) menu interface for designing of experiments accessible by selecting **Globals ➤ SAS/ASSIST ➤ PLANNING TOOLS ➤ DESIGN OF EXP** which is based on the PROC FACTEX procedure.

Minitab: In Minitab, the **Stat** ➢ **DOE** procedures provide access to two-level design and analysis facilities. To generate the output provided in Display 8.1, the necessary menu commands are as follows:

> Select **Stat** ➢ **DOE** ➢ **Create Factorial Design** ➢ for *Type of Design*, select **2-level factorial (specify generators)** ➢ for *Number of factors*, select **4**.
>
> Select **Designs** ➢ in *Designs* box, select **Full factorial** ➢ **Generators** ➢ for *Add factors to the base design by listing their generators*, enter **E = ABCD** ➢ click **OK** ➢ click **OK**.
>
> Select **Results** ➢ for *Printed Results*, select **Summary table, alias table, data table, defining relation** ➢ for *Content of Alias Table*, select **Interactions up through order** and choose **5** ➢ click **OK**.
>
> Select **Options** ➢ ensure *Fraction* set at **Use principal fraction** ➢ check **Randomize runs** ➢ click **OK** ➢ click **OK**.

The **specify generators** choice enables the user to design an experiment based on their choice of design generators rather than Minitab's default generators for the FFD planned. Selection of **Full factorial** provides the necessary basic design while use of **Generators** provides access to menu entry for the user-specified generators. Checking the **Randomize runs** option enables the design matrix to be presented in standard form, i.e. treatment combinations presented in logical order. Not checking would result in a randomised run order being presented for the design matrix which is how it should be implemented in practical design building. For this half-fraction design, choice of **2-level factorial (default generators)** could also have been selected as the Minitab default for a 2^{5-1} design is $I = ABCDE$.

For Exercise 8.3 (see Display 8.2), the menu commands would be as above except for specification of generators as **E = ABC F = BCD** and choice of interaction order of **6**. In Minitab, the default defining contrasts for a 2_{IV}^{6-2} design are $I = ABCE$ and $I = BCDF$ which are those required for the planned water filtration study. SAS's defaults differ and are $I = BCDE$ and $I = ACDF$.

Data entry

Data entry for two-level FFDs follows similar principles to that for two-level FDs. Generally, the actual levels of the factors can be used, e.g. low and high, though the −1/+1 system may also be necessary for analysis purposes. Data entry for Examples 8.3 and 8.4 would therefore comply with the information shown in Tables 8.11 and 8.13 in the text.

Exploratory analysis

Plots and summaries of two-level FFD data for exploratory analysis can be produced using similar procedures to those described in earlier Appendices.

Effect estimation and plotting

Production of effect estimates and associated data plots can be easily achieved within most software packages.

SAS: In SAS, ADX macros are the basis of the design and analysis procedures for two-level FFDs. They can be loaded into SAS by implementing the statements below based on macro storage in directory C:\SAS.

```
%INCLUDE 'C:\SAS\QC\SASMACRO\ADXGEN.SAS';
%INCLUDE 'C:\SAS\QC\SASMACRO\ADXFF.SAS';
%ADXINIT;
```

```
RUN;
```

Once the macros are loaded, we would then use the %ADXFFA macro to generate effect estimates and a normal plot of estimates using the following statements:

```
%ADXFFA(CH8.EX83, COLOUR, A B C D E, 5);
RUN;
```

The data to be analysed are contained in the data set CH8.EX83 and COLOUR specifies the response. The factors are entered next with design resolution entered as the last component. A similar set of statements could equally have been used to generate effect estimates for the coating quality study of Example 8.4. It should be noted, however, that use of this macro provides estimates which correspond to the coefficient estimates of the fitted multiple regression model of response as a function of all possible effects. These are half the true effect estimates though this does not affect the relative standing of the estimates within the generated normal plot. In addition, in two-level FFDs, SAS only prints effect estimates up to two factor interactions depending on design resolution. A macro, %ADXALIAS, is available for derivation of alias information but will only provide limited details by default.

Minitab: The **Stat** menu provides access to the effect estimation and plotting systems within Minitab. Displays 8.3 and 8.4 contain these presentations for the two-level product colour study of Example 8.3. The information contained in these displays was produced using the menu selections below. In the terms selection, choosing **5** enables the full alias structure to be printed.

Select **Stat** ➢ **DOE** ➢ **Analyze Custom Design** ➢ for *Design*, select **2-Level factorial** ➢ **Fit Model** ➢ for *Responses*, select **Colour** and click **Select** ➢ for *Factors*, select the box and enter **A B C D E**.

Select **Terms** ➢ for *Include terms in the model up through order*, ensure **5** is the entry ➢ click **OK**.

Select **Graphs** ➢ for *Effects Plots*, select **Normal** and **Pareto** ➢ for *Alpha*, enter **0.2** ➢ click **OK**.

Select **Storage** ➢ for *Model Information*, select **Effects** ➢ click **OK** ➢ click **OK**.

Data plots

Main effect plots, illustrated in Display 8.5, can be produced by the following menu selections though the illustrated plot has been edited using Minitab's graph edit facility **(Editor ➢ Edit)** before production in order to provide the range data and some annotation.

Select **Stat** ➢ **ANOVA** ➢ **Main Effects Plot** ➢ for *Factors*, enter **A B D E** ➢ for *Raw response data in*, click the box, select **Colour**, and click **Select** ➢ click **OK**.

The information for Example 8.4 was generated in a similar way. The interaction plots in Displays 8.12 and 8.13 were produced using the %Interact macro with plot editing used to incorporate the ranges and produce a more self-explanatory plot. For Display 8.12, the menu procedure was as follows:

Select **Stat** ➢ **ANOVA** ➢ **Interactions Plot** ➢ for *Factors*, enter **A B** ➢ for *Raw response data in*, click the box, select **CoatQual**, and click **Select** ➢ for *Title*, click the box and enter **Temperature x Force interaction** ➢ click **OK**.

Main effect and interaction plots can also be produced through the **Stat ➢ DOE ➢ Analyze Custom Design** menu after running the FFD analysis routine. The initial menu selection for this mode of data plot generation is

> Select **Stat ➢ DOE ➢ Analyze Custom Design ➢** ensure **Design** is set at **2-Level factorial ➢ Plot Response ➢** choose plot required **➢ Setup ➢** then follow menu to generate the required data plots.

ANOVA information
Derivation of ANOVA information for a proposed response model in two-level FFDs can be handled using software.

SAS: The PROC GLM procedure can provide the route to the ANOVA information required for a proposed two-level design model after screening. The statements below would produce the same ANOVA information as Display 8.6 for the product colour study of Example 8.3.

```
PROC GLM DATA = CH8.EX83;
    CLASS A B D E;
    MODEL COLOUR = D B E A;
    OUTPUT OUT = CH8.EX83A P = FITS R = RESIDS;
RUN;
```

For non-hierarchical response models, the general linear model (GLM) procedure can be used to provide the required ANOVA information and residual estimates for diagnostic checking by creating specific columns for the model interactions. For example, suppose the $A \times C$ interaction was required in Example 8.3. The following sequence of statements would produce the necessary numerical codes for this interaction effect to enable PROC GLM to generate the required ANOVA information.

```
DATA CH8.EX83;
    SET CH8.EX83;
    AC = A*C;
PROC GLM DATA = CH8.EX83;
    CLASS D B E A AC;
    MODEL COLOUR = D B E A AC;
    OUTPUT OUT = CH8.EX83A P = FITS R = RESIDS;
RUN;
```

Minitab: The general linear model (GLM) facilities in Minitab cannot handle a proposed response model if it is not hierarchical. For Exercise 8.7, the model was additive in the main effects so Minitab's GLM procedure could be used by default as such a model is not hierarchical.

> Select **Stat ➢ ANOVA ➢ General Linear Model ➢** for *Responses*, select **Colour** and click **Select ➢** for *Model*, select the box and enter **D B E A**.
> Select **Storage ➢ Fits ➢ Residuals ➢** click **OK**.
> Select **Graphs ➢** select the residual plots required **➢** click **OK ➢** click **OK**.

For non-hierarchical response models, the GLM procedure can be used to provide the required ANOVA information and residual estimates for diagnostic checking by creating specific columns for the model interactions. For example, suppose the $A \times C$ interaction was required in Example 8.3. A column of codes corresponding to this interaction, denoted AC, can be created using the **Calc ➢ Calculator** option selection through the multiplication of the codes for the A and C columns. For this, however, factor code settings must be numeric.

Diagnostic checking

Residual plots for diagnostic checking of a proposed response model using the stored residual and fits values can be produced using the procedures described in Appendix 3A.

9

Two-level Orthogonal Arrays

9.1 INTRODUCTION

So far, when discussing two-level designs for efficient multi-factor experimentation, we have introduced two-level FDs and two-level FFDs. For FFDs, we require to pre-specify at least one defining contrast which is to be sacrificed and from which the design's alias structure and resolution can be derived. Design applications can arise, however, where it may not be possible nor desirable to specify the defining contrast(s) or where the number of factors to be tested is large and deciding on sacrificial interactions is not straightforward. An alternative design selection procedure to the FFD approach may therefore be necessary.

Such an alternative is provided by **Orthogonal Array Designs** (OADs) (Clarke & Kempson 1997, Ross 1988). The array style of design presentation was first described by Rao (1946) and comprises a matrix form of design presentation where the columns are related to the factors and the rows represent the experimental runs. An illustration of an array structure, the $OA_8(2^7)$ array, is provided in Table 9.1. In fact, OADs are really the fraction of a FFD which contains the treatment combination (1), i.e. all factors at low level. This fraction may not coincide with the principal fraction of a FFD though as each fraction is essentially similar, any fraction could be used as a design base. This chapter will introduce some of the basic concepts and principles associated with two-level OADs. This will provide some of the statistical background to the design principles associated with Taguchi's quality engineering approach (see Chapter 10).

Table 9.1 — The $OA_8(2^7)$ array

Row	c_1	c_2	c_3	c_4	c_5	c_6	c_7
1	−	−	−	−	−	−	−
2	−	−	−	+	+	+	+
3	−	+	+	−	−	+	+
4	−	+	+	+	+	−	−
5	+	−	+	−	+	−	+
6	+	−	+	+	−	+	−
7	+	+	−	−	+	+	−
8	+	+	−	+	−	−	+

9.2 ARRAY STRUCTURES

The simplest practical two-level OAD structure is the $\boldsymbol{OA_8(2^7)}$ array presented in Table 9.1 where c_1, c_2, c_3, etc. are simply convenient column labels. Sometimes, factor levels in arrays are written as 1 and 2 where we have used − and +. Either mode of presentation is acceptable though the latter allows the orthogonality of array structures to be more readily apparent. In the specification $OA_8(2^7)$, the subscript 8 corresponds to the number of test runs (rows), 2 defines the number of levels tested for each controlled factor, and 7 specifies the number of columns. The patterning of symbols in the $OA_8(2^7)$ array listed in Table 9.1 closely resembles the contrast matrix listed in Table 9.2 for the 2^3 FD. This shows the similarity of the two design structures though the column ordering and sign ordering of some columns differ for the array structure.

The **general two-level Orthogonal Array** is written as $\boldsymbol{OA_r(2^c)}$ where $r = c + 1 = 2^m$ specifies the number of runs (rows), c is the number of columns, and m is the number of basic columns associated with a two-level r run design. These basic columns are designated by the column labels c_j where $j = 2^{i-1}$ and $i = 1, 2,..., m$. Basic columns contain $r/2^i$ repeats of each level in sequence and are the first constructed in the array. The other columns are constructed by pairwise column multiplication with the resultant column of signs multiplied by -1 if the first entry in the column is plus. The constructed array will be such that later columns show more change in factor levels than do the initial columns.

Table 9.2 — Contrast matrix for the 2^3 Factorial Design

Column	1	2	3	4	5	6	7
	−	−	+	−	+	+	−
	+	−	−	−	−	+	+
	−	+	−	−	+	−	+
	+	+	+	−	−	−	−
	−	−	+	+	−	−	+
	+	−	−	+	+	−	−
	−	+	−	+	−	+	−
	+	+	+	+	+	+	+

For the $OA_8(2^7)$ array, we have $r = 8$, i.e. eight entries per column, and $m = 3$ as a two-level eight run design has three basic columns, i.e. $8 = 2^m$. The basic columns are labelled c_1 ($i = 1, j = 1$), c_2 ($i = 2, j = 2$), and c_4 ($i = 3, j = 4$). Column c_1 contains $8/2^1 = 4$ repeats of each level (4 − followed by 4 +) while column c_2 is made up of $8/2^2 = 2$ repeats of each level. Column c_4 will contain $8/2^3 = 1$ repeat of each level, i.e. an alternating sequence of − and + signs. Columns c_1, c_2, and c_4 in Table 9.1 confirm this specification. Column c_3 is obtained by multiplying column c_1 by column c_2 which provides the signs sequence of

$$+ \; + \; - \; - \; - \; - \; + \; + \quad .$$

As the first entry is +, we require to multiply each entry by -1 to generate the sequence presented in column c_3 in Table 9.1. Column c_5 is $c_1 \times c_4$ with each entry multiplied by -1, c_6 is $c_2 \times c_4$ with each entry multiplied by -1, and c_7 is $c_1 \times c_2 \times c_4$.

For the $OA_8(2^7)$ array, an ideal column allocation is A to c_1, B to c_2, and C to c_4 which corresponds to the 2^3 design matrix though this can be altered to suit experimental objectives and constraints such as cost of factor level changes. The array structure resulting from this allocation is displayed in Table 9.3 with those columns corresponding to A, B, and C specifying the **design matrix**. The first row corresponds to each factor set at low to give treatment combination (1). The second row corresponds to A and B set low and C high to give treatment combination c while row 3 corresponds to A low, B high, and C low specifying treatment combination b. The other five rows can be similarly specified in respect of associated treatment combination. In each pair of columns, each factor level combination occurs the same number of times so satisfying the **balancing property** of two-level designs.

The **orthogonality property** of two-level designs is also satisfied as each column is a product of two other columns in the array though not a direct product, e.g. $c_1 \times c_2 = -c_3$, $c_1 \times c_4 = -c_5$, etc. On the basis of this column multiplication principle, columns c_3, c_5, c_6, and c_7 can be shown to be providing estimates for the interaction effects $-(A \times B)$, $-(A \times C)$, $-(B \times C)$, and $A \times B \times C$, respectively. The appearance of the minus sign reflects that the specification of $-$ and $+$ signs is opposite to that generated by multiplying together the columns associated with indicated factors. Thus, from column c_3, we would actually be estimating $-AB$ rather than AB though this is only a minor technicality. Similar comments apply to the estimates provided by columns c_5 and c_6 of the $OA_8(2^7)$ array shown in Table 9.3.

Table 9.3 — The $OA_8(2^7)$ array with A allocated to c_1, B to c_2, and C to c_4

	c_1	c_2	c_3	c_4	c_5	c_6	c_7	
Row	A	B	$-AB$	C	$-AC$	$-BC$	ABC	Combination
1	−	−	−	−	−	−	−	(1)
2	−	−	−	+	+	+	+	c
3	−	+	+	−	−	+	+	b
4	−	+	+	+	+	−	−	bc
5	+	−	+	−	+	−	+	a
6	+	−	+	+	−	+	−	ac
7	+	+	−	−	+	+	−	ab
8	+	+	−	+	−	−	+	abc

Basing a four factor experiment on the $OA_8(2^7)$ array, the ideal column allocation would be A to c_1, B to c_2, C to c_4, and D to c_7. By allocating the fourth factor D to column c_7, we associate it with the highest order interaction $A \times B \times C$, one of the prerequisites necessary for the provision of high design resolution and thereby efficient estimation (see Section 8.4.2). Columns c_3, c_5, and c_6 will each refer to two aliased two factor interactions. This structure is essentially a 2^{4-1}_{IV} FFD providing independent estimation of main effects and two factor interactions.

A larger version of the $OA_8(2^7)$ array is the $\boldsymbol{OA_{16}(2^{15})}$ array listed in Table 9.4 which has 16 rows (experimental runs) and 15 columns. This array corresponds to the contrast matrix for a 2^4 FD, as shown in Table 7.9, with similar column sign modifications and column re-ordering as for the $OA_8(2^7)$ array. For four factors, ideal basic column allocation is based on A to c_1, B to c_2, C to c_4, and D to c_8 to provide a design matrix identical to that of the 2^4 FD, row 1 providing treatment combination (1), row 2 combination d, row 3 combination c, and so on. The other columns would define associated two factor, three factor, and four factor interaction effects.

Table 9.4 — The $OA_{16}(2^{15})$ array

Row	c_1	c_2	c_3	c_4	c_5	c_6	c_7	c_8	c_9	c_{10}	c_{11}	c_{12}	c_{13}	c_{14}	c_{15}
1	–	–	–	–	–	–	–	–	–	–	–	–	–	–	–
2	–	–	–	–	–	–	–	+	+	+	+	+	+	+	+
3	–	–	–	+	+	+	+	–	–	–	–	+	+	+	+
4	–	–	–	+	+	+	+	+	+	+	+	–	–	–	–
5	–	+	+	–	–	+	+	–	–	+	+	–	–	+	+
6	–	+	+	–	–	+	+	+	+	–	–	+	+	–	–
7	–	+	+	+	+	–	–	–	–	+	+	+	+	–	–
8	–	+	+	+	+	–	–	+	+	–	–	–	–	+	+
9	+	–	+	–	+	–	+	–	+	–	+	–	+	–	+
10	+	–	+	–	+	–	+	+	–	+	–	+	–	+	–
11	+	–	+	+	–	+	–	–	+	–	+	+	–	+	–
12	+	–	+	+	–	+	–	+	–	+	–	–	+	–	+
13	+	+	–	–	+	+	–	–	+	+	–	–	+	+	–
14	+	+	–	–	+	+	–	+	–	–	+	+	–	–	+
15	+	+	–	+	–	–	+	–	+	+	–	+	–	–	+
16	+	+	–	+	–	–	+	+	–	–	+	–	+	+	–

For a five factor study using the $OA_{16}(2^{15})$ array, the main effects could be allocated as A to c_1, B to c_2, C to c_4, D to c_8, and E to c_{15}. This is the ideal column allocation for five factors because c_{15} will be associated with the four factor interaction $A \times B \times C \times D$. This association will provide highest design resolution. The remaining columns would correspond to at least two factor interactions and so the main effects and two factor interactions will not be aliased with each other as they will be estimated from different columns. Such a design structure is a resolution V 2^{5-1} FFD where each main effect is aliased with a four factor interaction and each two factor interaction with a three factor interaction.

Column allocation of factors in OADs can be modified to suit experimental constraints and objective. Additional factors should ideally be placed in columns corresponding to the highest order interactions in the basic factors that are available for confounding. Such an approach helps preserve as high a design resolution as possible as additional are included in the design structure.

9.3 EXPERIMENTAL PLANS

All two-level OADs have an associated resolution, as defined in Section 8.4.2, with higher resolution providing less restrictive alias structures and thereby more efficient effect estimation. A minimum of resolution V is necessary if all main effects and two factor interactions are to be separately estimable assuming third and higher order interactions have negligible effects. There is no simple link between the number of factors, the number of runs, and the resolution number of a plan stemming from array structures, though increasing the number of factors inevitably leads to a reduction in design resolution. Numerous design plans exist for each form of array.

9.3.1 Plans Based on the $OA_8(2^7)$ Structure

Table 9.5 outlines several plans associated with the $OA_8(2^7)$ array. The plans are developed incrementally with the initial three factor allocation of A to c_1, B to c_2, and C to c_4 as the base, i.e. the basic 2^3 FD. Each plan is obtained from the immediately previous one by adding the additional factor to an appropriate column. For example, the six factor plan is obtained using the five factor plan and adding the sixth factor F to column c_5. Main effect allocation for each incremental plan is specified in bold to signify the column allocation for that effect when added to the plan. Only main effects and two factor interactions are shown for four and more factors. Missing row/column entries correspond to column allocation for higher order interactions.

Table 9.5 — Plans generated from the $OA_8(2^7)$ array

No. of factors	Resolution	c_1	c_2	c_3	c_4	c_5	c_6	c_7
3 (Basic)	Full (VII)	***A***	***B***	$-AB$	***C***	$-AC$	$-BC$	ABC
4	IV			$-CD$		$-BD$	$-AD$	***D***
5	III	$-BE$	$-AE$	***E***	$-DE$			$-CE$
6	III	$-CF$	$-DF$		$-AF$	***F***	$-EF$	$-BF$
7	III	$-DG$	$-CG$	$-FG$	$-BG$	$-EG$	***G***	$-AG$

AB defines the $A \times B$ interaction, etc., main effect column allocation in bold

The three factor design in Table 9.5 is simply the 2^3 FD with all treatment combinations included. The four factor design is found from the three factor design by allocating the fourth factor D to column c_7 to relate it to the highest order interaction of the basic factors. To determine the columns associated with the interactions of D and the other three factors, consider the $OA_8(2^7)$ structure displayed in Table 9.3. Association, in this context, defines the column generating the contrast for aliased effects. Factor A is allocated to column c_1 and D to column c_7. The product of c_1 times c_7 produces the column pattern of – and + signs of

$$+ \quad - \quad - \quad + \quad + \quad - \quad - \quad +$$

corresponding to minus c_6. Therefore, the interaction $-(A \times D)$ will be associated with column c_6 and hence the assignment of $-AD$ to this column in Table 9.5. By similar reasoning, it is simple to show that the interaction $-(B \times D)$ is associated with column c_5 ($c_2 \times c_7 = -c_5$) and the interaction $-(C \times D)$ is associated with column c_3 ($c_4 \times c_7 = -c_3$). In fact, in array structures, interactions involving even numbers of factors will be associated with minus of a column of the array while interactions based on odd numbers of factors will be associated with a column of the array. Provided this sign reversal is understood, it is relatively straightforward to develop and discuss interaction associations within two-level OAD structures.

Using the column signs approach to determine column association with factor effects is relatively simple for the $OA_8(2^7)$ structure but less so for more complex arrays such as the $OA_{16}(2^{15})$ array. However, given the orthogonality property of array structures, it is easier to use a **triangular**, or **interaction**, **table** such as that displayed in Table 9.6. This table summarises the pairwise column multiplication results for the $OA_8(2^7)$ array and negates the necessity of determining column association from first principles. Referring to the four factor design discussed above, we can see that the $A \times D$ interaction ($c_1 \times c_7$) is associated with column c_6 as obtained from first principles though as this is an interaction involving two factors, we should note that it is strictly $-(A \times D)$ that is associated with column c_6. The $B \times D$ interaction ($c_2 \times c_7$) shows column c_5 as its associated column as before while $C \times D$

($c_4 \times c_7$) is indicated to be associated with column c_3 as before. Use of the triangular table can therefore greatly simplify column specification in multi-factor studies using OADs as the design base.

Table 9.6 — Triangular table for the $OA_8(2^7)$ array

	Column number					
Column number	2	3	4	5	6	7
1	3	2	5	4	7	6
2		1	6	7	4	5
3			7	6	5	4
4				1	2	3
5					3	2
6						1

The five factor design in Table 9.5 is obtained from the four factor design with the fifth factor E allocated to column c_3 to relate the new factor to the highest order interaction between basic factors that is still available for confounding. Allocating E to c_5 or c_6 would also have satisfied the same column allocation condition. Using Table 9.6, it is simple to ascertain column association for the interaction of E and the other four factors. With A allocated to column c_1 and E to column c_3, Table 9.6 states that the interaction $A \times E$ ($c_1 \times c_3$) will be allocated to column c_2 as shown though as with the four factor design, column c_2 is really estimating the $-(A \times E)$ interaction. Similarly, the $B \times E$ interaction, based on B to column c_2 and E to column c_3, will be associated with column c_1 as shown. Column associations for the $C \times E$ and the $D \times E$ interactions can be similarly derived. This approach can also be used to understand the structuring associated with the six and seven factor plans illustrated in Table 9.5.

9.3.2 Plans Based on the $OA_{16}(2^{15})$ Structure

Similar plans to those shown in Table 9.5 can be generated for the larger $OA_{16}(2^{15})$ structure. Generally, such plans are based on the initial four factor allocation of A to c_1, B to c_2, C to c_4, and D to c_8, i.e. the basic 2^4 FD, though this can be altered to suit experimental necessity. A triangular table of column multiplication results also exists for the $OA_{16}(2^{15})$ array as displayed in Table 9.7. Using this table and the specified basic factor allocation to the array columns, it can be shown that column c_3 refers to $-AB$, c_5 to $-AC$, c_6 to $-BC$, c_7 to ABC, c_9 to $-AD$, c_{10} to $-BD$, c_{11} to ABD, c_{12} to $-CD$, c_{13} to ACD, c_{14} to BCD, and c_{15} to $-ABCD$. Generally, additional factors are added to the array by relating them to high order interactions in the basic factors to generate designs with good resolution.

Example 9.1 Parts manufactured in an injection moulding process are showing excessive shrinkage. This is causing problems in the assembly operation upstream from the injection moulding area. A quality improvement team has decided to use a designed experiment to study the injection moulding process so that shrinkage effects could be investigated and, if possible, reduced. The objective of the experiment was to learn how each factor affects shrinkage and also, how the factors interact. The team decided to investigate six factors thought to have the most potential influence on shrinkage using an $OA_{16}(2^{15})$ array. The factors, each set at two levels, were mould temperature (A), screw speed (B), holding time

(*C*), cycle time (*D*), gate size (*E*), and holding pressure (*F*). Factor column allocation is planned to be based on *A* to c_1, *B* to c_2, *C* to c_4, *D* to c_8, *E* to c_{14}, and *F* to c_7. Columns for *A* to *D* provide the optimal basic design while *E* and *F* allocation enables these factors to be associated with three factor interactions in the basic factors to ensure design resolution is as high as possible.

Table 9.7 — Triangular table for the $OA_{16}(2^{15})$ array

Column number	Column number 2	3	4	5	6	7	8	9	10	11	12	13	14	15
1	3	2	5	4	7	6	9	8	11	10	13	12	15	14
2		1	6	7	4	5	10	11	8	9	14	15	12	13
3			7	6	5	4	11	10	9	8	15	14	13	12
4				1	2	3	12	13	14	15	8	9	10	11
5					3	2	13	12	15	14	9	8	11	10
6						1	14	15	12	13	10	11	8	9
7							15	14	13	12	11	10	9	8
8								1	2	3	4	5	6	7
9									3	2	5	4	7	6
10										1	6	7	4	5
11											7	6	5	4
12												1	2	3
13													3	2
14														1

Exercise 9.1 For the six factor injection moulding study of Example 9.1, we need to determine which columns are associated with which effects at least in respect of the main effects and two factor interactions.

Column allocation: Factor allocation was *A* to c_1, *B* to c_2, *C* to c_4, *D* to c_8, *E* to c_{14}, and *F* to c_7. Table 9.8 summarises the column allocation for each main effect and column association for each two factor interaction based on use of Table 9.7 for interaction column specification.

Table 9.8 — Column allocation for the six factor injection moulding study of Example 9.1

	Column c_1	c_2	c_3	c_4	c_5	c_6	c_7	c_8	c_9	c_{10}	c_{11}	c_{12}	c_{13}	c_{14}	c_{15}
(Basic)	***A***	***B***	−*AB*	***C***	−*AC*	−*BC*	*ABC*	***D***	−*AD*	−*BD*	*ABD*	−*CD*	*ACD*	*BCD*	−*ABCD*
						−*DE*				−*CE*		−*BE*		***E***	−*AE*
			−*CF*		−*BF*	−*AF*	***F***		−*EF*						−*DF*

AB defines the *A* × *B* interaction, etc., main effect column allocation in bold

For factor *E* allocated to column c_{14}, Table 9.7 specifies that the *A* × *E* interaction ($c_1 \times c_{14}$) is associated with column c_{15}, the *B* × *E* interaction ($c_2 \times c_{14}$) with column c_{12}, the *C* × *E* interaction ($c_4 \times c_{14}$) with column c_{10}, and the *D* × *E* interaction ($c_8 \times c_{14}$) with column c_6. By similar reasoning, allocation of the interactions of factor *F* with the other five factors can also be determined. For example, the *A* × *F* interaction ($c_1 \times c_7$) will be allocated to column c_6 while the *E* × *F* interaction ($c_{14} \times c_7$) will be associated with column c_9. Again, it must be remembered that these associations are really referring to minus the specified interaction estimate. □

9.3.3 Aliasing and Resolution in Orthogonal Arrays

Once effect column allocation is complete, it is necessary to check the array's alias structure and resolution to confirm its appropriateness for the study planned. The alias structure, at least as far as main and two factor interactions are concerned, can be found by reading down each column of the column allocation table until the plan required is reached. Each column will generally display a number of effects stemming from the basic factor plan and the plans associated with the additional factors. For each column, the effects obtained by this process are defined as **aliased effects**.

To illustrate, consider the five factor study based on the $OA_8(2^7)$ array (see Table 9.5) which is shown again in Table 9.9 to aid explanation. Reading down column c_1, we can see that A and $-BE$ share the contrast defined by this column and are therefore aliased. Checking column c_2 shows that B and $-AE$ are aliased while column c_3 specifies that $-AB$, $-CD$, and E form an alias string. The alias structure for main and two factor interaction effects for such a design is therefore

$$A \equiv -BE \quad B \equiv -AE \quad C \equiv -DE \quad D \equiv -CE \quad E \equiv -AB \equiv -CD \quad AC \equiv BD \quad BC \equiv AD$$

Aliasing of these effects with third and higher order interactions also occurs but is not readily available from the column allocation table. Aliasing in OADs can also be found using the column signs approach but this is more cumbersome to apply.

Table 9.9 — Column allocation for five factor design based on the $OA_8(2^7)$ array

	c_1	c_2	c_3	c_4	c_5	c_6	c_7
(Basic)	A	B	$-AB$	C	$-AC$	$-BC$	ABC
			$-CD$		$-BD$	$-AD$	$\mathbf{D}$
	$-BE$	$-AE$	$\mathbf{E}$	$-DE$			$-CE$

AB defines the $A \times B$ interaction, etc., factor column allocation in bold

Specification of defining contrasts for two-level OADs is based on relating the additional factors to the basic design effects which share the allocated column. The effect relationships this specifies are the **design generators**. For the five factor study being discussed, we can see from column c_7 that $D = ABC$ and from column c_3 that $E = -AB$. These represent the design generators for this array structure providing the **defining contrasts** $I = ABCD$ and $I = -ABE$. The additional contrast is $-CDE$ from multiplying these defining contrasts modulus 2. The full **defining relation** is therefore

$$I = ABCD = -ABE = -CDE$$

resulting in the design being of resolution III and therefore corresponding to a 2_{III}^{5-2} FFD. We can use the defining relation to determine the full alias structure for the five main effects as follows:

$$A \equiv BCD \equiv -BE \equiv -ACDE \quad B \equiv ACD \equiv -AE \equiv -BCDE \quad C \equiv ABD \equiv -ABCE \equiv -DE$$
$$D \equiv ABC \equiv -ABDE \equiv -CE \quad E \equiv ABCDE \equiv -AB \equiv -CD$$

confirming the hidden aliasing with three, four, and five factor interactions. Modifying factor column allocation cannot improve design resolution in such a study because at least one of D and E will be associated with a two factor interaction, the only effect would be changes to the aliasing structure and contrasts within the defining relation.

For a five factor study based on the $OA_{16}(2^{15})$ array, ideal column allocation would be A to c_1, B to c_2, C to c_4, D to c_8, and E to c_{15}, factor E allocation enabling it to be associated with the highest order interaction among the four basic factors. As this provides $E = -ABCD$, we therefore have a defining contrast $I = -ABCDE$ and a resolution V FFD. Each main effect would be aliased only with four factor interactions while two factor interactions would only be aliased with three factor interactions. Thus, such an array based structure would be capable of independent estimation of all main effects and all two factor interactions, the ideal estimation structure for multi-factor experiments.

Exercise 9.2 The column allocation for the injection moulding study of Example 9.1 was discussed in Exercise 9.1. We want to investigate the aliasing and resolution of the planned design. Table 9.8 is displayed again for convenience.

Table 9.8 — Column allocation for the six factor injection moulding study of Example 9.1

	Column														
	c_1	c_2	c_3	c_4	c_5	c_6	c_7	c_8	c_9	c_{10}	c_{11}	c_{12}	c_{13}	c_{14}	c_{15}
(Basic)	***A***	***B***	$-AB$	***C***	$-AC$	$-BC$	ABC	***D***	$-AD$	$-BD$	ABD	$-CD$	ACD	BCD	$-ABCD$
						$-DE$				$-CE$		$-BE$		***E***	$-AE$
			$-CF$		$-BF$	$-AF$	***F***		$-EF$						$-DF$

AB defines the $A \times B$ interaction, etc., main effect column allocation in bold

Aliasing: The alias structure of the proposed design can be determined by reading off the effects from columns in Table 9.8 with multiple entries. Beginning with column c_3, we can see that AB is aliased with CF while column c_5 provides AC aliased with BF. Column c_6 produces an alias string of $BC \equiv DE \equiv AF$. The full set of alias relationships for main effects E and F and the two factor interactions is as follows:

$$E \equiv BCD \quad F \equiv ABC$$
$$AB \equiv CF \quad AC \equiv BF \quad BC \equiv DE \equiv AF \quad AD \equiv EF \quad BD \equiv CE \quad CD \equiv BE \quad AE \equiv DF$$

No minus signs appear because they cancel out in the aliasing structure.

Again, aliasing with higher order interactions is hidden. In fact, those columns with only single entries specify that the named effect is aliased with third and higher order interactions. Thus, A (column c_1), B (column c_2), C (column c_4), and D (column c_8) are free of two factor interactions but will be aliased with at least one three factor interaction.

Resolution: From the columns in Table 9.8 associated with the two factors E and F additional to the four basic factors, we can see that the design generators are $E = BCD$ and $F = ABC$. From these, we have the defining contrasts $I = BCDE$ and $I = ABCF$ and full defining relation of

$$I = BCDE = ABCF = ADEF$$

where $ADEF$ is the additional defining contrast from multiplication of $BCDE$ and $ABCF$ modulus 2. As each contrast in the defining relation is of length 4, the design is therefore of resolution IV and so is a 2_{IV}^{6-2} FFD. From the defining relation, the main effect aliases can be shown to be

$$A \equiv ABCDE \equiv BCF \equiv DEF \quad B \equiv CDE \equiv ACF \equiv ABDEF \quad C \equiv BDE \equiv ABF \equiv ACDEF$$
$$D \equiv BCE \equiv ABCDF \equiv AEF \quad E \equiv BCD \equiv ABCEF \equiv ADF \quad F \equiv BCDEF \equiv ABC \equiv ADE$$

illustrating that the main effects are aliased with three and five factor interactions. This design separates main effects and two factor interactions and so it is acceptable for main effect estimation, provided third and higher order interactions can be assumed negligible. The aliasing of two factor interactions could cause some difficulties with data interpretation, however. □

Changing column allocation can modify the design structure and associated aliasing. For example, modifying column allocation of factor E to c_{15} for the planned study described in Example 9.1 produces a 2_{III}^{7-3} FFD with defining relation

$$I = -ABCDE = ABCF = -DEF$$

This differs from that shown in Exercise 9.2 and will consequently result in changes to the alias structure. Choice of column allocation, as stated earlier, can affect design resolution and thereby an array's ability to provide acceptable effect estimates for those effects of most interest, i.e. main effects and two factor interactions.

9.3.4 Saturated Designs

Saturated designs occur when every column in an array is allocated to a main effect. The seven factor design in Table 9.5 is an illustration of a saturated design as all columns of the array correspond to main effects only (A to c_1, B to c_2, C to c_4, D to c_7, E to c_3, F to c_5, G to c_6). Such a design could easily be constructed from the basic three factor design by associating the four additional factors, D to G, with the interactions of the original three factors (A to C) as follows: $D = ABC$ (column c_7), $E = -AB$ (column c_3), $F = -AC$ (column c_5), and $G = -BC$ (column c_6). The corresponding design generators are therefore $I = ABCD$, $I = -ABE$, $I = -ACF$, and $I = -BCG$ giving rise to an array structure corresponding to a 2_{III}^{7-4} FFD, i.e. a one-sixteenth fraction of a 2^7 design. Saturated designs are always resolution III as main effects are aliased with two factor interactions.

Larger arrays, such as the $\boldsymbol{OA_{32}(2^{31})}$ and $\boldsymbol{OA_{64}(2^{63})}$, also exist but obviously require considerably more experimental runs than the smaller $OA_8(2^7)$ and $OA_{16}(2^{15})$ arrays. When constructing OADs, if spare columns exist, it is possible to allocate pairs of them to measure certain two factor interactions. If estimation of such effects would be useful in a multi-factor investigation, this approach can enable these effects to be properly estimated free of aliasing. The method of specifically assigning interactions is often used in the construction of Taguchi designs within quality improvement studies (see Chapter 10). In addition, OADs can also be constructed with certain columns empty for error estimation. Such columns are referred to as **dummy columns**.

9.4 ANALYSIS COMPONENTS FOR TWO-LEVEL ORTHOGONAL ARRAYS

As a two-level OAD is essentially one fraction of a two-level FFD, the analysis procedures for data collected from OADs will closely resemble those described in Section 8.5 for FFDs. The methods of analysis are therefore exploratory (effect estimates, effect plots, data plots) and statistical modelling (proposed response model, ANOVA, diagnostic checking), the amount and nature of analysis again depending on the level of effect separation present in the collected data. Knowledge of the underlying alias structure and assumptions necessary concerning aliased effects must again be taken into account within the data analysis to ensure appropriate data interpretation.

Example 9.2 An experiment into microwave dissolution of selenium in biological samples by hydride generation atomic absorption spectroscopy (AAS) was carried out to ascertain which of seven factors and their associated effects most affected the accuracy of the collected measurements. The two-level factors tested were

Factors	*Levels*
Sample mass (A)	0.1 g and 0.25 g
Volume of nitric acid HNO_3 (B)	8 ml and 2 ml
Volume of sulphuric acid H_2SO_4 (C)	0 ml and 2 ml
Volume of percholic acid $HClO_4$ (D)	0.5 ml and 0 ml
Volume of hydrogen peroxide H_2O_2 (E)	0 ml and 0.5 ml
Microwave power (F)	100% and 50%
Time of dissolution (G)	5 min and 15 min

The experiment was based on the $OA_{16}(2^{15})$ array with A allocated to c_1, B to c_2, C to c_6, D to c_{10}, E to c_{14}, F to c_9, and G to c_{12}. The array structure and reported accuracy results (%), based on the ratio of recovered selenium and a certified reference value, are presented in Table 9.10.

Table 9.10 — Accuracy data for Example 9.2 based on $OA_{16}(2^{15})$

Combination	c_1 A	c_2 B	c_3	c_4	c_5	c_6 C	c_7	c_8	c_9 F	c_{10} D	c_{11}	c_{12} G	c_{13}	c_{14} E	c_{15}	Accuracy
(1)	−	−	−	−	−	−	−	−	−	−	−	−	−	−	−	67.4
defg	−	−	−	−	−	−	−	+	+	+	+	+	+	+	+	85.8
ceg	−	−	−	+	+	+	+	−	−	−	−	+	+	+	+	101.7
cdf	−	−	−	+	+	+	+	+	+	+	+	−	−	−	−	99.8
bcde	−	+	+	−	−	+	+	−	−	+	+	−	−	+	+	30.8
bcfg	−	+	+	−	−	+	+	+	+	−	−	+	+	−	−	48.5
bdg	−	+	+	+	+	−	−	−	−	+	+	+	+	−	−	41.6
bef	−	+	+	+	+	−	−	+	+	−	−	−	−	+	+	23.4
af	+	−	+	−	+	−	+	−	+	−	+	−	+	−	+	63.4
adeg	+	−	+	−	+	−	+	+	−	+	−	+	−	+	−	82.3
acefg	+	−	+	+	−	+	−	−	+	−	+	+	−	+	−	92.0
acd	+	−	+	+	−	+	−	+	−	+	−	−	+	−	+	78.1

(continued)

(continued)

abcdef	+	+	−	−	+	+	−	−	+	+	−	−	+	+	−	58.4
abcg	+	+	−	−	+	+	−	+	−	−	+	+	−	−	+	79.3
abdfg	+	+	−	+	−	−	+	−	+	+	−	+	−	−	+	48.5
abe	+	+	−	+	−	−	+	+	−	−	+	−	+	+	−	27.9

Reprinted from Lan, W. G., Wong, M. K., Chen, N. and Sin, Y. M. (1994) Orthogonal array design as a chemometric method for the optimisation of analytical procedures Part 1. Two-level design and its application in microwave dissolution of biological samples. *Analyst* **119** 1659–1667 with the permission of The Royal Society of Chemistry.

Exercise 9.3 For Example 9.2, we require to analyse the collected data fully to understand how the seven tested factors influence the accuracy of selenium recovery.

Alias information: From the factor column allocation specified, we can set up the column allocation for the two factor interactions. Table 9.11 provides this summary based on use of the triangular Table 9.7 for the $OA_{16}(2^{15})$ array.

Table 9.11 shows that the alias structure results in main effects and two factor interactions being aliased together, e.g. column c_2 provides $B \equiv -EG$, and some two factor interactions being aliased with others of the same order, e.g. column c_3 provides $AB \equiv DF$. Only main effect A is clear of two factor interactions and will thus be aliased with third and higher order interactions. Separation of aliased effects could be difficult but the investigators were able to specify that certain two factor interactions could be neglected according to experience. Those interactions coming into this category are underlined in Table 9.11. Inclusion of these assumptions greatly simplifies the effect of aliasing provided third and higher order interactions can also be assumed negligible.

Table 9.11 — Column allocation for the selenium recovery study of Example 9.2

Column														
c_1	c_2	c_3	c_4	c_5	c_6	c_7	c_8	c_9	c_{10}	c_{11}	c_{12}	c_{13}	c_{14}	c_{15}
$\mathbf{A}$	$\mathbf{B}$	$-AB$	$-BC$	ABC	$\mathbf{C}$	$-AC$	$-\underline{BD}$	ABD	$\mathbf{D}$	$-AD$	$-\underline{CD}$	ACD	BCD	$-ABCD$
			$-\underline{DE}$				$-\underline{CE}$				$-\underline{BE}$		$\mathbf{E}$	$-AE$
		$-\underline{DF}$				$-\underline{EF}$	$-\underline{AF}$	$\mathbf{F}$		$-\underline{BF}$				$-\underline{CF}$
	$-\underline{EG}$			$-FG$	$-\underline{DG}$				$-\underline{CG}$		$\mathbf{G}$	$-\underline{AG}$	$-\underline{BG}$	

AB defines the $A \times B$ interaction, main effect column allocation in bold, interactions underlined were assumed negligible by the authors

Resolution: From Table 9.11, comparison of the column allocation of the three additional factors E, F, and G with the interactions of the basic factors sharing the same column provides the design generators of $E = BCD$ (c_{14}), $F = ABD$ (c_9), and $G = -CD$ (c_{12}). The defining contrasts are therefore $I = BCDE$, $I = ABDF$, and $I = -CDG$ resulting in a defining relation of

$$I = BCDE = ABDF = -CDG = ACEF = -BEG = -ABCFG = -ADEFG$$

after inclusion of the additional defining contrasts. As the least complex interaction in the defining relation is of length 3, the $OA_{16}(2^{15})$ array implemented was therefore of resolution III and so corresponded to a 2^{7-3}_{III} FFD.

EDA: Initial plots and summaries with respect to factors, though not shown, highlighted that nitric acid (B) was the most important single factor. Changing levels of percholic acid (D) and hydrogen peroxide (E) appeared to have no significant effect on the reported accuracy figures. However, in all cases, the reported measurements were subject to wide variation indicating possible inadequacy of data precision.

Effect estimates: Display 9.1 provides the Minitab generated effect estimates for the seven main effects and the eight two factor interactions which can be estimated though each of these is aliased with other two factor interactions. Three main effects and two interactions appear important. These are, in numerical order of effect size, nitric acid (B), sulphuric acid (C), time of dissolution (G), the mass × nitric acid interaction ($A*B$), and the microwave power × time of dissolution interaction ($F*G$). Only percholic acid (D) and hydrogen peroxide (E) of the seven tested factors appear unimportant either independently or through an estimable interaction.

Display 9.1

Estimated Effects and Coefficients for Accuracy

Term	Effect	Coef	Term	Effect	Coef
Constant		64.31	A	3.86	1.93
B	−39.01	−19.51	C	18.54	9.27
D	2.71	1.36	E	−3.04	−1.52
F	1.34	0.67	G	16.31	8.16
A*B	13.59	6.79	A*C	2.89	1.44
A*D	−1.54	−0.77	A*E	0.86	0.43
A*F	−2.66	−1.33	A*G	2.26	1.13
B*C	0.36	0.18	F*G	−8.86	−4.43

Normal plot of effects: A normal plot of the derived effect estimates is shown in Display 9.2 for a significance level of $\alpha = 0.20$ for effect separation. The plot confirms the importance of the five highlighted effects which show up as distinctly different from the inactive effects. A Pareto chart provided the same conclusion. The aliasing assumptions provided enable us to be relatively sure that these effects are the most important within the analytical procedure under investigation.

Display 9.2

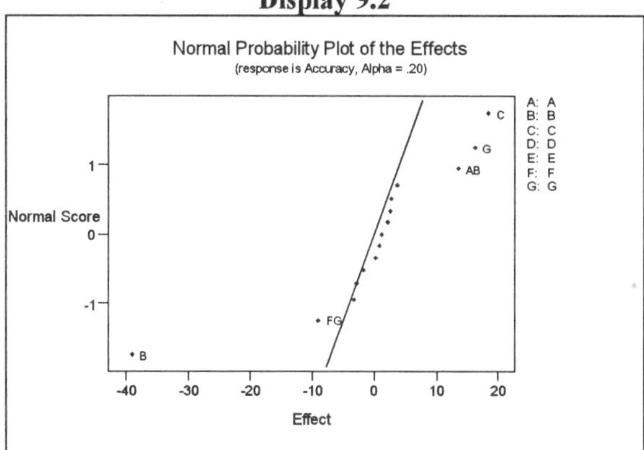

Data plots: Main effect plots for factors B, C, and G are shown in Display 9.3. Decreasing the volume of nitric acid (B) appears to reduce accuracy of selenium recovery. The opposite is the case for both sulphuric acid (C) and time of dissolution (G) which result in comparable increases in accuracy when level of contribution increases. The ranges shown indicate the wide variation in the reported accuracy figures. Initial indications are that B_-, C_+, and G_+ may be the optimum levels though the large ranges may affect this suggestion. Plots of the two interaction effects, though not shown, both exhibited distinct non-parallel trends indicative of significant interaction effect. Optimum levels appeared to be A_-B_- and F_-G_+ though, as with main effects analysis, large ranges were present.

Display 9.3

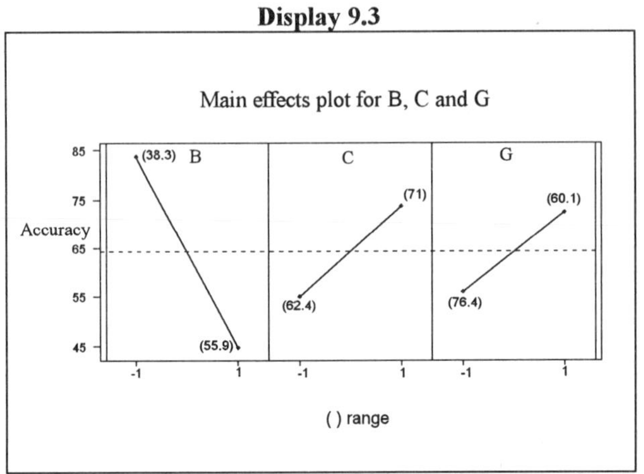

Analysis of proposed model: Effects analysis suggests a possible model of

$$\text{selenium recovered} = \text{nitric acid } (B) + \text{sulphuric acid } (C) + \text{time of dissolution } (G) + \text{mass} \times \text{nitric acid } (A \times B) + \text{microwave power} \times \text{time of dissolution } (F \times G) + \text{error}$$

to explain the experimental measurements. The error term comprises those effects deemed unimportant from the effects analysis. ANOVA showed that the microwave power × time of dissolution interaction was highly significant ($p < 0.01$) while the other four model effects were all very highly significant ($p < 0.001$) providing statistical back-up to the conclusions suggested by the effects analysis.

Diagnostic checking: A check of factor residual plots showed some column length differences by level indicating that variability in the reported results is affected by factor and level tested. In fact, for percholic acid (D) and hydrogen peroxide (E) which were deemed unimportant for mean accuracy, it was suggested that D_+ and E_+ could be best for reducing variability. Plots of the fitted responses showed no evidence of trend which could improve data understanding.

Summary: A summary of the findings from the analysis is provided in Table 9.12 and shows that treatment combination *cdeg* may be optimal. This combination was not tested in the experiment conducted but the near equivalent combination *ceg* was (see row 3 Table 9.10) and in fact produced the best recovery measurement so it may be that the combination indicated is best for optimising accuracy of selenium recovery. It would appear that, at the very least, optimum conditions for recovery of selenium by hydride generation AAS are 0.1 g sample mass (A_-), 8 ml of nitric acid (B_-), 2 ml of sulphuric acid (C_+), 0 ml of percholic acid (D_+), 0.5 ml of hydrogen peroxide (E_+), 100% microwave power (F_-), and 15 minutes dissolution time (G_+). Levels for percholic acid and hydrogen peroxide are based on reducing variability though both could be omitted as each appear to have negligible effect on mean accuracy. □

Table 9.12 — Summary of analysis of selenium recovery experiment of Example 9.2

Factors	A	B	C	D	E	F	G	Combination
Main effects		–	+				+	
$A \times B$ interaction	–	–						
$F \times G$ interaction						–	+	
Diagnostic checking				+	+		+	
Possible "best"	–	–	+	+	+	–	+	*cdeg*

The analysis discussed in Exercise 9.3 for the selenium recovery study of Example 9.2 represents an appropriate analysis for data collected from a two-level OAD. It is essentially comparable to that described for two-level FFDs and provides similar results.

9.5 THREE-LEVEL ORTHOGONAL ARRAYS

Two-level OADs can be extended to **three-level Orthogonal Arrays** very simply. The factor levels in such designs are specified as low '–', middle '0', and high '+'. Such arrays are particularly useful for quantitative factors where a measure of the curvature effect of the factor is required and where non-linear prediction equations are to be developed. Examples of such arrays include the $\boldsymbol{OA_9(3^4)}$ array (9 runs, 4 columns) shown in Table 9.13, the $\boldsymbol{OA_{16}(3^7)}$ array (16 runs, 7 columns), and the $\boldsymbol{OA_{27}(3^{13})}$ array (27 runs, 13 columns). The $OA_9(3^4)$ array occurs often as the parameter design layout in Taguchi experiments. The $OA_{16}(3^7)$ array is called **Stark's array** and differs from traditional arrays in that each factor level does not occur equally often, the centre levels appearing twice as many times compared to the low and high levels. As such designs are essentially one fraction of a three-level FFD, their associated alias structure can often be very complex with corresponding low design resolution (see Section 8.6). Data analysis for such designs generally consists of response plots by factor levels, interaction plots as appropriate, and ANOVA modelling of a proposed response model.

Table 9.13 — The $OA_9(3^4)$ array

Row	c_1	c_2	c_3	c_4
1	–	–	–	–
2	–	0	0	0
3	–	+	+	+
4	0	–	0	+
5	0	0	+	–
6	0	+	–	0
7	+	–	+	0
8	+	0	–	+
9	+	+	0	–

It is also possible to construct **mixed-level arrays** where factors are at different levels. For example, the $OA_{18}(2^1.3^7)$, shown in Table 9.14, specifies an eighteen run eight column array where one factor is at two levels and the other seven factors are at three levels. Conversion mechanisms to include three-level factors in two-level arrays and two-level factors in three-level arrays exist which result in unequal replication of factor levels (Grove & Davis 1992).

Table 9.14 — The $OA_{18}(2^1.3^7)$ array

Row	c_1	c_2	c_3	c_4	c_5	c_6	c_7	c_8
1	−	−	−	−	−	−	−	−
2	−	−	0	0	0	0	0	0
3	−	−	+	+	+	+	+	+
4	−	0	−	−	0	0	+	+
5	−	0	0	0	+	+	−	−
6	−	0	+	+	−	−	0	0
7	−	+	−	0	−	+	0	+
8	−	+	0	+	0	−	+	−
9	−	+	+	−	+	0	−	0
10	+	−	−	+	+	0	0	−
11	+	−	0	−	−	+	+	0
12	+	−	+	0	0	−	−	+
13	+	0	−	0	+	−	+	0
14	+	0	0	+	−	0	−	+
15	+	0	+	−	0	+	0	−
16	+	+	−	+	0	+	−	0
17	+	+	0	−	+	−	0	+
18	+	+	+	0	−	0	+	−

9.6 PLACKETT-BURMAN DESIGNS

Special array structures worth mentioning are those associated with **Plackett-Burman Designs** (PBDs) attributed to Placket & Burman (1946). They are two-level designs for the study of $k = n - 1$ factors in n experimental runs where n is a multiple of 4. When n is a power of 2, these arrays are identical to the structures discussed earlier. The Plackett-Burman arrays $\boldsymbol{OA_{12}(2^{11})}$, $\boldsymbol{OA_{20}(2^{19})}$, $\boldsymbol{OA_{24}(2^{23})}$, and $\boldsymbol{OA_{36}(2^{35})}$ are generally of most interest. PBDs enable many factors to be considered using few measurements for such a number of factors and therefore generally require fewer runs relative to comparable OADs.

PBDs are based on simple generators for appropriate values of n, the generators representing sequences of + and − signs. For $n = 12$, specifying the $OA_{12}(2^{11})$ array, the generator is

$$+ \; + \; - \; + \; + \; + \; - \; - \; - \; + \; -$$

This corresponds to the first row of the 12 run PBD. The second row is determined from the first by moving the first coefficient in row 1 to the last position in row 2 and assigning the coefficients of the first row one position to the left, i.e. cyclic rotation. Rows 3 to 11 are obtained in the same way. The final run, row 12, corresponds to − signs for all 11 columns, i.e. treatment combination (1). The results of the process are displayed in Table 9.15 and corresponds to one of several ways in which the $OA_{12}(2^{11})$ array can be presented. The array is not orthogonal as the product of any two columns is not itself a column of the array. The array is balanced as each combination of factor levels appears the same number of times in each pair of columns. Simple mechanisms exist for constructing the larger PBDs from the immediately previous array, i.e. the $OA_{20}(2^{19})$ array can be generated from the $OA_{12}(2^{11})$ array.

Table 9.15 — A Plackett-Burman Design for 11 factors in 12 runs, the $OA_{12}(2^{11})$ array

	Column										
Row	1	2	3	4	5	6	7	8	9	10	11
1	+	+	–	+	+	+	–	–	–	+	–
2	+	–	+	+	+	–	–	–	+	–	+
3	–	+	+	+	–	–	–	+	–	+	+
4	+	+	+	–	–	–	+	–	+	+	–
5	+	+	–	–	–	+	–	+	+	–	+
6	+	–	–	–	+	–	+	+	–	+	+
7	–	–	–	+	–	+	+	–	+	+	+
8	–	–	+	–	+	+	–	+	+	+	–
9	–	+	–	+	+	–	+	+	+	–	–
10	+	–	+	+	–	+	+	+	–	–	–
11	–	+	+	–	+	+	+	–	–	–	+
12	–	–	–	–	–	–	–	–	–	–	–

PBDs are particularly useful early in problem solving and when main effects are most of interest as they are classified as **main effects only designs**. The designs can have complex alias structures with correspondingly complicated confounding of interactions and main effects. They do not account well for the presence of two factor interactions and, if such are likely to affect a response, they should be used with extreme caution.

PROBLEMS

9.1 A four factor experiment is to be conducted using the $OA_8(2^7)$ array as the design base. Based on the column allocation shown in Table 9.5 for a four factor experiment, determine the alias structure for this design. What is the design generator? What type of FFD does this form of array structure correspond to?

9.2 A six factor experiment is to be conducted based on the $OA_8(2^7)$ array. Using the column allocation shown in Table 9.5 for a six factor experiment, determine the alias structure for this design. What are the design generators? What type of FFD does this form of array structure correspond to?

9.3 An electrical engineer was interested in assessing how six factors, each set at two levels, affected the start-up time of an electronic device, start-up time being measured in seconds. The engineer decided to base a simulation study on the $OA_{16}(2^{15})$ array. The factor column allocation A to c_1, B to c_2, C to c_4, D to c_8, E to c_{11}, and F to c_{15} was planned. Using the column allocation given and Table 9.7, obtain the aliases of the main effects and the two factor interactions for this design. What are the design generators? What is the design resolution? Suppose it was proposed to change the column allocation for factors E and F to columns c_7 and c_{14}, respectively. Does this change the design resolution and the ability of the design to estimate particular effects?

9.4 A ruggedness study is to be conducted by an analytical laboratory to investigate the resistance of a newly developed analytical procedure to deviations in the procedure factors. The experiment is to be based on seven two-level factors using the $OA_{16}(2^{15})$ array as the design base. The planned factor column allocation is to be A to c_1, B to c_2, C to c_4, D to c_8, E to c_{11}, F to c_{14}, and G to c_{15}. From the column allocation planned, determine the aliases of the main effects and the two factor interactions for this proposed experiment. What are the design generators? What is the design resolution?

9.5 Glass manufacture requires use of a number of additives in various amounts to produce a product with an appropriate combination of physical properties. The $OA_{16}(2^{15})$ array was being planned to be used to investigate the effect of 10 different two-level additives on the opacity (%) of the glass produced. The planned factor column allocation is to be A to c_1, B to c_2, C to c_4, D to c_8, E to c_5, F to c_6, G to c_3, H to c_9, J to c_{10}, and K to c_{12}. From the column allocation provided, determine the aliases of the main effects and the two factor interactions. What are the design generators? What is the design resolution?

9.6 In a study of capillary zone electrophoretic (CZE) analysis of heterocyclic amines (HCAs), five factors were studied as part of the optimisation of CZE separations. HCAs are mutagenic and carcinogenic compounds found in cooked meats. The five factors chosen for studying were pH (A) at 2.5 and 3.5, concentration of methanol MeOH (B) at 0% and 3.5%, concentration of electrolyte NaCl (C) at 0 mM and 30 mM, capillary temperature (D) at 35 °C and 25 °C, and applied potential (E) at 20 kV and 15 kV. Factors A and C were related to the buffer while factors D and E were related to the instrument used in the experiment. The study was based on the $OA_{16}(2^{15})$ array with column allocation of A to c_1, B to c_2, C to c_4, D to c_{11}, and E to c_{13}. The table below provides the array structure and response data. The response studied was a coded measurement representing the electrophoretic response function (ERF) minus 47. Determine the alias structure of the design. Taking into account the aliasing, draw appropriate conclusions from the presented data based on maximisation as the goal.

Row	c_1 A	c_2 B	c_3	c_4 C	c_5	c_6	c_7	c_8	c_9	c_{10}	c_{11} D	c_{12}	c_{13} E	c_{14}	c_{15}	Coded response
1	−	−	−	−	−	−	−	−	−	−	−	−	−	−	−	5.08
2	−	−	−	−	−	−	−	+	+	+	+	+	+	+	+	9.33
3	−	−	−	+	+	+	+	−	−	−	−	+	+	+	+	7.78
4	−	−	−	+	+	+	+	+	+	+	+	−	−	−	−	8.02
5	−	+	+	−	−	+	+	−	−	+	+	−	−	+	+	12.25
6	−	+	+	−	−	+	+	+	+	−	−	+	+	−	−	7.66
7	−	+	+	+	+	−	−	−	−	+	+	+	+	−	−	11.34
8	−	+	+	+	+	−	−	+	+	−	−	−	−	+	+	12.21
9	+	−	+	−	+	−	+	−	+	−	+	−	+	−	+	4.04
10	+	−	+	−	+	−	+	+	−	+	−	+	−	+	−	4.97
11	+	−	+	+	−	+	−	−	+	−	+	+	−	+	−	1.36
12	+	−	+	+	−	+	−	+	−	+	−	−	+	−	+	3.37
13	+	+	−	−	+	+	−	−	+	+	−	−	+	+	−	3.58
14	+	+	−	−	+	+	−	+	−	−	+	+	−	−	+	0.46

(continued)

(continued)

15	+	+	−	+	−	−	+	−	+	+	−	+	−	−	+	6.90
16	+	+	−	+	−	−	+	+	−	−	+	−	+	+	−	2.10

Reprinted from Wu, J., Wong, M. K., Lee, H. K. & Ong, C. N. (1996) Orthogonal array design for optimizing the capillary zone electrophoretic analysis of heterocyclic amines. *Journal of Chromatographic Science* **34** 139–145 by permission of Preston Publications, A Division of Preston Industries, Inc. and the authors.

9.7 The $OA_{16}(2^{15})$ array was used as a design base to study the effects of sample water content on the precision of gas chromatographic (GC) analysis of the pesticide methyl parathion. Six factors which can affect the water content of samples were studied based on knowledge of the GC analysis procedure. These were water concentration (A) at 0% and 30%, initial oven temperature (B) at 55 °C and 130 °C, concentration of cosolvent (C) at 0% and 10%, injector temperature (D) at 240 °C and 280 °C, injector volume (E) at 1 μL and 2 μL, and use of switch valve (F) at not used (−) and used (+). The array structure and responses are presented below. Because precision was being investigated, the response chosen for analysis was the CV (coefficient of variation) from the three replicate measurements at each treatment combination. Check the alias structure and design resolution. Carry out a relevant analysis of the presented data and draw conclusions.

Row	c_1 A	c_2 B	c_3	c_4 C	c_5	c_6	c_7	c_8 D	c_9	c_{10}	c_{11}	c_{12}	c_{13} E	c_{14} F	c_{15}	CV of methyl parathion
1	−	−	−	−	−	−	−	−	−	−	−	−	−	−	−	9.21
2	−	−	−	−	−	−	−	+	+	+	+	+	+	+	+	8.11
3	−	−	−	+	+	+	+	−	−	−	−	+	+	+	+	4.81
4	−	−	−	+	+	+	+	+	+	+	+	−	−	−	−	15.5
5	−	+	+	−	−	+	+	−	−	+	+	−	−	+	+	2.04
6	−	+	+	−	−	+	+	+	+	−	−	+	+	−	−	2.70
7	−	+	+	+	+	−	−	−	−	+	+	+	+	−	−	4.23
8	−	+	+	+	+	−	−	+	+	−	−	−	−	+	+	5.31
9	+	−	+	−	+	−	+	−	+	−	+	−	+	−	+	44.3
10	+	−	+	−	+	−	+	+	−	+	−	+	−	+	−	28.4
11	+	−	+	+	−	+	−	−	+	−	+	+	−	+	−	6.98
12	+	−	+	+	−	+	−	+	−	+	−	−	+	−	+	24.6
13	+	+	−	−	+	+	−	−	+	+	−	−	+	+	−	32.6
14	+	+	−	−	+	+	−	+	−	−	+	+	−	−	+	34.6
15	+	+	−	+	−	−	+	−	+	+	−	+	−	−	+	4.78
16	+	+	−	+	−	−	+	+	−	−	+	−	+	+	−	25.4

Reprinted from Wan, H. B., Wong, M. K. & Mok, C. Y. (1995) Use of statistically designed experiments to minimize the effect of water on gas chromatographic analysis. *Journal of Chromatographic Science* **33** 66–70 by permission of Preston Publications, A Division of Preston Industries, Inc. and the authors.

9.8 A screening experiment was set up to investigate the influence of seven two-level factors on the strength of a bond between two plastics within a bonding process. The experiment was based on the $OA_{16}(2^{15})$ array and the response measured was the bonding strength of the manufactured product with maximisation the objective. Column allocation of factors was based on A to c_1, B to c_2, C to c_4, D to c_8, E to c_{10}, F to c_{12}, and G to c_{15}. The treatment combinations run, in row order for the $OA_{16}(2^{15})$ array, and bonding strength responses, in kg cm^{-2}, were as follows:

$(1) = 95$	$defg = 42$	$cfg = 92$	$cde = 9$	$beg = 90$	$bdf = 65$	$bcef = 15$	$bcdg = 3$
$ag = 94$	$adef = 4$	$acf = 32$	$acdeg = 5$	$abe = 95$	$abdfg = 24$	$abcefg = 6$	$abcd = 3$

What is the alias structure for the main effects? Is it acceptable? By carrying out an analysis of these data, identify the important factors. Is an optimal combination suggested?

9A Appendix: Software Information for Two-level Orthogonal Array Designs

OAD construction

OADs are based on factor column allocation. These, in turn, provide design generators for the planned design structure. Once the generators are known, it is possible, in *Minitab*, to use the FFD construction facilities (see Appendix 8A) to generate the OAD structure. A trial and error approach is necessary by changing the fraction selection until the one with the required generators and design matrix showing first treatment combination as (1) is provided. This fraction selection can be achieved by modifying the **Options** selection thus

> Select **Options** ➤ for *Fraction*, select **Use fraction number**, select the empty box, and enter the **fraction number required** ➤ as Appendix 8A.

The advantage of checking design structure by software is that it can provide the full alias information and a design matrix.

Data entry

Data entry for two-level OADs follows similar principles to that for two-level FFDs. Generally, the actual levels of the factors can be used, e.g. low and high, though the −1/+1 system may also be necessary for analysis purposes. Data entry for Example 9.2 would therefore correspond to the design matrix for the seven factors and response measurements shown in Table 9.10 in the text.

Effect estimation and plotting

Production of effect estimates and associated data plots can be easily achieved within most software packages as indicated in Appendix 8A for FFDs.

SAS: In SAS, the ADX macros require to be used for analysis of OADs. They can be loaded into SAS by implementing the statements presented in Appendix 8A. Once the macros are loaded, we would then use the %ADXFFA macro to generate effect estimates and a normal plot of estimates using the following statements:

```
%ADXFFA(CH9.EX92, ACCURACY, A B C D E F G, 3);
RUN;
```

The data to be analysed are contained in the data set CH9.EX92 and ACCURACY specifies the reported accuracy responses. Explanation of structure of this statement is as in Appendix 8A for FFDs. Factor effect estimates generated are again half the correct value though relative positioning of the estimates in the normal plot is acceptable.

Minitab: The **Stat** menu provides access to the effect estimation and effect plotting facilities within Minitab. Displays 9.1 and 9.2 contain these presentations for the two-level selenium recovery study of Example 9.2. The information contained in these displays was produced using the following menu selections:

> Select **Stat** ➤ **DOE** ➤ **Analyze Custom Design** ➤ for *Design*, select **2-Level factorial** ➤ **Fit Model** ➤ for *Responses*, select **Accuracy** and click **Select** ➤ for *Factors*, select the box and enter **A B C D E F G**.
>
> Select **Terms** ➤ for *Include terms in the model up through order*, ensure **7** is the entry ➤ click **OK**.
>
> Select **Graphs** ➤ for *Effects Plots*, select **Normal** and **Pareto** ➤ for *Alpha*, enter **0.20** ➤ click **OK**.
>
> Select **Storage** ➤ for *Model Information*, select **Effects** ➤ click **OK** ➤ click **OK**.

Data plots

Main effect plots, illustrated in Display 9.3, and interaction plots can be produced as per the menu instructions provided in Appendices 7A and 8A.

ANOVA information

Derivation of ANOVA information for a proposed response model in two-level OADs can be handled using software in a similar way to that for two-level FDs and FFDs.

SAS: The PROC GLM procedure can provide the route to the ANOVA information required for a proposed response model after effect screening. The program below would produce the necessary ANOVA information for the model proposed for the selenium recovery study of Example 9.2 based on first generating specific interaction columns for the important interactions of $A \times B$ and $F \times G$.

```
DATA CH9.EX92;
    SET CH9.EX92;
    AB = A*B;
    FG = F*G;
PROC GLM DATA = CH9.EX92;
    CLASS B C G AB FG;
    MODEL ACCURACY = B C G AB FG;
    OUTPUT OUT = CH9.EX92A P = FITS R = RESIDS;
RUN;
```

Diagnostic checking plots can be produced as previously illustrated using the stored RESIDS and FITS values (see Appendix 3A).

Minitab: The ANOVA and general linear model (GLM) facilities in Minitab cannot handle a proposed response model if it is not hierarchical. However, GLM procedures can be used to provide the required ANOVA information and residual estimates for diagnostic checking by creating specific columns for the important interactions of $A \times B$ and $F \times G$ using Minitab's calculation facility within the **Calc** ➤ **Calculator** menu selection. For this, however, factor code settings must be numeric. The predictors **AB** and **FG** below refer to the appropriate column products of **A** and **B**, and **F** and **G**, respectively.

Select **Stat** ➢ **ANOVA** ➢ **General Linear Model** ➢ for *Response*, select **Accuracy** ➢ for *Model*, select the box and enter **B C G AB FG**.

Select **Storage** ➢ **Fits** ➢ **Residuals** ➢ click **OK**.

Select **Graphs** ➢ select the residual plots necessary ➢ click **OK** ➢ click **OK**.

10

Taguchi Methods

10.1 INTRODUCTION

So far, within this text, we have looked at experimental designs which have well defined statistical and analysis structures. Such designs have illustrated the importance of using cost-effective and easy to analyse design structures in **multi-factor** investigations. Since the mid 1980s, the search for quality within products and processes has led to increased usage of such methods. One of the forces behind this quality drive has been the work of **Genichi Taguchi**, a Japanese engineer. He has made use of design and analysis procedures for small run multi-factor experiments which have contributed significantly to the recent and continued success of Japanese industry. These techniques, generally called **Taguchi Methods**, have come to be applied in Western industry with some success. In fact, Taguchi methods are often synonymous with the implementation of experimental designs in many organisations.

Taguchi methods represent simple but novel applications of classical experimental designs. The underlying principles and methods are based on an integrated strategy of quality engineering principles and statistical methods and are commensurate with the ethos of other quality gurus such as Deming and Juran. With the recent push for quality improvement within British and European industry, Taguchi methods have become potentially useful tools to aid the production of high quality, reliable products. They provide a system to help develop product/process specifications, design those specifications into the product/process, and produce products/processes that continuously achieve target specifications, i.e. consistently work to target and are insensitive to extraneous, uncontrollable factors. In some cases, use of Taguchi methods has led to the developed products/processes often surpassing specifications.

Taguchi's philosophy is based on **building quality into the product/process as it is developed not during or after production**. This building-in of quality is called **off-line Quality Control** (QC) and is generally implemented in both the design and manufacturing phase of a product. It is used to assess how product quality is affected by controllable and uncontrollable factors in order to determine optimal levels for control factors in relation to product specification and usage requirements, the latter reflecting the factors outwith the control of the development or production process which could affect product performance. Traditional **Statistical Quality Control** (SQC) activities are based on control charts and process control and represent **on-line Quality Control**. They are primarily concerned with controlling product manufacture and process operation. Such techniques do not attempt to build quality into a product as they simply ensure that the product adheres to specified performance criteria which may not necessarily reflect acceptable product quality. Taguchi advocates that if off-line QC is used, then the reliance on SQC procedures for checking product specification can be lessened.

There are several basic principles on which Taguchi methods are based. The most important of these is that product quality must be **engineered in**. This involves experimenting at the product design and the process design stages to investigate how factors affect the operational quality of the product/process. The following points explain some of Taguchi's underlying approaches (Taguchi 1986, 1987a, b):

- **Good quality** begins with product design rather than with control of the manufacturing process as, in a competitive economy, continuous quality improvement and cost reduction are necessary for business viability.

- **Product design** must take account of manufacturing variability and of the environmental or usage conditions of the product within its working life. In other words, products must be **robust** with intrinsic quality and reliability characteristics which enable them to perform to, and possibly surpass, specification at all times under all conditions of usage.

- The **quality** of a manufactured product, however it is measured, is not determined by conformance to tolerances but by the total loss generated by that product to society. This loss can include customer dissatisfaction, added warranty costs to the producer, harmful side-effects, and company loss due to adverse publicity. The smaller the loss the better the quality.

- Customer satisfaction is inversely proportional to **performance variation**. This level of satisfaction, or "loss", is often related to the deviation of the product's performance characteristic from its target specification. Performance variation can be minimised by optimising the settings of the product/process parameters with respect to certain performance characteristics, e.g. high performance with low variability.

- **Statistically designed experiments** can be applied to identify product/process parameter settings that enable performance variation to be reduced. They can enable the optimal production/process settings for best quality product to be discerned and checked at the design/development phase prior to production.

Taguchi methods are therefore based on quality engineering principles. They concentrate on experiments at the product or process design stage rather than with subsequent optimisation of process operation as occurs with **Statistical Process Control** (SPC). They make use of techniques of **parameter design** to minimise a **loss function** with the emphasis on variability reduction of product/process performance. In most manufacturing situations, loss refers to the deviation of the product's functional characteristic from its desired target specification.

Successful application of Taguchi methods has taken place in many diverse areas. For example, glass manufacture to evaluate the levels required for various additives to optimise the physical properties of the glass, flow soldering of electronic components to printed circuit boards, the manufacture of car seat cushions, optimisation of the performance of load control software for the central processor units of a telephone exchange, optimising the yield of a magnetic card reader, the production of small plastic parts used in the carburettors of lawn mower engines, and the effects of various factors in electric light bulb production on usage characteristics of the bulbs. In spite of these successes, there is still considerable debate within the statistical community on the statistical appropriateness of Taguchi's design and analysis concepts. The debate has primarily focused on his design structures and the signal-to-noise characteristic on which Taguchi analysis is based. For more information on this, refer to Box *et al.* (1988), Box & Jones (1992), Kackar (1986), Myers *et al.* (1992), Nair (1992), Pignatiello & Ramberg (1991), Shoemaker & Tsui (1993), and Tribus & Szonyi (1989).

10.2 DESIGN STRUCTURES

Three design stages have been identified by Taguchi for determining factor effects on product and process characteristics such as nominal values, tolerances, or manufacturing settings. The three stages are:

1. **System Design** (*Primary Design, Functional Design*) This refers to the use of scientific, engineering, and technological knowledge to develop innovative products that consider the needs of customers and satisfy the requirements of the manufacturing environment. It involves surveying the pertinent technology and using what may be appropriate to aid product and process development.
2. **Parameter Design** (*Secondary Design*) This stage corresponds to the determination of the optimum combination of product or process parameter settings that help minimise **performance variation** and so reduce sensitivity of the product/process to potential sources of variation such as manufacturing variations, environmental factors, usage conditions, and product deterioration. The parameter design phase therefore attempts to understand and reduce the influence of specific sources of variation. This stage is based on **experimental design concepts and principles**.
3. **Tolerance Design** (*Tertiary Design*) This is the final stage of the quality improvement process and concerns determination of tolerances around the optimal settings identified within the parameter design phase. It is used to gauge the influence of these settings on the product/process characteristics so as to minimise manufacturing costs, operating costs, and lifetime costs, i.e. cost to the "customer". Cost calculations and the concept of quadratic loss functions play important roles in this stage.

Stages 2 and 3 can be used to understand the influence of controllable and uncontrollable factors on product quality to help reduce the effect of noise (uncontrolled variation) and so assure functional quality of the product/process being investigated. In particular, the parameter design phase of Taguchi's product quality design strategy is the phase which relies on the principles and methods of experimental design. The constructed experiments can be carried out through physical experimentation or through computer simulation trials. It is this phase of Taguchi methods which we will concentrate on in this chapter.

The parameter design structures advocated by Taguchi are based on **Orthogonal Arrays**, such as those described in Chapter 9, and therefore the column allocation and aliasing principles described for OADs will apply. In most instances,

Taguchi designs correspond to FFDs which are highly fractionated with low resolution and complex alias structures. Therefore, Taguchi's approach is based on the Fractional Factorial Design concept though there are several drawbacks. The design structures recommended by Taguchi enable each factor level to appear the same number of times in the design matrix and each level of each factor to appear with each level of any other factor the same number of times so retaining the balancing and orthogonality properties ideal in multi-factor investigations. In such designs, however, it is not always possible to carry out a full study of all interactions so the search for interactions among the controllable factors is de-emphasised. In fact, many of the design structures put forward by Taguchi lead to main effects aliased with two factor interactions. This inability to adequately estimate interaction effects is seen by many to be a serious deficiency in Taguchi designs.

To construct Taguchi arrays, we require to know the number of factors, number of levels of each factor, any specific two factor interactions requiring to be estimated, and any difficulties that may occur when running the experiment. The arrays are constructed so that the number of rows equals the number of experiments to be performed and the number of columns is equivalent to the maximum number of factors that can be studied in such an array. Instead of the –/+ approach in Chapter 9 for two-level OADs, Taguchi prefers to adopt a 1/2 system with 1 referring to low level and 2 high. For three-level arrays, the –/0/+ coding is replaced by a 1/2/3 system. Taguchi advocates use of the smallest possible orthogonal array commensurate with number of factors to be tested to help reduce experimental costs. Designs falling into this category include the $\boldsymbol{L_4(2^3)}$ array, the $\boldsymbol{L_8(2^7)}$ array shown in Table 10.1, and the $\boldsymbol{L_9(3^4)}$ array also displayed in Table 10.1. Taguchi specifies that the arrays shown in Table 10.1 be run in the order presented. He does not consider randomisation of run order as it could add to experimental costs. Other popular designs which occur often in Taguchi experiments are the $\boldsymbol{L_{16}(2^{15})}$ array, the $\boldsymbol{L_{18}(3^7)}$ array, the $\boldsymbol{L_{16}(4^5)}$ array, and the $\boldsymbol{L_{27}(3^{13})}$ array. To help in array construction, Taguchi developed the principle of **linear graphs** to aid column allocation particularly in cases where interactions need to be included. Bisgaard (1996) and Grove & Davis (1992) contain more details on this approach to Taguchi design construction.

Table 10.1 — Taguchi's $L_8(2^7)$ and $L_9(3^4)$ arrays

$L_8(2^7)$ array							$L_9(3^4)$ array			
1	2	3	4	5	6	7	1	2	3	4
1	1	1	1	1	1	1	1	1	1	1
1	1	1	2	2	2	2	1	2	2	2
1	2	2	1	1	2	2	1	3	3	3
1	2	2	2	2	1	1	2	1	2	3
2	1	2	1	2	1	2	2	2	3	1
2	1	2	2	1	2	1	2	3	1	2
2	2	1	1	2	2	1	3	1	3	2
2	2	1	2	1	1	2	3	2	1	3
							3	3	2	1

Often, the chosen array is a **saturated** one where all columns of the array are allocated to tested factors (see Section 9.3.4). Nearly saturated designs with the majority of columns allocated to factors are also common. For such design structures, all two factor interactions will be confounded with main effects and other higher order interactions (resolution III design) and no degrees of freedom will be specifically reserved for measuring experimental error. Taguchi methods tend to explore as many control factors as possible, thus "saturating" the array providing great economy in experimental effort but little information on factor interactions unless specifically included in the array.

Example 10.1 Consider an investigation into an integrated circuit fabrication process where thickness problems are affecting circuit viability. Seven two-level control factors are to be tested using Taguchi's $L_8(2^7)$ array as the parameter design base. As the array contains seven columns and there are seven factors, all columns of the array will be assigned to a factor as illustrated in Table 10.2 resulting in a saturated design. Each row represents an experimental run corresponding to a particular treatment combination. Six replications of each combination are planned. We want to check the associated alias structure of the planned design to confirm it is appropriate to the needs of the study.

Table 10.2 — Planned $L_8(2^7)$ array structure for Example 10.1

A 1	***B*** 2	***C*** 3	***D*** 4	***E*** 5	***F*** 6	***G*** 7	Combination	Thickness response replicates
1	1	1	1	1	1	1	(1)	x x x x x x
1	1	1	2	2	2	2	*defg*	x x x x x x
1	2	2	1	1	2	2	*bcfg*	x x x x x x
1	2	2	2	2	1	1	*bcde*	x x x x x x
2	1	2	1	2	1	2	*aceg*	x x x x x x
2	1	2	2	1	2	1	*acdf*	x x x x x x
2	2	1	1	2	2	1	*abef*	x x x x x x
2	2	1	2	1	1	2	*abdg*	x x x x x x

factor column allocation in bold, x denotes an observation

Column allocation: Based on the structure presented in Table 10.2, we have factor column allocation of A to column 1, B to 2, C to 3, D to 4, E to 5, F to 6, and G to 7. Column association for the two factor interactions can be determined using the triangular Table 9.6 for an eight run seven column array. Table 10.3 provides the outcome to this process.

Table 10.3 — Column allocation for the planned seven factor Taguchi study for Example 11.1

1	2	3	4	5	6	7
A	***B***	***C***	***D***	***E***	***F***	***G***
$-BC$	$-AC$	$-AB$		$-AD$	$-BD$	$-CD$
$-DE$			$-AE$		$-CE$	$-BE$
	$-DF$	$-EF$	$-BF$	$-CF$		$-AF$
$-FG$	$-EG$	$-DG$	$-CG$	$-BG$	$-AG$	

factor column allocation in bold, AB denotes the $A \times B$ interaction, etc.

Aliasing: This design is in fact a 2^{7-4}_{III} FFD, i.e. a one-sixteenth fraction of a 2^7 FD. As $2^{-4} = 1/2^4$, we have, from Section 8.4, that the level of fractionation is $p = 4$. From expression (8.5), we can therefore state that each effect will have $4 + (2^4 - 4 - 1) = 15$

aliases. Reading down the columns in Table 10.3 shows main effects aliased with three of the two factor interactions and two factor interactions aliased with two effects of the same order confirming that the design resolution is III. Following similar reasoning to that applied in Chapter 9 and switching C and D for tractability, we can obtain the defining relation as

$$I = -ABC = -ADE = -BDF = ABDG = BCDE = ACDF = -CDG = ABEF$$
$$= -BEG = -AFG = -CEF = ACEG = BCFG = DEFG = -ABCDEFG$$

The details of this have been discussed in Exercise 9.2. By multiplying the full defining relation by a main effect modulus 2, e.g. factor A, as in Exercise 8.2, main effects can be shown to be aliased with not only two factor interactions but also three (BDG, CDF, BEF, CEG), four, five, and six factor interactions.

Providing we can accept that interactions will be negligible, the planned design structure can be considered acceptable. The number of contrasts specified in the defining relation highlights the complexity of the aliasing inherent in this proposed Taguchi design. In fact, complexity of alias structure is often a feature of Taguchi designs. □

Performance variation, or functional operation (usage), of products or manufacturing processes can arise from two primary sources according to Taguchi, **design parameter variables** and **noise variables**. Design parameter variables represent **control**, or **external**, **factors** corresponding to product or process parameters integral to product construction or manufacturing process specification, e.g. raw material used, cycle time, temperature of manufacturing process, input voltage. Such factors are varied across different levels, with the Taguchi array to which they are assigned called the **inner array** or **parameter design matrix**. Noise variables represent **uncontrollable**, or **nuisance**, **factors** that are either difficult, impossible, or expensive to control such as customer usage, component deterioration, and product operating conditions. They correspond to factors that can affect product/process performance and cause it to differ from that desired. Such factors are assigned to what Taguchi calls an **outer array** or **noise matrix**. Thus, in Taguchi experiments, the design can consist of two parts, an inner array and an outer array. Essentially, the outer array is **nested** within the inner array meaning that each treatment combination in the parameter design matrix is run at each combination within the noise matrix.

An illustration of this approach is provided in Table 10.4 for a quality improvement study involving four three-level control factors, A to D, and three two-level noise factors, E to G. The four control factors have been assigned to Taguchi's $L_9(3^4)$ array to form the inner array. In this array, the design parameter variables (factors) are listed in its columns with the rows specifying the factor combinations to be tested. The outer array, based on the $L_4(2^3)$ array, consists of the three two-level noise variables and is presented in transpose form for convenience. The noise variables are listed in the rows and the appropriate combinations of settings in the columns. Each experimental run, comprising a combination of control factors, would be carried out four times, once for each combination of noise factors. The replication provided by the inclusion of the outer array enables the response to be measured over a range of operating conditions for each combination of control factor settings. Taguchi then uses the replicate data to generate a performance statistic, defining process optimisation, which becomes the basis of the analysis procedures (see Section 10.3).

Table 10.4 — Example of a Taguchi design with inner and outer arrays

					Outer array (L_4)				
					E	1	1	2	2
Inner array (L_9)					*F*	1	2	1	2
Run	***A***	***B***	***C***	***D***	*G*	1	2	2	1
1	1	1	1	1		x	x	x	x
2	1	2	2	2		x	x	x	x
3	1	3	3	3		x	x	x	x
4	2	1	2	3		x	x	x	x
5	2	2	3	1		x	x	x	x
6	2	3	1	2		x	x	x	x
7	3	1	3	2		x	x	x	x
8	3	2	1	3		x	x	x	x
9	3	3	2	1		x	x	x	x

control factor column allocation in bold, x denotes an observation

Product development and manufacturing processes can be influenced by many factors, both controllable and uncontrollable. Such factors may work independently or in combination and could strongly influence response variation. Taguchi advocates that as many sources of variation as possible be identified. These sources must then be varied within an experiment to investigate their influence to determine the nominal values of the design parameter variables that provide the least impact on product/process performance over the chosen noise variables. Selection of appropriate variables to include in the design and noise matrices is not trivial. Values for the controllable variables should be selected to make the product/process least sensitive to changes in the noise variables, i.e. robust. Thus, instead of finding and eliminating causal noise variables, Taguchi methods attempt simply to remove or reduce the impact of these variables on product performance. By deliberately investigating noise factors, Taguchi believes that their effect on performance variation can be significantly reduced.

When both design and noise matrices are included in a Taguchi experiment, the design structure is called an **inner × outer Taguchi design** or **crossed array design**. The structure illustrated in Table 10.4 is therefore an $L_9(3^4) \times L_4(2^3)$ Taguchi design. The design has 9×4 = 36 combinations of control and noise factors and so it will be based on a total of 36 experiments, nine treatment combinations replicated four times. The corresponding full FD based on the same number of control and noise factors would consist of 648 ($3^4 \times 2^3$) factor combinations illustrating the saving in both experimental effort and cost available through the use of Taguchi's parameter design approach. However, implementing such an experiment will not provide information on all control factor interactions.

In many experimental situations, it may be impractical or inappropriate to use the same number of levels for all factors and so **mixed-level arrays** may need to be used. For example, the $L_{18}(2^1.3^7)$ array, shown in Table 9.14, could be a suitable structure for an eight factor study where one factor is to be set at two levels and the other seven at three levels. The **idle column** method developed by Taguchi enables a two-level array to be modified into a mixed-level array which is not strictly orthogonal. Grove & Davis (1991) discuss this technique further. It is possible also not to incorporate a noise matrix within a Taguchi designed experiment. This may correspond to a simple screening experiment designed to try to minimise the number of design parameter variables to include in a fuller Taguchi design, or a Taguchi experiment in which data are collected at various points in the process, e.g. points on the production line, or to simple replication of each experimental run.

Example 10.2 To reduce the carbon monoxide (CO) content of exhaust gas of an engine, seven two-level factors were studied. The parameter design base was Taguchi's $L_8(2^7)$ array which, as seven factors were tested, was a saturated design. Measured CO data, in g/m^3, were obtained for three different driving modes. The design and data are illustrated in Table 10.5 where 1 refers to low level and 2 high level.

Table 10.5 — CO emission data from Taguchi's $L_8(2^7)$ array for Example 10.2

Parameter design matrix (L_8)									
A	*B*	*C*	*D*	*E*	*F*	*G*	CO responses		
1	1	1	1	1	1	1	1.04	1.20	1.54
1	1	1	2	2	2	2	1.42	1.76	2.10
1	2	2	1	1	2	2	1.01	1.23	1.52
1	2	2	2	2	1	1	1.50	1.87	2.25
2	1	2	1	2	1	2	1.28	1.34	2.05
2	1	2	2	1	2	1	1.14	1.26	1.88
2	2	1	1	2	2	1	1.33	1.42	2.10
2	2	1	2	1	1	2	1.33	1.52	2.13

Reprinted from Taguchi, G. (1986) *Introduction to Quality Engineering*. Asian Productivity Organisation, Japan with the permission of the Asian Productivity Organisation.

Another important aspect of Taguchi designs is the principle of **loss**. According to Taguchi, loss is defined as not only the unnecessary development and production costs of the manufacturer but also loss arising from product usage (financial, functional, environmental) ranging from inconvenience to monetary loss and physical injury. It can occur not only when a product is outside the specification, but also when a product falls within specification. Taguchi advocates, at the very least, that losses to the manufacturer and the next organisation or person in the chain should be considered as part of a quality improvement study.

Three basic types of **loss function** have been proposed by Taguchi, depending on the quality criterion applicable. Each reflects that undesirable consequences increase in proportion to some function of the discrepancy between a product's measurable characteristic and its target specification. Taguchi proposes the quadratic functions shown in Table 10.6 for each of three possible quality criterion. For n items, the functions essentially measure the average loss per item. The loss functions, for n items, can also be written as

$$L(y) = c\text{MSD}$$

where MSD refers to the sample mean square deviation. All the functions displayed in Table 10.6 reflect quadratic trend. For nominal is best, the minimum is at the target value and small variations from the target are penalised.

Table 10.6 — Taguchi's loss functions for each quality criterion

Quality criterion	For one item	For n items
Nominal is best	$L(y) = c(y - m)^2$	$L(y) = c\Sigma(y - m)^2/n$
(Dimensions, electrical outputs, colour matching, product sizing)		
Large is best	$L(y) = c/y^2$	$L(y) = c\Sigma(1/y^2)/n$
(Material strength, yield, amplification, adhesion)		
Small is best	$L(y) = cy^2$	$L(y) = c(\Sigma y^2)/n$
(Distortion, impurities, contamination, friction)		

y is the measured response, c is a constant reflecting maximum loss and product specification, m is the target value

10.3 PERFORMANCE STATISTICS

Most classical experimental designs are concerned with response optimisation or with assessing factor influences on the average response. Taguchi methods, by contrast, are primarily designed to assess how factors affect response variation. To this end, Taguchi uses a performance statistic called a **signal-to-noise ratio** (*SN* ratio) to reflect the quality of the product performance with

$$SN = -10\log_{10}(\text{MSD}) \qquad (10.1)$$

where $\log_{10}$ is the preferred transformation of Taguchi to produce the 'decibel' style of *SN* ratio measure though $\log_e$ could equally be used. Choice of this statistic as the quality characteristic to be analysed is based on the need to control both the mean level of the process (signal) and the variation around this mean (noise). The *SN* ratio, expressed in decibels (Db), is an objective measure that combines both of these components in a single measure. It can be used to help determine factor levels that best cope with the noise factors and can be adapted to investigate the statistical properties of the factors being experimented on.

Based on expression (10.1), the three *SN* ratios recommended by Taguchi for the quality criteria in Table 10.6 become:

$$\text{Nominal is best:}\ \ SN_N = 20\log_{10}(\bar{y}/s) \qquad (10.2)$$

$$\text{Large is best:}\ \ SN_L = -10\log_{10}(\Sigma(1/y^2)/n) \qquad (10.3)$$

$$\text{Small is best:}\ \ SN_S = -10\log_{10}(\Sigma y^2/n) \qquad (10.4)$$

based on response replication either through inclusion of an outer array or simple combination replication. y is used to refer to the measured response while n, $\bar{y}$, and s refer, respectively, to the number of replications, mean response, and standard deviation of responses for each treatment combination. Specification of expressions (10.3) and (10.4) is straightforward given the form of the associated MSD. For the "Nominal is best" quality criterion, however, it is not possible to directly relate the

MSD term to the SN_N expression (10.2) (Maghsoodloo 1990). The format presented is an alternative put forward by Taguchi.

These SN ratios are such that they will be maximum when the optimal treatment combination occurs irrespective of quality criterion. According to Taguchi, this will produce response optimisation while at the same time minimising response variability and corresponds to achieving an average response close to target with minimal variation. For the "Nominal is best" criterion, the SN ratio (10.2) is a function of the logarithm of the inverse of the coefficient of variation (CV, see expression (1.4)). The presented SN ratios can also be derived from the expression

$$SN = 10\log c - 10\log[L(y)]$$

where c is the constant associated with the loss function. Thus, we can see that maximising the SN ratio will also lead to minimising the corresponding loss function $L(y)$.

Taguchi's use of SN ratio as the performance measure has generated considerable debate in respect of its validity as a statistical measure on which to base data analysis. From expression (10.2), we can see that

$$SN_N = 20\log_{10}(\bar{y}) - 20\log_{10}(s)$$

and so, if the mean is fixed at a target estimated by $\bar{y}$, maximising SN_N will be equivalent to minimising $\log_{10}(s)$. Using $\log(s)$ provides a clearer connection with an important response characteristic, i.e. variability, and so could aid understanding of how factors affect process variability. The SN ratios (10.3) and (10.4) confound both location and dispersion effects, the former referring to factors affecting mean response and the latter factors influencing response variability. This means that use of such statistics may not provide a clear picture of factor influences. Some believe the SN_L and the SN_S ratios should not be used as part of the data analysis because of the confounding effect.

Given the apparent statistical inefficiency of SN ratios as data analysis measures, it is recommended that parallel analysis of **means** and **standard deviations**, through the natural logarithm of the standard deviation s, $\ln(s)$, should also be carried out to augment the analysis of the SN ratios. This will help separate location and dispersion effects for a more complete data analysis. Analysing the mean response enables average performance to be assessed while $\ln(s)$ analysis will provide information of how performance variability is affected by the tested factors, $\ln(s)$ being adopted to improve residual normality. By conducting such a comprehensive analysis, it is possible to gain a better understanding and insight into the process and thereby come to more appropriate conclusions on the factors and factor levels having most influence on the measured response.

A Taguchi experiment based on the inner × outer design structure illustrated in Table 10.4, for example, consists of running experiments for every row in the parameter design matrix under the conditions specified in every row of the noise matrix. This would produce four response measurements for each combination tested. From these replicate observations, each type of performance statistic could be computed. Each would be assessed to determine factor influences with a view to seeing if a consensus emerges in respect of the optimal control factor levels. In many instances, this combination of control factors is not actually run within the Taguchi experiment and as Taguchi experiments are highly fractionated FFDs, this is not unexpected.

Exercise 10.1 For the CO emission study of Example 10.2, we want to discuss the alias structure and illustrate performance statistic derivation.

Alias structure: As the implemented design was a saturated $L_8(2^7)$ array comparable to the planned design for the integrated circuit study described in Example 10.1, the alias structure will be identical to that discussed. Therefore, main effects will be aliased with two, three, four, five, and six factor interactions providing a resolution III design.

Manual derivation: Given that three ($n = 3$) replicate CO content measures were made for each treatment combination, calculation of the associated performance statistics is relatively straightforward. Consider experimental run 1 (row 1 of Table 10.5) with measurements 1.04, 1.20, and 1.54. As minimisation is the goal, the *SN* ratio to be used is that of expression (10.4). For the data highlighted, this becomes

$$SN_S = -10\log_{10}[(1.04^2 + 1.20^2 + 1.54^2)/3] = -10\log_{10}[4.8932/3] = -2.12472,$$

the mean, expression (1.2), is

$$\bar{y} = (1.04 + 1.20 + 1.54)/3 = 3.78/3 = 1.26,$$

and the standard deviation, expression (1.3), is

$$s = \sqrt{\frac{4.8932 - 3.78^2/3}{3-1}} = \sqrt{0.0652} = 0.255343$$

providing a ln(s) value of −1.36515. Similar calculations can be applied to provide the performance statistics for each of the other seven factor combinations tested.

Software derivation: As manual derivation of performance statistics can be cumbersome, it is best to use software to determine the required figures. Table 10.7 shows the Minitab generated performance statistics using simple calculation procedures (see Appendix 10A.1). The summaries shown for run 1 (row 1) agree with those derived manually. □

Table 10.7 — Performance statistic summaries for CO emission in Taguchi study of Example 10.2

Parameter design matrix (L_8)							Performance statistics			
A	*B*	*C*	*D*	*E*	*F*	*G*	*SN*	Mean	*s*	Ln(*s*)
1	1	1	1	1	1	1	−2.12472	1.26000	0.255343	−1.36515
1	1	1	2	2	2	2	−5.01698	1.76000	0.340000	−1.07881
1	2	2	1	1	2	2	−2.08029	1.25333	0.255799	−1.36336
1	2	2	2	2	1	1	−5.56680	1.87333	0.375011	−0.98080
2	1	2	1	2	1	2	−4.05773	1.55667	0.428291	−0.84795
2	1	2	2	1	2	1	−3.30522	1.42667	0.397157	−0.92342
2	2	1	1	2	2	1	−4.36444	1.61667	0.420991	−0.86514
2	2	1	2	1	1	2	−4.58195	1.66000	0.417971	−0.87234

10.4 DATA ANALYSIS COMPONENTS FOR TWO-LEVEL TAGUCHI PARAMETER DESIGNS

Statistical analysis of two-level Taguchi designs involves first determining the *SN* ratio, mean, and ln(s) performance statistics for each experimental run. These measures are then analysed using similar procedures to those considered for the two-level designs in Chapters 7, 8, and 9, i.e. effects plots, data plots, and ANOVA modelling. As previously, it is advisable to screen the factors for importance before constructing an ANOVA model for each performance statistic.

Use of statistical software such as SAS and Minitab can enable Taguchi calculations to take place easily and can expedite the data presentation and analysis. Such software packages are not totally dedicated to Taguchi analysis methods but contain sufficient commands and procedures to enable a Taguchi analysis to be undertaken.

10.4.1 Estimation of Factor Effects

For Taguchi's two-level arrays, when r test runs have been carried out, factor effect estimation for all performance statistics is provided by

$$\textit{Effect estimate} = [(\Sigma \text{ level 2 results}) - (\Sigma \text{ level 1 results})]/(r/2) \qquad (10.5)$$

Interpretation of such estimates is as before (see Sections 7.3.2 and 8.5.2). Important factors, in respect of each type of performance statistic, will be those with high numerical estimates distinctly separated from those of the unimportant factors.

Exercise 10.2 Exercise 10.1 illustrated derivation of the performance statistic summaries for the CO emission study of Example 10.2. Minitab generated effect estimates for each type of performance statistic are presented in Display 10.1.

Display 10.1

Estimated Effects and Coefficients

	SN ratio		Mean		Ln(s)	
Term	Effect	Coef	Effect	Coef	Effect	Coef
Constant		−3.887		1.55083		−1.037
A	−0.380	−0.190	0.02833	0.01417	0.320	0.160
B	−0.522	−0.261	0.10000	0.05000	0.033	0.017
C	0.270	0.135	−0.04667	−0.02333	0.016	0.008
D	−1.461	−0.730	0.25833	0.12917	0.147	0.073
E	−1.728	−0.864	0.30167	0.15083	0.188	0.094
F	0.391	0.196	−0.07333	−0.03667	−0.041	−0.021
G	−0.094	−0.047	0.01333	0.00667	−0.007	−0.003

Effect estimation: We know that $r = 8$. From Table 10.7, we can determine the necessary totals for derivation of effect estimate expression (10.5). To illustrate, consider the effect estimate for factor A for the *SN* ratio. From Table 10.7, we can obtain Σ level 1 results = −14.78879 and Σ level 2 results = −16.30934. Thus, expression (10.5) becomes

$$\text{estimate}_A = [(-16.30934) - (-14.78879)]/(8/2) = -0.380$$

which agrees with the result presented in Display 10.1. The other presented effect estimates for each performance statistic can be evaluated similarly. □

10.4.2 Factor Effect Plots

Plots of effect estimates can be used in two-level Taguchi experiments, as they were in the two-level designs of Chapters 7, 8, and 9, to screen the factors for those that are potentially significant and those that are not as regards their influence on the performance statistic analysed. A **normal plot**, **half-normal plot**, and **Pareto chart** are the commonly used graphical displays for effect estimates generated in two-level Taguchi designs. Such plots require to be constructed for each of the performance statistics *SN*, mean, and ln(*s*).

Interpretation of the effects plots conforms to the general rules introduced in Section 7.3.2 with important factors differing in position from unimportant factors. For the normal plot, this corresponds to points lying to the left (large negatives) of the line through the unimportant factors or to the right (large positives) of this line (see Section 7.3.2.1). For the Pareto chart, important factors will appear in the top right corner of the plot with a distinct elbow effect signifying the jump from unimportant to important factors, the unimportant factors clustering to form a near vertical line parallel to the *Y* axis of the plot (see Section 7.3.2.3). As with previous illustrations, both plots will generate comparable information and thereby similar interpretation.

Exercise 10.3 Graphical presentation and analysis of the effect estimates shown in Display 10.1 is now necessary to begin the data interpretation. Display 10.2 shows the Pareto charts for all three performance statistics.

> *Pareto chart of effect estimates*: Factors *D* and *E* show through as important to all three performance statistics with factor *A* only important to ln(*s*). This would suggest that factors *D* and *E* play significant roles in affecting location and dispersion. Factor *A* only appears to affect dispersion and therefore could be considered as solely a tuning parameter. Such conclusions are obviously dependent on assuming that all two factor interactions are unimportant (see aliasing discussed in Example 10.1). □

Display 10.2

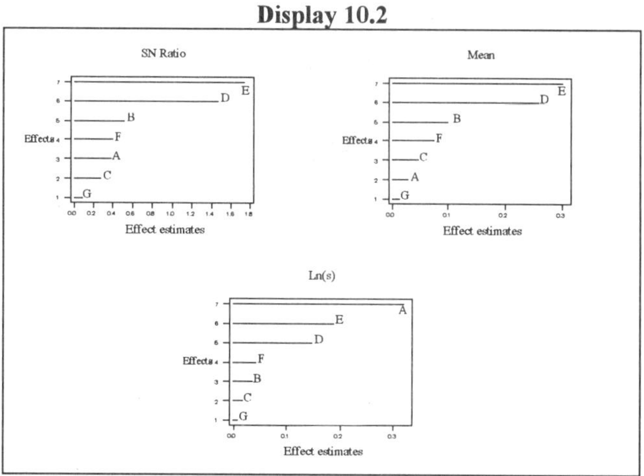

10.4.3 Data Plots

Data plots of the "marginal means" (average of each performance statistic for each factor level) provide, as before, visual mechanisms for displaying factor influences. Plots worth considering include **main effect plots** and **interaction plots**. The latter are appropriate if factor interaction has been specifically included in the parameter design matrix and for assessing control × noise interaction when a noise matrix is used to generate the necessary replicate response measurements. For the interactions between control and noise factors, it is advisable to assess all such interactions fully to gauge how the control factors perform in the presence of the noise factors. Assessment of the latter form of interaction is seen by Taguchi to be inappropriate if full analysis of the *SN* ratio performance statistic is carried out. However, by including such analysis, additional information on the performance of control factors in the presence of noise factors can be forthcoming and may provide further evidence of the influence of the control factors on the measured response.

In the plot interpretation, it is hoped that optimal factor levels will be shown to influence all performance characteristics equally. Optimal levels will correspond to the end-points of the line, lower for minimisation and upper for maximisation. On occasions, conflict may arise when a factor level is optimal for one aspect but not another. In such cases, a **trade-off** between performance statistics may be necessary depending on whether optimisation of the response or minimisation of response variability would be most beneficial for the problem being investigated.

Exercise 10.4 Main effect plots for the seven tested factors in the CO emission study of Example 10.2 are provided in Displays 10.3 to 10.5. We want to interpret these plots for the information they contain on factor effects and optimal factor levels.

SN ratio: The main effect plots for the *SN* ratio are shown in Display 10.3. They indicate that all factors have some influence on *SN* ratio but that factors *D* and *E* show greatest effect with a distinct downward trend apparent as tested level increases. Factors *A*, *B*, *C*, and *F* also have some effect though very much less so than *D* and *E*. Only *G* appears to have no real influence on *SN* ratio. Thus for maximisation of *SN* ratio, the optimal combination would appear to be $A_1B_1C_2D_1E_1F_2G_1$.

Display 10.3

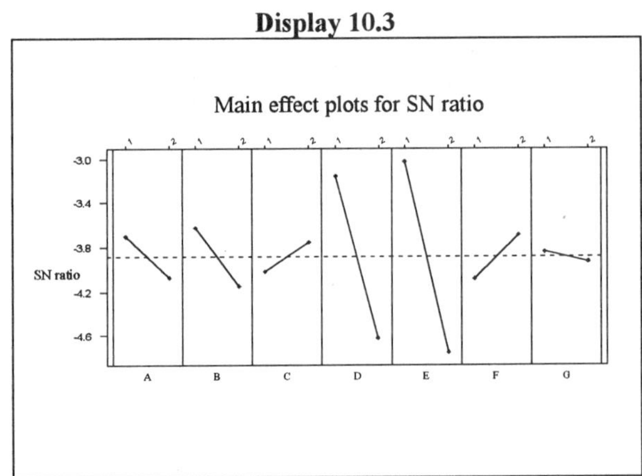

Mean: The main effect plots for the mean response in Display 10.4 provide a similar picture of overall factor effect with *D* and *E* again showing through as the most important. An upward trend appears for both of these factors. Factors *B*, *C*, and *F* have lesser effect while *A* and *G* appear to have negligible effect. As minimising CO content is the goal, the optimal combination of factors to achieve this would appear to be $A_1B_1C_2D_1E_1F_2G_1$ which corresponds to the optimal suggested from the *SN* ratio analysis.

Display 10.4

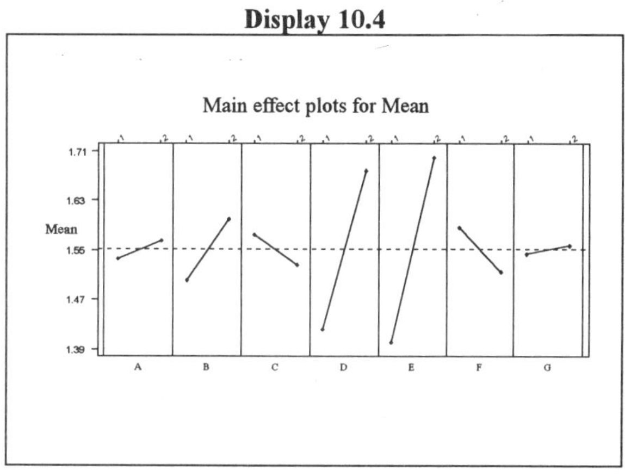

Ln(s): The ln(*s*) main effect plots, presented in Display 10.5, confirm that factors *A*, *D*, and *E* have most influence on CO content variability. They show greatest change for changes in the factor levels. The other four factors look to have little or no influence on CO content variability. Since minimisation of variability is ideal, factor optimisation would be achieved through choice of the lower end points. This suggests that the combination $A_1B_1C_1D_1E_1F_2G_2$ may be optimal for minimising variability. □

Display 10.5

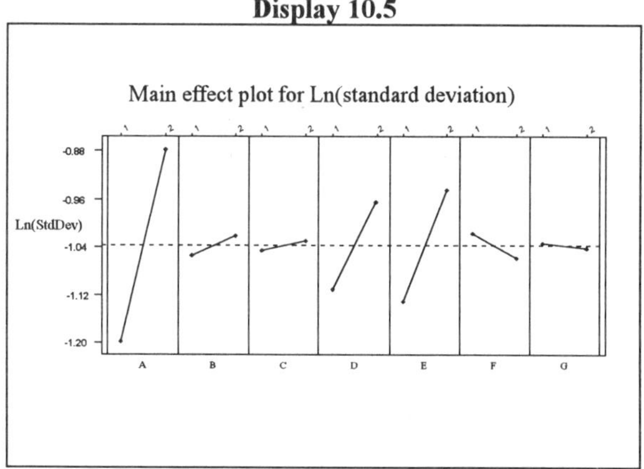

A further plot to be considered is that of $\ln(s^2)$ versus $\ln(\bar{y})$ for each replicated treatment combination. Such a plot can be used to investigate the form of the relationship between the variance and the mean. A trend in the plot may signify that variance is a function of the mean and that a data transformation may be appropriate. A simple horizontal trend would be indicative of the presence of location effects only.

10.4.4 ANOVA Modelling

After analysis of the effects, the next step is generally to develop and assess a model for each performance statistic as occurred in the two-level screening designs discussed earlier. Using the **ANOVA principle**, such a model can be assessed to help confirm the statistical significance of those effects thought to have an important influence on the corresponding summary measure. As before, ANOVA is based on splitting response variation, the total sum of squares *SS*Total, into sums of squares corresponding to the important effects. The total sum of squares is given by

$$SS\text{Total} = \Sigma y_i^2 - (\Sigma y_i)^2/r \tag{10.6}$$

with $r - 1$ degrees of freedom where y_i is the appropriate performance characteristic (*SN* ratio, mean, ln(s)) for the ith experimental run of the parameter design matrix and r is the number of test runs. For each estimable effect, we have

$$SS\text{Effect} = [(\Sigma \text{ level 1 results})^2 + (\Sigma \text{ level 2 results})^2]/(r/2) - (\Sigma y_i)^2/r \tag{10.7}$$

based on one degree of freedom, i.e. number of levels minus 1. If effect estimates are available, expression (10.7) can be re-written as

$$SS\text{Effect} = (\textit{Effect estimate})^2 \times (r/4) \tag{10.8}$$

As each factor sum of squares is based on only one degree of freedom, we have *MS*Effect = *SS*Effect.

Exercise 10.5 From the effects and plots analysis of the CO emission data of Example 10.2, the following models can be suggested for the performance statistics:

$$SN \text{ ratio} = E + D + \text{error}$$

$$\text{Mean} = E + D + \text{error}$$

$$\text{Ln}(s) = A + E + D + \text{error}$$

ANOVA analysis: Table 10.8 shows a summary of the ANOVA results for each of the aforementioned models. For the *SN* ratio associated with factor D, we have Σ level 1 results = −12.62718 and Σ level 2 results = −18.47095. From the design run, we have $r = 8$ and $\Sigma y_i = -31.09813$. Expression (10.7) therefore becomes

$$SSD = [(-12.62718)^2 + (-18.47095)^2]/(8/2) - (-31.09813)^2/8 = 4.269.$$

agreeing with the value presented in Table 10.8. All other *SS* components can be similarly computed.

The figures show that all specified model effects have statistically significant influences on the associated performance statistic. All factors, except factor D for the *SN* ratio, are highly significant ($p < 0.01$). The results confirm the findings generated from the analysis of the effects.

Table 10.8 — ANOVA summary for Exercise 10.5

Source	*SN* df	MS	F	p	Mean df	MS	F	p	Ln(s) df	MS	F	p
A									1	0.2046	132.00	< 0.001
E	1	5.975	22.89	< 0.01	1	0.1820	24.59	< 0.01	1	0.0706	45.55	< 0.01
D	1	4.266	16.36	< 0.05	1	0.1335	18.04	< 0.01	1	0.0430	27.74	< 0.01
Error	5	0.261			5	0.0074			4	0.00155		

Summary of findings: A summary of the findings from the data analysis for the CO emission study of Example 10.2 is presented in Table 10.9. Technically, only factors A, D, and E appear important as they were the only ones which showed up as such in the data analysis carried out. Consensus on best level is apparent for all factors except C and G which differ for the ln(s) performance statistic. A trade-off between minimising mean or minimising variability is therefore necessary as both cannot be satisfied by choice of level for each of these factors. A check of the main effect plots in Displays 10.4 and 10.5 may help in the decision making. For factor C, the mean appears more affected by choice of level than variability suggesting C_2 as possibly best. By contrast, factor G has minimal effect on both characteristics so choice of level can be based on other considerations such as cost and ease of implementation. It would appear that $A_1B_1C_2D_1E_1F_2$ with either G_1 or G_2 may be the optimal combination for minimisation of CO content. □

Table 10.9 — Summary of optimal results for CO emission study of Example 10.2

Performance statistic	Factor and level *A*	*B*	*C*	*D*	*E*	*F*	*G*
SN ratio	1	1	2	1*	1*	2	1
Mean	1	1	2	1*	1*	2	1
Ln(s)	1*	1	1	1*	1*	2	2

* significant using effects analysis and ANOVA

The aliasing present in the Taguchi design used for the CO emission study of Example 10.2 (see Exercise 10.1) states that each control factor is aliased with three two factor interactions and other higher order interactions. It is possible that any one of these two factor interactions may be the effect generating the associated estimate, not the main effect that the analysis suggests. In fact, for the ln(s) characteristic, it could be that the significant effects are D, E, and their interaction because of the aliasing of A with $D \times E$. Thus, it is possible that several explanations could be put forward to explain the occurrence of important effects, any one of which could provide a better description of the system being investigated. Failure to investigate other possibilities can be one of the failings of Taguchi practitioners. Overlooking these cases, therefore, could lead to erroneous conclusions.

To illustrate this, consider the Minitab generated matrix of interaction plots for the mean performance statistic presented in Display 10.6. For the mean, the analysis provided in Exercises 10.3 to 10.5 suggested that factors D and E were the important control factors with both ideally set at low level. From Table 10.3, the associated aliases of these effects are

$$D \equiv -AE \equiv -BF \equiv -CG \qquad E \equiv -AD \equiv -CF \equiv -BG$$

The plots in Display 10.6 show that these aliased two factor interactions all exhibit strong interaction effect with non-parallel lines evident in each of their plots. For $A \times D$, which is aliased with E, the plot indicates that low A causes a major change in CO emission as D increases suggesting that A_1D_1 may be optimal. For $A \times E$, which is aliased with D, the same patterning emerges providing comparable interpretation for increasing levels of E and suggesting an optimal of A_1E_1. Each interaction plot for the aliased two factor interactions can be interpreted similarly. A summary of the results of this analysis is provided in Table 10.10. Carrying out such analysis for each performance statistic showed that $A_1D_1E_1$ appeared optimal across all, but that consensus on levels for the other four factors could not be achieved. This highlights that factors A, D, and E appear most important in determining the levels of CO emission of the tested engine. Further experimentation on these factors around the suggested optimal levels should be considered to enhance the conclusion.

Display 10.6

Interaction plots for Mean

Table 10.10 — Summary of interaction plots analysis for CO emission study of Example 10.2

		A	B	C	D	E	F	G
	D				1			
D aliases	$A \times E$	1				1		
	$B \times F$		1				1	
			2				2	
	$C \times G$			1				1
				2				2
	E					1		
E aliases	$A \times D$	1			1			
	$C \times F$			2			2	
	$B \times G$		1					1

1 and 2 denote low and high levels

Though the Taguchi experiment in Example 10.2 has highlighted some interesting features, it is possible that design construction and the consequent effect estimation could have been improved. The experiment was based on 8×3 = 24 separate experiments and had an extremely complex alias structure with each main effect aliased minimally with three two factor interactions. A better way of approaching the parameter design construction to help separate main effects and two factor interactions could have been to use the $L_{16}(2^{15})$ array with column allocation of A to column 1, B to 2, C to 4, D to 8, E to 5, F to 9, and G to 14. This would generate a resolution IV design with defining relation

$$I = ABCE = ABDF = BCDG = CDEF = ADEG = ACFG = BEFG$$

and alias structure up to three factor interactions shown in Table 10.11. With single replication, this structure would require fewer experimental runs but at least effect estimation would provide main effect estimates clear of two factor interactions and a more comprehensive rangc of treatment combinations would be tested. A second possibility would be to use a 2^{7-2} FFD (32 runs) with defining contrasts $I = -ABCDF$ and $I = -BCDEG$ which would also provide a resolution IV design though the increase in the number of runs may mitigate against its adoption.

Table 10.11 — Partial alias structure of alternative $L_{16}(2^{15})$ parameter design structure for CO emission study of Example 10.2

$A \equiv BCE \equiv BDF \equiv DEG \equiv CFG$	$B \equiv ACE \equiv ADF \equiv CDG \equiv EFG$
$C \equiv ABE \equiv BDG \equiv DEF \equiv AFG$	$D \equiv ABF \equiv BCG \equiv CEF \equiv AEG$
$E \equiv ABC \equiv CDF \equiv ABG \equiv BFG$	$F \equiv ABD \equiv CDE \equiv ACG \equiv BEG$

$G \equiv BCD \equiv ADE \equiv ACF \equiv BEF$

$AB \equiv CE \equiv DF$ $\quad AC \equiv BE \equiv FG$ $\quad AD \equiv BF \equiv EG$ $\quad AE \equiv BC \equiv DG$

$AF \equiv BD \equiv CG$ $\quad AG \equiv DE \equiv CF$ $\quad BG \equiv CD \equiv EF$

$ABG \equiv CEG \equiv DFG \equiv ACD \equiv BDE \equiv BCF \equiv AEF$

Taguchi's initial recommendations on ANOVA modelling were not wholly appropriate. If there is no obvious error term for the ANOVA table, Taguchi recommended combining the smaller effect mean squares (*MS*) and working out their average to obtain an "error" mean square (*MSE*) for testing the remaining factors against. To help determine the smallest *MS*s, Taguchi advised ranking the effect *MS*s in order of magnitude with the smallest ones chosen to make up the *MSE* estimate. There is no clear guidance from Taguchi on where to stop pooling the smallest *MS*s except F statistics greater than 1 were thought indicative, whereas F statistics greater than 2 were thought to be evidence of real effects. This, however, can increase the probability of Type I error ("false positives") and so could bias the analysis results. Carrying out the type of analysis described in this section avoids many of Taguchi's controversial methods and will help to generate more statistically valid conclusions.

Once analysis of the performance characteristic data has taken place, Taguchi advises that a **confirmation experiment** be run. This involves using the optimal levels of the tested factors to confirm that the suggested improvement in product quality and consequent reduction in variability can be achieved and sustained.

10.5 DATA ANALYSIS COMPONENTS FOR THREE-LEVEL TAGUCHI PARAMETER DESIGNS

Three-level arrays such as the $\mathbf{L_9(3^4)}$ and the $\mathbf{L_{27}(3^{13})}$ are also common in Taguchi experiments as the basis for the parameter design matrix. They provide the added benefits of curvature assessment unlike two-level arrays. Again, such structures are highly fractionated FFDs with, often, even more complex aliasing than occurs in two-level Taguchi arrays. Analysis procedures for experiments based on these arrays are similar to those illustrated for two-level Taguchi parameter designs.

Example 10.3 In an investigation into the assembly of an elastomeric connector to a nylon tube, a Taguchi-based experiment was conducted to try to maximise pull-off force to ensure the manufactured product was suitable for use in automotive engine components. Elastomeric connector means a rubber-based connector which returns to its original shape after the deforming force is removed. Four controllable factors and three noise factors were identified as shown. The parameter design matrix was based on Taguchi's $L_9(3^4)$ array while the three noise factors were placed in an $L_8(2^7)$ array. The design was therefore an $L_9(3^4) \times L_8(2^7)$ Taguchi design with saturated inner array. Noise factor column allocation was based on E to column 1, F to 2, and G to 4 for the L_8 outer array. The collected pull-off force data, in lbs, are presented in Table 10.12. The eight measurements presented for each run correspond to the eight rows of the noise matrix. In other words, the value 15.6 was obtained with E at 24h, F at 72 °F, and G at 25%, i.e. column 1 of the presented noise matrix, while 9.5 was obtained with E at 24h, F at 72 °F, and G at 75%, i.e. column 2 of the presented noise matrix.

Factors		*Levels*
Control	Interference (A)	low, medium, and high
	Connector wall thickness (B)	thin, medium, and thick
	Insertion depth (C)	shallow, medium, and deep
	Percent adhesive in connector pre-dip (D)	low, medium, and high
Noise	Conditioning time (E)	24 hours and 120 hours
	Conditioning temperature (F)	72 °F and 150 °F
	Conditioning relative humidity (G)	25% and 75%

Table 10.12 — Pull-off force data from Taguchi's $L_9(3^4) \times L_8(2^7)$ design for Example 10.3

				Outer array (L_8)								
				E	1	1	1	1	2	2	2	2
Inner array (L_9)				F	1	1	2	2	1	1	2	2
A	B	C	D	G	1	2	1	2	1	2	1	2
1	1	1	1		15.6	9.5	16.9	19.9	19.6	19.6	20.0	19.1
1	2	2	2		15.0	16.2	19.4	19.6	19.7	19.8	24.2	21.9
1	3	3	3		16.3	16.7	19.1	15.6	22.6	18.2	23.3	20.4
2	1	2	3		18.3	17.4	18.9	18.6	21.0	18.9	23.2	24.7
2	2	3	1		19.7	18.6	19.4	25.1	25.6	21.4	27.5	25.3
2	3	1	2		16.2	16.3	20.0	19.8	14.7	19.6	22.5	24.7
3	1	3	2		16.4	19.1	18.4	23.6	16.8	18.6	24.3	21.6
3	2	1	3		14.2	15.6	15.1	16.8	17.8	19.6	23.2	24.4
3	3	2	1		16.1	19.9	19.3	17.3	23.1	22.7	22.6	28.6

10.5.1 Data Plots

As with two-level Taguchi designs, data plotting of performance statistics is an integral part of data analysis in three-level designs. **Main effect plots** and **interaction plots** again form the analysis base. In the case of main effect plots, all factors levels should be included.

Exercise 10.6 The performance statistics for the pull-off force data of Example 10.3 are presented in Table 10.13 with the *SN* results based on the "Large is best" expression (10.3). Data plots for the "marginal means" (averages) for each performance statistic now require analysis. They are provided in Displays 10.7 to 10.9.

Data plots: For the *SN* ratio, factors *A* and *C* show greater variation than factors *B* and *D* and thus, larger effect. From the plots, the optimal combination for maximising *SN* ratio would appear to be $A_{\text{med}}B_{\text{med}}C_{\text{deep}}D_{\text{low}}$ though C_{med} is not too different from C_{deep}. For the mean, the effect of each factor is similar to that for the *SN* ratio performance statistic with $A_{\text{med}}B_{\text{med}}C_{\text{deep}}D_{\text{low}}$ again suggested as the optimal combination. The ln(*s*) plots in Display 10.9, however, highlight a different trend. Choice of optimal (minimal variability) is more complicated with two levels producing similar variability characteristics for each of factors *A*, *C*, and *D*. Only factor *B*, where B_{thin} appears best for reducing variability, provides a simple conclusion though for *D*, it could be argued that high is best. Checking Display 10.9 for ln(*s*), we could suggest that A_{med} and C_{deep} look best for minimising variability producing a potential optimal combination of $A_{\text{med}}B_{\text{thin}}C_{\text{deep}}D_{\text{high}}$. This analysis also shows that no factor is unimportant for any performance statistic as they all appear to play some role in determining the level of the statistic.

Table 10.13 — Performance statistic summaries for the $L_9(3^4) \times L_8(2^7)$ study of Example 10.3

Parameter design matrix (L_9)				Performance statistics			
A	*B*	*C*	*D*	*SN*	Mean	*s*	Ln(*s*)
1	1	1	1	24.0253	17.5250	3.61258	1.28442
1	2	2	2	25.5216	19.4750	2.90652	1.06696
1	3	3	3	25.3348	19.0250	2.88333	1.05894
2	1	2	3	25.9043	20.1250	2.59766	0.95461
2	2	3	1	26.9075	22.8250	3.42751	1.23184
2	3	1	2	25.3257	19.2250	3.37967	1.21778
3	1	3	2	25.7108	19.8500	2.98472	1.09351
3	2	1	3	24.8323	18.3375	3.77470	1.32832
3	3	2	1	26.1520	21.2000	3.94787	1.37318

Thus, initially from this analysis, it would appear that A_{med} and C_{deep} may be best for *A* and *C* respectively. For factor *B*, B_{thin} appears to improve consistency and though providing lower average pull-off force compared to the medium level, was chosen as best by the experimenters on cost grounds. For factor *D*, smaller variations are associated with mid and high levels though these, in turn, produce lower average pull-off force compared to the low level.

Display 10.7

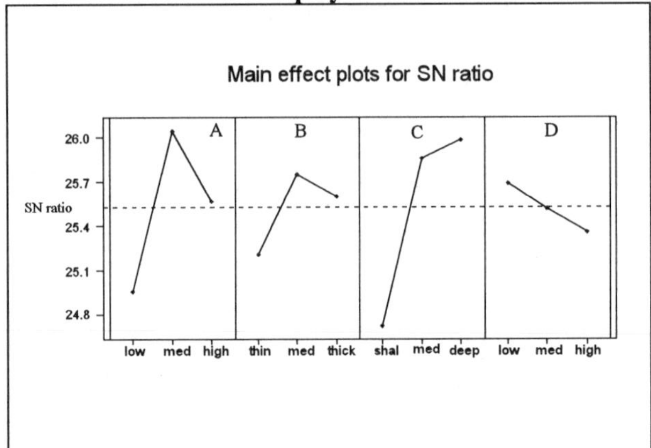

Display 10.8

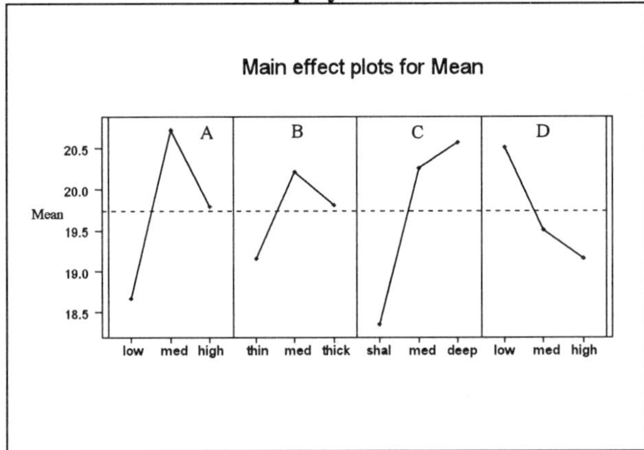

Display 10.9

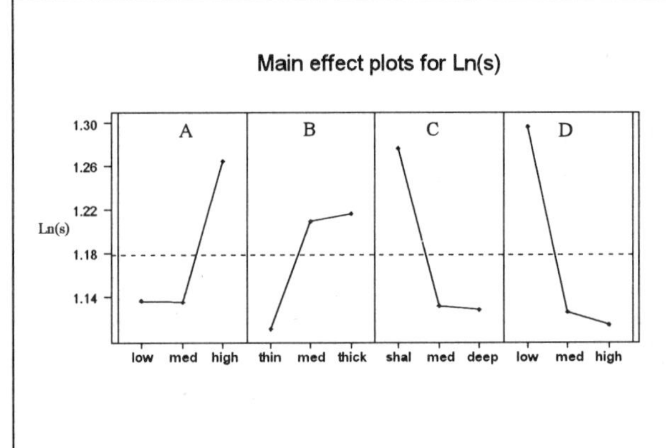

Control × noise interaction plots: Additional analysis through control × noise factor interactions are also worth considering as specific noise factors were included in the outer array. In such cases, assessment of such interactions is advised though not all will necessarily provide interesting interpretations. The authors discussed two control × noise interactions which generated interesting patterns, $A \times G$ shown in Display 10.10 and $D \times E$ in Display 10.11.

The $A \times G$ interaction, representing interference × conditioning relative humidity, shows evidence of interaction through the non-parallel lines. The plot shows that A_{med} appears least sensitive to humidity changes (almost horizontal line) while also providing highest average pull-off force. Thus, medium level looks best for the interference factor A. The plot of the adhesive × conditioning time ($D \times E$) interaction in Display 10.11 shows a distinctive interaction effect. D_{med} has the least slope and so appears least influenced by conditioning time. D_{low}, however, appears to provide best pull-off force and may be best level for the adhesive factor. Combining the analysis of all the control × noise interactions, it appeared that the combination $A_{med}B_?C_{deep}D_{low}$ may be best for maximising pull-off force where $B_?$ signifies that no definite decision on best level for factor B was forthcoming.

Display 10.10

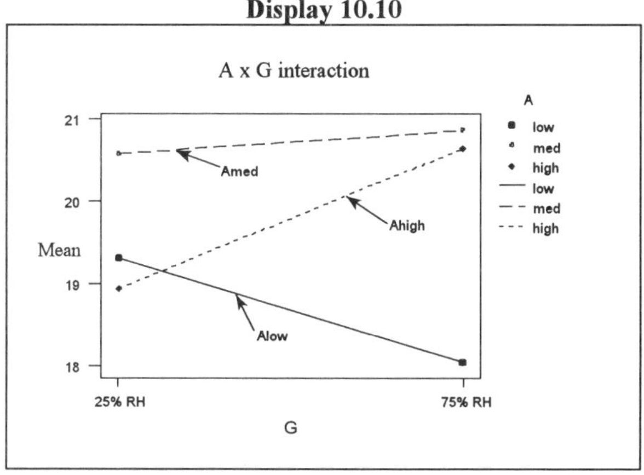

Display 10.11

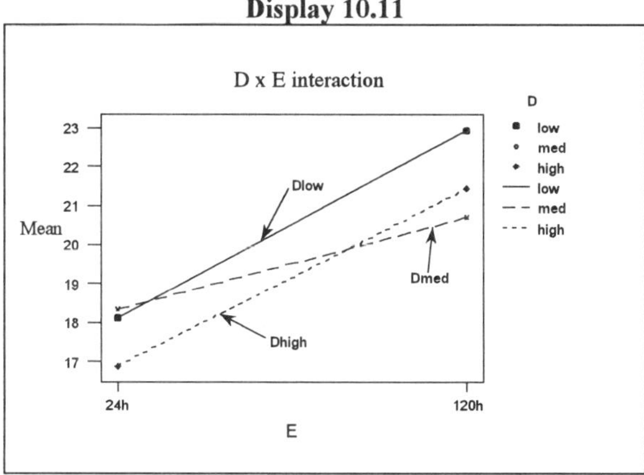

Summary of findings: Table 10.14 provides a summary of the findings. It shows that consensus appears for choice A_{med}, C_{deep}, and D_{low}. For factor B, no consensus appears as the interaction plots fail to help clarify best level. The experimenters chose thin from a cost perspective arguing that a thin wall was least expensive. Thus, it would appear that the combination $A_{med}B_{thin}C_{deep}D_{low}$ may be optimal. As with FFDs, this combination, represented by $A_2B_1C_3D_1$, was not run within the experiment (see inner array in Table 10.12). The experimenters also conducted a cost and assembly effort analysis which suggested that C_{med} was possibly better. They therefore concluded that $A_{med}B_{thin}C_{med}D_{low}$ was the optimal combination for maximising pull-off force. A confirmation experiment was run and acceptable results reported. □

Table 10.14 — Summary for pull-off force study of Example 10.3

		Factor and level			
		A	*B*	*C*	*D*
Factor plots	*SN* ratio	med	med	deep	low
	Mean	med	med	deep	low
	Ln(*s*)	low/med	thin	med/deep	high
Interaction plots	$A \times G$	med			
	$D \times E$				low
	overall	med	?	deep	low

10.5.2 ANOVA Modelling

Using data plots enables factor screening for each performance statistic to be carried out. On the basis of these findings, we can again consider developing a model of the performance statistic to assess the statistical significance of the screened factors. The total sum of squares, *SS*Total, is based on expression (10.6) as before. Effect sum of squares for three-level factors are determined as

$$SS\text{Effect} = [(\Sigma \text{ level 1 results})^2 + (\Sigma \text{ level 2 results})^2 + (\Sigma \text{ level 3 results})^2]/(r/3) - (\Sigma y_i)^2/r \qquad (10.9)$$

based on two degrees of freedom, i.e. number of levels minus 1. Using the usual ratio of sum of squares over degrees of freedom produces the effect mean square expressions of

$$MS\text{Effect} = SS\text{Effect}/2 \qquad (10.10)$$

ANOVA table construction is as before.

From the plots analysis of the performance statistics associated with the pull-off force data of Example 10.3, it is difficult to suggest models for each performance statistic. This is because of the small number of control factors tested and the lack of any that appear unimportant. Only the *SN* ratio appeared capable of being modelled as *SN* ratio = $C + A$ + error with ANOVA specifying that *C* was significant at the 5% level ($p = 0.033$) while *A* was significant at the 10% level ($p = 0.070$).

10.6 OTHER DATA ANALYSIS COMPONENTS

10.6.1 Exponential/Weibull Plots

Exponential/Weibull plots correspond to plots of ordered effect *MS*s and are used similarly to half-normal plots in two-level designs. They are most appropriate for three-level Taguchi designs. The *Y* axis of a Weibull plot refers to the values of the expression $100(i - 0.5)/k$ where *i* is the rank value of the ordered *MS* and *k* is the number of control factors. The X axis corresponds to the ordered MS terms. Plot interpretation is similar to that associated with a half-normal plot (see Section 7.3.2.2). The unimportant effects will cluster together to form a nearly vertical line almost parallel to the Y axis. The important effects will be distinctly separate from this line generally positioned to the top right of the plot. Such a plot will only be useful if the numbers of factors is large and there is distinct separation of *MS*s, the **effects sparsity** principle.

10.6.2 Test of Homogeneity of the Factor Mean Squares

This provides an alternative to ANOVA modelling when determining the statistical significance of factor effects in respect of each performance statistic. The procedure involves calculating the mean (m) and standard deviation (sd) of the mean squares of the k factors. Under the null hypothesis of no real effects, the expression

$$[(k-1)\times(\nu/2)](sd/m)^2 \qquad (10.11)$$

follows a χ^2 distribution with $k-1$ degrees of freedom where ν refers to the degrees of freedom of the associated active effect. For two-level factors, $\nu = 1$ while $\nu = 2$ for three-level factors.

If the test indicates significant differences between the *MS*s, one or more of the larger *MS*s may be deleted and the test repeated until a non-significant result (corresponding to a p value between 0.1 and 0.5) is obtained. The remaining *MS*s may then be pooled to provide a provisional estimate of the error variance, i.e. *MS*Error estimate, for construction of appropriate F tests for testing the larger effect *MS*s. Use of the percentage points of the χ^2 distribution (Table A.2) provides an adequate approximation though tables for direct comparison of sd/m could also be used (Bissell 1992). Using this procedure to detect important factors requires that the factors exhibit distinct separation such that sd will exceed m significantly in all comparisons but the final one.

10.7 FINAL COMMENTS

Since Taguchi methods first appeared in the West, debate on their validity has been ongoing. Several improvements to his methods have been suggested which make use of more appropriate statistical experimental designs specifically developed for multi-factor investigations.

Tribus & Szonyi (1989), for instance, suggest replacing Taguchi's arrays with a sequential test plan consisting of a FFD followed by a Central Composite Design (see Chapter 11) to give more information per unit of cost. They also propose using Response Surface Method techniques (see Chapter 11) to replace ANOVA and Decision Analysis techniques to find the best combination of location and dispersion parameters. Box & Jones (1992) examine how split-plot designs could be used in preference to Taguchi's parameter designs advocating that use of such designs could reduce the amount of experimentation as well as providing more efficient parameter estimation. Myers *et al.* (1992) describe augmentations of the Taguchi approach by combining it with Response Surface Method techniques to exploit the best of both optimisation methods. They advocate using a dual response surface approach involving designing an experiment with the control and noise factors in the same structure. Instead of *SN* ratio analysis, methods are suggested for fitting models for both response mean and variance. They also point out the possibility of using Central Composite Designs instead of two- or three-level factorial arrays to cut down the number of experimental runs. An article by Tuck *et al.* (1993) provides an illustration of this approach from the milling industry. Shoemaker & Tsui (1993), on the other hand, advocate methods for fitting models for response variance only using variability measures in conjunction with control × noise interaction plots to determine which levels of the control factors should be chosen to produce the most robust product/process.

From the outline provided, it can be seen that Taguchi methods provide an alternative experimental approach to traditional experimental design. They can be used to identify those design parameter settings (factor levels) at which the effect of noise on the response is minimised and which reduce cost without sacrificing quality. Though the philosophy is sound, it is the underlying statistical aspects which cause the controversy surrounding the application of Taguchi methods. Implementation of these methods may provide useful results but because of interaction aliasing, we may not fully understand how the improvement has occurred. We may therefore have a solution but our knowledge of the overall process would not be extensive enough to avoid future problems.

Many successful applications of Taguchi methods have been reported. Most have occurred in high volume manufacturing processes where previous uses of experimental design strategies were not common. Best guess and OFAT techniques were often the approaches adopted. Such techniques are very much hit or miss and so the application of Taguchi methods, with their strong factorial basis, were likely to lead to the identification of improvements.

Taguchi methods can also lead to large experiments. For instance, Example 10.3 illustrated one requiring 72 runs though only seven factors were assessed. This makes such design structures potentially useful for high volume processes but very much less so for low volume, specialised processes such as occur in the chemical and process industries, the electronics industry, and in instrument optimisation in an analytical laboratory. Given the amount of experimentation often associated with Taguchi methods, use of small run FFDs and Response Surface designs would provide more cost effective and efficient alternatives for robust product and process development.

PROBLEMS

10.1 An electrical engineer conducted an experiment based on Taguchi's $L_8(2^7)$ array into the output voltage of an electronic device. Four two-level factors were identified to be the control factors with column allocation based on A to column 1, B to 2, C to 4, and D to 7. Through this allocation, estimation of three interaction effects is possible with columns 3, 5, and 6 providing the means of measuring the $A \times B$, $A \times C$, and $B \times C$ interactions, respectively. Three noise factors were also identified and varied with an $L_4(2^3)$ array. The parameter design matrix and the four replicate voltage measurements, in kV, are presented. Check out the alias structure for main effects and two factor interactions. Carry out a full and relevant analysis of the presented data for all three possible performance statistics based on "nominal is best" quality criterion where target voltage should be in the range 24.5 to 25 kV. Is there an optimal combination suggested?

Run	1 A	2 B	3	4 C	5	6	7 D	Voltage data
1	1	1	1	1	1	1	1	20.5 21.7 22.6 21.5
2	1	1	1	2	2	2	2	21.4 22.2 20.8 21.5
3	1	2	2	1	1	2	2	26.8 27.6 28.4 27.3
4	1	2	2	2	2	1	1	24.4 25.7 24.8 25.1
5	2	1	2	1	2	1	2	22.6 23.3 22.9 23.7
6	2	1	2	2	1	2	1	24.7 24.5 26.4 25.6
7	2	2	1	1	2	2	1	25.3 23.7 24.8 24.4
8	2	2	1	2	1	1	2	21.8 23.7 22.6 24.5

10.2 A Taguchi experiment based on a saturated $L_9(3^4)$ array was conducted to investigate the effect of four factors on the bonding strength of an epoxy resin. The factors chosen for experimentation were composition of bond (A), bonding method (B), surface treatment (C), and operator (D). All factors were tested at three levels with two pieces bonded in each experiment. Measured data for bonding strength, in tons, are given below with maximum considered ideal. If the present levels are all level 2, i.e. centre level, is there an improvement with the suggested optimal?

Factor	A	B	C	D	Bonding strength data
	1	1	1	1	6.80 2.27
	1	2	2	2	2.49 3.43
	1	3	3	3	2.17 1.57
	2	1	2	3	1.79 1.33
	2	2	3	1	1.98 2.57
	2	3	1	2	2.93 2.72
	3	1	3	2	1.70 2.12
	3	2	1	3	4.24 1.91
	3	3	2	1	1.50 4.05

Reprinted from Taguchi, G. (1986) *Introduction to Quality Engineering*. Asian Productivity Organisation, Japan with the permission of the Asian Productivity Organisation.

10.3 The bonding strength of a plastic product was assessed using Taguchi's $L_{27}(3^{13})$ array in order to ascertain optimum bonding conditions. The factors chosen for experimentation were: etch time (A) at 5 min, 10 min, and 15 min; etch temperature (B) at 60 °C, 65 °C, and 80 °C; etchant composition (C) at C_1, C_2, and C_3; preprocessing (D) at none, solvent, and warm water; accelerator (E) at E_1, E_2, and E_3; catalyst (F) at present, proposed$_1$, and proposed$_2$; and neutralising method (G) at G_1, G_2, and G_3. All of the main effects and the two factor interactions among the factors A, B, and C were assigned to the $L_{27}(3^{13})$ array. Factor allocation was A to column 1, B to 2, C to 5, D to 9, E to 10, F to 12, and G to 13. Columns 3 and 4 were used for the $A \times B$ interaction, columns 6 and 7 for the $A \times C$ interaction, and columns 8 and 11 for the $B \times C$ interaction. For the $A \times B$ interaction, this meant that all nine interaction combinations were tested, these being, in order from the parameter design matrix, A_1B_1, A_2B_2, A_3B_3, A_2B_3, A_3B_1, A_1B_2, A_3B_2, A_1B_3, and A_2B_1. The same applies for the other two interactions. Two test pieces were manufactured for each set of conditions and a pull-off test performed yielding the presented bonding strength, measurede in kg cm^{-2}. Maximum bond strength is considered optimal. Carry out a relevant analysis of these data. Is an optimal combination of factors forthcoming?

Factors	*A*	*B*			*C*				*D*	*E*		*F*	*G*		
Column	1	2	3	4	5	6	7	8	9	10	11	12	13	Strength data	
	1	1	1	1	1	1	1	1	1	1	1	1	1	6	5
	1	1	1	1	2	2	2	2	2	2	2	2	2	10	8
	1	1	1	1	3	3	3	3	3	3	3	3	3	10	12
	1	2	2	2	1	1	1	2	2	2	3	3	3	3	10
	1	2	2	2	2	2	2	3	3	3	1	1	1	18	18
	1	2	2	2	3	3	3	1	1	1	2	2	2	23	18
	1	3	3	3	1	1	1	3	3	3	2	2	2	9	13
	1	3	3	3	2	2	2	1	1	1	3	3	3	33	30
	1	3	3	3	3	3	3	2	2	2	1	1	1	29	29
	2	1	2	3	1	2	3	1	2	3	1	2	3	6	8
	2	1	2	3	2	3	1	2	3	1	2	3	1	7	11
	2	1	2	3	3	1	2	3	1	2	3	1	2	23	24
	2	2	3	1	1	2	3	2	3	1	3	1	2	1	1
	2	2	3	1	2	3	1	3	1	2	1	2	3	31	31
	2	2	3	1	3	1	2	1	2	3	2	3	1	32	35
	2	3	1	2	1	2	3	3	1	2	2	3	1	16	20
	2	3	1	2	2	3	1	1	2	3	3	1	2	32	35
	2	3	1	2	3	1	2	2	3	1	1	2	3	29	32
	3	1	3	2	1	3	2	1	3	2	1	3	2	1	1
	3	1	3	2	2	1	3	2	1	3	2	1	3	37	34
	3	1	3	2	3	2	1	3	2	1	3	2	1	33	28
	3	2	1	3	1	3	2	2	1	3	3	2	1	13	16
	3	2	1	3	2	1	3	3	2	1	1	3	2	37	35
	3	2	1	3	3	2	1	1	3	2	2	1	3	31	33
	3	3	2	1	1	3	2	3	2	1	2	1	3	28	28
	3	3	2	1	2	1	3	1	3	2	3	2	1	35	38
	3	3	2	1	3	2	1	2	1	3	1	3	2	36	35

Reprinted from Taguchi, G. (1986) *Introduction to Quality Engineering*. Asian Productivity Organisation, Japan with the permission of the Asian Productivity Organisation.

10.4 A Taguchi experiment was conducted into the breaking strength of a moulded plastic part used in the carburettor of a lawn mower engine. The plastic must withstand the solvent effects of fuel and the pressures of the return spring in the automatic choke mechanism. The parameter design matrix was based on the $L_{27}(3^{13})$ array with the noise matrix based on the $L_{18}(3^6)$ array. The parameter design matrix contained six factors based on the column allocation of feed rate to column 1, first screw speed to 2, second screw speed to 3, gate size to 9, first temperature to 10, and second temperature to 12. The noise matrix consisted of six three-level noise factors. The parameter design matrix and *SN*, mean, and standard deviation (s) summaries from the 18 replications of each treatment combination are presented. Assess each performance statistic and draw relevant conclusions.

Feed rate (gms/min)	First screw speed (rpm)	Second screw speed (rpm)	Gate size (000s from nominal)	First temp (°F)	Second temp (°F)	*SN*	Mean	*s*
1000	400	850	−30	280	320	35.9	87.4	31.5
1000	400	900	0	320	360	40.5	115.6	24.6
1000	400	950	+30	360	400	36.8	106.2	36.3
1000	440	850	0	320	400	37.3	101.5	34.6
1000	440	900	+30	360	320	39.1	117.6	35.4
1000	440	950	−30	280	360	40.5	115.2	24.3
1000	480	850	+30	360	360	41.4	131.1	30.7
1000	480	900	−30	280	400	36.8	93.9	35.4
1000	480	950	0	320	320	40.8	134.5	35.5
1200	400	850	0	360	360	40.5	111.6	21.9
1200	400	900	+30	280	400	39.2	108.6	30.5
1200	400	950	−30	320	320	40.0	111.9	29.3
1200	440	850	+30	280	320	39.3	105.7	28.0
1200	440	900	−30	320	360	41.1	118.3	20.1
1200	440	950	0	360	400	41.5	133.1	34.0
1200	480	850	−30	320	400	38.8	104.1	33.2
1200	480	900	0	360	320	42.3	144.5	35.4
1200	480	950	+30	280	360	42.2	133.5	21.4
1400	400	850	+30	320	400	33.2	82.5	41.6
1400	400	900	−30	360	320	29.5	85.8	42.0
1400	400	950	0	280	360	40.4	120.4	33.5
1400	440	850	−30	360	360	38.0	99.3	36.7
1400	440	900	0	280	400	36.2	99.1	41.5
1400	440	950	+30	320	320	38.8	115.3	41.1
1400	480	850	0	280	320	34.1	96.2	40.9
1400	480	900	+30	320	360	40.4	121.6	36.7
1400	480	950	−30	360	400	39.2	120.8	46.4

Reprinted from Barker, T. B. (1986) Quality engineering by design: Taguchi's philosophy. *Quality Progress* 19 32–42 with the permission of the American Society for Quality Control.

10A.1 Appendix: Software Information for Two-level Taguchi Designs

Data entry

Data entry for two-level Taguchi designs follows similar principles to that for the two-level designs discussed throughout this text though the 1/2 level coding system is used for specification of factor levels. Data entry for Example 10.2 would therefore look exactly as that shown in Table 10.5. The control factors will comprise seven variables and the replicate data three variables. Thus, the input data set contained 10 variables. Provision of full replicate data is necessary for performance statistic calculation with the replicate observations referred to as R1, R2, and R3. If a noise matrix is included, the recorded replicate data would be entered in a similar manner for each treatment combination.

Performance statistic calculation

Generally, software have sufficient in-built commands to carry out the necessary performance statistic calculations. Irrespective of availability of specific Taguchi features, performance statistic derivation is useful to enable summaries of the performance statistics to be derived for use in the data analysis.

SAS: For Example 10.2, the following statements will generate the statistics required:

```
DATA CH10.EX102;
     SET CH10.EX102;
     SN = –10*LOG10(USS(OF R1-R3)/3);
     XBAR = MEAN(OF R1-R3);
     S = STD(OF R1-R3);
     LNS = LOG(S);
RUN;
```

The statements use the DATA step with the SET option to calculate and store the required performance statistics. SET CH10.EX102 specifies that the additional calculated data are to be appended to the original data set giving a data set of 14 (10 + 4) variables. The term USS(OF R1-R3) in the SN statement corresponds to $\sum y^2$ in expression (10.4) and therefore provides the sum of the squares of the three replicate observations across each experimental run. The other three statements involving summary statistic calculations can be similarly explained. In SAS, LOG will generate the natural logarithm while LOG10 provides log to base 10. The statistics generated will comply with those in Table 10.7.

Example 11.2 was concerned with minimisation and so the SN_S expression (10.4) for SN ratio evaluation was used. For SN_N (expression (10.2)), assuming three replicate observations, we would use the statement

```
SN = 20*LOG10(MEAN(OF R1-R3)/STD(OF R1-R3));
```

while for SN_L (expression (10.3)), we would need to use the Data step statements

```
R4 = 1/R1;
R5 = 1/R2;
R6 = 1/R3;
SN = –10*LOG10(USS(OF R4-R6)/3);
```

The first three statements are necessary to generate the reciprocal $1/y$ of each response measurement to enable the $\sum(1/y^2)$ component of expression (10.3) to be determined through use of USS(OF R4-R6).

Minitab: In Minitab, the **Calc** menu provides access to Minitab's calculation procedures for performance statistic derivation. Columns C1 to C10 will contain the experimental information with columns C11 to C14 used for performance statistic storage. For the *SN* ratio (10.4), we would use

Select **Calc** ➤ **Row Statistics** ➤ for *Statistic*, select **Sum of Squares** ➤ select the **Input variables** box, highlight **R1 to R3**, and click **Select** ➤ select the **Store Results in** box and enter **C15** ➤ click **OK**.

Select **Calc** ➤ **Calculator** ➤ for *Store result in variable*, enter **SN** ➤ select the **Expression** box and enter **–10*LOGT(C15/3)** ➤ click **OK**.

Similar use of the **Calc ➢ Row Statistics** with change of *Statistic* selection to **Mean** and **Standard Deviation** and change of storage column to Mean and *s*, respectively, will provide the other response summaries. The ln(*s*) statistic is then found using

> Select **Calc ➢ Calculator ➢** for *Store result in variable*, enter **Ln(s) ➢** select the **Expression** box and enter **LOGE(S) ➢** click **OK**.

In Minitab, LOGE will generate the natural logarithm while LOGT provides log to base 10.

For SN_N (expression (10.2), based on three replicate observations, we would use the **Calc ➢ Calculator** selection with storage entry of **SN** and change of **Expression** entry to

20*LOGT(MEAN/S)

where the MEAN and S columns contain the row mean and row standard deviation for each tested treatment combination. For SN_L (expression (10.3)), based on three replications, we would first use the selection

> Select **Calc ➢ Calculator ➢** for *Store result in variable*, enter **R4 ➢** select the **Expression** box and enter **1/R1 ➢** click **OK**.

This would be repeated for replications R2, producing R5, and R3, producing R6. The required *SN* ratio could then be evaluated using the **Calc ➢ Calculator** selection with storage entry of **SN** and change of **Expression** entry to

−10*LOGT(R4**2 + R5**2 + R6**2)/3)

Effect estimation and plotting

SAS: In SAS, there is no direct procedure for evaluation of effect estimates and generation of effect plots for Taguchi designs. However, by developing a simple sequence of statements along the lines explained in Appendix 7A for two-level FDs, it is possible to generate the associated effect estimates and normal plots for each performance statistic. The only modifications required to the statements provided in Appendix 7A would be to the MODEL and KEEP statements to suit the factors accounted for in the Taguchi design run. For Example 10.2, the MODEL statement would read

```
MODEL SN = A B C D E F G;
```

while the KEEP statement would read

```
KEEP A B C D E F G EFFEST;
```

By changing the response SN in the MODEL statement, the same sequence of statements could be used to generate effects information for the mean (XBAR) and ln(*s*) (LNS) performance statistics.

Minitab: The **Stat ➢ DOE ➢ Analyze Custom Design** menu provides access to data plotting procedures for two-level Taguchi designs. The selection used to produce Displays 10.3 to 10.6 was:

Select **Stat** ➢ **DOE** ➢ **Analyze Custom Design** ➢ for *Design*, select **Inner/outer array** ➢ **Fit Model** ➢ for *Response data are in*, select **R1 to R3** and click **Select** ➢ for *Inner array data are in*, select the box and enter **A B C D E F G**.

Select **Options** ➢ for *Signal to Noise Ratio*, select **Smaller is better** ➢ click **OK**.

Select **Graphs** ➢ **Main effects and interactions for Signal to Noise ratios** ➢ **Main effects and interactions for means** ➢ **Main effects and interactions for standard deviations** ➢ ensure that the box for **Use ln(s) in graphs for standard deviations** is active (box crossed) ➢ click **OK**.

Select **Storage** ➢ for *Store the following items*, select **Signal to Noise ratios**, **Means across rows of response data**, and **Standard deviations across rows of response data** ➢ click **OK** ➢ click **OK**.

The **Storage** selection can be used as an alternative to the performance statistic calculation method described earlier though the procedure for calculation of ln(s) still requires to be implemented as this Minitab selection does not store values for ln(s).

Data plots

Main effect plots can also be produced as explained in Appendices 7A, 8A, and 9A using the input factor levels and calculated performance statistics. In addition to main effect plot derivation, the **Stat** ➢ **ANOVA** ➢ **Interactions Plot** menu in Minitab also provides access to interaction plotting facilities which can generate a matrix of interaction plots such as shown in Display 10.6 to assess how aliased interaction effects affect the performance statistics.

ANOVA modelling

Derivation of ANOVA information, such as that displayed in Table 10.8, is as previously explained in Appendices 7A, 8A, and 9A for two-level screening experiments.

10A.2 Appendix: Software Information for Three-level Taguchi Designs

Data entry

Data entry for three-level Taguchi designs requires to be provided in two forms. The first will correspond to the parameter design matrix and observation presentation provided in the data tables such as Table 10.12. This form of data entry is useful for performance statistic calculation and for control factor assessment. The second form of data entry is necessary if control × noise interaction plots are to be included in the data analysis. It involves entering the parameter design matrix a number of times according to the number of run replications carried out. The noise matrix would then be entered for each series of repetitions with a single response corresponding to each combination of control and noise factors. For Example 10.3, this would produce

Control factors				Noise factors			Pull-off force
A	B	C	D	E	F	G	response
1	1	1	1	1	1	1	15.6
1	1	1	1	1	1	2	9.5
1	1	1	1	1	2	1	16.9
.	.	.	.	.	.	.	.
1	2	2	2	1	1	1	15.0

(continued)

(continued)

1	2	2	2	1	1	2	16.2
1	2	2	2	1	2	1	19.4
.	.	.	.	.	.	.	.

Data plots

The main effect plots of the performance statistics, provided in Displays 10.7 to 10.9, used the first form of data entry and were produced using Minitab's macro for main effects plotting (see Appendices 7A, 8A, and 9A) with plot editing before presentation. Inclusion of control × noise interaction assessment utilised the second form of data entry as its base. The plots presented in Displays 10.10 and 10.11 were produced using Minitab's interaction plotting facility (see Appendices 7A and 8A) and contain some plot editing.

ANOVA modelling

Derivation of ANOVA information, if feasible, will be as previously explained for two-level screening experiments (see Appendices 7A, 8A, and 9A).

11

Response Surface Methods

11.1 INTRODUCTION

Response Surface Methods (**RSM**) are powerful experimental design tools applicable to product development, process improvement, and optimisation. They represent a collection of experimental design and multiple regression based techniques that can be used to analyse problems where several factors may influence a response, and where the goal is either to optimise or understand system performance. Their use within optimisation may follow a sequence of FFDs (see Chapter 8) through which the most important factors have been identified. The response model within such designs can be either first- or second-order, the latter being particularly useful if interaction or curvature within a response is expected.

Response surface techniques were developed in the 1950s for the purpose of determining optimal operating conditions for production processes in the chemical industry (Box & Wilson 1951). Later work concentrated on the analysis principles for RSM strategies together with aspects of design structure such as rotatability, robustness, and optimality.

Application areas for RSM are diverse. Chemists have used them to help optimise analytical procedures such as those associated with gas chromatography and assays. The food industry has made use of them to optimise the properties of foodstuffs and beverages. RSM strategies have also found application in the biological and clinical sciences within pollution studies and cancer research. Industrial usage of RSM techniques is widespread in both research and development and in optimisation of processes particularly in relation to improving product quality.

RSM techniques are based on sequential procedures using two- and three-level FDs to identify important factors through characterising the response surface by a polynomial model and using the model to indicate the direction of future exploration for the likely optimum. Subsequent experiments on specified treatment combinations are then used to home in on a region of factor levels that appear most likely to produce optimum response. Such applications of experimental design and analysis can lead to the vicinity of the optimum by rapid and systematic means.

The relationship between the response variable Y and the factors $X_1, X_2, \ldots, X_k$ in RSM applications is generally expressed as

$$Y = f(X_1, X_2, \ldots, X_k) + \varepsilon \qquad (11.1)$$

where ε represents the noise or error observed in the response Y. The function f in expression (11.1) is called the **response surface** and corresponds to an approximation of the response function for the measured response as the exact nature of this functional relationship is generally unknown. Using experimental data, we attempt to predict the direction of movement of the factor levels to achieve the optimal response. This can result in unimodal (one optimum) or multi-modal response surfaces (local and global optima) which can be represented graphically by means of surface or contour plots.

Experimental design structures associated with RSM studies include **Central Composite Design**, **face-centred cube design**, and **Box-Behnken design**, all of which stem from the two-level multi-factor FDs described earlier. Such designs are appropriate for estimation of main effects, two factor interactions, and the quadratic effect of experimental factors. Replication through inclusion of centre points to aid trend and error estimation is a key feature.

Choice of design structure for data collection in RSM studies, referred to as a **Response Surface Design**, is of fundamental importance. Most are based on two- and three-level FDs. When selecting a design for a response surface experiment, it is desirable that it has at least some of the following features:

- the treatment combinations provide good coverage of the experimental region;
- the experiment can be carried out using as few runs as possible;
- the data collected provide statistically valid estimates of all model parameters and enable model fit to be assessed (lack-of-fit);
- the structure must enable an adequate estimate of error to be obtained, generally through use of replication (inclusion of centre points);
- the design must enable specific variance criteria for parameter and response estimates to be met.

The last feature is best served by using orthogonal two-level designs which minimise the variance of the estimated regression coefficients and enable independent estimation of all model parameters. There are also several **robustness** features worthy of consideration within design choice. These include robustness to outliers, robustness to errors in factor levels, and robustness to model extrapolation under conditions of model misspecification. It may not always be possible, however, for a proposed Response Surface (RS) design to have all these desirable features. Box & Draper (1987), Khuri & Cornell (1987), and Myers & Montgomery (1995) provide further background on design choice criteria and aspects of analysis for RSM studies.

11.2 FIRST-ORDER DESIGN STRUCTURES

Orthogonal designs associated with fitting first-order response surface models include **two-level Factorial Designs**, **two-level Fractional Factorial Designs** where main effects are not aliased with each other, and **Simplex Designs**. These designs satisfy different combinations of the desirable properties of RS designs.

11.2.1 Two-level Design

For two factors, a **two-level FD** would be a useful design base but without replication of certain runs, no estimation of experimental error can be forthcoming. Use of **centre points** (see Section 7.5.3) is the common way in which replication is included in RS designs to provide for error estimation. Centre points refer to treatment combinations corresponding to factor levels equi-distant from the "low" and "high" settings of the factors. Parameter estimation and design orthogonality are unaffected by inclusion of centre points.

In most RS designs, factor levels are generally **coded** by means of the expression

$$\text{coded value} = (\text{factor level} - M)/S \qquad (11.2)$$

where M is the average of the "low" and "high" levels (the centre) and S is half the difference between the "high" and "low" levels. Based on this form of coding, −1 will represent "low" level, 0 centre level, and +1 "high" level. A possible orthogonal RS design structure for two factors, based on this form of coding, is shown in Table 11.1 and corresponds to a 2^2 FD augmented with five centre points. The number of points in the factorial portion is generally denoted n_f with the number of centre points given by n_c. For the illustration in Table 11.1, we have $n_f = 4$ and $n_c = 5$. A pictorial display of this design is presented in Fig. 11.1 and shows that the experimental region is a square with the centre points corresponding to points at the centre of the design.

Table 11.1 — Two factor first-order Response Surface Design with centre point replication

	Factorial portion		*Centre portion*	
Coded factors	X_1	X_2	X_1	X_2
	−1	−1	0	0
	+1	−1	0	0
	−1	+1	0	0
	+1	+1	0	0
			0	0

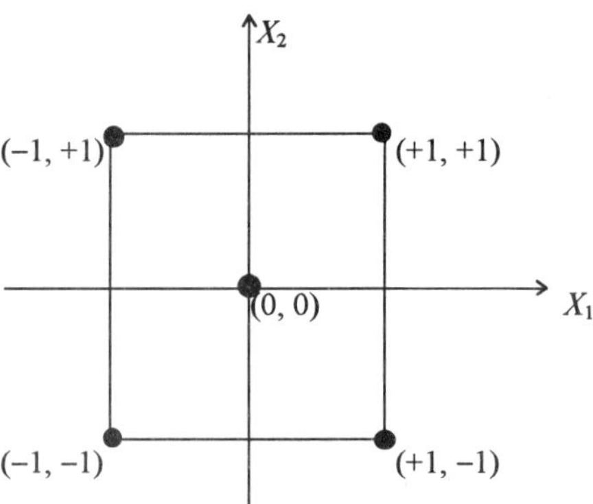

Fig. 11.1 — First-order Response Surface Design for two factors

Example 11.1 A chemist wished to maximise the percent purity of yield from a chemical reaction. Two controllable factors were identified: temperature, in °C, and pH. To investigate the problem, a first-order design was based on a region of exploration of temperature (47.5 °C, 52.5 °C) and pH (3.5, 4.0). The experimental design is shown in

Table 11.2 and corresponds to a 2^2 FD ($n_f = 4$) augmented with five centre points ($n_c = 5$). The coded variables are expressed as

$$X_1 = (\text{temperature} - 50)/2.5 \text{ and } X_2 = (\text{pH} - 3.75)/0.25 \quad .$$

Table 11.2 — Purity data from first-order Response Surface design for Example 11.1

Factorial portion					*Centre portion*				
Temperature	pH	X_1	X_2	Purity, Y	Temperature	pH	X_1	X_2	Purity, Y
47.5	3.5	−1	−1	82.8	50	3.75	0	0	84.7
52.5	3.5	+1	−1	85.8	50	3.75	0	0	85.0
47.5	4.0	−1	+1	84.1	50	3.75	0	0	85.4
52.5	4.0	+1	+1	86.9	50	3.75	0	0	84.5
					50	3.75	0	0	85.2

11.2.2 Simplex Design

The **Simplex Design** is an orthogonal structure of particular use when seeking an optimum for k experimental factors. It represents a geometrical figure with $(k + 1)$ vertices in k dimensions. For $k = 2$ variables, the experimental region is an equilateral triangle while for $k = 3$ variables, it represents a regular tetrahedron (Montgomery 1997).

11.3 RESPONSE SURFACE ANALYSIS FOR FIRST-ORDER DESIGNS

As stated earlier, RSM techniques were developed for the purpose of determining optimal factor conditions for process operation. RS analysis incorporates flexible and sequential steps to achieve this goal. The first step involves fitting the simplest polynomial model, generally a first-order model, to a set of data collected at the points of a first-order design. If the fitted first-order model is adequate, the information provided by it is used to locate areas within, or outside, the experimental region, but within the boundaries of the operability region, where more desirable values of the response are suspected to be.

In the new region, the cycle is repeated in that the first-order model is fitted and tested for adequacy of fit. If at any stage of this process lack-of-fit is detected, a test for interaction and/or curvature can be conducted. If significance of curvature is found, a second-order model can be fitted and, if it is adequate, used to map or describe the shape of the surface in the experimental region. If the optimal or most desirable response values are found, then locating the best values as well as the settings of the input variables that produce the best response values should be the next step in the analysis.

11.3.1 First-order Response Model

The response model (10.1) for a first-order RS study is expressed in the form of a multiple linear model of the response Y as a function of the k experimental factors. This can be represented as

$$Y = \beta_0 + \beta_1 X_1 + \beta_2 X_2 + \ldots + \beta_k X_k + \varepsilon = \beta_0 + \sum_{i=1}^{k} \beta_i X_i + \varepsilon \qquad (11.3)$$

where k is the number of experimental factors, β_0 is the constant term, β_i is the coefficient associated with factor i, and ε is the random error. Such a model therefore assumes that the coded variables X_1, X_2,..., and X_k have an additive and independent effect on the response.

Consider an experiment involving two factors X_1 and X_2 where the simplest model we can propose is a plane

$$Y = \beta_0 + \beta_1 X_1 + \beta_2 X_2 + \varepsilon.$$

Providing the tested levels of the two factors are sufficiently close, such a model will be a reasonable approximation to the true response surface for the measured response. As the model contains three unknown parameters β_0, β_1, and β_2, then the associated design structure for data collection must contain at least three treatment combinations in order to be able to furnish statistically feasible parameter estimates.

For three factors, a first-order response model would be

$$Y = \beta_0 + \beta_1 X_1 + \beta_2 X_2 + \beta_3 X_3 + \varepsilon$$

containing four unknown parameters. A minimum of four treatment combinations would therefore need to be included in the associated RS design to estimate such a response surface.

11.3.2 Parameter Estimation

Least squares estimation, as in regression modelling, is the technique used to derive estimates b_i of the β_i parameters within the first-order response model (11.3). Inclusion of centre points does not change the parameter estimates obtained as least squares estimation only uses the factorial portion of the design to estimate model parameters.

Parameter estimation for the model (11.3) is based on using effect contrasts comparable to those outlined in Section 7.2.1 for two-level designs. Based on the specified coding of factor levels (see expression (11.2)), this means that $b_0 = \overline{Y}$ (mean response) and

$$b_i = \tfrac{1}{2}(\text{effect of factor } i) = \tfrac{1}{4}(\text{contrast}_i) = \Sigma X_i Y / \Sigma X_i^2 \tag{11.4}$$

where $\Sigma X_i Y$ refers to the product of the code values for the ith variable X_i and the recorded response Y at these settings.

For the first-order RS design illustrated in Table 11.1, a model of the form

$$Y = \beta_0 + \beta_1 X_1 + \beta_2 X_2 + \varepsilon$$

would be proposed. Parameter estimates would be $b_0 = \overline{Y}$,

$$b_1 = \tfrac{1}{4}(\text{contrast}_A) = \tfrac{1}{4}(a + ab - (1) - b) = \Sigma X_1 Y / \Sigma X_1^2$$

and

$$b_2 = \tfrac{1}{4}(\text{contrast}_B) = \tfrac{1}{4}(b + ab - (1) - a) = \Sigma X_2Y / \Sigma X_2^2 \quad .$$

Exercise 11.1 The chemical purity study of Example 11.1 represents a first-order RS design application. We must first specify the response model that will be fitted to the data and use the methods described above to estimate the parameters of the model.

Response model: As the design is of first-order type, then the response model for purity will be

$$Y = \beta_0 + \beta_1 X_1 + \beta_2 X_2 + \varepsilon$$

where β_0 is the constant term, β_i is the regression coefficient of factor i (i = 1 temperature, i = 2 pH), X_i is the coded form of factor i, and ε defines the random error. We wish to use the collected data to estimate this model and use the features of the fitted model to assess what the data specify on the influence of the factors on chemical purity.

Manual derivation: From Table 11.2, we have (1) = 82.8, a = 85.8, b = 84.1, and ab = 86.9. In addition, we have

$$\Sigma X_1Y = (-1)\times 82.8 + (+1)\times 85.8 + (-1)\times 84.1 + (+1)\times 86.9 + (0)\times 84.7 + \ldots + (0)\times 85.2 = 5.8,$$

$\Sigma X_1^2 = 4$, $\Sigma X_2Y = 2.4$, $\Sigma X_2^2 = 4$, and $\overline{Y} = 84.933$. Thus, $b_0 = 84.933$,

$b_1 = \tfrac{1}{4}(85.8 + 86.9 - 82.8 - 84.1) = 1.45$, and $b_2 = \tfrac{1}{4}(84.1 + 86.9 - 82.8 - 85.8) = 0.6$

so the fitted model reads

$$\text{purity} = 84.933 + 1.45\,X_1 + 0.6\,X_2 \quad .$$

Coefficient estimates suggest unit change in coded temperature (X_1) causes greater change in purity than unit change in coded pH (X_2) and thus temperature appears to be exerting a greater influence on chemical purity.

Software generation: As with previous design applications, software usage for RS studies is desirable not only to simplify the calculation process but also to provide access to RS data analysis features. Display 11.1 shows the Minitab generated first-order RS output for this study. Parameter estimation, specified in the 'Coef' column, is identical to that obtained above. The other aspects of the output obtained will be discussed later. □

11.3.3 ANOVA Table

As RS models are essentially special cases of regression models, we can use the **ANOVA principle** to split response variability into controlled and uncontrolled components, the former defining the variation due to the response model fitted and the latter the error variation. Sum of squares (*SS*) and mean square (*MS*) terms again underpin this approach with the overall *SS* term, *SS*Total, expressed as

$$SS\text{Total } (SST) = SS\text{Regression } (SSRegn) + SS\text{Error } (SSE) \qquad (11.5)$$

Display 11.1

```
Term              Coef       StDev         T        P
Constant       84.9333      0.1009   742.490    0.000
X1              1.4500      0.1514     9.578    0.000
X2              0.6000      0.1514     3.963    0.007

S = 0.3028       R-Sq = 94.7%      R-Sq(adj) = 92.9%

Analysis of Variance for Purity
Source           DF      Seq SS      Adj SS       Adj MS       F       P
Regression        2      9.8500     9.85000      4.92500   53.73   0.000
  Linear          2      9.8500     9.85000      4.92500   53.73   0.000
Residual Error    6      0.5500     0.55000      0.09167
  Lack-of-Fit     2      0.0180     0.01800      0.00900    0.07   0.936
  Pure Error      4      0.5320     0.53200      0.13300
Total             8     10.4000
```

The calculation aspects and general **ANOVA table** for the first-order RS model (11.3) are presented in Table 11.3. The statistical components for a first-order model are, as usual, based on F test procedures using appropriate ratios of mean squares. These will be discussed more fully in Sections 11.3.4 and 11.3.5. For the purity study of Example 11.1, the associated ANOVA information is provided in Display 11.1 in a similar format.

Table 11.3 — ANOVA table for first-order Response Surface model of a response Y as a function of k experimental factors

Source	df	SS	MS
Regression	k	$SSRegn = \underline{b}'X'Y - (\sum y)^2/N$	$MSRegn = SSRegn/df$
Error	$N - k - 1$	SSE	$MSE = SSE/df$
Lack-of-fit	$N - k - n_r$	$SSLF = SSE - SSPE$	$MSLF = SSLF/df$
Pure error	$n_r - 1$	$SSPE = \sum y_c^2 - (\sum y_c)^2/n_c$	$MSPE = SSPE/df$
Total	$N - 1$	$SST = \sum y^2 - (\sum y)^2/N$	

k is the number of experimental factors, N is the number of data points, Y is an $n\times1$ column vector of responses, X is an $n\times(k + 1)$ matrix of the X settings (first column being n 1s), $\underline{b}$ is a $(k + 1)\times1$ column vector of the unknown regression coefficients, y are the experimental responses, n_r is the number of treatment combination replications, y_c are the responses at the n_c centre points, df defines degrees of freedom, SS defines sum of squares, MS defines mean square

11.3.4 Adequacy of Fit

Once a first-order model is fitted to the response data, an important aspect of its assessment involves checking the adequacy of its fit. This is carried out first at a general level and then, if necessary, at a more specific level corresponding to trend effects, e.g. interaction and curvature. These checks are designed to investigate how the fitted equation both affects and is affected by the experimental factors.

11.3.4.1 Lack-of-fit

Use of centre points provides the replication necessary for experimental error estimation. This, in turn, allows model **lack-of-fit** to be assessed through the partitioning of the error sum of squares (SSE) into two components: lack-of-fit (LF) and pure error (PE), i.e. $SSE = SSLF + SSPE$. If a fitted model is an adequate fit, $SSLF$ will be low relative to $SSPE$ signifying that more of the error appears to be due to experimental uncertainty rather than an accountable trend. If $SSLF$ is high relative to *SSPE*, then the adequacy of the fitted model is in doubt since most of the error appears due to a definitive trend.

A simple statistical test can be performed to assess for lack-of-fit. The test is based on the null hypothesis of H_0: response surface first-order. In a first-order RS model, we have k model components (the factors) and N responses. If any of the treatment combinations are replicated, then n_r is the number of data associated with these points. In RS studies, often only the centre point is replicated and so $n_r = n_c$. The lack-of-fit test statistic is given by

$$F = MSLF/MSPE \tag{11.6}$$

with degrees of freedom $df_1 = N - k - n_r$ and $df_2 = n_r - 1$.

A low value of test statistic (11.6) is indicative of model adequacy. A high value, by contrast, suggests model adequacy has not been demonstrated and that checks to ascertain how lack-of-fit is occurring either through the presence of interaction or curvature or both is necessary. Regression diagnostics could also be useful if the adequacy of the model is in doubt.

Exercise 11.2 A first-order response surface model was fitted to the purity data of Example 11.1 (see Exercise 11.2). The first stage of model analysis concerns the assessment of lack-of-fit.

Lack-of-fit check: Display 11.1 contains the information relevant to this check. The implemented first-order design provides $k = 2$, $N = 9$, and since the centre point is replicated five times, $n_c = 5 = n_r$. From Display 11.1, we have $MSLF = 0.009$, $df_1 = 2$, $MSPE = 0.133$, and $df_2 = 4$. The F test statistic (11.6) is estimated as 0.07 with p value 0.936. The high p value indicates there appears no evidence to say lack-of-fit is occurring so the fitted first-order model would appear to be an acceptable fit ($p > 0.05$). □

11.3.4.2 Interaction

If lack-of-fit is significant, it could be due to the presence of **interaction**, or **crossproduct**, effect between the factors. The number of possible interaction components is defined as n_i. For $k = 2$ factors, only one interaction term X_1X_2 is possible so $n_i = 1$ while for $k = 3$ factors, three interaction terms X_1X_2, X_1X_3, and X_2X_3 are possible so $n_i = 3$. Estimation of the parameters of interaction terms follows comparable rules to those outlined in Section 11.3.2 for the linear parameters of the first-order RS model (11.3), i.e.

$$b_{ij} = ½(\text{effect of interaction of factors } i \text{ and } j) = ¼(\text{contrast}_{ij}) \tag{11.7}$$

The sum of squares of the interaction component, *SS*Intn, is expressed as $n_f\Sigma\Sigma b_{ij}^2$ where n_f defines the number of design points in the factorial portion of the RS design. The associated mean square term is *MS*Intn = *SS*Intn/n_i. A simple test for assessing whether lack-of-fit is due to interaction effects can therefore be based on the test statistic

$$F = MS\text{Intn}/MSPE \tag{11.8}$$

with degrees of freedom $df_1 = n_i$ and $df_2 = n_r - 1$. High values of this test statistic will be indicative of presence of interaction specifying that the detected lack-of-fit may appear to be due to this trend.

As lack-of-fit was not significant for the purity study of Example 11.1, there is no need to assess for the presence of interaction effects. If it is necessary to assess for this trend, then modifying the choice of type of RS model to **Linear + interactions** within Minitab's menu procedure will provide the necessary ANOVA modification and test information generation. Display 11.2 illustrates this case for Example 11.1 and confirms that the contribution of the interaction term to model fit is not significant ('Interaction', $b_{12} = -0.05$, $n_f = 4$, $SS\text{Intn} = 0.01$, $n_i = 1$, $F = 0.09$, $p = 0.773$).

Display 11.2

```
Term              Coef       StDev        T        P
Constant       84.9333      0.1095   684.045   0.000
X1              1.4500      0.1643     8.824   0.000
X2              0.6000      0.1643     3.651   0.015
X1*X2          -0.0500      0.1643    -0.304   0.773

S = 0.3286       R-Sq = 94.8%     R-Sq(adj) = 91.7%

Analysis of Variance for Purity
Source            DF     Seq SS      Adj SS       Adj MS       F       P
Regression         3     9.8600     9.86000      3.28667   30.43   0.001
  Linear           2     9.8500     9.85000      4.92500   45.60   0.001
  Interaction      1     0.0100     0.01000      0.01000    0.09   0.773
Residual Error     5     0.5400     0.54000      0.10800
  Lack-of-Fit      1     0.0080     0.00800      0.00800    0.06   0.818
  Pure Error       4     0.5320     0.53200      0.13300
Total              8    10.4000
```

11.3.4.3 Curvature

Another possible reason for lack-of-fit of a first-order model could be the presence of curvature (quadratic effect) within the data which the fitted model is not explaining adequately. The number of quadratic terms possible is generally denoted n_q. For $k = 2$ variables ($n_f = 4$), there are only three, i.e. $n_f - 1$, degrees of freedom for parameter estimation. As two linear components require to be estimated, this leaves only one degree of freedom for other trend effects and so $n_q = 1$, either X_1^2 or X_2^2. For $k = 3$ variables ($n_f = 8$), there are seven degrees of freedom for parameter estimation so all quadratic terms X_1^2, X_2^2, and X_3^2 are possible for inclusion and $n_q = 3$.

A simple test of whether lack-of-fit is due to quadratic effects is based on the test statistic

$$F = MS\text{Quad}/MSPE \qquad (11.9)$$

with degrees of freedom $df_1 = n_q$ and $df_2 = n_r - 1$. MSQuad is calculated as $SS\text{Quad}/n_q$ where

$$SS\text{Quad} = n_f n_c(\bar{y}_f - \bar{y}_c)^2/(n_f + n_c)$$

using $\bar{y}_f$ and $\bar{y}_c$ to define the mean response for the factorial and centre portions, respectively. Again, high values of the test statistic would be suggestive that lack-of-fit of the fitted model appears due to a quadratic trend.

11.3.5 Statistical Validity of Fitted Model

Checking the statistical validity of a fitted k component RS model can be achieved by carrying out a simple hypothesis test. This test, based on the use of ANOVA principles, assesses the null hypothesis H_0: model inadequate against an alternative H_1: model adequate. Based on response variation partitioning into controlled and uncontrolled elements, we can specify the test statistic as the ratio of regression mean square to error mean square, i.e.

$$F = MSRegn/MSE \qquad (11.10)$$

with derivation of the MS terms as shown in Table 11.3. This test statistic is based on degrees of freedom $df_1 = k$ and $df_2 = N - k - 1$. Additionally, we should also assess the adjusted coefficient of determination R^2_{adj} of the fitted RS model, expressed as

$$R^2_{adj} = 100[(N-1)SSRegn - (k)SS\text{Total}]/[(N-k-1)SS\text{Total}]\% \qquad (11.11)$$

to help assess the model's explanation of response variation. In general, we are looking for the p value of test statistic (11.10) to be less than 0.05 and R^2_{adj} to exceed 80%.

For Example 11.1, Display 11.1 provides the test statistic of $F = 53.73$, p value of approximately 0.000, and R^2_{adj} of 92.9% [R-Sq(adj)] for the fitted first-order RS model. All confirm the statistical validity of the fitted model. Addition of the interaction term X_1X_2 to the first-order RS model appears to reduce R^2_{adj} by 1.2% (see Display 11.2) reinforcing the unimportant nature of the interaction effect to explanation of the purity response.

11.3.6 Additional Aspects of Analysis

Simple tests of each linear component of the RS model (11.3) may also be of interest. This can be achieved through either

$$t = b_i/\text{se}(b_i) \qquad (11.12)$$

with degrees of freedom $N - k - 1$ where the standard error $\text{se}(b_i) = \sqrt{(MSE/\Sigma X_i^2)}$ or

$$F = MS(b_i)/MSE \qquad (11.13)$$

with degrees of freedom $df_1 = 1$ and $df_2 = N - k - 1$ where $MS(b_i) = b_i^2 \Sigma X_i^2$. Diagnostic checking of model residuals can also be considered to assess for the presence of surface trends unaccounted for by the fitted model which could help indicate the direction of future experimentation. Use of main effect plots for each factor tested can also be useful to help understand how the response varies as a result of changes to the factor levels.

Exercise 11.2 It would be useful to consider assessment of main effect plots for the purity study of Example 11.1.

Main effects analysis: The main effect plots for temperature and pH are presented in Display 11.3. Both show the positive effect on purity of increasing factor levels. The temperature effect appears the greater of the two as indicated earlier in model assessment. Variability is comparable at all temperature levels but differs with pH quite markedly at the high level. pH therefore may have greater influence on consistency of purity than the level of purity. □

Display 11.3

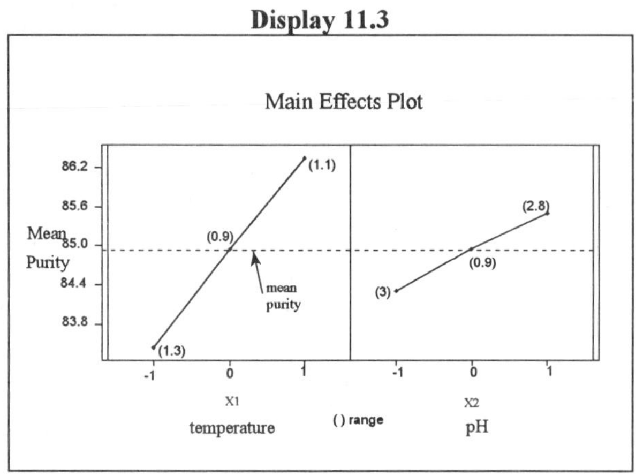

11.3.7 Contour Plot

Contour plots of the estimated linear response for pairings of the experimental factors can be used to provide a visual insight into which direction to follow in pursuit of the optimum. This entails considering how the contours change as the levels of the suggested factor settings change and using this knowledge to suggest the direction for future experimentation. For first-order models, contour plots will simply be a series of parallel lines joining factor levels that produce equal response values. The direction for future experimentation will be indicated by the line perpendicular to the contour lines.

Exercise 11.3 The statistical validity of the fitted first-order RS model for the purity study of Example 11.1 has been established. We must now assess the associated contour plot which is provided in Display 11.4.

Contour plot: The contour plot, as expected, exhibits the parallel lines feature associated with first-order RS models. The nature of the patterning suggests that increasing both factors would appear to increase purity, the change caused by temperature is again shown to be greater than that due to pH. It would appear that the next stage of experimentation should be directed toward testing higher settings of these factors than those currently used. □

11.4 METHOD OF STEEPEST ASCENT

The objective of RSM studies is to try to characterise the region of optimal response by determining the combination of factor levels providing this optimal. The **method of steepest ascent** is a "hill-climbing" procedure for moving upwards in the direction of the optimum until the maximum response is reached. For minimisation, the direction would be downwards and the technique called the **method of steepest descent**.

Display 11.4

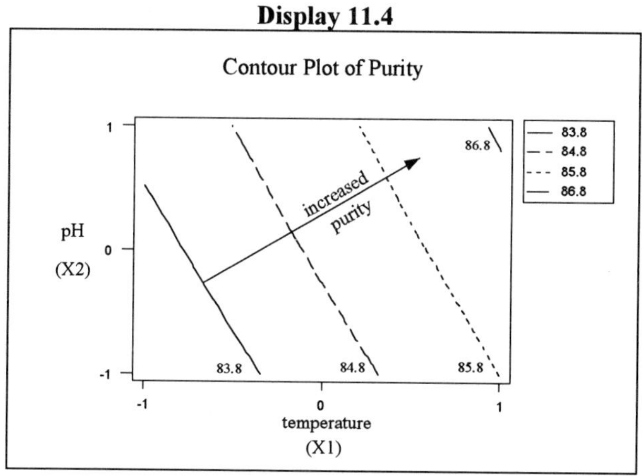

Fig. 11.2 illustrates a contour plot for a first-order RS model dealing with process maximisation. The **path of steepest ascent** begins at the centre point of the experimental region tested in the first-order RS design, i.e. the point (0, 0). The objective is to move along this path until an optimum response is reached by carrying out experiments at points on the path. Experiments are continued until no further increase in the response is detected. At this point, further experimentation can be considered to investigate the response surface around this suggested optimum. Such additional experimentation generally takes the form of second-order RS modelling to account for any curvature that may be present in the response.

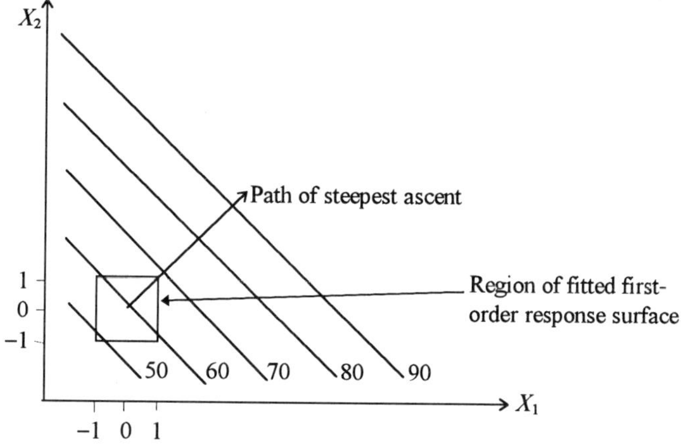

Fig. 11.2 — First-order Response Surface and path of steepest ascent

For two factors, as in Fig. 11.2, the estimated linear equation relating the factors to the measured response is

$$\hat{Y} = b_0 + b_1X_1 + b_2X_2.$$

For the path of steepest ascent, an incremental step size of ΔX_1 is chosen for the coded variable X_1 (factor A) based on factor knowledge or the fact that it has largest absolute regression coefficient (largest effect on response for unit change in level). The step size in X_2 (factor B) is then chosen as $\Delta X_2 = (b_2/b_1)\Delta X_1$, i.e. proportional to the regression coefficients of the fitted model. For a unit change in X_1, this would mean X_2 changing by (b_2/b_1) units for each experiment.

Based on the first-order model for the purity study of Example 11.1, we have $b_1 = 1.45$ and $b_2 = 0.6$ so improvement along the path of steepest ascent involves 1.45 units in the X_1 direction for every 0.6 units in the X_2 direction. The path of steepest ascent would begin at ($X_1 = 0$, $X_2 = 0$) with slope b_2/b_1 of 0.41. The chemist opted for 1 °C as the basic step size for temperature. This is equivalent to a unit change in X_1 so $\Delta X_1 = 1$ and $\Delta X_2 = 0.41$ specifying a pH increase of 0.41 for every one Celsius degree increase in temperature. Experiments were conducted along this path until a drop in purity was observed.

The series of experiments based on the path of steepest ascent produced the results shown in Table 11.4 which are depicted graphically in Fig. 11.3. The results indicate a steady increase in purity along the path until setting (temperature 54 °C, pH 5.39) after which a decrease begins to occur. This would suggest that maximum purity may well be occurring in the vicinity of this setting which should be the basis of the next optimisation experiment.

Table 11.4 — Experiments on the path of steepest ascent for purity study of Example 11.1

		Coded variables		Factors		
Experiment	Steps	X_1	X_2	Temperature	pH	Purity, Y
1	Origin	0	0	50	3.75	85.1
2	Origin + Δ	1	0.41	51	4.16	87.4
3	Origin + 2Δ	2	0.82	52	4.57	90.6
4	Origin + 3Δ	3	1.23	53	4.98	92.8
5	Origin + 4Δ	4	1.64	54	5.39	94.1
6	Origin + 5Δ	5	2.05	55	5.80	92.7
7	Origin + 6Δ	6	2.46	56	6.21	91.5

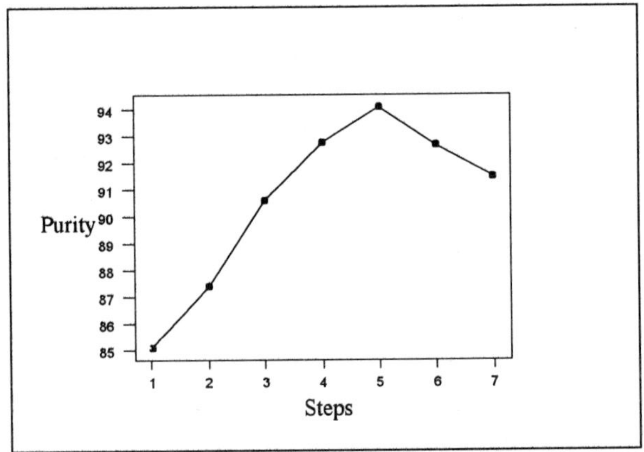

Fig. 11.3 — Levels of purity against steps for path of steepest ascent for purity study of Example 11.1

11.5 SECOND-ORDER DESIGN STRUCTURES

One drawback of first-order RS designs is their inability to account for interaction and quadratic effects within the factors tested. To cater for this, it is advisable to use a **second-order Response Surface design** which is specifically oriented to enabling such effects to be estimated and properly assessed. Orthogonal designs associated with fitting of second-order RS models include **Central Composite Designs, three-level Factorial Designs, three-level Fractional Factorial Designs**, and **Box-Behnken Designs**. CCDs were introduced by Box & Wilson (1951) as an alternative to three-level FDs and, as such, are often referred to as **Box-Wilson Designs**.

11.5.1 Central Composite Design

Orthogonal **Central Composite Designs** (CCDs) are the most widely used design framework for second-order RS modelling within k factor experiments. These design structures are based on full, or fractional, two-level FDs augmented by centre point replication and inclusion of an **axial portion**. Thus, the structure is split into three portions: a factorial portion providing $\boldsymbol{n_f}$ **points**, a centre portion of $\boldsymbol{n_c}$ **points**, and an axial portion of $\boldsymbol{n_a}$ **axial** (or **star**) **points**. In CCDs, $n_a = 2k$. Combined, this results in a total of $N = n_f + n_c + n_a$ design points.

Axial points refer to points beyond the "low" and "high" settings chosen for the factorial portion of the design structure which are equi-distant from the centre level for each factor. By appropriate choice of axial points, such designs can be made **rotatable**, a further desirable property of RS designs. Rotatability means that the variance of the predicted response at a particular point depends only on the distance of the point from the centre and not its direction, i.e. the design provides parameter estimates with equal precision in all directions. Axial points are set at treatment combinations corresponding to axial levels for one factor and centre levels for the others. Based on coding variables of −1 "low", 0 "centre", and +1 "high", this means the axial points will be ($\pm\alpha$, 0, 0,...), (0, $\pm\alpha$, 0,...), (0, 0, $\pm\alpha$,...), etc. where α will exceed 1 numerically.

The **orthogonality property** of a CCD is governed by the choice of axial spacing according to the expression

$$\alpha = \sqrt{\frac{\sqrt{n_f(n_f + n_c + n_a)} - n_f}{2}} \tag{11.14}$$

The **rotatability property** of a CCD is also governed by the choice of axial spacing through the expression

$$\alpha = (n_f)^{1/4} = (2^k)^{1/4} \tag{11.15}$$

and only depends on the factorial portion of the planned RS design structure. When these two properties provide the same α value, and therefore axial spacing, the associated design structure is classified as an **orthogonal rotatable CCD**. A more general form of (11.15) can be provided as

$$\alpha = (r_f n_f / r_a)^{1/4} \tag{11.16}$$

where r_f and r_a refer to the respective number of response replications measured in the factorial and axial portions of the RS design. Expression (11.15) refers to the simple case of single replication of both factorial and axial portions, i.e. $r_f = r_a = 1$.

For two factors ($k = 2$), consider a two-level RS design. For such a structure, the factorial portion will provide $n_f = 4$ design points while the axial portion will have $n_a = 2k = 4$ design points at ($\pm\alpha$, 0) and (0, $\pm\alpha$). From the rotatability specification (11.15), we must have α equal to ±1.414, i.e. $(2^2)^{1/4} = \sqrt{2}$. To ensure the design is also orthogonal, we need to choose the number of centre points n_c to ensure expression (11.14) provides an answer of ±1.414, i.e.

$$1.414 = \sqrt{\frac{\sqrt{4 \times (4 + n_c + 4)} - 4}{2}}$$

Solving this produces $n_c = 8$ so we therefore must use eight centre points to ensure orthogonality of design structure which is displayed in Table 11.5 and graphically in Fig. 11.4. As the two design specifications coincide, the planned second-order structure is therefore an orthogonal rotatable CCD with axial design points set at (±1.414, 0) and (0, ±1.414). It can be seen from Fig. 11.4 that the experimental region, excluding centre point replication, represents a circle of radius 1 passing through the axial points and the vertices of the factorial region. It is this circular patterning which specifies design rotatability.

Reducing the number of centre points to five ($n_c = 5$) produces α equal to ±1.27 for an orthogonal CCD and α equal to ±1.414 for a rotatable CCD. Two second-order CCDs are therefore possible in this case with choice of which to use dependent on which of the orthogonality or rotatability properties is more desirable. To convert the axial point co-ordinate into a factor level, we reverse expression (11.2) to produce the expression

$$\text{axial level} = M + \alpha S \qquad (11.17)$$

where M and S are as defined for the general level coding expression (11.2).

Table 11.5 — Two factor orthogonal rotatable Central Composite Design

Factorial portion		*Centre portion*		*Axial portion*	
X_1	X_2	X_1	X_2	X_1	X_2
−1	−1	0	0	−1.414	0
+1	−1	0	0	+1.414	0
−1	+1	0	0	0	−1.414
+1	+1	0	0	0	+1.414
		0	0		
		0	0		
		0	0		
		0	0		

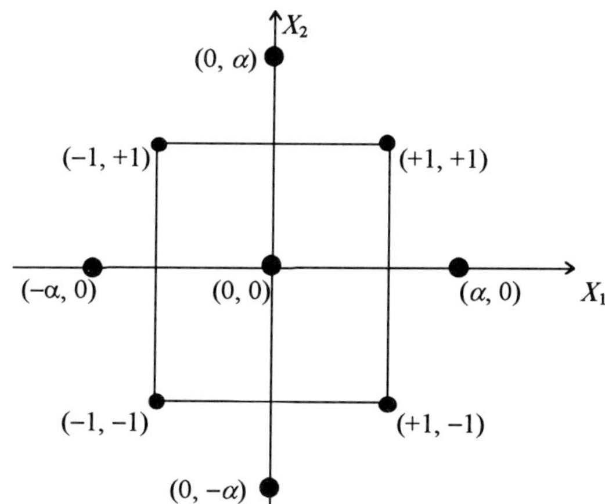

Fig. 11.4 — Orthogonal rotatable Central Composite Design for two factors

Choice of number of centre points n_c therefore controls the properties of a CCD. Depending on the choice made, a CCD structure may be made **orthogonal** or **uniform-precision**. A uniform-precision CCD refers to a design structure where the variance of the predicted response is the same at the design centre and a unit distance from the centre, i.e. invariant to position within the experimental region. Such a design provides more protection against estimation bias of the regression coefficients and would be useful if exploration of the response surface near the centre was desirable. Table 11.6 summarises the possible orthogonal and uniform-precision rotatable CCDs for RS experiments incorporating up to five factors. It can be seen that the centre point replication is the primary source of difference between these two types of CCD.

Table 11.6 — Orthogonal and uniform-precision rotatable Central Composite Designs for up to five factors

Number of factors (k)		2	3	4	5	5 (half replicate)
	n_f	4	8	16	32	16
	n_a (2k)	4	6	8	10	10
	α	±1.414	±1.682	±2	±2.378	±2
orthogonal	n_c	8	9	12	17	10
	N	16	23	36	59	36
uniform-precision	n_c	5	6	7	10	6
	N	13	20	31	52	32

Example 11.2 The chemist extended the investigation into the yield purity introduced in Example 11.1. Based on the results from the path of steepest ascent experiments, the chemist conducted further experiments using the suggested optimal setting of temperature 54 °C and pH 5.4 as the base with a view to trying to optimise the process settings. In the factorial portion, temperature was varied from 53 °C to 55 °C while pH was varied between 5.0 and 5.8. The design chosen as the design base was a uniform-precision rotatable CCD with axial spacing set at $\alpha = \pm 1.414$. The findings are shown in Table 11.7.

Table 11.7 — Purity data from uniform-precision rotatable Central Composite Design for Example 11.2

	X_1	X_2	Temperature	pH	Purity, Y
Factorial portion	−1	−1	53	5.0	90.1
	+1	−1	55	5.0	91.8
	−1	+1	53	5.8	90.7
	+1	+1	55	5.8	93.6
Centre portion	0	0	54	5.4	94.1
	0	0	54	5.4	94.6
	0	0	54	5.4	94.2
	0	0	54	5.4	93.9
	0	0	54	5.4	94.0
Axial portion	−1.414	0	52.59	5.4	89.0
	+1.414	0	55.41	5.4	92.3
	0	−1.414	54	4.83	90.7
	0	+1.414	54	5.96	92.5

11.5.2 Three-level Design

Three-level factorials can also be considered for second-order RS studies but often result in more experimental runs being required. For instance, for a three factor RS study, an orthogonal CCD would require 23 experimental runs (see Table 11.6), a uniform-precision CCD 20 runs (see Table 11.6), and a three-level factorial 27 runs showing the increased experimentation necessary for implementation of a three-level FD. Three-level factorials and their fractions are not rotatable and therefore may not be good choices for second-order RS designs if rotatability is deemed an important feature of a proposed design structure.

11.5.3 Box-Behnken Design

Box-Behnken designs represent three-level Composite Designs for fitting response surfaces across three-level factors and was proposed by Box & Behnken (1960). They can be constructed by combining two-level factorials within incomplete block designs to produce small run designs which are rotatable, or nearly so. An illustration of a Box-Behnken design structure for three factors is shown in Table 11.8 and visually in Fig. 11.5. This structure enables each factor level to be tested four times with only 13 of the 27 possible treatment combinations of a formal three-level factorial considered. Combinations involving only the "low" and "high" levels are not included so a Box-Behnken design maps out an experimental region that is more spherical than the cube shape prevalent in factorial structures. Omission of the corners of the cube can be advantageous if such combinations are difficult within the process being assessed or are too costly to perform.

11.5.4 Face-Centred Design

CCDs require that each factor be tested at five levels $-\alpha$, -1, 0, $+1$, and α. Often, this may be too costly or time consuming to consider or it may be that excessive changing of factor levels makes implementation of the experiment more difficult. The **face-centred CCD** or **face-centred cube** represents a variation on the CCD structure where α is set equal to 1 resulting in factors being tested at three levels only. For an experiment in three factors, the axial points will be located on the centres of the faces of the cube defining the experimental region as illustrated in Fig. 11.6. Structuring the RS design in this way requires fewer treatment combinations than the three-level FD but generates a structure which is not rotatable. In practice, centre point replication in face-centred CCDs is generally less than would be necessary for the comparable rotatable CCD.

Table 11.8 — Box-Behnken design for three factors

Factorial portion						*Centre portion*		
X_1	X_2	X_3	X_1	X_2	X_3	X_1	X_2	X_3
−1	−1	0	−1	0	+1	0	0	0
+1	−1	0	+1	0	+1	0	0	0
−1	+1	0	0	−1	−1	0	0	0
+1	+1	0	0	+1	−1			
−1	0	−1	0	−1	+1			
+1	0	−1	0	+1	+1			

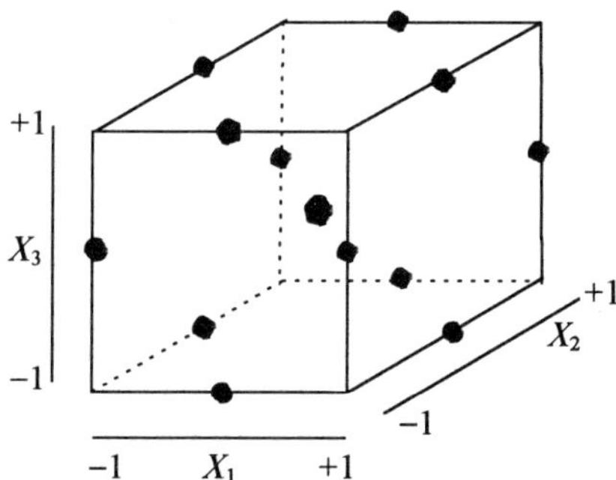

Fig.11.5 — Box-Behnken design for three factors

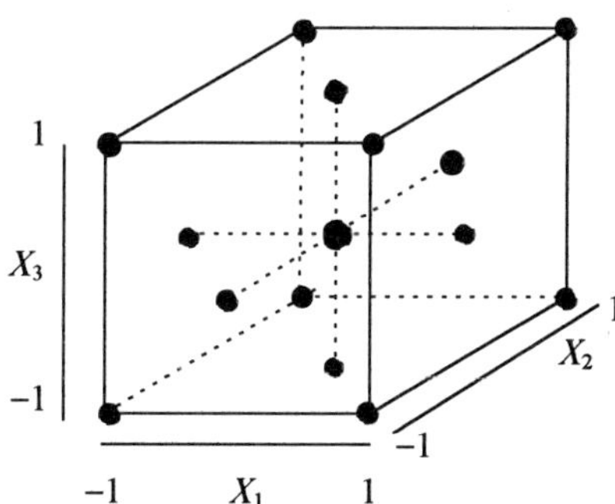

Fig.11.6 — Face-centred Central Composite Design for three factors

11.5.5 Blocking Design

Blocking in RS designs can be appropriate if it is necessary to eliminate the effect of nuisance factors such as time lapses between the implementation of an initial first-order design and the running of further experiments to produce a second-order design. As experiments are run at different times, the response could be influenced by changes to the test conditions. Blocking according to the time of experimentation would help to account for the influence of such a variable.

Orthogonality of design structure can be retained by assigning treatment combinations to the blocks so as not to affect parameter estimation. Generally, centre points are split equally between the blocks. In second-order designs, each block must be a first-order or orthogonal design. The first block generally comprises the factorial portion (n_f points) plus half the centre points (n_{cf} points) and the second block the axial points (n_a points) and remaining centre points (n_{ca} points). Table 11.9 provides an illustration of such a structure for a two factor RS study.

Table 11.9 — Blocked two factor orthogonal Central Composite Design

Block 1		Block 2	
X_1	X_2	X_1	X_2
−1	−1	−1.414	0
+1	−1	+1.414	0
−1	+1	0	−1.414
+1	+1	0	+1.414
0	0	0	0
0	0	0	0
0	0	0	0
0	0	0	0

To provide for orthogonal blocking, the axial spacing must satisfy the expression

$$\alpha = \sqrt{\frac{n_f(n_a + n_{ca})}{2(n_f + n_{cf})}} \qquad (11.18)$$

The blocked design of Table 11.9 results in $\alpha = \pm 1.414$ for expression (11.18). This corresponds to the α value necessary for orthogonality (see Table 11.6) so the illustrated blocked RS structure is orthogonal. Expression (11.18) and the rotatability expression (11.15) are not always compatible in blocked designs so a relaxation of the necessity of satisfying both orthogonality and rotatability properties is often necessary. Box & Draper (1987), Khuri & Cornell (1987), and Myers & Montgomery (1995) contain more details on blocking concepts for RS designs.

11.6 RESPONSE SURFACE ANALYSIS FOR SECOND-ORDER DESIGNS

The analysis principles for fitted second-order RS models are very similar to those adopted for first-order models in Section 11.4. They involve RS model specification, statistical assessment of the fitted model, assessment of contour plots, and analysis of the outcome to assess if the optimum has been reached.

11.6.1 Second-order Response Model

A **second-order**, or **quadratic**, **response model** for an experimental response Y is expressed in the form

$$Y = \beta_0 + \sum_{i=1}^{k} \beta_i X_i + \sum_{i=1}^{k} \beta_{ii} X_i^2 + \sum_{i<j}\sum \beta_{ij} X_i X_j \qquad (11.19)$$

where k is the number of experimental factors, β_0 is the constant term, β_i is the linear coefficient for factor i, β_{ii} is the quadratic coefficient for factor i, β_{ij} is the interaction coefficient for factor i and factor j, and ε is the random error. A second-order model is therefore a multiple non-linear regression model of the response containing p components of which k are linear, $n_q = k$ are quadratic, and $n_i = k(k - 1)/2$ are interaction terms whereby $p = k(k + 3)/2$.

For two factors X_1 and X_2, a second-order RS model for a response Y would take the form

$$Y = \beta_0 + \beta_1 X_1 + \beta_2 X_2 + \beta_{11} X_1^2 + \beta_{22} X_2^2 + \beta_{12} X_1 X_2 + \varepsilon$$

requiring estimation of six parameters. A minimum of six treatment combinations would therefore need to be tested to estimate such a RS model though error estimation would require more than six to be considered. For three factors, a second-order response model would be expressed as

$$Y = \beta_0 + \beta_1 X_1 + \beta_2 X_2 + \beta_3 X_3 + \beta_{11} X_1^2 + \beta_{22} X_2^2 + \beta_{33} X_3^2 + \beta_{12} X_1 X_2 + \beta_{13} X_1 X_3 + \beta_{23} X_2 X_3 + \varepsilon$$

requiring estimation of ten parameters using a minimum of ten treatment combinations.

11.6.2 Parameter Estimation

We use **least squares estimation**, as in first-order model fitting, to derive the estimates b_i, b_{ii}, and b_{ij} of the β parameters of the second-order response model (11.19). Parameter estimation is unaffected by the centre points as least squares estimation only uses the factorial and axial portions of the design to estimate model parameters. Linear and interaction component estimation is based on the same estimation procedure as outlined in Sections 11.3.2 and 11.3.4.2. Estimation of the parameters of the quadratic terms is by least squares methods.

Exercise 11.4 The extended chemical purity study of Example 11.2 represents a second-order RS design application. We must first specify the response model that will be fitted to the data and use software to provide the model specification.

Response model: As the design is of second-order type, then the response model for purity will be

$$\text{purity} = \beta_0 + \beta_1 X_1 + \beta_2 X_2 + \beta_{11} X_1^2 + \beta_{22} X_2^2 + \beta_{12} X_1 X_2 + \varepsilon$$

where β_0 is the constant term, β_i is the regression coefficient of factor i ($i = 1$ temperature, $i = 2$ pH), X_i is the coded form of factor i, β_{ii} refers to the coefficients of the quadratic components, β_{12} refers to the coefficient of the interaction component X_1X_2, and ε defines the random error. We wish to use the collected data to estimate this model and use the features of the fitted model to assess what the data specify for the influence of the factors on chemical purity and to determine if an optimal combination of factors can be reached.

Fitted model: As with Example 11.1, software generation of the RS model is required to not only simplify the calculation process but also to provide summary information to aid the analysis of the model. Display 11.5 shows the SAS generated second-order RS output for the optimisation of the yield purity investigation. The fitted model reads, using the column headed 'Parameter Estimate',

$$\text{purity} = 94.160 + 1.158\ \text{temp} + 0.618\ \text{pH} - 1.649\ \text{temp}^2 - 1.174\ \text{pH}^2 + 0.3\ \text{temp} \times \text{pH}$$

Coefficient estimates indicate unit change in coded temperature causes roughly twice the change in purity of unit change in coded pH thus suggesting that temperature produces the greater influence. The quadratic elements provide for similar interpretation while the interaction looks the least important of the three trend elements. The remaining aspects of the output presented will be discussed in the following sections. □

Display 11.5

Response Surface for Variable Purity

Response Mean	92.423077
Root MSE	0.306579
R-Square	0.9836
Coef. of Variation	0.3317

Regression	Degrees of Freedom	Type I Sum of Squares	R-Square	F-Ratio	Prob > F
Linear	2	13.791766	0.3429	73.368	0.0000
Quadratic	2	25.413376	0.6318	135.2	0.0000
Crossproduct	1	0.360000	0.0090	3.830	0.0912
Total Regress	5	39.565142	0.9836	84.189	0.0000

Residual	Degrees of Freedom	Sum of Squares	Mean Square	F-Ratio	Prob > F
Lack of Fit	3	0.365935	0.121978	1.671	0.3091
Pure Error	4	0.292000	0.073000		
Total Error	7	0.657935	0.093991		

Parameter	Degrees of Freedom	Parameter Estimate	Standard Error	T for H0: Parameter=0	Prob > |T|	Parameter Estimate from Coded Data
INTERCEPT	1	94.159949	0.137106	686.8	0.0000	94.159949
Temp	1	1.158450	0.108400	10.687	0.0000	1.638048
pH	1	0.618243	0.108400	5.703	0.0007	0.874196
Temp*Temp	1	-1.648993	0.116263	-14.183	0.0000	-3.296990
pH*Temp	1	0.300000	0.153290	1.957	0.0912	0.599819
pH*pH	1	-1.173850	0.116263	-10.097	0.0000	-2.346990

Factor	Degrees of Freedom	Sum of Squares	Mean Square	F-Ratio	Prob > F
Temp	3	30.002255	10.000752	106.4	0.0000
pH	3	12.998734	4.332911	46.099	0.0001

Canonical Analysis of Response Surface (based on coded data)

Factor	Critical Value Coded	Critical Value Uncoded
Temp	0.268477	0.379627
pH	0.220545	0.311851

Predicted value at stationary point 94.476238

Eigenvalues	Eigenvectors Temp	Eigenvectors pH
-2.260233	0.277883	0.960615
-3.383747	0.960615	-0.277883

Stationary point is a maximum.

11.6.3 Adequacy of Fit

Again, **ANOVA principles** underpin the fitting procedure for a second-order RS model as the **ANOVA table** in Table 11.10 demonstrates. Total variation, *SS*Total, is split into regression and error elements as in first-order modelling. These components are further split to reflect the aspects of surface fitting that affect them. In the case of the regression elements, this means linear, quadratic, and interaction (crossproduct) to reflect the three trend effects accounted for in the RS model (11.19). Error split is as for first-order modelling.

Table 11.10 — ANOVA table for second-order Response Surface model of a response Y as a function of k factors

Source	*df*	*SS*	*MS*
Regression	p	$SSRegn = \underline{b}'X'Y - (\Sigma y)^2/N$	$MSRegn = SSRegn/df$
Linear	k	$SS\text{Lin} = \Sigma b_i^2 \Sigma X_i^2$	$MS\text{Lin} = SS\text{Lin}/df$
Interaction	n_i	$SS\text{Intn} = n_f \Sigma\Sigma\, b_{ij}^2$	$MS\text{Intn} = SS\text{Intn}/df$
Quadratic	n_q	$SS\text{Quad}$	$MS\text{Quad} = SS\text{Quad}/df$
Error	$N - p - 1$	SSE	$MSE = SSE/df$
Lack-of-fit	$N - p - n_r$	$SSLF = SSE - SSPE$	$MSLF = SSLF/df$
Pure error	$n_r - 1$	$SSPE = \Sigma y_c^2 - (\Sigma y_c)^2/n_c$	$MSPE = SSPE/df$
Total	$N - 1$	$SST = \Sigma y^2 - (\Sigma y)^2/N$	

$p = k + n_i + n_q$ is the number of model components, $\underline{b}$ is a $(k + 1)\times 1$ column vector of the unknown regression coefficients with $\underline{b}'$ referring to the transpose of this vector, X is an $n\times(k + 1)$ matrix of the X settings (first column being n 1s) with X' defining its transpose, Y is an $n\times 1$ column vector of responses, y are the experimental responses, N is the number of data points, n_r is the number of treatment combination replications, y_c are the responses at the n_c centre points, *df* defines degrees of freedom, *SS* defines sum of squares, *MS* defines mean square

The first aspect of analysis, as was the case with first-order RS modelling, must be to check **lack-of-fit** of the fitted model. This can be carried out using the lack-of-fit test statistic (11.6) but with modified numerator degrees of freedom $df_1 = N - p - n_r$.

11.6.4 Statistical Validity of Fitted Model

General testing of model validity uses the usual F test and R^2_{adj} for model appropriateness (see Section 11.3.5). However, for second-order model fitting, we need to replace k in the R^2_{adj} expression (11.11) by p to account for all the model components.

Exercise 11.5 A second-order RS model was fitted to the purity data of Example 11.2 (see Exercise 11.4). The first stage of model analysis concerns the assessment of lack-of-fit.

Lack-of-fit check: Display 11.5 contains the information relevant to this check. The implemented second-order design provides $p = 5$, $N = 13$, $k = 2$, $n_q = 2$, $n_i = 1$, and $n_r = 5$. From Display 11.5, the lack-of-fit F test statistic (11.6) is presented as 1.671 with p value 0.3091. The high p value indicates no evidence to say lack-of-fit is occurring so the fitted second-order model for yield purity would appear to be an acceptable fit ($p > 0.05$) and appears to be providing a statistically acceptable explanation of the response surface.

Overall validity: Display 11.5 provides the test statistic of F = 84.189 ('Total Regress'), p value of approximately 0.0000 (less than 0.00005), and a calculated value of R^2_{adj} of 97.2% using expression (11.11). Not surprisingly, all confirm the statistical validity of the fitted model ($p < 0.05$). The increase of 5.5% in R^2_{adj} from the fitted first-order model provides further evidence that the response surface is second-order near the optimum. □

11.6.5 Additional Analysis

Provided model fit is statistically adequate, we should check the importance of each trend to the response model fitted. This entails checking the importance of the linear, interaction, and quadratic components individually. The linear component can be tested by the test statistic

$$F = MS\text{Lin}/MSPE \tag{11.20}$$

based on degrees of freedom $df_1 = k$ and $df_2 = n_r - 1$. The interaction and quadratic elements can be tested using test statistics (11.8) and (11.9) respectively. In the case of the quadratic component, if the test is significant, then this indicates that we may be near the optimum for the response surface. Testing of each individual factor contribution to each trend type can also be carried using the procedures mentioned in Section 11.3.6.

Exercise 11.6 To complete the statistical assessment of the fitted second-order RS model for the purity data of Example 11.2, we should assess the significance of each trend component separately to assess its importance to the explanation of the purity response surface.

Linear trend: The linear trend information (row 'Linear') in Display 11.5 confirms the statistical importance of this aspect to the fitted response surface (F = 73.368, p = 0.0000, i.e. less than 0.00005). Test information for the individual linear components is provided in the 'Parameters' section and shows that each is highly significant with temperature being more important (higher t value).

Quadratic trend: General information for the quadratic trend check is provided in the row 'Quadratic' of Display 11.5. The F value of 135.2 and the p value of 0.0000 (less than 0.00005) provide confirmation of the statistical importance of the quadratic components to the fitted response surface ($p < 0.05$). The sum of squares of the quadratic component is larger than those associated with the other trend components suggesting the response surface is quadratic. The individual quadratic elements for both temperature and pH are also highly significant ($p < 0.01$).

Interaction: Display 11.5 confirms that interaction between temperature and pH appears not to be present ('Crossproduct', F = 3.830, p = 0.0912) using a 5% significance level. At the 10% significance level, however, it would be declared significant though only just. It could be, therefore, that an element of interaction is present but not one that provides a major contribution to the response surface.

From this analysis of the individual model components, only the interaction element appears to be unimportant to the explanation of the fitted response surface. The result for the quadratic components point to the conclusion that we may be near the optimum of the response surface and therefore the optimal settings for maximum yield purity. □

11.6.6 Contour Plot

Again, a contour plot for factor pairings should be produced to visually display the region of optimal factor settings. For second-order response surfaces, such a plot can be more complex than the simple series of parallel lines that can occur with first-order models. Several possible contour patterns could occur with some illustrated in Figs 11.7 to 11.10.

The pattern displayed in Fig. 11.7 is a symmetrical mound shape with maximum response within the central contour. Minimisation would generate a surface with a depression within the central contour. When a contour patterning similar to Fig. 11.7 occurs, circular shaped contours tend to suggest independence of factor effects while elliptical contours may indicate factor interactions. Fig. 11.8, by contrast, illustrates a falling ridge with the minimum outside the experimental region. In Fig. 11.9, a stationary ridge pattern is presented with response decrease either side of the maximum response. A saddle point, or minimax, feature is also possible as Fig 11.10 illustrates. In such a case, the optimum is neither a maximum nor a minimum, but such that the centre is maximum in one direction and minimum in the other. Thus, depending on the direction of travel from the centre, the response can either increase or decrease.

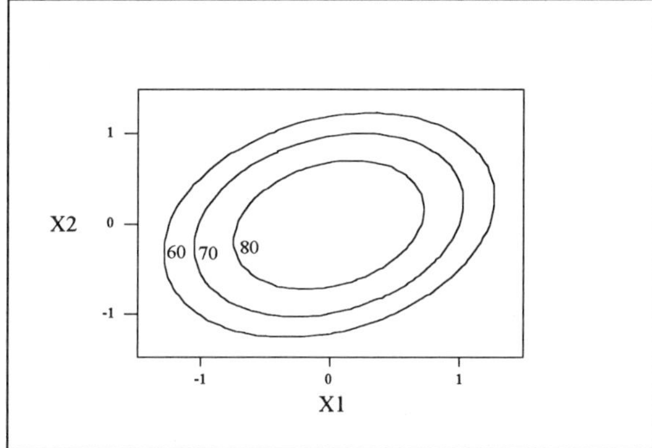

Fig. 11.7 — Response surface contour plot exhibiting a maximum surface

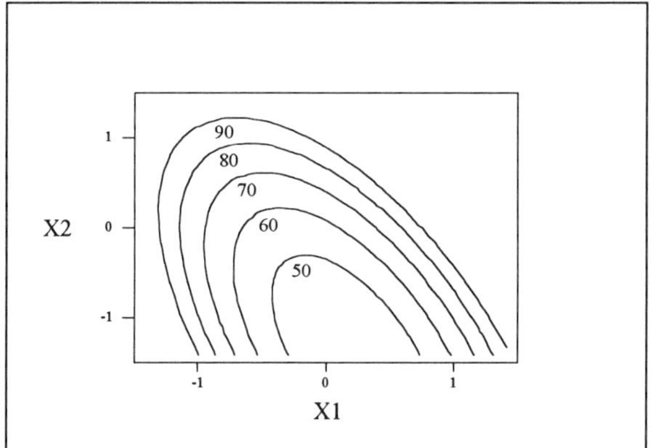

Fig. 11.8 — Response surface contour plot exhibiting a falling ridge

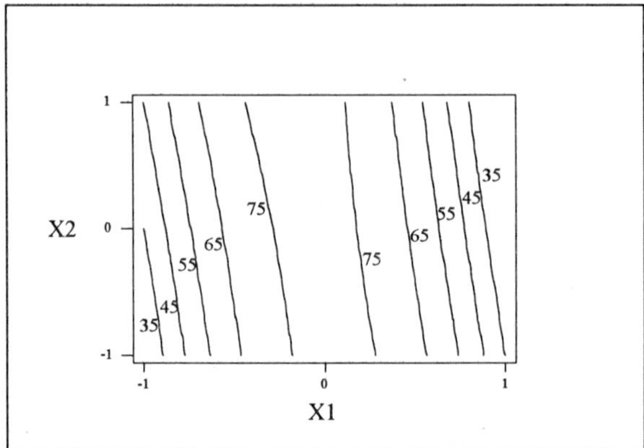

Fig. 11.9 — Response surface contour plot exhibiting a stationary ridge

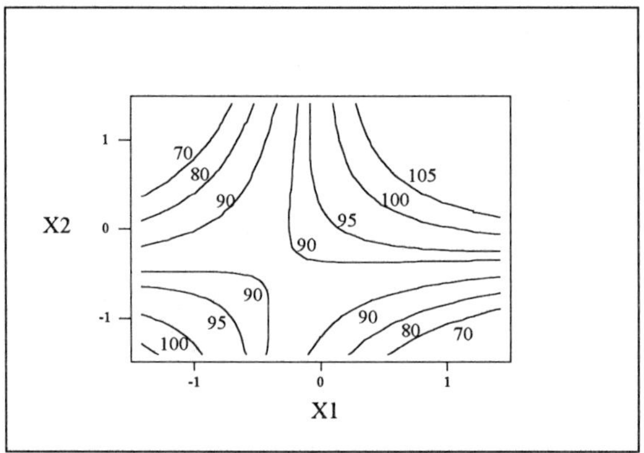

Fig. 11.10 — Response surface contour plot exhibiting a minimax feature

Exercise 11.7 For the purity optimisation experiment of Example 11.2, the statistical validity of the fitted model has been confirmed. To enhance the analysis, we need to produce a contour plot and assess what information it provides in respect of process optimisation. The contour plot for the second-order RS study of Example 11.2 is presented in Display 11.6.

Contour plot: The plot shows a distinctive circular mound shape indicative of possible independence of factors with response maximisation within the central contour. The optimum appears near the middle of the plot within the contour of 94. This suggests that a setting of temperature between 54 °C (0) and 55 °C (1) and pH between 5.4 (0) and 5.8 (1) may be optimal. It would appear, therefore, that process optimisation has been achieved. □

11.6.7 Exploration of Response Surface

Most of the analysis methods discussed so far provide primarily statistical measures of the suitability of a fitted second-order response surface. A useful additional analysis, if optimisation is attainable, would be to estimate the stationary point on the surface and its nature, i.e. is it a minimum, maximum, or minimax? Through this, we can investigate the response surface to assess if we are near the optimum and determine how sensitive the response may be to changes in the factor levels.

Display 11.6

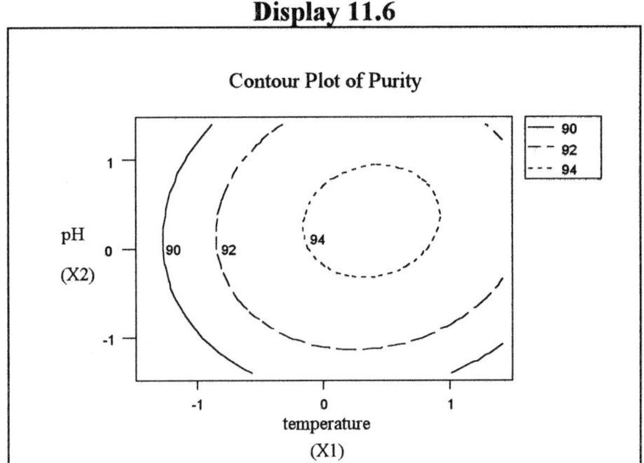

A formal approach to this aspect of RS analysis is based on the use of calculus (partial differentiation) and matrix algebra (eigenvalues) through consideration of the **canonical form**

$$Y = y_0 + \lambda_1 w_1^2 + \lambda_2 w_2^2 + \ldots + \lambda_k w_k^2 \qquad (11.21)$$

of the fitted response equation where y_0 is the predicted response at the optimal settings. The λ_i's are constants, or eigenvalues, and the w_i terms refer to transformed variables generated by rotating the X_i variables to place the origin of the new co-ordinate system at the specified stationary point. The sign and magnitude of the λ_i coefficients provide the information on the nature of the response surface. If the λ_i's are positive for all i, we declare a minimum while λ_i negative for all i suggests maximum. If the λ_i values differ, we have a minimax or saddle point representing differing optima in respect of each design factor and possible inadequacy of fit of the RS model. If any are zero, the surface is a stationary ridge since the response does not appear to change along the associated factor axis. SAS can provide this information as Exercise 11.8 will demonstrate.

Exercise 11.8 To finalise the analysis of the yield purity data from Example 11.2, we want to assess how the response surface appears near the optimum and to determine the position of this optimal treatment combination. The information necessary for this aspect of analysis is provided in Display 11.5 under 'Canonical Analysis of Response Surface'.

RS exploration: The output suggests that the optimal settings can provide a yield purity of 94.48% ('Predicted value at stationary point'). This is very close to the highest value recorded within the experimental results shown in Table 11.7 indicating that the fitted RS model looks to be providing a suitable explanation of the region of optimisation for yield purity. Additionally, the output specifies that the optimum is a maximum ('Stationary point is a maximum'). This also indicates that the response surface is rising to a maximum from both the temperature and pH directions as would be hoped for when trying to maximise a response.

The predicted response at the stationary point is y_0 = 94.48%. This, together with the quoted eigenvalues ('Eigenvalues' from Display 11.5), the λ_i terms, means the canonical form of the fitted model is

$$\text{purity} = 94.48 - 2.2602\,w_1^2 - 3.3837\,w_2^2$$

Since both the eigenvalues are negative and the suggested stationary point lies within the experimental region tested, we have further evidence that the stationary point is a maximum.

Determination of the optimal settings from the canonical analysis uses the factor coding principle of expression (11.2) but in reverse as

$$\text{factor level} = M + \text{optimal code level} \times S$$

where M represents the centre point level and S is half the difference between the high and low levels. For temperature, $M = 54$ and $S = 1$ while for pH, $M = 5.4$ and $S = 0.4$. Using the 'Critical Value Uncoded' results in Display 11.5, we can obtain the optimal settings for each factor as

$$\text{temperature: level} = 54 + (0.379627)1 = 54.38\ ^\circ\text{C}$$

$$\text{pH: level} = 5.4 + (0.311851)(0.4) = 5.53$$

Therefore, in conclusion, we can suggest that the optimal setting for maximising yield purity appears to be temperature 54.38 °C and pH 5.53. Use of this setting would appear to be able to improve yield purity from the values first obtained in the study described in Example 11.1. □

We have seen, through Examples 11.1 and 11.2, how to analyse RS design data for the information they contain. Not all aspects of RS analysis have been covered but what has been illustrated is sufficient for most practical purposes. Obviously, it is hoped that optimal factor settings will be reached at the end of the experimentation. Additional experiments, based on factor levels suggested from the canonical form (11.21), could be considered if further exploration of the response surface in the vicinity of the optimum was appropriate.

11.7 MIXTURE DESIGNS

In RSM applications, the factor levels tested for each factor are considered independent of those for every other factor. Cases can arise where the experimental factors correspond to components or ingredients which are mixed to form a product, e.g. a fruit juice, a vaccine, or a liquid fertiliser. The percentage of ingredients must sum to 100% to produce the required product resulting in choice of factor levels not being independent. The response associated with such products, which could be a measure of product quality or potency, therefore depends on the proportions of ingredients rather than their amounts with variations in the proportions affecting the final product. To understand the relationship between the measured response and the ingredient proportions, it would be necessary to use **Mixture Designs** (MDs) which are specifically oriented to optimising processes involving the mixing of components. Many of the analysis principles discussed for RS designs apply in MD analysis. A comprehensive discussion of MDs is provided in Cornell (1990) and Myers & Montgomery (1995).

11.8 FURTHER ASPECTS OF RESPONSE SURFACE METHODS

11.8.1 Multiple Responses and Non-parametric Methods

Most practical applications of RSM are concerned with single response experiments. Often, optimisation experiments may require multiple responses across similar factors to be optimised. Some multi-response methods exist (Khuri & Cornell 1987). Further developments in this area in respect of design and statistical analysis approaches are being developed. Use of overlaid contour plots can also be considered. In addition, development of non-parametric techniques for RS modelling and analysis to enable non-quantitative types of response data to be analysed by RS strategies are also under review (Myers *et al.* 1989).

11.8.2 Optimal Design Theory

The development of RSM slowed down in the 1970s and 80s. During this time, optimal design theory concepts were advanced and incorporated into RS strategies. Although the contribution and impact of optimal-design theory cannot be denied, there are some who believe that its application in RS design work should be made with extreme caution. Theoretical considerations of design optimality produce several forms of optimal design. These designs go under titles such as A-optimal, D-optimal, E-optimal, and G-optimal which correspond to different optimality criterion in respect of validity of parameter estimates. The reader needs to consult other texts such as Khuri & Cornell (1987) for further details.

PROBLEMS

11.1 A laboratory test program was undertaken to examine the effect of curing conditions on the strength of a composite material product. Three factors were identified as most probably contributing to product strength. The factors were autoclave cure time (hours), air cure time (hours), and autoclave cure temperature (°F). Air cure time refers to the amount of time elapsed between product formation and autoclaving. The first-order RS design chosen for the study was a 2^3 FD augmented with two centre points. The low and high factor levels represented the extremes of the range normally used in production. The response selected to be measured was the modulus of rupture which is known to characterise product strength, high being considered optimal. The response data presented below refer to the mean modulus of rupture for five specimens for each treatment combination in the factorial portion and ten specimens at the centre points. Assess the fit of a first-order RS model to these data. In what direction do the results suggest future experimentation should follow?

	Factors						
	Autoclave cure time	Air cure	Autoclave cure temperature	*Coded factors* X_1	X_2	X_3	Average modulus of rupture (psi)
Factorial	4	4	300	−1	−1	−1	1520
	12	4	300	+1	−1	−1	2340
	4	12	300	−1	+1	−1	1670
	12	12	300	+1	+1	−1	2230
	4	4	330	−1	−1	+1	2450

(continued)

(continued)

	12	4	330	+1	−1	+1	2900
	4	12	330	−1	+1	+1	2540
	12	12	330	+1	+1	+1	3230
Centre	8	8	315	0	0	0	2270
	8	8	315	0	0	0	2620

Reprinted from Kinzer, G. R. (1985). Application of two-cubed factorial designs to process studies. In: Snee, R. D., Hare, L. B. & Trout, J. R. (eds.), *Experiments in industry: Design, analysis, and interpretation of results*, American Society for Quality Control, Milwaukee, pp. 37–45 with the permission of the American Society for Quality Control.

11.2 A plastics manufacturer wished to optimise the weight yield from a new manufacturing process. Three factors were thought to have most effect on yield: time of processing, in hours, temperature, in °C, and the quantity, in g, of catalyst added near the end of the manufacturing process. The manufacturer used a first-order RS design based on a 2^3 FD with five centre point replications. The design and process yield results, in kg, are shown below. Assess the fit of a first-order RS model to these data. What does the analysis suggest is a possible direction for future experimentation?

	Factors			*Coded factors*			
	Time	Temp	Catalyst	X_1	X_2	X_3	Yield response
Factorial	1	115	10	−1	−1	−1	56.8
	4	115	10	+1	−1	−1	70.8
	1	145	10	−1	+1	−1	60.5
	4	145	10	+1	+1	−1	67.5
	1	115	20	−1	−1	+1	74.1
	4	115	20	+1	−1	+1	87.8
	1	145	20	−1	+1	+1	70.8
	4	145	20	+1	+1	+1	79.2
Centre	2.5	135	15	0	0	0	72.1
	2.5	135	15	0	0	0	69.4
	2.5	135	15	0	0	0	76.8
	2.5	135	15	0	0	0	73.2
	2.5	135	15	0	0	0	75.4

11.3 A screening experiment, based on a two-level Plackett-Burman design, was conducted into the effect of nine factors on the measurement of phenol in soils using a supercritical fluid extraction (SFE) process. Two factors, the extraction cell temperature (A), in °C, and level of acetic anhydride (B), in µL, were suggested to be the most important of those tested. Based on this, an orthogonal CCD experiment was conducted to optimise phenol measurement. The CCD chosen was based on a 2^2 factorial augmented with two centre points and four axial points. Axial spacing was set at $\alpha = \pm 1.0781$ to provide an orthogonal CCD structure. The design and results for total phenol recovered are shown below. Fit a second-order model to these data and analyse its fit. What are the suggested optimal factor levels?

	Run	*Factors* *A* (°C)	*B* (μL)	*Coded factors* X_1	X_2	Total phenol recovery (%)
Factorial	7	90	80	−1	−1	71.23
	3	110	80	+1	−1	88.70
	5	90	130	−1	+1	82.24
	8	110	130	+1	+1	90.09
Centre	1	100	105	0	0	81.57
	10	100	105	0	0	84.31
Axial	6	89.22	105	−1.0781	0	73.83
	9	110.78	105	+1.0781	0	91.97
	4	100	83.83	0	−1.0781	85.32
	2	100	116.17	0	+1.0781	83.68

Reprinted from Llompart, M. P., Lorenzo, R. A. & Cela, R. (1996) Multivariate optimization of supercritical fluid derivatization and extraction of phenol in soil samples. *Journal of Chromatographic Science* **34** 43–51 by permission of Preston Publications, A Division of Preston Industries, Inc. and the authors.

11.4 Abnormal levels of thyroxine in serum are indicative of hyper or hypothyroidism. Direct measurements of the concentration of the hormone can be misleading if not corrected for the binding capacity of a primary binding protein, thyroxine binding globulin (TBG). Three process conditions required to be co-optimised for accuracy of response and a face-centred cube design was chosen as the design base. The conditions tested were enzyme volume, conjugate volume, and serum sample volume, the first two factors corresponding to reagent concentrations. Five patient serum pools with different TBG binding capacities were assayed four times each for each treatment combination. The overall mean TBG binding capacity, in ma/min, for each treatment combination are presented below. By fitting a second-order model to these data, carry out a full and relevant assessment of the collected data.

	Factors Enzyme vol (μl)	Conjugate vol (μl)	Sample vol (μl)	*Coded factors* X_1	X_2	X_3	Mean TBG binding capacity
Factorial	20	20	60	−1	−1	−1	295.82
	40	20	60	+1	−1	−1	679.43
	20	60	60	−1	+1	−1	64.44
	40	60	60	+1	+1	−1	131.30
	20	20	120	−1	−1	+1	331.72
	40	20	120	+1	−1	+1	731.47
	20	60	120	−1	+1	+1	100.59
	40	60	120	+1	+1	+1	217.97
Centre	30	40	90	0	0	0	291.30
	30	40	90	0	0	0	287.83
	30	40	90	0	0	0	293.16
Axial	20	40	90	−1	0	0	180.67
	40	40	90	+1	0	0	412.57
	30	20	90	0	−1	0	501.33
	30	60	90	0	+1	0	127.58

(continued)

(continued)

30	40	60	0	0	−1	244.57
30	40	120	0	0	+1	321.26

Reprinted from Myers, G. C. (1985). Use of response surface methodology in clinical chemistry. In: Snee, R. D., Hare, L. B. & Trout, J. R. (eds.), *Experiments in industry: Design, analysis, and interpretation of results*, American Society for Quality Control, Milwaukee, pp. 59–73 with the permission of the American Society for Quality Control.

11.5 Photosensitive drugs such as riboflavin are often used for medical or cosmetic reasons. Formulation of these drugs must be carried out properly so as to achieve maximum photostability. An investigation into the amount of riboflavin (R, vitamin B_2) recovered was carried out using a three -level Box-Behnken design in respect of three factors: light absorbers oil red O (mmol) and sulisobenzene, S (mmol), and cholesterol, C (mmol). The collected response was the entrapment value of R, as a percent, whereby high values suggest high photostability. The design and response data are shown below. Fit a second-order model. What conditions lead to optimal recovery of riboflavin?

	Factors			*Coded variables*			
	O	S	C	X_1	X_2	X_3	Entrapment (%)
Factorial	0.0	0.0	1.0	−1	−1	0	47
	0.2	0.0	1.0	+1	−1	0	38
	0.0	0.2	1.0	−1	+1	0	13
	0.2	0.2	1.0	+1	+1	0	14
	0.0	0.1	0.0	−1	0	−1	8
	0.2	0.1	0.0	+1	0	−1	9
	0.0	0.1	2.0	−1	0	+1	23
	0.2	0.1	2.0	+1	0	+1	22
	0.1	0.0	0.0	0	−1	−1	14
	0.1	0.2	0.0	0	+1	−1	4
	0.1	0.0	2.0	0	−1	+1	35
	0.1	0.2	2.0	0	+1	+1	12
Centre	0.1	0.1	1.0	0	0	0	25
	0.1	0.1	1.0	0	0	0	24.5
	0.1	0.1	1.0	0	0	0	27

Reprinted from Loukas, Y. L. (1996) Formulation optimization of novel multicomponent photoreceptive liposomes by using response surface methodology. *Analyst* **121** 279–284 with the permission of The Royal Society of Chemistry.

11A Appendix: Software Information for Response Surface Designs

Design construction facilities

Most statistical software have default design build facilities for the construction of RS designs together with creation of a data sheet for data recording.

SAS: In SAS, RS design construction facilities are accessible through the ADX macros which contain CCD macros. The statements necessary for loading these macros are, based on macro storage in directory C:\SAS:

```
%INCLUDE 'C:\SAS\QC\SASMACRO\ADXGEN.SAS';
%INCLUDE 'C:\SAS\QC\SASMACRO\ADXFF.SAS';
```

```
%INCLUDE 'C:\SAS\QC\SASMACRO\ADXCC.SAS';
%ADXINIT;
RUN;
```

Once the macros have been loaded, the %ADXPCC(*k*) procedure can be used to check which *k* factor RS designs are available with the %ADXCCD procedure available to generate the design structure.

Minitab: In Minitab, the **Stat** ➢ **DOE** ➢ **Create RS Design** procedures provide access to RS design and analysis facilities for CCDs and Box-Behnken designs.

Data entry
Data entry for RS designs conforms to the same principles as those for two-level FFDs. Factor codes are based on use of the coded variables as RS analysis procedures utilise regression procedures and so require numerical code data to be available. Data entry for Example 11.1 would therefore conform to the data presentation below with that for Example 11.2 similarly specified.

Temp	pH	Purity	Temp	pH	Purity
−1	−1	82.8	0	0	84.7
1	−1	85.8	0	0	85.0
−1	1	84.1	0	0	85.4
1	1	86.9	0	0	84.5
			0	0	85.2

Response surface fitting
Response surface fitting can be easily achieved within most software packages.

SAS: In SAS, PROC RSREG is the RS modelling procedure for response surface data. Display 11.5 was obtained through implementing the statements

```
PROC SORT DATA = CH11.EX112;
     BY TEMP PH;
PROC RSREG DATA = CH11.EX112;
     MODEL PURITY = TEMP PH / LACKFIT;
RUN;
```

In the MODEL statement, the LACKFIT option requests a printing of lack-of-fit information for the fitted second-order response model.

Minitab: The **Stat** menu provides access to the RS surface fitting procedure within Minitab. The information contained in Display 11.1 was produced using the following menu selections:

Select **Stat** ➢ **DOE** ➢ **Analyze Custom Design** ➢ for *Design*, select **Response surface** ➢ **Fit Model** ➢ for *Responses*, select **Purity** and click **Select** ➢ for *Factors*, select the box and enter **X1 X2**.

Select **Terms** ➢ for *Include the following terms*, select **Linear** ➢ click **OK** ➢ click **OK**.

Display 11.2 was generated using the same procedure except for a change to the *Include the following terms* selection to **Linear + interactions**. Comparable information to that shown in Display 11.5 could be generated using this same procedure except for a change to the *Include the following terms* selection to **Full quadratic**. The menu commands for Display 11.3 are as shown in Appendix 7A. Menu commands for Displays 11.4 and 11.6 were as follows, based on first running the RS analysis procedure shown above:

> Select **Stat** ➤ **DOE** ➤ **Analyze Custom Design** ➤ for *Design Type*, ensure **Response surface** selected ➤ **Plot Response** ➤ for *Response Surface Plots*, select **Contour plot** and click **Setup** ➤ for *Response*, entry should read **Purity** ➤ for *Select a pair of factors for a single plot*, the entry for *X Axis* should read **X1** and the entry for *Y Axis* should read **X2**.
>
> Select **Contours** ➤ for *Contour Levels*, select **Values**, click the empty box, and enter **83.8 84.8 85.8 86.8** ➤ for *Line Styles*, select **Use different types** ➤ for *Line Colors*, select **Make all lines black** ➤ click **OK** ➤ click **OK** ➤ click **OK**.

For RS designs with more than two factors, contour plots for each factor pairing may be necessary. This can be achieved by following the menu procedure above except that selection of plot information in **Setup** should be **Generate plots for all pairs of factors**.

Bibliography

Barker, T. B. (1986) Quality engineering by design: Taguchi's philosophy. *Quality Progress* **19** 32–42.

Bayne, C. K. & Rubin, I. B. (1986) *Practical experimental design and optimisation methods for chemists.* VCH Publishers Inc., Deerfield Beach, Florida.

Bisgaard, S. (1996) A comparative analysis of the performance of Taguchi's linear graphs for the design of two-level fractional factorials. *Applied Statistics* **45** 311–322.

Bissell, A. F. (1992) Mean squares in saturated fractional designs revisited. *Journal of Applied Statistics* **19** 351–366.

Blom, G. (1958) *Statistical Estimates and Transformed Beta Variables.* John Wiley & Sons, Inc., New York.

Box, G. E. P. (1954) Some theorems on quadratic forms applied in the study of analysis of variance problems. *Annals of Mathematical Statistics* **25** 290–302.

Box, G. E. P. (1990) George's column: A simple way to deal with missing observations from designed experiments. *Quality Engineering* **3** 249–254.

Box, G. E. P. & Behnken, D. W. (1960) Some new three level designs for the study of quantitative variables. *Technometrics* **2** 455–476.

Box, G. E. P., Bisgaard, S. & Fung, C. (1988) An explanation and critique of Taguchi's contribution to quality engineering. *Quality and Reliability Engineering Journal* **4** 123–131.

Box, G. E. P. & Cox, D. R. (1994) An analysis of transformations. *Journal of the Royal Statistical Society (B)* **26** 211–252.

Box, G. E. P. & Draper, N. R. (1959) A basis for the selection of a response surface design. *Journal of the American Statistical Association* **54** 622–654.

Box, G. E. P. & Draper, N. R. (1987) *Empirical model-building and response surfaces.* Wiley, New York.

Box, G. E. P. & Hunter, J. S. (1957) Multifactor experimental designs for exploring response surfaces. *Annals of Mathematical Statistics* **28** 622–654.

Box, G. E. P. & Jones, S. (1992) Split-plot designs for robust product experimentation. *Journal of Applied Statistics* **19** 3–26.

Box, G. E. P. & Wilson, K. B. (1951) On the experimental attainment of optimum conditions. *Journal of the Royal Statistical Society (B)* **13** 1–45.

Byrne, D. M. & Taguchi, S. (1987) The Taguchi approach to experimental design. *Quality Progress* **20** 19–26.

Cheung, C. -S. & Jacroux, M. (1988) The construction of trend-free run orders of two-level factorial designs. *Journal of the American Statistical Association* **83** 1152–1158.

Clarke, G. M. & Kempson, R. E. (1997) *Introduction to the design and analysis of experiments.* Arnold, London.

Connor, W. S. & Zelen, M. (1959) *Fractional factorial experimental designs for factors at three levels.* National Bureau of Standards, Washington, D. C., Applied Mathematics Series, No. 54.

Cornell, J. A. (1990) *Experiments with mixtures: Designs, models and the analysis of mixture data.* 2nd ed. Wiley, New York.

Cox, D. R. (1958) *Planning of experiments.* Wiley, New York.

Daniel, C. (1959) Use of half-normal plots in interpreting factorial two-level experiments. *Technometrics* **1** 311–341.

Daniel, C. (1976) *Applications of statistics to industrial experimentation.* Wiley, New York.

Daniel, W. W. (1990) *Applied nonparametric statistics.* 2nd ed. PWS-Kent Publishing Company, Boston, Massachusetts.

Davies, L. (1993) *Efficiency in research, development and production: The statistical analysis of chemical experiments.* The Royal Society of Chemistry, Cambridge.

Everitt, B. S. & Der, G. (1996) *A handbook of statistical analyses using SAS.* Chapman and Hall, London.

Fisher, R. A. (1970) *Statistical methods for research workers.* 14th ed. Hafner Publishing Company, Darien, Connecticut.

Fowlkes, W. Y. & Creveling, C. M. (1995) *Engineering methods for robust product design.* Addison-Wesley, Reading, Massachusetts.

Gardiner, W. P. (1997) *Statistical analysis methods for chemists: A software-based approach.* The Royal Society of Chemistry, Cambridge.

Grove, D. M. & Davis, T. P. (1991) Taguchi's idle column method. *Technometrics* **33** 349–353.

Grove, D. M. & Davis, T. P. (1992) *Engineering quality and experimental design.* Longman, Harlow, Essex.

Haaland, P. D. (1989) *Experimental design in biotechnology.* Marcel Dekker, Inc., New York.

Hicks, C. R. (1982) *Fundamental concepts in the design of experiments.* 3rd ed. Holt, Rinehart and Winston, New York.

Hocking, R. R. (1973) A discussion of the two-way mixed model. *The American Statistician* **27** 148–152.

Huynh, H. & Feldt, L. S. (1970) Conditions under which mean square ratios in repeated measurements designs have exact F-distributions. *Journal of the American Statistical Association* **65** 1582–1589.

Kackar, R. N. (1986) Taguchi's quality philosophy: Analysis and commentary. *Quality Progress* **19** 21–29.

Kay, R. (1995) The p-value in statistics: Its value and its drawbacks. *Clinical Research Focus* **6** 4–5.

Khuri, A. I. & Cornell, J. A. (1987) *Response surfaces: Designs and analyses.* Marcel Dekker, Inc., New York.

Kuehl, R. O. (1994) *Statistical principles in research design and analysis.* Duxbury Press, Belmont, California.

Lipsey, M. W. (1990) *Design sensitivity: Statistical power for experimental research.* Sage Publications, Newbury Park, London.

Maghsoodloo, S. (1990) The exact relation of Taguchi's signal-to-noise ratio to his quality loss function. *Journal of Quality Technology* **22** 57–67.

McGovern, J. L. (1994a) A critique of the Taguchi approach — Part I: A presentation of some deficiencies and how these limit its efficiency and validity. *Journal of Coatings Technology* **66** No. 830 65–70.

McGovern, J. L. (1994b) A critique of the Taguchi approach — Part II: An alternative that is more efficient. *Journal of Coatings Technology* **66** No. 831 55–61.

Mendenhall, W. (1968) *Introduction to linear models.* Duxbury Press, Belmont, California.

Miller, J. C. & Miller, J. N. (1993) *Statistics for analytical chemistry.* 3rd ed. Ellis Horwood, Chichester.

Milliken, G. A. (1990) Analysis of repeated measures. In: Berry, D. A. (ed.), *Statistical methodology in the pharmaceutical sciences*, Marcel Dekker, Inc., New York, pp. 83–116.

Milliken, G. A. & Johnson, D. E. (1992) *Analysis of messy data: Volume 1 - Designed experiments.* Chapman and Hall, New York.

Milliken, G. A. & Johnson, D. E. (1989) *Analysis of messy data: Volume 2 - Non-replicated experiments.* Van Nostrand Reinhold, New York.

Montgomery, D. C. (1997) *Design and analysis of experiments.* 4th ed. Wiley, New York.

Myers, R. H. (1991) Response surface methodology in quality improvement. *Communications in Statistics - Theory and Methods* **20** 457–476.

Myers, R. H., Khuri, A. I. & Carter, W. H. (1989) Response surface methodology: 1966-1988. *Technometrics* **31** 137–157.

Myers, R. H., Khuri, A. I. & Vining, G. (1992) Response surface alternatives to the Taguchi robust parameter design approach. *The American Statistician* **46** 131–139.

Nair, V. N. (ed) (1992) Taguchi's parameter design: A panel discussion. *Technometrics* **34** 127–161.

Ott L. (1993) *An introduction to statistical methods and data analysis.* 4th ed. Duxbury Press, Belmont, California.

Phadke, M. S. (1989) *Quality engineering using robust design.* Prentice-Hall, Englewood Cliffs, New Jersey.

Pignatiello, J. J. & Ramberg, J. S. (1991) Top ten triumphs and tragedies of Genichi Taguchi. *Quality Engineering* **4** 211–225.

Plackett, R. L. & Burman, J. P. (1946) The design of optimum multi-factorial experiments. *Biometrika* **33** 305–325.

Ross, P. J. (1988) *Taguchi techniques for quality engineering.* McGraw-Hill, New York.

Searle, S. R. (1997) *Linear models.* Wiley, New York.

Shaw, R. G. & Mitchell-Olds, T. (1993) ANOVA for unbalanced data. *Ecology* **74** 1638–1645.

Shoemaker, A. C. & Tsui, K-L. (1993) Response model analysis for robust design experiments. *Communications in Statistics - Simulation* **22** 1037–1064.

Siegel, S. & Castellan, N. J. (1988) *Non-parametric statistics for the behavioral sciences.* 2nd ed. McGraw-Hill, New York.

Snedecor, G. W. & Cochran, W. G. (1980) *Statistical methods.* 7th ed. Iowa State University Press, Ames, Iowa.

Stone, R. A. & Veevers, A. (1994) The Taguchi influence on designed experiments. *Journal of Chemometrics* **8** 103–110.

Taguchi, G. (1986) *Introduction to quality engineering*, Asian Productivity Organisation, Tokyo, Japan.

Taguchi, G. (1987a) *System of experimental design, Volume 1.* UNIPUB/Kraus International Publications, White Plains, New York.

Taguchi, G. (1987b) *System of experimental design, Volume 2.* UNIPUB/Kraus International Publications, White Plains, New York.

Tribus, M. & Szonyi, G. (1989) An alternative view of the Taguchi approach. *Quality Progress* **22** 46–52.

Walpole, R. E. & Myers, R. H. (1993) *Probability and statistics for engineers and scientists.* 5th ed. Macmillan, New York.

Wang, P. C. & Jan, H. W. (1995) Designing two-level factorial experiments using orthogonal arrays when the run order is important. *The Statistician* **44** 379–388.

Winer, B. J., Brown, D. R. & Michels, K. M. (1991) *Statistical principles in experimental design.* 3rd ed. McGraw-Hill, New York.

Yates, F. W. (1934) The analysis of multiple classifications with unequal numbers in the different classes. *Journal of the American Statistical Association* **29** 51–66.

Zar, J. H. (1996) *Biostatistical analysis.* 3rd ed. Prentice Hall, Englewood Cliffs, New Jersey.

APPENDIX A

Statistical Tables

Table A.1 *Critical Values of the Student's t Distribution*

df	$t_{0.10}$	$t_{0.05}$	$t_{0.025}$	$t_{0.01}$	$t_{0.005}$	*df*
1	3.078	6.314	12.710	31.820	63.660	1
2	1.886	2.920	4.303	6.965	9.925	2
3	1.638	2.353	3.182	4.541	5.841	3
4	1.533	2.132	2.776	3.747	4.604	4
5	1.476	2.015	2.571	3.365	4.032	5
6	1.440	1.943	2.447	3.143	3.707	6
7	1.415	1.895	2.365	2.998	3.499	7
8	1.397	1.860	2.306	2.896	3.355	8
9	1.383	1.833	2.262	2.821	3.250	9
10	1.372	1.812	2.228	2.764	3.169	10
11	1.363	1.796	2.201	2.718	3.106	11
12	1.356	1.782	2.179	2.681	3.055	12
13	1.350	1.771	2.160	2.650	3.012	13
14	1.345	1.761	2.145	2.624	2.977	14
15	1.341	1.753	2.131	2.602	2.947	15
16	1.337	1.746	2.120	2.583	2.921	16
17	1.333	1.740	2.110	2.567	2.898	17
18	1.330	1.734	2.101	2.552	2.878	18
19	1.328	1.729	2.093	2.539	2.861	19
20	1.325	1.725	2.086	2.528	2.845	20
21	1.323	1.721	2.080	2.518	2.831	21
22	1.321	1.717	2.074	2.508	2.819	22
23	1.319	1.714	2.069	2.500	2.807	23
24	1.318	1.711	2.064	2.492	2.797	24
25	1.316	1.708	2.060	2.485	2.787	25
26	1.315	1.706	2.056	2.479	2.779	26
27	1.314	1.703	2.052	2.473	2.771	27
28	1.313	1.701	2.048	2.467	2.763	28
29	1.311	1.699	2.045	2.462	2.756	29
30	1.310	1.697	2.042	2.457	2.750	30
32	1.309	1.694	2.037	2.449	2.738	32
34	1.307	1.691	2.032	2.441	2.728	34
36	1.306	1.688	2.028	2.434	2.719	36
38	1.304	1.686	2.024	2.429	2.712	38
40	1.303	1.684	2.021	2.423	2.704	40
50	1.299	1.676	2.009	2.403	2.678	50
60	1.296	1.671	2.000	2.390	2.660	60
90	1.291	1.662	1.987	2.369	2.632	90

The critical values presented are appropriate for both two-tail and one-tail testing.

For a *two-tail test* at the 5% ($\alpha = 0.05$) significance level, $\alpha/2 = 0.025$ and the value is read from column $t_{0.025}$.

For a *one-tail test* at the 5% significance level, $\alpha = 0.05$ and the value is read from column $t_{0.05}$.

Computed with Minitab® by W. P. Gardiner.

Table A.2 *Critical Values of the χ^2 Distribution*

df	$\chi^2_{0.995}$	$\chi^2_{0.99}$	$\chi^2_{0.975}$	$\chi^2_{0.95}$	$\chi^2_{0.90}$	$\chi^2_{0.10}$	$\chi^2_{0.05}$	$\chi^2_{0.025}$	$\chi^2_{0.01}$	$\chi^2_{0.005}$	df
1	0.00	0.00	0.00	0.00	0.02	2.71	3.84	5.02	6.63	7.88	1
2	0.01	0.02	0.05	0.10	0.21	4.61	5.99	7.38	9.21	10.60	2
3	0.07	0.11	0.22	0.35	0.58	6.25	7.81	9.35	11.34	12.84	3
4	0.21	0.30	0.48	0.71	1.06	7.78	9.49	11.14	13.28	14.86	4
5	0.41	0.55	0.83	1.15	1.61	9.24	11.07	12.83	15.09	16.75	5
6	0.68	0.87	1.24	1.64	2.20	10.65	12.59	14.45	16.81	18.55	6
7	0.99	1.24	1.69	2.17	2.83	12.02	14.07	16.01	18.48	20.28	7
8	1.34	1.65	2.18	2.73	3.49	13.36	15.51	17.53	20.09	21.95	8
9	1.73	2.09	2.70	3.33	4.17	14.68	16.92	19.02	21.67	23.59	9
10	2.16	2.56	3.25	3.94	4.87	15.99	18.31	20.48	23.21	25.19	10
11	2.60	3.05	3.82	4.57	5.58	17.28	19.68	21.92	24.72	26.76	11
12	3.07	3.57	4.40	5.23	6.30	18.55	21.03	23.34	26.22	28.30	12
13	3.57	4.11	5.01	5.89	7.04	19.81	22.36	24.74	27.69	29.82	13
14	4.07	4.66	5.63	6.57	7.79	21.06	23.68	26.12	29.14	31.32	14
15	4.60	5.23	6.26	7.26	8.55	22.31	25.00	27.49	30.58	32.80	15
16	5.14	5.81	6.91	7.96	9.31	23.54	26.30	28.85	32.00	34.27	16
17	5.70	6.41	7.56	8.67	10.09	24.77	27.59	30.19	33.41	35.72	17
18	6.26	7.01	8.23	9.39	10.87	25.99	28.87	31.53	34.81	37.16	18
19	6.84	7.63	8.91	10.12	11.65	27.20	30.14	32.85	36.19	38.58	19
20	7.43	8.26	9.59	10.85	12.44	28.41	31.41	34.17	37.57	40.00	20
21	8.03	8.90	10.28	11.59	13.24	29.62	32.67	35.48	38.93	41.40	21
22	8.64	9.54	10.98	12.34	14.04	30.81	33.92	36.78	40.29	42.80	22
23	9.26	10.20	11.69	13.09	14.85	32.01	35.17	38.08	41.64	44.18	23
24	9.89	10.86	12.40	13.85	15.66	33.20	36.42	39.36	42.98	45.56	24
25	10.52	11.52	13.12	14.61	16.47	34.38	37.65	40.65	44.31	46.93	25
26	11.16	12.20	13.84	15.38	17.29	35.56	38.89	41.92	45.64	48.29	26
27	11.81	12.88	14.57	16.15	18.11	36.74	40.11	43.19	46.96	49.65	27
28	12.46	13.56	15.31	16.93	18.94	37.92	41.34	44.46	48.28	50.99	28
29	13.12	14.26	16.05	17.71	19.77	39.09	42.56	45.72	49.59	52.34	29
30	13.79	14.95	16.79	18.49	20.60	40.26	43.77	46.98	50.89	53.67	30

Computed with Minitab® by W. P. Gardiner.

Table A.3 *Critical Values of the F Distribution*

Values of $F_{0.1}$ ($\alpha = 0.1$)

df_1 = degrees of freedom of numerator variance df_2 = degrees of freedom of denominator variance

				df_1					
df_2	1	2	3	4	5	6	7	8	df_2
2	8.53	9.00	9.16	9.24	9.29	9.33	9.35	9.37	2
3	5.54	5.46	5.39	5.34	5.31	5.28	5.27	5.25	3
4	4.54	4.32	4.19	4.11	4.05	4.01	3.98	3.95	4
5	4.06	3.78	3.62	3.52	3.45	3.40	3.37	3.34	5
6	3.78	3.46	3.29	3.18	3.11	3.05	3.01	2.98	6
7	3.59	3.26	3.07	2.96	2.88	2.83	2.78	2.75	7
8	3.46	3.11	2.92	2.81	2.73	2.67	2.62	2.59	8
9	3.36	3.01	2.81	2.69	2.61	2.55	2.51	2.47	9
10	3.29	2.92	2.73	2.61	2.52	2.46	2.41	2.38	10
11	3.23	2.86	2.66	2.54	2.45	2.39	2.34	2.30	11
12	3.18	2.81	2.61	2.48	2.39	2.33	2.28	2.24	12
14	3.10	2.73	2.52	2.39	2.31	2.24	2.19	2.15	14
16	3.05	2.67	2.46	2.33	2.24	2.18	2.13	2.09	16
18	3.01	2.62	2.42	2.29	2.20	2.13	2.08	2.04	18
20	2.97	2.59	2.38	2.25	2.16	2.09	2.04	2.00	20
22	2.95	2.56	2.35	2.22	2.13	2.06	2.01	1.97	22
26	2.91	2.52	2.31	2.17	2.08	2.01	1.96	1.92	26
30	2.88	2.49	2.28	2.14	2.05	1.98	1.93	1.88	30
40	2.84	2.44	2.23	2.09	2.00	1.93	1.87	1.83	40
50	2.81	2.41	2.20	2.06	1.97	1.90	1.84	1.80	50
60	2.79	2.39	2.18	2.04	1.95	1.87	1.82	1.77	60

				df_1				
df_2	9	10	12	15	20	24	30	df_2
2	9.38	9.39	9.41	9.42	9.44	9.45	9.46	2
3	5.24	5.23	5.22	5.20	5.18	5.18	5.17	3
4	3.94	3.92	3.90	3.87	3.84	3.83	3.82	4
5	3.32	3.30	3.27	3.24	3.21	3.19	3.17	5
6	2.96	2.94	2.90	2.87	2.84	2.82	2.80	6
7	2.72	2.70	2.67	2.63	2.59	2.58	2.56	7
8	2.56	2.54	2.50	2.46	2.42	2.40	2.38	7
9	2.44	2.42	2.38	2.34	2.30	2.28	2.25	8
10	2.35	2.32	2.28	2.24	2.20	2.18	2.16	9
11	2.27	2.25	2.21	2.17	2.12	2.10	2.08	10
12	2.21	2.19	2.15	2.10	2.06	2.04	2.01	11
14	2.12	2.10	2.05	2.01	1.96	1.94	1.91	12
16	2.06	2.03	1.99	1.94	1.89	1.87	1.84	16
18	2.00	1.98	1.93	1.89	1.84	1.81	1.78	18
20	1.96	1.94	1.89	1.84	1.79	1.77	1.74	20
22	1.93	1.90	1.86	1.81	1.76	1.73	1.70	22
26	1.88	1.86	1.81	1.76	1.71	1.68	1.65	26
30	1.85	1.82	1.77	1.72	1.67	1.64	1.61	30
40	1.79	1.76	1.71	1.66	1.61	1.57	1.54	40
50	1.76	1.73	1.68	1.63	1.57	1.54	1.50	50
60	1.74	1.71	1.66	1.60	1.54	1.51	1.48	60

Table A.3 (continued)

Values of $F_{0.05}$ ($\alpha = 0.05$)

df_1 = degrees of freedom of numerator variance df_2 = degrees of freedom of denominator varaince

	df_1								
df_2	1	2	3	4	5	6	7	8	df_2
2	18.51	19.00	19.16	19.25	19.30	19.33	19.35	19.37	2
3	10.13	9.55	9.28	9.12	9.01	8.94	8.89	8.85	3
4	7.71	6.94	6.59	6.39	6.26	6.16	6.09	6.04	4
5	6.61	5.79	5.41	5.19	5.05	4.95	4.88	4.82	5
6	5.99	5.14	4.76	4.53	4.39	4.28	4.21	4.15	6
7	5.59	4.74	4.35	4.12	3.97	3.87	3.79	3.73	7
8	5.32	4.46	4.07	3.84	3.69	3.58	3.50	3.44	8
9	5.12	4.26	3.86	3.63	3.48	3.37	3.29	3.23	9
10	4.96	4.10	3.71	3.48	3.33	3.22	3.14	3.07	10
11	4.84	3.98	3.59	3.36	3.20	3.09	3.01	2.95	11
12	4.75	3.89	3.49	3.26	3.11	3.00	2.91	2.85	12
14	4.60	3.74	3.34	3.11	2.96	2.85	2.76	2.70	14
16	4.49	3.63	3.24	3.01	2.85	2.74	2.66	2.59	16
18	4.41	3.55	3.16	2.93	2.77	2.66	2.58	2.51	18
20	4.35	3.49	3.10	2.87	2.71	2.60	2.51	2.45	20
22	4.30	3.44	3.05	2.82	2.66	2.55	2.46	2.40	22
26	4.23	3.37	2.98	2.74	2.59	2.47	2.39	2.32	26
30	4.17	3.32	2.92	2.69	2.53	2.42	2.33	2.27	30
40	4.08	3.23	2.84	2.61	2.45	2.34	2.25	2.18	40
50	4.03	3.18	2.79	2.56	2.40	2.29	2.20	2.13	50
60	4.00	3.15	2.76	2.53	2.37	2.25	2.17	2.10	60

	df_1							
df_2	9	10	12	15	20	24	30	df_2
2	19.38	19.40	19.41	19.43	19.45	19.45	19.46	2
3	8.81	8.79	8.74	8.70	8.66	8.64	8.62	3
4	6.00	5.96	5.91	5.86	5.80	5.77	5.75	4
5	4.77	4.74	4.68	4.62	4.56	4.53	4.50	5
6	4.10	4.06	4.00	3.94	3.87	3.84	3.81	6
7	3.68	3.64	3.57	3.51	3.44	3.41	3.38	7
8	3.39	3.35	3.28	3.22	3.15	3.12	3.08	8
9	3.18	3.14	3.07	3.01	2.94	2.90	2.86	9
10	3.02	2.98	2.91	2.85	2.77	2.74	2.70	10
11	2.90	2.85	2.79	2.72	2.65	2.61	2.57	11
12	2.80	2.75	2.69	2.62	2.54	2.51	2.47	12
14	2.65	2.60	2.53	2.46	2.39	2.35	2.31	14
16	2.54	2.49	2.42	2.35	2.28	2.24	2.19	16
18	2.46	2.41	2.34	2.27	2.19	2.15	2.11	18
20	2.39	2.35	2.28	2.20	2.12	2.08	2.04	20
22	2.34	2.30	2.23	2.15	2.07	2.03	1.98	22
26	2.27	2.22	2.15	2.07	1.99	1.95	1.90	26
30	2.21	2.16	2.09	2.01	1.93	1.89	1.84	30
40	2.12	2.08	2.00	1.92	1.84	1.79	1.74	40
50	2.07	2.03	1.95	1.87	1.78	1.74	1.69	50
60	2.04	1.99	1.92	1.84	1.75	1.70	1.65	60

Table A.3 (continued)

Values of $F_{0.025}$ ($\alpha = 0.025$)

df_1 = degrees of freedom of numerator variance df_2 = degrees of freedom of denominator variance

df_2	df_1: 1	2	3	4	5	6	7	8	df_2
2	38.51	39.00	39.17	39.25	39.30	39.33	39.36	39.37	2
3	17.44	16.04	15.44	15.10	14.88	14.73	14.62	14.54	3
4	12.22	10.65	9.98	9.60	9.36	9.20	9.07	8.98	4
5	10.01	8.43	7.76	7.39	7.15	6.98	6.85	6.76	5
6	8.81	7.26	6.60	6.23	5.99	5.82	5.70	5.60	6
7	8.07	6.54	5.89	5.52	5.29	5.12	4.99	4.90	7
8	7.57	6.06	5.42	5.05	4.82	4.65	4.53	4.43	8
9	7.21	5.71	5.08	4.72	4.48	4.32	4.20	4.10	9
10	6.94	5.46	4.83	4.47	4.24	4.07	3.95	3.85	10
11	6.72	5.26	4.63	4.28	4.04	3.88	3.76	3.66	11
12	6.55	5.10	4.47	4.12	3.89	3.73	3.61	3.51	12
14	6.30	4.86	4.24	3.89	3.66	3.50	3.38	3.29	14
16	6.12	4.69	4.08	3.73	3.50	3.34	3.22	3.12	16
18	5.98	4.56	3.95	3.61	3.38	3.22	3.10	3.01	18
20	5.87	4.46	3.86	3.51	3.29	3.13	3.01	2.91	20
22	5.79	4.38	3.78	3.44	3.22	3.05	2.93	2.84	22
26	5.66	4.27	3.67	3.33	3.10	2.94	2.82	2.73	26
30	5.57	4.18	3.59	3.25	3.03	2.87	2.75	2.65	30
40	5.42	4.05	3.46	3.13	2.90	2.74	2.62	2.53	40
50	5.34	3.97	3.39	3.05	2.83	2.67	2.55	2.46	50
60	5.29	3.93	3.34	3.01	2.79	2.63	2.51	2.41	60

df_2	df_1: 9	10	12	15	20	24	30	df_2
2	39.39	39.40	39.41	39.43	39.45	39.46	39.46	2
3	14.47	14.42	14.34	14.25	14.17	14.12	14.08	3
4	8.90	8.84	8.75	8.66	8.56	8.51	8.46	4
5	6.68	6.62	6.52	6.43	6.33	6.28	6.23	5
6	5.52	5.46	5.37	5.27	5.17	5.12	5.07	6
7	4.82	4.76	4.67	4.57	4.47	4.42	4.36	7
8	4.36	4.30	4.20	4.10	4.00	3.95	3.89	8
9	4.03	3.96	3.87	3.77	3.67	3.61	3.56	9
10	3.78	3.72	3.62	3.52	3.42	3.37	3.31	10
11	3.59	3.53	3.43	3.33	3.23	3.17	3.12	11
12	3.44	3.37	3.28	3.18	3.07	3.02	2.96	12
14	3.21	3.15	3.05	2.95	2.84	2.79	2.73	14
16	3.05	2.99	2.89	2.79	2.68	2.63	2.57	16
18	2.93	2.87	2.77	2.67	2.56	2.50	2.44	18
20	2.84	2.77	2.68	2.57	2.46	2.41	2.35	20
22	2.76	2.70	2.60	2.50	2.39	2.33	2.27	22
26	2.65	2.59	2.49	2.39	2.28	2.22	2.16	26
30	2.57	2.51	2.41	2.31	2.20	2.14	2.07	30
40	2.45	2.39	2.29	2.18	2.07	2.01	1.94	40
50	2.38	2.32	2.22	2.11	1.99	1.93	1.87	50
60	2.33	2.27	2.17	2.06	1.94	1.88	1.82	60

Table A.3 (continued)

Values of $F_{0.01}$ ($\alpha = 0.01$)

df_1 = degrees of freedom of numerator variance df_2 = degrees of freedom of denominator variance

df_2	df_1 1	2	3	4	5	6	7	8	df_2
2	98.50	99.00	99.17	99.25	99.30	99.33	99.36	99.37	2
3	34.12	30.82	29.46	28.71	28.24	27.91	27.67	27.49	3
4	21.20	18.00	16.69	15.98	15.52	15.21	14.98	14.80	4
5	16.26	13.27	12.06	11.39	10.97	10.67	10.46	10.29	5
6	13.75	10.92	9.78	9.15	8.75	8.47	8.26	8.10	6
7	12.25	9.55	8.45	7.85	7.46	7.19	6.99	6.84	7
8	11.26	8.65	7.59	7.01	6.63	6.37	6.18	6.03	8
9	10.56	8.02	6.99	6.42	6.06	5.80	5.61	5.47	9
10	10.04	7.56	6.55	5.99	5.64	5.39	5.20	5.06	10
11	9.65	7.21	6.22	5.67	5.32	5.07	4.89	4.74	11
12	9.33	6.93	5.95	5.41	5.06	4.82	4.64	4.50	12
14	8.86	6.51	5.56	5.04	4.69	4.46	4.28	4.14	14
16	8.53	6.23	5.29	4.77	4.44	4.20	4.03	3.89	16
18	8.29	6.01	5.09	4.58	4.25	4.01	3.84	3.71	18
20	8.10	5.85	4.94	4.43	4.10	3.87	3.70	3.56	20
22	7.95	5.72	4.82	4.31	3.99	3.76	3.59	3.45	22
26	7.72	5.53	4.64	4.14	3.82	3.59	3.42	3.29	26
30	7.56	5.39	4.51	4.02	3.70	3.47	3.30	3.17	30
40	7.31	5.18	4.31	3.83	3.51	3.29	3.12	2.99	40
50	7.17	5.06	4.20	3.72	3.41	3.19	3.02	2.89	50
60	7.08	4.98	4.13	3.65	3.34	3.12	2.95	2.82	60

df_2	df_1 9	10	12	15	20	24	30	df_2
2	99.39	99.40	99.42	99.43	99.45	99.46	99.46	2
3	27.35	27.23	27.05	26.87	26.69	26.60	26.50	3
4	14.66	14.55	14.37	14.20	14.02	13.93	13.84	4
5	10.16	10.05	9.89	9.72	9.55	9.47	9.38	5
6	7.98	7.87	7.72	7.56	7.40	7.31	7.23	6
7	6.72	6.62	6.47	6.31	6.16	6.07	5.99	7
8	5.91	5.81	5.67	5.52	5.36	5.28	5.20	8
9	5.35	5.26	5.11	4.96	4.81	4.73	4.65	9
10	4.94	4.85	4.71	4.56	4.41	4.33	4.25	10
11	4.63	4.54	4.40	4.25	4.10	4.02	3.94	11
12	4.39	4.30	4.16	4.01	3.86	3.78	3.70	12
14	4.03	3.94	3.80	3.66	3.51	3.43	3.35	14
16	3.78	3.69	3.55	3.41	3.26	3.18	3.10	16
18	3.60	3.51	3.37	3.23	3.08	3.00	2.92	18
20	3.46	3.37	3.23	3.09	2.94	2.86	2.78	20
22	3.35	3.26	3.12	2.98	2.83	2.75	2.67	22
26	3.18	3.09	2.96	2.81	2.66	2.58	2.50	26
30	3.07	2.98	2.84	2.70	2.55	2.47	2.39	30
40	2.89	2.80	2.66	2.52	2.37	2.29	2.20	40
50	2.78	2.70	2.56	2.42	2.27	2.18	2.10	50
60	2.72	2.63	2.50	2.35	2.20	2.12	2.03	60

Computed with Minitab® by W. P. Gardiner.

Table A.4 *Table of z Scores for the Standard Normal Distribution*

α	z_α	α	z_α	α	z_α	α	z_α
0.20	0.8416	0.10	1.2816	0.0333	1.8344	0.0036	2.6875
0.19	0.8779	0.09	1.3408	0.025	1.9600	0.0033	2.7164
0.18	0.9154	0.08	1.4051	0.0167	2.1272	0.0025	2.8070
0.17	0.9542	0.07	1.4758	0.0125	2.2414	0.0024	2.8202
0.16	0.9945	0.06	1.5548	0.0083	2.3954	0.0017	3.9290
0.15	1.0364	0.05	1.6449	0.0075	2.4324	0.0012	3.0357
0.14	1.0803	0.04	1.7507	0.0067	2.4730	0.001	3.0902
0.13	1.1264	0.03	1.8808	0.005	2.5758		
0.12	1.1750	0.02	2.0537	0.0048	2.5899		
0.11	1.2265	0.01	2.3263	0.0042	2.6356		

Values refer to $P(z > z_\alpha) = \alpha$ where z corresponds to the standard, or unit, normal distribution, mean $\mu = 0$ and variance $\sigma^2 = 1$. The values represent the 100α percent critical values of the standard normal distribution.

Computed with Minitab® by W. P. Gardiner.

Table A.5 *5% Critical Values for the Studentised Range Statistic*

r = number of steps between ordered means df = "error" degrees of freedom

	r														
df	2	3	4	5	6	7	8	9	10	11	12	13	14	15	*df*
1	18.00	27.00	32.80	37.10	40.40	43.10	45.40	47.40	49.10	50.60	52.00	53.20	54.30	55.40	1
2	6.09	8.30	9.80	10.90	11.70	12.04	13.00	13.50	14.00	14.40	14.70	15.01	15.40	15.70	2
3	4.50	5.91	6.82	7.50	8.04	8.48	8.85	9.18	9.46	9.72	9.95	10.20	10.40	10.50	3
4	3.93	5.04	5.76	6.29	6.71	7.05	7.35	7.60	7.83	8.03	8.21	8.37	8.52	8.66	4
5	3.64	4.60	5.22	5.67	6.03	6.33	6.58	6.80	6.99	7.17	7.32	7.47	7.60	7.72	5
6	3.46	4.34	4.90	5.31	5.63	5.89	6.12	6.32	6.49	6.65	6.79	6.92	7.03	7.14	6
7	3.34	4.16	4.69	5.06	5.36	5.61	5.82	6.00	6.16	6.30	6.43	6.55	6.66	6.76	7
8	3.26	4.04	4.53	4.89	5.17	5.40	5.60	5.77	5.92	6.05	6.18	6.29	6.39	6.48	8
9	3.20	3.95	4.42	4.76	5.02	5.24	5.43	5.60	5.74	5.87	5.98	6.09	6.19	6.28	9
10	3.15	3.88	4.33	4.65	4.91	5.12	5.30	5.46	5.60	5.72	5.83	5.93	6.03	6.11	10
11	3.11	3.82	4.26	4.57	4.82	5.03	5.20	5.35	5.49	5.61	5.71	5.81	5.90	5.99	11
12	3.08	3.77	4.20	4.51	4.75	4.95	5.12	5.27	5.40	5.51	5.62	5.71	5.80	5.88	12
13	3.06	3.73	4.15	4.45	4.69	4.88	5.05	5.19	5.32	5.43	5.53	5.63	5.71	5.79	13
14	3.03	3.70	4.11	4.41	4.64	4.83	4.99	5.13	5.25	5.36	5.46	5.55	5.64	5.72	14
16	3.00	3.65	4.05	4.33	4.56	4.74	4.90	5.03	5.15	5.26	5.35	5.44	5.52	5.59	16
18	2.97	3.61	4.00	4.28	4.49	4.67	4.82	4.96	5.07	5.17	5.27	5.35	5.43	5.50	18
20	2.95	3.58	3.96	4.23	4.45	4.62	4.77	4.90	5.01	5.11	5.20	5.28	5.36	5.43	20
24	2.92	3.53	3.90	4.17	4.37	4.54	4.68	4.81	4.92	5.01	5.10	5.18	5.25	5.32	24
30	2.89	3.49	3.84	4.10	4.30	4.46	4.60	4.72	4.83	4.92	5.00	5.08	5.15	5.21	30
40	2.86	3.44	3.79	4.04	4.23	4.39	4.52	4.63	4.74	4.82	4.91	4.98	5.05	5.11	40
60	2.83	3.40	3.74	3.98	4.16	4.31	4.44	4.55	4.65	4.73	4.81	4.88	4.94	5.00	60

$r = j + 1 - i$ is the number of steps the treatments in positions i and j are apart in the ordered list of treatment mean responses.

Table A.6 *1% Critical Values for the Studentised Range Statistic*

r = number of steps between ordered means df = error degrees of freedom

	r														
df	2	3	4	5	6	7	8	9	10	11	12	13	14	15	df
2	14.0	19.0	22.3	24.7	26.6	28.2	29.5	30.7	31.7	32.6	33.4	34.1	34.8	35.4	2
3	8.26	10.6	12.2	13.3	14.2	15.0	15.6	16.2	16.7	17.1	17.5	17.9	18.2	18.5	3
4	6.51	8.12	9.17	9.96	10.6	11.1	11.5	11.9	12.3	12.6	12.8	13.1	13.3	13.5	4
5	5.70	6.97	7.80	8.42	8.91	9.32	9.67	9.97	10.2	10.5	10.7	10.9	11.1	11.2	5
6	5.24	6.33	7.03	7.56	7.97	8.32	8.61	8.87	9.10	9.30	9.49	9.65	9.81	9.95	6
7	4.95	5.92	6.54	7.01	7.37	7.68	7.94	8.17	8.37	8.55	8.71	8.86	9.00	9.12	7
8	4.74	5.63	6.20	6.63	6.96	7.24	7.47	7.68	7.87	8.03	8.18	8.31	8.44	8.55	8
9	4.60	5.43	5.96	6.35	6.66	6.91	7.13	7.32	7.49	7.65	7.78	7.91	8.03	8.13	9
10	4.48	5.27	5.77	6.14	6.43	6.67	6.87	7.05	7.21	7.36	7.48	7.60	7.71	7.81	10
11	4.39	5.14	5.62	5.97	6.25	6.48	6.67	6.84	6.99	7.13	7.26	7.36	7.46	7.56	11
12	4.32	5.04	5.50	5.84	6.10	6.32	6.51	6.67	6.81	6.94	7.06	7.17	7.26	7.36	12
13	4.26	4.96	5.40	5.73	5.98	6.19	6.37	6.53	6.67	6.79	6.90	7.01	7.10	7.19	13
14	4.21	4.89	5.32	5.63	5.88	6.08	6.26	6.41	6.54	6.66	6.77	6.87	6.96	7.05	14
16	4.13	4.78	5.19	5.49	5.72	5.92	6.08	6.22	6.35	6.46	6.56	6.66	6.74	6.82	16
18	4.07	4.70	5.09	5.38	5.60	5.79	5.94	6.08	6.20	6.31	6.41	6.50	6.58	6.65	18
20	4.02	4.64	5.02	5.29	5.51	5.69	5.84	5.97	6.09	6.19	6.29	6.37	6.45	6.52	20
24	3.96	4.54	4.91	5.17	5.37	5.54	5.69	5.81	5.92	6.02	6.11	6.19	6.26	6.33	24
30	3.89	4.45	4.80	5.05	5.24	5.40	5.54	5.56	5.76	5.85	5.93	6.01	6.08	6.14	30
40	3.82	4.37	4.70	4.93	5.11	5.27	5.39	5.50	5.60	5.69	5.77	5.84	5.90	5.96	40
60	3.76	4.28	4.60	4.82	4.99	5.13	5.25	5.36	5.45	5.53	5.60	5.67	5.73	5.79	60

$r = j + 1 - i$ is the number of steps the treatments in positions i and j are apart in the ordered list of treatment mean responses.

Table A.7 *Othogonal Polynomials*

Number of levels	Order	c_1	c_2	c_3	c_4	c_5	c_6	c_7	c_8	c_9	c_{10}	$\sum c_j^2$
3	1	−1	0	1								2
	2	1	−2	1								6
4	1	−3	−1	1	3							20
	2	1	−1	−1	1							4
	3	−1	3	−3	1							20
5	1	−2	−1	0	1	2						10
	2	2	−1	−2	−1	2						14
	3	−1	2	0	−2	1						10
	4	1	−4	6	−4	1						70
6	1	−5	−3	−1	1	3	5					70
	2	5	−1	−4	−4	−1	5					84
	3	−5	7	4	−4	−7	5					180
	4	1	−3	2	2	−3	1					28
	5	−1	5	−10	10	−5	1					252
7	1	−3	−2	−1	0	1	2	3				28
	2	5	0	−3	−4	−3	0	5				84
	3	−1	1	1	0	−1	−1	1				6
	4	3	−7	1	6	1	−7	3				154
	5	−1	4	−5	0	5	−4	1				84
	6	1	−6	15	−20	15	−6	1				924
8	1	−7	−5	−3	−1	1	3	5	7			168
	2	7	1	−3	−5	−5	−3	1	7			168
	3	−7	5	7	3	−3	−7	−5	7			264
	4	7	−13	−3	9	9	−3	−13	7			616
	5	−7	23	−17	−15	15	17	−23	7			2184
	6	1	−5	9	−5	−5	9	−5	1			264
9	1	−4	−3	−2	−1	0	1	2	3	4		60
	2	28	7	−8	−17	−20	−17	−8	7	28		2772
	3	−14	7	13	9	0	−9	−13	−7	14		990
	4	14	−21	−11	9	18	9	−11	−21	14		2002
	5	−4	11	−4	−9	0	9	4	−11	4		468
	6	4	−17	22	1	−20	1	22	−17	4		1980
10	1	−9	−7	−5	−3	−1	1	3	5	7	9	330
	2	6	2	−1	−3	−4	−4	−3	−1	2	6	132
	3	−42	14	35	31	12	−12	−31	−35	−14	42	8580
	4	18	−22	−17	3	18	18	3	−17	−22	18	2860
	5	−6	14	−1	−11	−6	6	11	1	−14	6	780
	6	3	−11	10	6	−8	8	6	10	−11	3	660

Table A.8 *Critical Values of the Ryan-Joiner Correlation Test of Normality*

	One-tail areas			
N	0.10	0.05	0.01	N
10	0.9347	0.9179	0.8804	10
12	0.9422	0.9275	0.8947	12
14	0.9481	0.9351	0.9061	14
16	0.9529	0.9411	0.9153	16
18	0.9567	0.9461	0.9228	18
20	0.9600	0.9503	0.9290	20
22	0.9627	0.9538	0.9343	22
24	0.9651	0.9569	0.9387	24
26	0.9672	0.9595	0.9426	26
28	0.9690	0.9618	0.9460	28
30	0.9707	0.9639	0.9490	30
32	0.9721	0.9657	0.9516	32
34	0.9734	0.9674	0.9539	34
36	0.9746	0.9689	0.9561	36
38	0.9757	0.9702	0.9580	38
40	0.9767	0.9715	0.9597	40
42	0.9776	0.9726	0.9613	42
44	0.9785	0.9737	0.9627	44
46	0.9792	0.9746	0.9640	46
48	0.9800	0.9756	0.9652	48
50	0.9807	0.9764	0.9664	50
52	0.9813	0.9772	0.9674	52
54	0.9819	0.9779	0.9684	54
56	0.9825	0.9786	0.9693	56
58	0.9830	0.9792	0.9702	58
60	0.9835	0.9799	0.9709	60
62	0.9840	0.9804	0.9717	62
64	0.9844	0.9810	0.9724	64
66	0.9848	0.9815	0.9731	66
68	0.9852	0.9820	0.9737	68
70	0.9856	0.9824	0.9743	70
72	0.9860	0.9829	0.9748	72
74	0.9863	0.9833	0.9755	74
76	0.9867	0.9837	0.9759	76
78	0.9870	0.9841	0.9763	78
80	0.9873	0.9845	0.9768	80

N represents the number of observations in the data set being tested.

The Ryan-Joiner test statistic R must be GREATER THAN the critical value for acceptance of normality (accept H_0).

Table A.9 *Power Values for Treatment F Test Based on the Non-central $F(f_1, f_2, \lambda)$ Distribution for 5% Significance Testing*

f_1 = degrees of freedom of numerator *MS* f_2 = degrees of freedom of denominator *MS*

$f_1 = 2$

f_2	ϕ 0.5	1.0	1.2	1.4	1.6	1.8	2.0	2.2	2.6	3.0	f_2
6	8.7	21.1	28.7	37.4	47.0	56.7	66.0	74.5	87.4	94.8	6
8	9.2	23.3	31.9	41.8	52.3	62.7	72.3	80.5	91.8	97.3	8
10	9.5	24.8	34.1	44.6	55.7	66.4	75.9	83.8	93.9	98.3	10
12	9.7	25.8	35.6	46.6	58.0	68.8	78.3	85.8	95.1	98.7	12
14	9.9	26.6	36.8	48.1	59.7	70.6	79.9	87.2	95.8	99.0	14
16	10.0	27.3	37.7	49.2	61.0	71.9	81.1	88.2	96.3	99.2	16
18	10.1	27.8	38.4	50.1	62.0	72.9	82.0	88.9	96.7	99.3	18
20	10.2	28.2	38.9	50.8	62.8	73.7	82.7	89.5	96.9	99.4	20
22	10.3	28.5	39.4	51.4	63.4	74.3	83.2	89.9	97.2	99.4	22
24	10.3	28.8	39.8	51.9	64.0	74.8	83.7	90.3	97.3	99.5	24
26	10.4	29.0	40.2	52.3	64.4	75.3	84.1	90.6	97.4	99.5	26
28	10.4	29.2	40.4	52.7	64.8	75.7	84.4	90.8	97.5	99.5	28
30	10.5	29.4	40.7	53.0	65.2	76.0	84.7	91.0	97.6	99.6	30
32	10.5	29.6	40.9	53.3	65.5	76.3	84.9	91.2	97.7	99.6	32
34	10.5	29.7	41.1	53.5	65.7	76.5	85.2	91.4	97.8	99.6	34
36	10.6	29.9	41.3	53.7	66.0	76.7	85.3	91.5	97.8	99.6	36
38	10.6	30.0	41.5	53.9	66.2	76.9	85.5	91.7	97.9	99.6	38
40	10.6	30.1	41.6	54.1	66.3	77.1	85.7	91.8	97.9	99.6	40

$f_1 = 3$

f_2	ϕ 0.5	1.0	1.2	1.4	1.6	1.8	2.0	2.2	2.6	3.0	f_2
6	8.6	20.9	28.7	37.7	47.6	57.7	67.4	75.9	88.7	95.6	6
8	9.1	23.6	32.7	43.2	54.3	65.2	75.0	83.1	93.6	98.2	8
10	9.4	25.5	35.5	46.9	58.7	69.8	79.4	86.9	95.8	99.0	10
12	9.7	26.9	37.5	49.5	61.7	72.9	82.2	89.3	97.0	99.4	12
14	9.9	27.9	39.1	51.5	63.9	75.1	84.2	90.8	97.6	99.6	14
16	10.1	28.8	40.3	53.0	65.6	76.7	85.6	91.8	98.0	99.7	16
18	10.2	29.5	41.3	54.2	66.9	78.0	86.6	92.6	98.3	99.8	18
20	10.4	30.0	42.1	55.2	67.9	78.9	87.4	93.2	98.5	99.8	20
22	10.4	30.5	42.7	56.0	68.8	79.7	88.0	93.6	98.7	99.8	22
24	10.5	30.9	43.3	56.7	69.5	80.4	88.6	94.0	98.8	99.8	24
26	10.6	31.3	43.8	57.2	70.1	80.9	89.0	94.3	98.9	99.9	26
28	10.7	31.6	44.2	57.7	70.6	81.4	89.4	94.5	99.0	99.9	28
30	10.7	31.8	44.6	58.2	71.1	81.8	89.7	94.7	99.0	99.9	30
32	10.8	32.1	44.9	58.6	71.5	82.1	89.9	94.9	99.1	99.9	32
34	10.8	32.3	45.2	58.9	71.8	82.4	90.2	95.1	99.1	99.9	34
36	10.8	32.4	45.4	59.2	72.1	82.7	90.4	95.2	99.2	99.9	36
38	10.9	32.6	45.7	59.5	72.4	83.0	90.5	95.3	99.2	99.9	38
40	10.9	32.8	45.9	59.7	72.6	83.2	90.7	95.4	99.2	99.9	40

Table A.9 (continued)

f_1 = degrees of freedom of numerator *MS* f_2 = degrees of freedom of denominator *MS*

$f_1 = 4$	ϕ										
f_2	0.5	1.0	1.2	1.4	1.6	1.8	2.0	2.2	2.6	3.0	f_2
6	8.5	20.9	28.9	38.2	48.4	58.8	68.6	77.2	89.7	96.2	6
8	9.1	24.0	33.5	44.5	56.1	67.3	77.1	85.1	94.9	98.7	8
10	9.5	26.2	36.8	48.9	61.2	72.6	82.1	89.2	97.0	99.4	10
12	9.8	27.9	39.3	52.1	64.8	76.2	85.2	91.7	98.0	99.7	12
14	10.0	29.2	41.3	54.5	67.5	78.7	87.3	93.2	98.6	99.8	14
16	10.2	30.3	42.8	56.4	69.5	80.6	88.8	94.2	98.9	99.9	16
18	10.4	31.2	44.0	57.9	71.0	82.0	89.9	95.0	99.1	99.9	18
20	10.6	31.9	45.0	59.1	72.3	83.1	90.7	95.5	99.3	99.9	20
22	10.7	32.5	45.9	60.1	73.3	83.9	91.4	95.9	99.4	99.9	22
24	10.8	33.0	46.6	60.9	74.1	84.7	91.9	96.2	99.5	100	24
26	10.9	33.5	47.2	61.7	74.8	85.3	92.3	96.5	99.5	100	26
28	10.9	33.9	47.7	62.3	75.5	85.8	92.7	96.7	99.6	100	28
30	11.0	34.2	48.2	62.8	76.0	86.2	93.0	96.9	99.6	100	30
32	11.1	34.5	48.6	63.3	76.4	86.6	93.3	97.0	99.6	100	32
34	11.1	34.8	49.0	63.7	76.8	86.9	93.5	97.2	99.7	100	34
36	11.2	35.0	49.3	64.1	77.2	87.2	93.7	97.3	99.7	100	36
38	11.2	35.3	49.6	64.4	77.5	87.4	93.8	97.4	99.7	100	38
40	11.3	35.5	49.9	64.7	77.8	87.7	94.0	97.5	99.7	100	40

$f_1 = 5$	ϕ										
f_2	0.5	1.0	1.2	1.4	1.6	1.8	2.0	2.2	2.6	3.0	f_2
6	8.5	21.0	29.1	38.7	49.1	59.7	69.6	78.2	90.5	96.7	6
8	9.1	24.4	34.3	45.7	57.6	69.0	78.9	86.6	95.7	99.0	8
10	9.5	26.9	38.1	50.7	63.4	74.9	84.2	90.9	97.7	99.6	10
12	9.9	28.9	41.0	54.3	67.5	78.8	87.5	93.4	98.6	99.8	12
14	10.2	30.4	43.2	57.1	70.4	81.6	89.7	94.8	99.1	99.9	14
16	10.4	31.7	45.0	59.3	72.7	83.5	91.2	95.8	99.4	99.9	16
18	10.6	32.7	46.5	61.0	74.4	85.0	92.3	96.5	99.5	100	18
20	10.8	33.6	47.7	62.5	75.8	86.2	93.1	97.0	99.6	100	20
22	10.9	34.4	48.7	63.6	77.0	87.1	93.7	97.3	99.7	100	22
24	11.0	35.0	49.6	64.6	77.9	87.9	94.2	97.6	99.7	100	24
26	11.1	35.5	50.3	65.5	78.7	88.5	94.6	97.8	99.8	100	26
28	11.2	36.0	51.0	66.2	79.4	89.0	94.9	98.0	99.8	100	28
30	11.3	36.5	51.6	66.9	80.0	89.4	95.2	98.1	99.8	100	30
32	11.4	36.8	52.1	67.4	80.5	89.8	95.4	98.3	99.8	100	32
34	11.4	37.2	52.5	67.9	80.9	90.2	95.6	98.4	99.9	100	34
36	11.5	37.5	52.9	68.3	81.3	90.4	95.8	98.4	99.9	100	36
38	11.6	37.8	53.3	68.7	81.7	90.7	96.0	98.5	99.9	100	38
40	11.6	38.0	53.6	69.1	82.0	90.9	96.1	98.6	99.9	100	40

Table A.9 (continued)

f_1 = degrees of freedom of numerator *MS* f_2 = degrees of freedom of denominator *MS*

$f_1 = 6$					ϕ						
f_2	0.5	1.0	1.2	1.4	1.6	1.8	2.0	2.2	2.6	3.0	f_2
6	8.4	21.1	29.4	39.1	49.7	60.4	70.4	79.0	91.1	97.0	6
8	9.1	24.8	34.9	46.7	58.9	70.4	80.2	87.8	96.3	99.2	8
10	9.6	27.6	39.1	52.2	65.2	76.7	85.8	92.2	98.2	99.7	10
12	10.0	29.8	42.4	56.3	69.6	80.9	89.2	94.6	99.0	99.9	12
14	10.3	31.5	44.9	59.4	72.9	83.8	91.4	96.0	99.4	99.9	14
16	10.5	33.0	47.0	61.8	75.3	85.9	92.9	96.9	99.6	100	16
18	10.8	34.2	48.7	63.8	77.2	87.4	93.9	97.5	99.7	100	18
20	10.9	35.2	50.1	65.4	78.8	88.6	94.7	97.9	99.8	100	20
22	11.1	36.1	51.3	66.7	80.0	89.5	95.3	98.2	99.8	100	22
24	11.2	36.8	52.3	67.8	81.0	90.3	95.8	98.4	99.9	100	24
26	11.4	37.5	53.1	68.8	81.8	90.9	96.1	98.6	99.9	100	26
28	11.5	38.0	53.9	69.6	82.6	91.4	96.4	98.7	99.9	100	28
30	11.6	38.5	54.6	70.3	83.2	91.9	96.7	98.9	99.9	100	30
32	11.7	39.0	55.2	70.9	83.7	92.2	96.9	99.0	99.9	100	32
34	11.7	39.4	55.7	71.5	84.2	92.5	97.1	99.0	99.9	100	34
36	11.8	39.8	56.2	72.0	84.6	92.8	97.2	99.1	99.9	100	36
38	11.9	40.1	56.6	72.4	85.0	93.1	97.3	99.1	100	100	38
40	11.9	40.4	57.0	72.8	85.3	93.3	97.4	99.2	100	100	40

Entries refer to estimated power for the planned treatment F test based on (f_1, f_2) degrees of freedom and non-centrality parameter $\lambda = \phi^2(f_1 + 1)$ where $\phi = \sqrt{[(nES^2)/(2k)]}$. Definitions of the parameters n, ES, and k can be found in Sections 3.6 and 5.6. The values presented refer to percentages.

Computed with SAS® by W. P. Gardiner.

APPENDIX B

Answers to Selected Problems

Problem 2.1

This is a two sample experiment involving assessment of the effectiveness of two independent coating procedures. For both mean and variability, interest lies in ascertaining if there is a difference between the two coating procedures so of hypotheses for means and variances will be required. The null hypothesis for both cases can be expressed as H_0: no difference in coating deposition with method, while the alternative for both would be two-sided and expressed as H_1: difference between the methods.

Problem 2.3

For both means and variances, interest lies in assessing for specific difference (one-tail tests). The null hypothesis in both cases is H_0: no difference between the two groups. For the mean, we have H_1: mean glucose level for control group less than that for treatment group ($\mu_{contr} < \mu_{treat}$) while for the variability, we have H_1: variability in glucose level lower for treatment group ($\sigma^2_{contr} > \sigma^2_{treat}$). The data suggest that blood glucose levels may be higher but not more variable after treatment. Testing the difference in means using the pooled t test indicates that there appears sufficient evidence to suggest treatment increases blood glucose ($t = -3.7862$, $p = 0.0007$). The 95% confidence interval for the difference in mean blood glucose levels between the control and treatment groups of (−6.00, −1.72) μg l^{-1} provides further evidence that treatment appears to be able to increase blood glucose levels by anything from 1.72 μg l^{-1} to 6 μg l^{-1}, on average. For the variability assessment, the conclusion is no evidence of difference in the variability in blood glucose levels as a result of drug treatment ($F = 1.53$, $p = 0.2864$).

Problem 2.6

For this paired samples study, we determine the difference D as washed – control. For the mean difference, we want to test the data for general difference so the alternative will be two-sided. Similarly for the variability in extensibility, we want to also test for general difference the alternative will also be two-sided. Initial impressions suggest washing can increase extensibility of the thread though with only six out of the eight differences being positive, statistical difference may not be substantiated. Testing the mean difference leads to the conclusion that the evidence suggests no difference in the extensibility of the thread ($t = 1.31$, $p = 0.2306$). The 95% confidence interval for the mean difference in extensibility of (−0.70, 2.45)% covers more positive values than negative indicating that washing can increase extensibility but not to a level that is statistically significant. The variability test results in acceptance of no difference in the variance of the extensibility of the thread with washing ($t = -1.10$, $p > 0.05$).

Problem 2.9

We want to estimate sample size $n = n_1 = n_2$ for a two-tail test of the proposed methods of storing distilled water. We have $d = 0.1$ and $\sigma = 0.05$ so $ES = 2$. The two sample t test is planned to be carried out at the 5% significance level so $z_{0.025} = 1.96$. With power of 95%, $\beta = 0.05$ and $z_{0.05} = 1.6449$. Sample size n is estimated as 7.

Problem 3.1

The response model for the cadmium measurements is $\mu + \tau_j + \varepsilon_{ij}$. EDA highlights that cadmium level varies with species, with *F. vesiculosus* highest and *Laminaria* lowest. *Pelvetia* and *Ascophyllum* appear similar though the latter does have higher values. Statistically, the species appear to differ ($F = 9.25$, $p = 0.0009$) and use of the SNK multiple comparison provided the result *Lam Pelv Asco Ves* signifying that *F. vesiculosus* differs from the other species and that *Laminaria*, *Pelvetia*, and *Ascophyllum* appear not to differ. Treatment effect confidence intervals for each species show that *F. vesiculosus* and *Laminaria* are different from the other two as they are the only ones exhibiting significant treatment effect. The contrast of $L = Ves - (Pelv + Asco + Lam)/3$ has a 95% confidence interval of (0.082, 0.220) μg g^{-1} dry weight which does not contain 0 suggesting that mean cadmium level in *F. vesiculosus* differs significantly from that of the other three species ($F = 21.15$, $p = 0.0003$). Diagnostic checking suggests that the cadmium data do conform to normality ($R = 0.99401$, $p > 0.1$) but not to equal variability. A variance stabilising transformation of such data may be advisable in future experiments when such a response is measured.

Problem 3.5

The proposed CRD structure provides the following information: $k = 4$, $n = ?$, $\delta = 0.7$ mg/ml, and $\sigma^2 = 0.14$ so $ES = 1.8708$. The F test for comparison of biscuit types is planned for the 5% significance level with degrees of freedom $f_1 = 3$ and $f_2 = 4(n - 1)$. From expression (3.15), using $n = 4$, ϕ is estimated to be 1.32 with Table A.9 ($f_2 = 12$) providing a power estimate of around 44%. This is well below the ideal so increasing n is necessary. Changing the number of biscuits sampled to six ($n = 6$) increases power to approximately 67% ($f_2 = 18$, $\phi = 1.62$) while increasing to eight ($n = 8$) provides a more acceptable power estimate of approximately 85% ($f_2 = 24$, $\phi = 1.87$). This would suggest that at least eight biscuits of each type be tested.

Problem 4.1

The response model for the plot yield data is $\mu + \beta_i + \tau_j + \varepsilon_{ij}$. Control and carbon disulphide show similar trend with low yield and wide variability. The other two sterilants are also similar to one another exhibiting high yield with low variability. Mean yield agrees with this interpretation of two sterilant pairings with the *CV*s highlighting the differing yield variability for the sterilants. Statistically, the sterilants appear to differ ($F = 28.73$, $p = 0.0006$) with the eelworm intensity p value of 0.0353 signifying that blocking according to eelworm intensity was beneficial. Application of the SNK multiple comparison generated the result

contr carbdis meth chlor

indicating that control and carbon disulphide differ from methamsodium and chloropicrin but that each pair of sterilants is similar. Treatment effect confidence intervals for each sterilant showed that all sterilants exhibit significant treatment effect with each showing wide variation in their effect, with chloropicrin possibly most beneficial. Diagnostic checking suggests that the plot yield data may require transformation in future experiments to stabilise variability.

Problem 4.5

The proposed RBD structure provides the following information: $k = 4$, $n = 5$, $\delta = 6.5$, and $\sigma^2 = 2.25$ so $ES = 2.8889$. The methods F test is planned for the 5% significance level with degrees of freedom $f_1 = 3$ and $f_2 = 12$. From expression (3.15), ϕ is estimated to be 2.04 with

Table A.9 ($f_2 = 12$) providing a power estimate of around 83.5%. This is acceptable so five machines appear sufficient.

Problem 4.8

From the study information provided, we have that $b = 6$, $t = 4$, $k = 2$, $bk = 12 = rt \Rightarrow r = 3$, and $\lambda = 3\times(2-1)/(4-1) = 1$. From these values, the following BIBD structure is possible, though randomisation of treatment order would be necessary for practical implementation.

		Control	A	B	C
	1	x	x	-	-
Twin	2	x	-	x	-
	3	x	-	-	x
lambs	4	-	x	x	-
	5	-	x	-	x
	6	-	-	x	x

Problem 4.11

The response model for the octane number data within this LS design is $\mu + \alpha_i + \beta_j + \tau_k + \varepsilon_{ijk}$. EDA highlighted octane number differences for the four tested petroleum spirits. Spirit A appears highest with B lowest and variability in octane number also differing with spirit type. The petroleum spirit F test result provides evidence pointing to octane number differences with type of petroleum spirit ($F = 26.46$, $p = 0.0007$). The day p value of 0.0397 and the tester group p value of 0.0138 signify that blocking on both factors was beneficial. Use of the SNK multiple comparison highlights that A differs from all the others and that C differs from B. Treatment effect confidence intervals show that all spirits tested exhibit significant treatment effect with A and B exhibiting most treatment effect and C least. Diagnostic checking shows no problem with normality for octane number but does suggest that, in future experiments, the octane number data may require transformation to stabilise variability.

Problem 5.1

The experimental structure is that of a two factor FD so the response model for the sales data will be $\mu + \alpha_i + \beta_j + \alpha\beta_{ij} + \varepsilon_{ijk}$. EDA indicated that sales differ with shelf position with bottom least and most consistent and middle highest but most variable. Sales difference was also apparent for the aisle position factor with sales very much higher for the nearest level but with wide variation. Initially, it would appear that position of product can markedly affect sales in respect of both volume and variation, as might be expected. Statistically, the shelf position × aisle position interaction is significant ($F = 14.46$, $p = 0.0006$). The interaction plot showed three different patterns depending on shelf position, with sales for the top shelf position lower at furthest compared to nearest aisle position. This difference was less marked for the middle shelf position and did not appear for the bottom shelf position. Variation in sales differs little for each combination of product position though for the aisle position factor, sales appear more consistent though lower for the furthest position. Diagnostic checking showed no particular problems.

Problem 5.6

From the two factor FD planned, the following information is available: $a = 3$, $b = 2$, $k' = ab = 6$, $n' = n = ?$, $f_1 = 2$, $f_2 = 6(n-1)$, $\delta = 0.86$, and $\sigma^2 = 0.0622$ so $ES = 3.4483$. For the statistical test of the primer × application method interaction, 5% is the proposed significance level. A start point of $n = 6$ has been suggested. This provides a power estimate of approximately 95% ($f_2 = 30$, $\phi = 2.4$) which is very much higher than the required power.

Changing n to 5 decreases power to 90.5% ($f_2 = 24$, $\phi = 2.23$) while $n = 4$ provides a power estimate of around 81.5% ($f_2 = 18$, $\phi = 1.90$). These calculations suggest at least four replications would be advised to satisfy the power constraint of at least 80%. This would result in a total of 24 test specimens being required.

Problem 5.9

This corresponds to a three factor FD with factors of diet (A), level of vitamin A (B), and level of vitamin E (C). The planned design structure provides the following information: $a = 3 = b = c$, $n = 5$, A fixed, B random, and C random. For the restricted model approach, the nature of the factor effects provides $D_a = 0$, $D_b = 1$, and $D_c = 1$. Checking the associated *EMS* expressions for both types of mixed model and using the usual *EMS* ratio principle, the test statistics are as follows:

Effect	*Restricted*	*Unrestricted*
A	$MS/(MSAB + MSAC - MSABC)$	$MSA/(MSAB + MSAC - MSABC)$
B	$MSB/MSBC$	$MSB/(MSAB + MSBC - MSABC)$
$A{\times}B$	$MSAB/MSABC$	$MSAB/MSABC$
C	$MSC/MSBC$	$MSC/(MSAC + MSBC - MSABC)$
$A{\times}C$	$MSAC/MSABC$	$MSAC/MSABC$
$B{\times}C$	$MSBC/MSE$	$MSBC/MSABC$
$A{\times}B{\times}C$	$MSABC/MSE$	$MSABC/MSE$

Problem 5.11

The experimental structure is that of a three factor FD so the response model for the operating temperature data is $\mu + \alpha_i + \beta_j + \alpha\beta_{ij} + \gamma_k + \alpha\gamma_{ik} + \beta\gamma_{jk} + \alpha\beta\gamma_{ijk} + \varepsilon_{ijkl}$. EDA showed that operating temperature differs with the settings of all factors but most obviously with start condition and line voltage. Initially, it would appear that all factors exert some influence on operating temperature. The three factor interaction is not significant ($F = 0.97$, $p = 0.3834$). The start condition × ambient temperature interaction effect shows statistical significance ($F = 103.47$, $p = 0.0001$) while the ambient temperature × line voltage interaction is only just not significant ($F = 2.57$, $p = 0.0811$). The plot for the start condition × ambient temperature interaction showed two different patterns. For the hot start condition, ambient temperature appeared to have no effect with operating temperature staying high while the cold start condition resulted in an obvious increase in operating temperature as ambient temperature increased. The plot for the ambient temperature × line voltage interaction shows only minimal interaction effect with a general increase in operating temperature as ambient temperature and line voltage increased. Variation in operating temperature is large for all treatment combinations. Diagnostic checking highlighted only a clustering pattern in the fits plot.

Problem 6.1

Based on the presented information, we can deduce that the site effect is fixed, the area of sampling is random, and the areas sampled are particular to the site of testing so the area effect defines a nested factor. The response model, for this mixed model, is lead measurement = $\mu + \alpha_i + \beta_{j(i)} + \varepsilon_{k(ij)}$. From the factor information and the fact that a mixed model is being applied, we have test statistics of MSS/MSA(S) for site and MSA(S)/MSE for area. Lead levels show site differences with site B appearing highest and least variable. The site effect is significant ($F = 7.37$, $p = 0.005$) as is the area effect ($F = 39.99$, $p = 0.000$). Application of the SNK follow-up to the fixed site effect showed site B to be significantly different from sites C, D, and E ($p < 0.05$) and site A to be significantly different from sites C and D ($p < 0.05$). Testing for significant nested factor effects showed that statistically significant differences in lead levels between sampling areas were occurring in sites A, B, C, and D ($p < 0.001$) with only site E differing from this trend ($p > 0.1$). Variance estimates generated percentage contribution of 95.1% for the area sampled indicating that most of the differences in lead measurement variation appear associated with the areas sampled, much as might be expected. Changing n to 3 marginally reduced the treatment variance but not sufficiently to suggest design improvement. Diagnostic checking suggested a variance stabilising transformation of lead response may be necessary in future studies. Normality for the lead response appeared acceptable ($R = 0.963$, $p > 0.1$).

Problem 6.6
From the information supplied, we can deduce that the supplier factor is fixed, the regulator factor is random, and the time points of measurement are fixed resulting in a mixed model. Additionally, the regulators tested from each supplier are independent so the regulator effect refers to a nested factor. The response model for operating voltage in this RMD structure is $\mu + \alpha_i + \beta_{j(i)} + \gamma_k + \alpha\gamma_{ik} + \varepsilon_{ijk}$. The supplier effect showed A and B with similar average voltages but very different levels of variation, A exhibiting considerably more variation. Time trends varied with regulator as did the associated operating voltage with the trends found not sufficient to suggest that a significant supplier × time interaction appeared likely. Test statistic construction for the supplier effect gives rise to a test statistic of MSSuppl/MSReg(Suppl) with the other three test statistics all having denominator MSE. The supplier × time interaction was not statistically significant ($F = 0.69$, $p = 0.5617$) but the supplier effect was ($F = 9.70$, $p = 0.0089$). In addition, both the nested regulator effect ($F = 9.09$, $p = 0.0001$) and the time effect ($F = 2.98$, $p = 0.0441$) were statistically significant. The differences in the regulators does give cause for concern as it would appear suppliers cannot provide comparable regulators which can work to the same specifications. Variance estimation generated variation accountability of 66.9% for the regulator effect and 33.1% for the error indicative that most variation appears to stem from the regulators tested. The high value for the regulator effect provides further evidence of the differences apparent in the operating voltage of the tested regulators. Assessment of linear trend for each regulator showed only that the intercepts differed statistically ($F = 8.72$, $p = 0.012$). This further backs-up the horizontal nature of the trend in operating voltage for each regulator but highlights that regulators generate different operating voltages. Consideration of a quadratic trend for each regulator was less successful and not significant.

Problem 6.9
Given that the design was a COD, then the response model for the biochemical measurements was $\mu + \alpha_i + \beta_{j(i)} + \gamma_k + \tau_{m(i,k)} + \lambda_{m(i,k-1)} + \varepsilon_{ijk}$. $\tau_{m(i,k)}$ defines the direct effect of the mth treatment measured at period k of treatment sequence i while $\lambda_{m(i,k-1)}$ refers to the carryover effect of the mth treatment administered in period $(k - 1)$ of treatment sequence i. Initial examination of the data showed no sequence effect though there was a suggestion of a period effect with period 4 having the least variable measurements. Summary statistics for the nested patient effect showed no evidence of patient differences with treatment sequence. The sequence test is not significant ($F = 0.69$, $p = 0.6034$) and neither is the nested patient effect ($F = 0.87$, $p = 0.5059$). The period effect was also not significant ($F = 0.76$, $p = 0.5328$). These results confirm the exploratory analysis findings of comparability of biochemical measurements across the treatment sequences, between the patients within the treatment sequences, and across the periods of measurement. Carryover treatment effects adjusted for the presence of direct effects was not significant ($F = 0.46$, $p = 0.7164$) while the test of direct treatment effects accounting for carryover effects was close to significance at the 5% level ($F = 3.15$, $p = 0.0560$). The latter result suggests difference in the biochemical measurements with the different treatments indicating that treatment used could strongly influence the level of breathing abnormality and thus, the relief of patients.

Problem 7.1

EDA indicated that date of planting appeared to affect yield most though spacing was also important. Variability in yield is comparable for all factor levels tested. The normal plot of effect estimates exhibits an "ideal" pattern with distinct separation of date of sowing, spacing, and the variety × spacing interaction indicated. Spacing is the only negative estimate and is least of the three numerically. A Pareto chart exhibited comparable characteristics of effects sparsity. Main effect plots indicated that date of sowing affects timber yield markedly with late sowing generating higher yields. A downward trend from narrow to wide occurs for spacing suggesting narrow is best though range of yield is high but similar in all cases. The variety × spacing interaction plot exhibited a different pattern for the two varieties assessed with markedly diminishing timber yield for variety X and little change for variety Y. Fitting a model of timber yield = date of sowing (C) + variety × spacing ($A \times B$) + spacing (B) + error generated significant effects, date of sowing with $p = 0.003$, variety × spacing interaction with $p = 0.008$, and spacing with $p = 0.019$. Checking the residual plots showed evidence of a distinct trend for spacing and a possible quadratic trend in the fits plot. Analysis suggests that the optimal factor combination may be variety X, narrow spacing, and late sowing.

Problem 7.4

EDA indicated that pressure and stirring appear to affect yield most though temperature has some influence also. The normal plot of effect estimates separated the effects temperature × time (positive), pressure (positive), stirring (negative), pressure × time (negative), and temperature (positive) from the others. The associated Pareto chart provided a comparable interpretation. The main effects plot highlights the significant influence of both pressure and stirring on yield (steeply sloping line) and the marginal influence of temperature. The temperature × time interaction plot showed distinctly different trends with low temperature reducing yield while high temperature increased it as time increased. The pressure × time interaction plot showed that as time increases, low pressure produces an upward trend in the yield while high pressure generates a downward trend. Modelling yield as temperature × time + pressure + stirring + pressure × time + temperature + error proved useful with four model effects significant at the 0.1% level and only temperature significant at the 5% level ($p = 0.020$). The residual plot for the time factor showed evidence of a trend as did the fits plot. Normality of the yield response was in doubt ($R = 0.952$, $p < 0.1$). Overall, the analysis suggested that the optimal factor combination may be temperature 40 °C, pressure 1.5 atmospheres, time 20 minutes, and no stirring though the temperature × time interaction suggested two possible combinations as best for its effect.

Problem 8.1

The design is to be a 2^4 FD confounded in two blocks of eight runs each with $ABCD$ confounded with the blocks. Any block assignment method generates the following block structure:

Block 1:	(1)	*ab*	*ac*	*ad*	*bc*	*bd*	*cd*	*abcd*
Block 2:	*a*	*b*	*c*	*abc*	*d*	*abd*	*acd*	*bcd*

Problem 8.3

The design is to be a 2^{5-1} FFD ($p = 1$) in 16 runs with defining relation $I = ACDE$. Solving the defining relation $I = ACDE$ for the additional factor E provides the design generator $E =$

ACD. The treatment combinations within this FFD will be (1), *ae*, *b*, *abe*, *ce*, *ac*, *bce*, *abc*, *de*, *ad*, *bde*, *abd*, *cd*, *acde*, *bcd*, and *abcde*. The corresponding alias structure for this design is $A \equiv CDE$, $B \equiv ABCDE$, $C \equiv ADE$, $D \equiv ACE$, $E \equiv ACD$, $AB \equiv BCDE$, $AC \equiv DE$, $AD \equiv CE$, $AE \equiv CD$, $BC \equiv ABDE$, $BD \equiv ABCE$, and $BE \equiv ABCD$. Thus, main effects are aliased with mostly three factor interactions while two factor interactions are aliased with others of the same order and four factor interactions. Main effects and two factor interactions are therefore estimable independently of each other. The design resolution is therefore IV and provides acceptable effect estimation.

Problem 8.5

The design run was a 2^{4-1} FFD with defining contrast $I = ABCD$ providing a design of resolution IV with independent estimation of the main effects but not the two factor interactions. EDA suggested some trend in the response data with both belt angle (A) and belt speed (B) appearing to have significant effect on the number of damaged items with high levels increasing the number of damaged letters. Variability in the number of damaged items is large and differs for both belt angle and belt speed. The normal plot and Pareto chart (α = 0.2) indicate that the belt angle and belt speed effects stand out with the interaction belt angle × number of letters sorted also important. This interaction is aliased with the belt speed × belt material interaction. As belt angle appeared more important than belt speed, only the belt angle × number of letters sorted interaction has been assessed. Main effect plots for the belt angle and belt speed highlight the strong positive effect of these factors with minimal damage apparent at the low level. The belt angle × number of letters sorted interaction suggests minimal damage at belt angle 10° and high number of letters sorted. Modelling the number of damaged items as belt angle (A) + belt speed (B) + belt angle × number of letters sorted ($A \times D$) + error generated significant effects, belt angle and belt speed at the 0.1% level ($p = 0.000$) and the interaction at the 5% level ($p = 0.014$). Assessment of the residual plots showed that high level of belt material (C) and number of letters sorted (D) could reduce variability in the number of damaged items. It would appear that a possible optimal treatment combination is belt angle 10°, belt speed slow, belt material type 2, and number of letters sorted high. This refers to combination *cd* which was run in the experiment and had the minimal response.

Problem 8.8

The design run was a 2^{5-2} FFD with defining contrasts $I = ABD$ and $I = ACE$, and design generators of $D = AB$ and $E = AC$. The full defining relation is $I = ABD = ACE = BCDE$ providing a design of resolution III where main effects are aliased with two factor interactions. Only N (B), CM (C), and CR (D) appear to have any real effect on retention time of simazine with low level generating higher responses. Variation in retention time measurements is high at all factor levels suggesting precision of measurement response may be in doubt. The normal plot and Pareto chart (α = 0.2) confirm that CM and CR are the most important effects. However, it must be remembered that CM (C) is aliased with the pH × F ($A \times E$) interaction while CR (D) is aliased with the pH × N ($A \times B$) interaction. Since both pH and F generate small effect estimates, it could be argued that their interaction may be negligible and thus, the effect estimate of −12.673 fully explains the CM effect. Since N also produces a small effect estimate, it could equally be argued that its interaction with pH may be negligible and thus, the effect estimate of −7.427 associated with the CR effect could be deemed to correspond to that effect only. The main effect plots for CM and CR show distinctly negative trends indicative of higher retention time at the low factor level. Modelling retention time by CM (C) + CR (D) + error provided statistically significant effects, CM at the 0.1% level ($p = 0.000$) and CR at the 1% level ($p = 0.001$). The residual plots showed patterning for both N and F suggesting that high level of each could reduce retention time variability while the fits and predicted values plots both showed evidence of patterning. Normality of retention time was also in doubt. Assuming maximising retention time is best, optimal levels for the tested factors would appear to be pH either 5.8 or 7.2, N 8, CM 35% methanol, CR 2.75, and F 1.3. These two combinations were tested in the experiment carried out and produced the highest retention time measurements.

Problem 9.1

Factor column allocation is A to c_1, B to c_2, C to c_4, and D to c_7. For factor D, its associated interactions will be allocated as $AD \rightarrow c_6$, $BD \rightarrow c_5$, and $CD \rightarrow c_3$. This generates aliasing of $AB \equiv CD$ from c_3, $AC \equiv BD$ from c_5, and $BC \equiv AD$ from c_6. The factor D allocation leads to the design generator of $D = ABC$, defining contrast of $I = ABCD$, and defining relation of $I = ABCD$. Thus, the proposed OAD is a resolution IV FFD.

Problem 9.3

Factor column allocation is A to c_1, B to c_2, C to c_4, D to c_8, E to c_{11}, and F to c_{15}. This gives rise to the column association of $AE \rightarrow c_{10}$, $BE \rightarrow c_9$, $CE \rightarrow c_{15}$, $DE \rightarrow c_3$, $AF \rightarrow c_{14}$, $BF \rightarrow c_{13}$, $CF \rightarrow c_{11}$, $DF \rightarrow c_7$, and $EF \rightarrow c_4$. This structuring results in aliasing of $AB \equiv DE$ (c_3), $C \equiv -EF$ (c_4), $ABC \equiv -DF$ (c_7), $AD \equiv BE$ (c_9), $BD \equiv AE$ (c_{10}), $ABD \equiv E \equiv -CF$ (c_{11}), $ACD \equiv -BF$ (c_{13}), $BCD \equiv -AF$ (c_{14}), and $-ABCD \equiv F \equiv -CE$ (c_{15}). The corresponding design generators are $E = ABD$ and $F = -ABCD$ providing defining contrasts of $I = ABDE$ and $I = -ABCDF$ with defining relation of $I = ABDE = -ABCDF = -CEF$. The proposed OAD is a resolution III FFD.

Changing column allocation of E to c_7 and F to c_{14} alters the design generators to $E = ABC$ and $F = BCD$. The defining contrasts change to $I = ABCE$ and $I = BCDF$ which, in turn, change the defining relation to $I = ABCE = BCDF = ADEF$. This is an OAD with resolution IV which is an improvement on the initial design resolution of III and thus can provide improved effect estimation.

Problem 9.6

Factor column allocation is A to c_1, B to c_2, C to c_4, D to c_{11}, and E to c_{13}. From this, we can obtain the two factor interaction allocations of $AE \rightarrow c_{12}$, $BE \rightarrow c_{15}$, $CE \rightarrow c_9$, and $DE \rightarrow c_6$. The aliasing of main effects and two factor interactions is $BC \equiv DE$, $BD \equiv CE$, $ABCD \equiv AE$, $BCD \equiv E$, and $CD \equiv BE$. This provides the design generator of $E = BCD$ and defining relation of $I = BCDE$. The design implemented corresponds to a resolution IV FFD with main effects aliased with three factor interactions and two factor interactions aliased with interactions of the same order. pH (A), MeOH (B), and NaCl (C) appear to have important influence on the response with response variation large for all levels tested suggesting measurement precision may be in doubt. Effect estimate plots ($\alpha = 0.2$) suggest that pH is an important factor as are the interactions MeOH × capillary temperature ($B \times D$), pH × MeOH ($A \times B$), MeOH × applied potential ($B \times E$), and MeOH × NaCl ($B \times C$). From the aliasing, these interactions are aliased with four factor interactions so interpretation of their effects is possible. The main effect plots highlight the obvious effect of pH on the response and the lesser effect of MeOH. All the interaction plots show distinct interaction properties indicative of non-additive response effect. For MeOH × capillary temperature, MeOH 3.5% and capillary temperature 35 °C appears best while for pH × MeOH, pH 2.5 and MeOH 3.5% may be best. For MeOH × applied potential, MeOH 3.5% and applied potential 20 kV appears best while MeOH × NaCl suggests MeOH 3.5% and NaCl 30 mM as best. Modelling the coded response as pH + MeOH × capillary temperature + pH × MeOH + MeOH + MeOH × applied potential + MeOH × NaCl + error provided significant effects, pH at the 0.1% level (p = 0.000), MeOH × capillary temperature at the 1% level (p = 0.003), and all other effects at the 5% level. Residual plots provided no additional useful information. As maximising the coded response was the goal, analysis suggested that optimisation could be achieved with the factor combination of pH 2.5, MeOH 3.5%, NaCl concentration 30 mM, capillary temperature 35 °C, and applied potential 20 kV. This combination, corresponding to *bc*, was tested and produced the second highest response measurement with only combination *bcd* higher. The suggested optimal factor combination, therefore, looks to be acceptable for optimising the electrophoretic response function.

Problem 10.1

The design generator, from column allocation of D, is $D = ABC$ resulting in a defining relation of $I = ABCD$ and a resolution of IV for the Taguchi design conducted. The aliases are $A \equiv BCD$, $B \equiv ACD$, $C \equiv ABD$, $D \equiv ABC$, $AB \equiv CD$, $AC \equiv BD$, and $AD \equiv BC$. For the *SN* ratio, effects $A \times C$ and $A \times B$ showed through as most important with B and D appearing unimportant. Maximising *SN* ratio occurred at the levels A_1, AB_2, C_1, AC_2, BC_2, and D_2, the "2" for interaction effects suggesting opposite levels of each factor best. Interaction plotting suggested A_1B_2, A_2C_1 or A_1C_2, and B_2C_1 to be the factor combinations maximising the *SN* ratio. Modelling the *SN* ratio by $AC + AB + C + A + BC$ + error showed AC and AB significant at the 1% level (p = 0.006 and p = 0.009, respectively), C and A at the 5% level (p = 0.041 and p = 0.048, respectively), and BC at the 10% level (p = 0.090). The optimal combination for maximising the *SN* ratio could be $A_1B_2C_1D_2$. For the mean performance statistic, effects $A \times B$, B, $B \times C$, and $A \times C$ look the most important with A and D unimportant. Average voltage nearest to target range occurs at the levels A_2, B_2, AB_2, C_1, AC_1, BC_2, and D_1. Plots of the interactions suggest A_1B_2, A_1C_1, and B_2C_1 best for achieving target voltage. Fitting a model of mean = $AB + B + BC + AC$ + error showed AB and B significant at the 1% level (p = 0.004 and p = 0.007, respectively), BC at the 5% level (p = 0.019), and AC at the 10% level (p = 0.079). For optimising voltage, combination $A_1B_2C_1D_2$ may be best. For the variability performance statistic ln(s), effect $A \times C$ stands out though $A \times B$, A, and C also look to have some influence. Minimum ln(s) occurs at the levels A_1, B_1, AB_2, C_1, AC_2, BC_2, and D_2. Plots of the interactions suggest A_1B_2, A_1C_2, and B_1C_1 appear the optimal factor combinations. Fitting a model of ln(s) = $AC + AB + A + C$ + error showed AC significant at the 1% level (p = 0.007), AB at the 5% level (p = 0.038), and A and C with p values of 0.102 and 0.125, respectively. The optimal combination for minimising voltage variability could be $A_1B_1C_1D_2$. From the overall analysis, the consensus on important factors and effects is not readily apparent because of the differing conclusions from each performance statistic. By examining all results, however, it would appear that combination $A_1B_2C_1D_2$ may be best for optimising output voltage.

Problem 11.1

The RS model is average modulus of rupture = $\beta_0 + \beta_1X_1 + \beta_2X_2 + \beta_3X_3 + \varepsilon$. The fitted model is average modulus of rupture = −7188 + 79 X_1 + 14 X_2 + 28 X_3. The lack-of-fit test is not significant (F = 0.22, p = 0.913) and overall model fit is acceptable (F = 34.50, p = 0.000, R^2_{adj} = 91.8%). Of the individual model components, only that for air cure was not significant at the 1% level (p = 0.310). Inclusion of the possible interaction terms still results in insignificant lack-of-fit (F = 0.35, p = 0.767) as the inclusion of the interaction terms is not significant (F = 0.24, p = 0.862). In fact, inclusion of the interaction terms reduces R^2_{adj} by 5% providing further evidence of the non-significant influence of the interaction components. The main effect plots show that both cure time and cure temperature exhibit positive sloping lines. The contour plots for each pair of factors exhibit the expected parallel lines feature of first-order RS models. The plot for cure time and air cure suggests higher levels may be best. The other two contour plots exhibit different trends but ones that suggest high levels of the factors may be best. From the plots, it appears changing cure time and cure temperature have most effect as the plot trends show greater change for these factors. It would appear that the next stage of experimentation should be directed toward testing higher settings of the three factors than those initially tested.

Problem 11.3

The RS model is total phenol recovery = $\beta_0 + \beta_1 X_1 + \beta_2 X_2 + \beta_{11} X_1^2 + \beta_{22} X_2^2 + \beta_{12} X_1 X_2 + \varepsilon$. The fitted model is total phenol recovery = $83.281 + 7.096\ X_1 + 1.681\ X_2 - 0.678\ X_1^2 + 0.699\ X_2^2 - 2.405\ X_1 X_2$. Both quadratic coefficients are low suggesting little influence of these trend effects on phenol recovery. The interaction estimate is larger suggesting interaction important to phenol recovery. The lack-of-fit test is not significant ($F = 2.612$, $p = 0.4201$) and the fitted model is statistically acceptable ($F = 8.732$, $p = 0.0283$, $R^2_{adj} = 81.1\%$). Of the individual model components, the linear component is statistically significant ($F = 20.281$, $p = 0.0081$) but the quadratic components ($p = 0.8618$) and the interaction component ($p = 0.1701$) appear not significant. The test of X_1 (extraction cell temperature) was statistically significant ($t = 6.197$, $p = 0.0034$) showing the importance of this factor to phenol recovery levels. Main effect plots show distinct change in recovery with increase in extraction cell temperature but no significant change for acetic anhydride. The contour plot exhibits a rising ridge pattern indicative of the optimum lying outside the experimental region. High levels of each factor look likely to provide optimal response. The canonical form of the fitted model is total phenol recovery = $91.34 + 1.6224 w_1^2 - 1.5985 w_2^2$ with the predicted response at the stationary point being 91.35%. The different signs of the eigenvalues indicates that the suggested stationary point is a saddle point indicating minimum for extraction cell temperature (positive eigenvalue) and maximum for acetic anhydride (negative eigenvalue). Presence of a minimax solution suggests possible inadequacy of model explanation given that response is only maximised for one factor and not both. Optimal factor levels are suggested to be extraction cell temperature 118.18 °C and acetic anhydride 153.14 μL.

Subject Index

Ellis Horwood Series in
Mathematics and its Applications

MATHEMATICAL ANALYSIS AND PROOF

DAVID S. G. STIRLING, Senior Lecturer in Mathematics, University of Reading

ISBN 1-898563-36-5 256 pages 1997

This introductory first and second year degree mathematics course texts emphasises mathematical proof and the need for it, developing associated technical and logical skills. This is brought to bear on core material of analysis. The development reads naturally and progressively, by indicating how proofs are constructed. It emphasises the need for familiarity with long mathematical arguments and manipulation, and the ability to construct proofs in analysis.

The author writes in good mathematical style and elegant prose. Helpful to students for its "easy" language, usefully explained with clarity, logic and brevity. Over 100 worked examples and some 200 problem exercises (and their solutions/ hints). Index of Notation and Subject Index

Contents: Setting the scene; Logic and deduction; Mathematical induction; Sets and numbers; Order and inequalities; Decimals; Limits; Infinite series; Structure of real number system; Continuity; Differentiation; Functions defined by power series; Integration; Functions of several variables.

"**A fine line between accuracy and exactitude. David Stirling treads it carefully ... a thorough a comprehensive introduction ... very much in the classical mould but written in a chatty style with the common student misunderstandings in mind. It should be in your undergraduate reference library**" - *The Mathematical Gazette* (K.L. McAvaney, Deakin University, Geelong, Australia)

"**Concise and straightforward ... an excellent textbook for beginning undergraduates and polytechnic students**" - *Zentralblatt für Mathematik und ihre Grenzgebiete/Mathematics Abstracts,* Germany (J.A. Adepoju)

STOCHASTIC DIFFERENTIAL EQUATIONS & APPLICATIONS

XUERONG MAO, Department of Statistics & Modelling Science, University of Strathclyde, Glasgow

ISBN 1-898563-26-8 360 pages 1997

This advanced undergraduate and graduate text covers basic principles and applications of various types of stochastic systems which now play an important role in many branches of science and industry. It is a source book for pure and applied mathematicians, statisticians and probabilists, and engineers in control and communications, information scientists, physicists and economists.

It emphasises the analysis of stability in stochastic modelling and illustrates the practical use of stochastic stabilisation and destabilization, stochastic oscillators, stochastic stock models and stochastic neural networks in pragmatic, real life situations.

Contents: Generalised Gronwall inequality and Bihari inequality; Introduces the Brownian motions and Stochastic integrals; Analyses the classical Ito formula and the Feynman-Kac formula; Demonstrates the manifestations of the Lyapunov method and the Ruzumikhin technique; Discusses Cauchy-Marayama's and Carathedory's approximate solutions to stochastic differential equations; and more.

TEACHING AND LEARNING MATHEMATICAL MODELLING:
Innovation, Investigation and Application

Editors: S.K. HOUSTON, The University of Ulster, Northern Ireland; W.BLUM, The University of Kassel, Germany; IAN HUNTLEY, The University of Bristol, England; N.T.NEILL, The University of Ulster, Northern Ireland

ISBN- 1-898563-29-2 416 pages 1997

Sponsored by the organising committee of International Conferences on the Teaching of Mathematical Modelling and Applications (ICTMA), this book contributes to teaching and learning mathematical modelling in universities throughout degree study, colleges of technology, teachers' training colleges, high schools and sixth form colleges.

International mathematicians from Austria, Australia, Germany, Holland, Italy, Japan, Russia, Spain, UK, USA reflect current knowledge and development. Philosophically and creatively they discuss innovation and assessment, and teaching and study, at all levels. The interdisciplinary topics reflect their use in such areas of application as mechanics and engineering, medicine, patient flow in hospitals, computing science, traffic control, business studies, and mathematics (fractals and analysis), all pointing to a wide choice of future careers.

Contents: Reflections and investigations; Assessment at undergraduate level; Secondary courses and case studies; Tertiary case studies; Tertiary courses.

"**Deals with assessment, particularly at undergraduate level, secondary education and case studies of good practice with examples of courses and how they are taught in a variety of countries including Russia, the Netherlands, the USA and the UK. Concludes with descriptions of ideas relating to undergraduate modelling courses**" - *Zentralblatt für Mathematik und ihre Grenzgebiete/Mathematics Abstracts, (Germany)*

DECISION AND DISCRETE MATHEMATICS

prepared by THE SPODE GROUP with Ian Hardwick, Truro School, Cornwall

ISBN 1-898563-27-6 240 pages 1996

A complete coverage in the decision mathematics (or Discrete Mathematics) module of A-level examination syllabuses. Also suitable for first year undergraduate courses in qualitative studies or operational research, or for access courses. Reflects the combined teaching skills and experience of authors within The Spode Group. The text is modular, explaining concepts used in decision mathematics and related operational research, and electronics. Emphasises techniques and algorithms in real life situations and working problems. Clear diagrams; plentiful worked examples; Exam-standard questions; Many exercises.

Contents: Introduction to networks; Recursion; Shortest route; Dynamic programming; Network flows; Critical path analysis; Linear programming: Graphical; Linear programming: Simplex method; The transportation problem; Matching and assignment problems; Game theory; Recurrence relations; Simulation; Iterative processes; Sorting; Algorithms; Appendices; Answers; Glossary.

"**Topics include networks, recursive and iterative processes, critical path analysis, linear programming, the transporation problem, recurrence, game theory simulation, and sorting/packing algorighms ... explores both key concepts and fundamental algorithms, but also relates the area of decision mathematics to real-life situations for lower-division undergraduates and two-year technical program students**" - *Choice, American Library Association*

MATHEMATICAL MODELLING:
Teaching and Assessment in a Technology-Rich World

P. GALBRAITH, University of Queensland, Brisbane, Australia; W. BLUM, The University of Kassel, Germany; G. BOOKER, Griffith University, Brisbane, Australia; IAN D .HUNTLEY, University of Bristol, England

ISBN 1-898563-42-X 368 pages Hardback 1997

Very few projects in science or industry are attempted without the safeguard of prior mathematical modelling. This book contributes to the teaching, learning and assessing of mathematical modelling in this era of rapidly expanding technology. It addresses all levels of education, from secondary schools through teacher training colleges, colleges of technology, universities, and state and national departments of mathematical education and research groups.

Sponsored by the International Conferences on Teaching of Mathematical Modelling and Applications (ICTMA), it reflects the very latest ideas and methods contributed by specialists from some fifteen countries in Africa, the Americas, Asia, Australia, Continental Europe, and the United Kingdom. These contributions reflect common issues shared globally, and also those that represent emergent or on-going challenges in particular communities.

The broad range of topics considered may be classified in terms of identifiable themes. Issues and alternatives which improve the quality of performance continue to attract attention, as also do practical examples of modelling in action. They encompass medical, engineering, social or sporting applications. The theme of technology dominates throughout, and extends beyond the theme which bears its name. In a range of applications the use of graphics calculators, spreadsheets, symbolic manipulator software, and special purpose programs all feature both as aids to mathematical processing, and also in relation to educational implications. Examples of the use of models to achieve particular goals are given, while more generally the role and purpose of mathematical modelling at particular system or national level, are examined from within a range of national contexts.

Contents: A Theme; Issues and alternatives in assessing modelling; B Theme: Technologically enriched mathematical modelling; C Theme: Real world: Models and applications; D Theme: Applications and modelling in teaching and learning; E Theme: Applications and modelling in a system or national context.

FUNDAMENTALS OF UNIVERSITY MATHEMATICS

COLIN McGREGOR, JOHN NIMMO and WILSON W. STOTHERS, Department of Mathematics, University of Glasgow

ISBN: 1-898563-09-8 540 pages Hardback 1994
ISBN: 1-898563-10-1 200 diagrams Paperback

A unified course for first year mathematics, bridging the school/university gap, suitable for pure and applied mathematics courses, and those leading to degrees in physics, chemical physics, computing science, or statistics. The treatment is careful, thorough and unusually clear, and the slant and terminology are modern, fresh and original, in parts sophisticated and demanding some student commitment. Fresh ideas for teachers, students, and tutorials. 300 worked examples, rigorous proofs for most theorems, 750 exercises with answers. Problems and solutions for all topics.

Contents: Preliminaries, functions & inverse functions; Polynomials & rational functions; Induction & the binomial theorem; Trigonometry; Complex numbers; limits & continuity; Differentiation - fundamentals; Differentiation - applications; Curve sketching; Matrices & linear equations; Vectors & three dimensional geometry; Products of vectors; Integration - fundamentals; Logarithms & exponentials; Integration - methods & applications; Ordinary differential equations; Sequences & series; Numerical methods.

Engineering (Electronic and Electrical)

GEOMETRY OF SURFACE NAVIGATION

ROY WILLIAMS, Master Mariner, BSc, PhD, FRIN, AFIMA

ISBN: 1-898563-46-2 144 pages Hardback 1999

There is no written exposition which offers a clear and understandable mathematical approach to navigation. This book offers a treatment of the Earth as an ellipsoid of revolution, pictured as a sphere, as has served navigators for centuries. Now that the science of navigation has entered the electronic age, orbital man-made satellites have replaced the stars. Many problems in navigation which, because of the true shape of the Earth, seemed numerically and algebraically impractical, can now be resolved by computer. Nevertheless, traditional methods provide important back-up for which geometry is invaluable. The scope of this book applies to surface navigation, and also to the surfaces of planets. It also includes methods for fixing a position by astronomical observation, and also a method of fixing position from the tracking of the angular co-ordinates of an astronomical body.

It reviews mathematical analysis for developing the methods in computing navigation by setting out the proof in terms of the geometry of differentials and methods of the calculus. In this way the differential geometry makes for a full analysis of the methods of navigation as applied to the ellipsoid of revolution. The computed numerical solutions of the resulting equations introduces new numerical methods of application for navigation, which make a contribution to mathematical theory by the uses of geometry for route planning and position fixing. The author assumes familiarity with the elements of navigation, terminology and definitions.

Contents: Geometrical representation of the earth: Mathematics of chart projections: Navigating along Rhumb lines: Shortest paths on the surface of a sphere: Shortest paths on the surface of an ellipsoid: Paths between nearly antipodean points: Great ellipse on the surface of an ellipsoid: Navigating along the arc of a small ellipse: Surface position from astronomical observation: Surface position from satellite data: Appendices: Table of latitude parts (meridian distance): Transformation between equations: Direct cubic spline approximation.

SIGNAL PROCESSING IN ELECTRONIC COMMUNICATIONS

MICHAEL J. CHAPMAN, DAVID P. GOODALL, and NIGEL C. STEELE, School of Mathematics and Information Sciences, University of Coventry

ISBN 1-898563-30-6 288 pages 1997

This text for advanced undergraduates reading electrical engineering, applied mathematics, and branches of computer science involved with signal processing (speech synthesis, computer vision and robotics). Serves also as a reference source in academia and industry. Signal processing is an important aspect of electronic communications in its role of transmitting information, and the mathematical language of its expression is developed here in an interesting and informative way, imparting confidence to the reader.

The first part of the book focuses on continuous-time models, with chapters on signals and linear systems, and system responses. Fourier methods, vital in information theory, are developed prior to a discussion of methods for the design of analogue filters. The second part covers discrete-time signals and systems. There is full development of *z*- and discrete Fourier transforms to support the digital filter design chapter. The final chapter draws together all preceding material on important aspects of speech processing as an up-to-date example of the theory. Topics considered include a speech production model, linear predictive filters, lattice filters and cepstral analysis, with application to recognition of non-nasal voiced speech and formant estimation.

*Contents: Sig*nal and linear system fundamentals; System responses; Fourier methods; Analogue filters; Discrete-time signals and systems; Discrete-time system responses; Discrete-time Fourier analysis; The design of digital filters; Aspects of speech processing; Appendices: The complex exponential; Linear predictive coding algorithms; Answers.

GAME THEORY: Mathematical Models of Conflict

A.J. JONES

ISBN: 1-898563-14-4 *ca.* 300 pages 1998

A modern, up-to-date text for senior under-graduate and graduate students (and teachers and professionals) of mathematics, economics, sociology; and operational research, psychology, defence and strategic studies, and war games. Engagingly written with agreeable humour, this account of game theory can be understood by non-mathematicians. It shows basic ideas of extensive form, pure and mixed strategies, the minimax theorem, non-cooperative and co-operative games, and a "first class" account of linear programming, theory and practice. The book is self-contained with comprehensive references from source material.

DELTA FUNCTIONS: A Fundamental Introduction to Generalised Functions

R.F. HOSKINS, Department of Mathematical Sciences, De Montfort University, Leicester

ISBN: 1-898563-44-6 *ca.* 250 pages 1999

This book provides a text for final year and postgraduate courses in applied mathematics, and electrical and mechanical engineering. It may be used as a working manual in its own right, as well as serving as a preparation for the study of more advanced treatises.

It is a successor book to the author's *Generalised Functions* (Ellis Horwood Limited, 1979) following numerous world-wide enquiries since its disappearance. It is updated, modernised and restructured, particularly so to address the continual problem of the transition from elementary delta functions to Schwartz distributions. After a detailed account of the formal properties of delta functions and of various applications, the book gives a conceptually satisfying interpretation of delta functions and other generalised functions, using concepts and techniques from a simple form of Non-Standard Analysis (NSA).